Flora of Telangana

The 29th State of India

THE AUTHOR

Prof. T.Pullaiah obtained his M.Sc. (1973) and Ph.D. (1976) degrees in Botany from Andhra University. He was a Post Doctoral Fellow at Moscow State University, Russia during 1976-78. He traveled widely in Europe and USA and visited Universities and Botanical Gardens in about 17 countries. Professor Pullaiah joined Sri Krishnadevaraya University as Lecturer in 1979 and became Professor in 1993. He held several positions in the University, which include Dean, Faculty of Life Sciences, Head of the Department of Botany, Chairman, BOS in Botany, Head of the Department of Sericulture, Co-ordinator and Chairman, BOS in Biotechnology, Vice Principal and Principal, S.K.University College. He retired from active service on 31st May 2011. He was selected by UGC as UGC-BSR Faculty Fellow and is working in Sri Krishnadevaraya University. Prof. Pullaiah has published 60 books, 303 research papers and 35 popular articles. He is Principal Investigator of 20 major Research Projects totaling more than a Crore of Rupees funded by DBT, DST, CSIR, UGC, BSI, WWF, GCC etc. Under his guidance 53 students obtained their Ph.D. degrees and 35 students their M.Phil. degrees. He is recipient of Best Teacher Award from Government of Andhra Pradesh, Prof. P. Maheswari Gold Medal and Prof. G.Panigrahi Memorial Lecture of Indian Botanical Society and Prof. Y.D.Tyagi Gold medal of Indian Association for Angiosperm Taxonomy. He was Past President of Indian Association for Angiosperm Taxonomy and presently he is President of Indian Botanical Society. He was Member of Species Survival Commission of International Union for Conservation of Nature and Natural Resources (IUCN).

Flora of Telangana

The 29th State of India

Volume 3

Author
T. Pullaiah
Department of Botany
Sri Krishnadevaraya University
Anantapur – 515 003
Andhra Pradesh
India

2021
Regency Publications
A Division of
Astral International Pvt. Ltd.
New Delhi – 110 002

Flora of Telangana-*The 29th State of India*
(3 Volume Set)

Volume 1: Page 0001-0446
Volume 2: Page 0447-0892
Volume 3: Page 0893-1306

Cataloging in Publication Data--DK
Courtesy: D.K. Agencies (P) Ltd. <docinfo@dkagencies.com>

Pullaiah, T., author.
Flora of Telangana : the 29th state of India / author, T. Pullaiah.
3 volumes cm
Includes bibliographical references and index.
ISBN 9789352220021 (Vol.3)
ISBN 9788189233983 (Set)

1. Plants--India--Telengana--Classification. 2. Flowers--India--Telengana--Classification. I. Title.

DDC 582.13095485 23

Published by : **Regency Publications**
A Division of
Astral International Pvt. Ltd.
– ISO 9001:2015 Certified Company –
4736/23, Ansari Road, Darya Ganj
New Delhi-110 002
Ph. 011-43549197, 23278134
E-mail: info@astralint.com Website:
www.astralint.com

Laser Typesetting : **Classic Computer Services**, Delhi - 110 035

Printed at : **Replika Press Pvt. Ltd.**

Preface

Flora of Former Andhra Pradesh was published in 1997 and Additions in 2008. The book went out of print in 2010. The long awaited decision to divide the state of Andhra Pradesh coincidentally happened when I had started revising the book sitting in Missouri Botanical Garden, U.S.A. It was then the idea of separate Flora for Telangana and Andhra Pradesh began. I continued my work using the facilities of Missouri Botanical Garden Herbarium and Library. I would like to highlight the warmth and affection which I received there and was inspired by their hard work and dedication. I thank authorities of Missouri Botanical Garden, Herbarium and Library for the permission to consult the Library and Herabrium.

Out of ten districts of Telangana we already published Flora of Five districts and I have taken liberally information from these books. The contributions of the scholars in these projects who are presently well placed in various esteemed institutions include Dr. P.V. Prasanna, Prof. B. Ravi Prasada Rao, Dr. C. Prabhakar, Dr. M.Silar Mohammad and Dr. B.V.Raghava Swamy. I thank all these scholars for their contributions which enabled me to write this book.

I thank the Scientists and scholars of Botanical Survey of India, Deccan Circle, Hyderabad, especially Dr. P.Venu, Dr. P.V.Prasanna, Scientists, Dr. L.Rasingam, Dr. Shankara Rao, Alok Chorghe, P.S.Annamma, besides ministerial staff. Many a times I had stimulating discussion with them. I am confident that this centre will grow. Dr. G.V.S.Murthy, Joint Director of Botanical Survey of India, Sourthern Circle, Coimbatore was always encouraging, supporting and extending facilities at Madras Herbarium for which I am grateful to him. Dr. M.V.Ramana, Assistant Professor, Osmania University gave his thesis on Flora of Hyderabad and many a times had animated discussion on the subject. This young boy will contribute in future much to the science of Plant Taxonomy. I also thank Prof. K. Seshagiri Rao, of University of Hyderabad, for giving the details of Flora of their campus. Librarian of Kakatiya University extended Library facilities during my visit and I am grateful to them for the same. Dr.

M.Sanjappa, former Director of Botanical Survey of India was critical about my work and always wanted best from my hands and I am grateful to him for the same. May his tribe grow. Taxonomists at Calicut University helped me in nomenclature and other aspects of different groups of plants, I thank especially Prof. M.Sabu, Prof. Santhosh Nampy and Dr. A.K.Pradeep for their help in various ways.

Authorities of Sri Krishnadevaraya University were graceful enough to provide facilities for carrying out this work. I thank Prof. R.R.Venkata Raju, Prof. C.Sudhakar, Prof. B.Ravi Prasad Rao, Prof. S. Thimma Naik, Heads of Department during different periods for their support and extending the facilities in the Department. Dr. S. Sandhya Rani, the latest entrant in the Botany Department needs special mention who read the manuscript, gave suggestions for its improvement, corrected the manuscript and contributed in many ways to bring out this book. Dr. S,.Karuppusamy Assitant Professor at The Madura College and Dr. K.Sri Rama Murthy, Professor and Head, Department of Biotechnology, Montessori Mahila Kalasala, Vijayawada have also contributed much for this book.

University Grants Commission gave me BSR Faculty Fellowship which enabled me to carry out this work after superannuation and the same is gratefully acknowledged.

I thank Anil Mittal and Dinesh Arora at Astral International, New Delhi for their patience, support and encouragement to bring out his book.

I tried to make this work as perfect as possible. Had I tried for perfection I could never complete this work. I request the readers of this book to point out errors and omissions which can be rectified in future.

T. Pullaiah

Contents

MONOCOTYLEDONS

HYDROCHARITACEAE

1. Acaulescent herbs; spathes peduncled:
    2. Spathes winged .... **Ottelia**
    2. Spathes unwinged:
        3. Perianth lobes 6; peduncle straight .... **Blyxa**
        3. Perianth lobes 3; peduncle coiled .... **Vallisneria**
1. Caulescent herbs; spathes sessile:
    4. Leaves whorled .... **Hydrilla**
    4. Leaves not whorled .... **Nechamandra**

BLYXA Noronha ex Thouars

1. Leaves broad at base, gradually attenuate; flowers bisexual; spathes 1-flowered; stamens 3 .... **B. aubertii**
1. Leaves narrow at ends; flowers unisexual; spathes several – flowered; stamens 8-12 .... **B. octandra**

Blyxa aubertii Rich, Mem. Cl. Sci. Math. Inst. Natl. France 1811(2): 19. 1812. *Diplosiphon oryzetorum* Decne. in Jacq. Voy. Bot. 167. 1844. *Blyxa oryzetorum* (Decne.) Hook. f., Fl. Brit. India 5: 661. 1888. *B. griffithii* Planch. ex Hook. f., Fl. Brit. India 5: 661. 1888. *p.p. B. ceylanica* Hook. f., Fl. Brit. India 5: 661. 1888; Fischer, Fl. Madras 3: 1397. 1928.

An acaulescent aquatic herb. Leaves linear, 20-30 × 0.5-1 cm, membranous, glabrous, base broad, margin hyaline, apex gradually tapering, setaceous. Spathe to 6 cm, 1-flowered, bisexual; shorter than leaves, sepals 3, oblong; petals 3, linear; stamens 3; ovary as long as spathe, rostrum subterete, rigid. Fruit linear; seeds not tailed, obscurely tubercled or smooth.

In stagnant waters of Hyderabad and Warangal Districts.

Fl. & Fr.: June – October.

Thammalakunta (HYD), *TR* 884 (HH); Warangal (WGL), *P. Satyanarayana Reddy* 128 (HH); Narsapur (MDK), *RG* 116455 (BSID).

Blyxa octandra (Roxb.) Planch. ex Thwaites, Enum. Pl. Zeyl. 332. 1864; Fischer, Fl. Madras 3: 1397. 1928; Hartog in Steenis, Fl. Males. 1.5:392. 1957. *Vallisneria octandra* Roxb., Pl. Coromandel t. 165. 1802 & Fl. Ind. 3: 752. 1832. *Blyxa roxburghii* Rich., Mem. Sci. Math. Inst. Natl. France 1811(2): 19. 1812; Hook.f., Fl. Brit. India 5: 660. 1888.

An acaulescent aquatic herb. Leaves linear, 18-45 × 0.2-1.2 cm, membranous, glabrous, base broad, margin entire, apex acuminate; petioles sheathing at base. Flowers unisexual, on 5-7 cm long spathes; sepals 3; petals 3. Male flowers: 6-9 per spathe, spathes to 7 × 0.3 cm, stamens 8-12, filaments unequal, pistillodes 3, basally connate, Female flowers: solitary in spathes, sessile. Fruits linear, to 6 × 1.5 cm, torulose, glabrescent; seeds many, very shortly tailed, distinctly tubercled, in 6-8 rows.

In stagnant waters in Warangal and Mahabubnagar districts.

Fl. & Fr.: June – October.

Pakhal (WGL), *ANH* 15909 (MH); Mukilayru, (MBNR), *SRR* 1126 (HH).

HYDRILLA L.C. Rich.

Hydrilla verticillata (L.f.) Royle, Ill. Bot. Himal. Mts. t. 376. 1839; Hook. f., Fl. Brit. India 5: 659. 1888; Fischer, Fl. Madras 3: 1396. 1928; Hartog in Steenis Fl. Males 1, 5. 385. 1957. *Serpicula verticillata* L. f., Suppl. Pl. 416. 1781.

Acaulescent submerged herb; branchlets 15-75 cm long; basal nodes distant. Leaves linear-lanceolate, whorled, 3-6 per node, basal ones linear-oblong, upper ones ovate, 0.6-1.5 × 0.2-0.4 cm, membranous, glabrous, base rounded, margin entire-sharply serrulate, scaberulous on indentations, apex apiculate. Flowers unisexual, monoecious or dioecious; sepal 3; petals 3. Male flowers: Spathe solitary, pedicellate, stamens 3; Female flower: solitary, sessile, ovary oblong, beaked, to 3 cm, style 3-fid. Fruit subulate, subterete, to 5-seeded; seeds oblong.

Very common in stagnant waters, freshwater ponds, puddles, tanks, lakes and rivers in all districts.

Fl. & Fr.: September-January.

Sone (ADB), *TP & PVP* 4045; Sri Ramsagar project (NZB), *BR & GO* 9227; Pathur (MDK), *TP* & *CP* 12123; Dharmapur (RR), *MSM* 10485; Narsapur (MDK), *KMS* 6801 (MH); Pakhal (WGL), *ANH* 15904 (MH); Dindi river near Bollarum (NLG), *BVRS* 167; Dimmagudem (KMM), *RCS* 99061 (BSID).

NECHAMANDRA Planch.

Nechamandra alternifolia (Roxb.) Thwaites, Enum. Pl. Zeyl. 332. 1864; Subramanyam & N.P. Balak., Bull. Bot. Surv. India 3: 23. 1962. *Vallisneria alternifolia* Roxb. ex Wight, Bot. Misc. 2: 344. 1831. *Nechamandra roxburghii* Planch., Ann. Sci. Nat. Bot. Ser. 3. 11: 78. 1849. *Lagarosiphon roxburghii* (Planch.) Benth., Gen. Pl. 3: 451. 1880; Hook.f., Fl. Brit. India 5: 659. 1888. *L. alternifolia* (Roxb.) Druce, Bot. Soc. Exch. Club. Brit. Isles 4: 630. 1917; Govind., Proc. Ind. Acad. Sci. (Plant Sci.) 93: 15. 1984; Fischer, Fl. Madras 3: 1396. 1928.

A fresh water submerged herb; branchlets 20-50 cm, glabrous. Leaves alternate, basally décussate, linear, 1-9 × 0.3-1 cm, membranous, glabrous, 3-5-nerved, base amplexicaul, margin entire, apex acuminate. Flowers dioecious, in axillary, solitary, sessile spathes; sepals 3; petals 3. Male flowers: stamens 2, divergent, anthers ovate. Female flowers: Spathe subulate, turgid, to 8 mm; ovary obovoid, beaked, styles 3, notched; stigma papillose. Fruit ovoid, included in spathe, many-seeded; seeds ellipsoid, closely pitted in longitudinal rows.

Common in stagnant waters, ponds, tanks, lakes, rivers in Nizamabad, Medak and Warangal districts.

Fl. & Fr.: September-February.

Kammarpalli (NZB), *TP* & *BR* 6238; Uppalwai (NZB), *BR* & *PSPB* 9566; Medak (MDK), *CP* 11798, *KMS* 6754 (CAL); Pakhal lake (WGL), *KMS* 11681 (MH & CAL); Pakhal (WGL), *ANH* 15906 (CAL).

OTTELIA Pers.

Ottelia alismoides (L.) Pers., Syn. Pl. 1: 400. 1805; Hook.f., Fl. Brit. India 5: 662. 1888; Fischer, Fl. Madras 3: 1398. 1928; Hartog in Steenis Fl. Males 1, 5: 398. 1957. *Stratiotes alismoides* L., Sp. Pl. 535. 1753.

An acaulescent submerged herb. Leaves submerged and floating; submerged leaves oblong, 4-9 × 3-18 cm, chartaceous, base cuneate, truncate, margin entire, apex rounded, obtuse. Flowers unisexual or/and bisexual. Spathe elliptic-ovate, greenish, 6-winged, apex 2-lobed; peduncle 10-50 cm, 3-angled. Flowers solitary, sessile; sepals 3, linear, persistent; petals 3, white with yellow spots, oblong, stamens 9-15, ovary oblong, beaked, 6-9-celled, ovules arranged on parietal placentation, styles 10-15, bifid. Fruit oblong, crowned by sepals; many-seeded

Common in tanks and slugglish streams in all districts.

Fl. & Fr.: May – November.

Jannaram (ADB), *TP* & *PVP* 4192; Kammarpally (NZB), *TP* & *BR* 6249; Pathur (MDK), *TP* & *CP* 12110; Malkalcheruvu (RR), *MSM* & *KH* 10540; Hajipur hills (MBNR), *TDCK* 13650; Pakhal lake, (WGL), *KMS* 11655 (MH); Lakavarapucheruvu (KMM), *MHR* & *MCK* 17014; Ramagiricheruvu (NLG), *NVN* & *KI* 18750; Ramancheru (MDK), *RG* 96270 (BSID);

VALLISNERIA L.

Vallisneria natans (Lour.) H. Hara, J. Jap. Bot. 49: 136. 1974. *Physkium natans* Lour., Fl. Cochinch. 663. 1790. *Vallisneria spiralis* auct. non. L. 1753; Hook.f., Fl. Brit. India 5: 660. 1888. *p.p;* Fishcer, Fl. Madras 3: 1369. 1928.

Acaulescent, stoloniferous freshwater herbs; stolons to 5.5 cm, terete. Leaves radical, very long and linear, ribbon like, 8-36 × 0.4-1.2 cm, narrow at base, margin denticulate, apex obtuse. Flowers unisexual (dioecious). Male flowers: minute, many together in a shortly peduncled ovoid, 3-lobed spathe; spathes oblong, 6 × 4 mm. Female flowers: solitary, in a tubular 3-toothed spathe at the end of a very long filiform spirally coiled scape, 50-60 cm long, sepals 3; petals absent; ovary narrow, ovules arranged on marginal placentation, styles 3, 2-lobed, stigmas 3, broad, notched. Fruit linear, to 2 cm, included in the spathe, many-seeded.

A common weed rooting at the bottom of pools and tanks in all districts.

Fl. & Fr.: September-March.

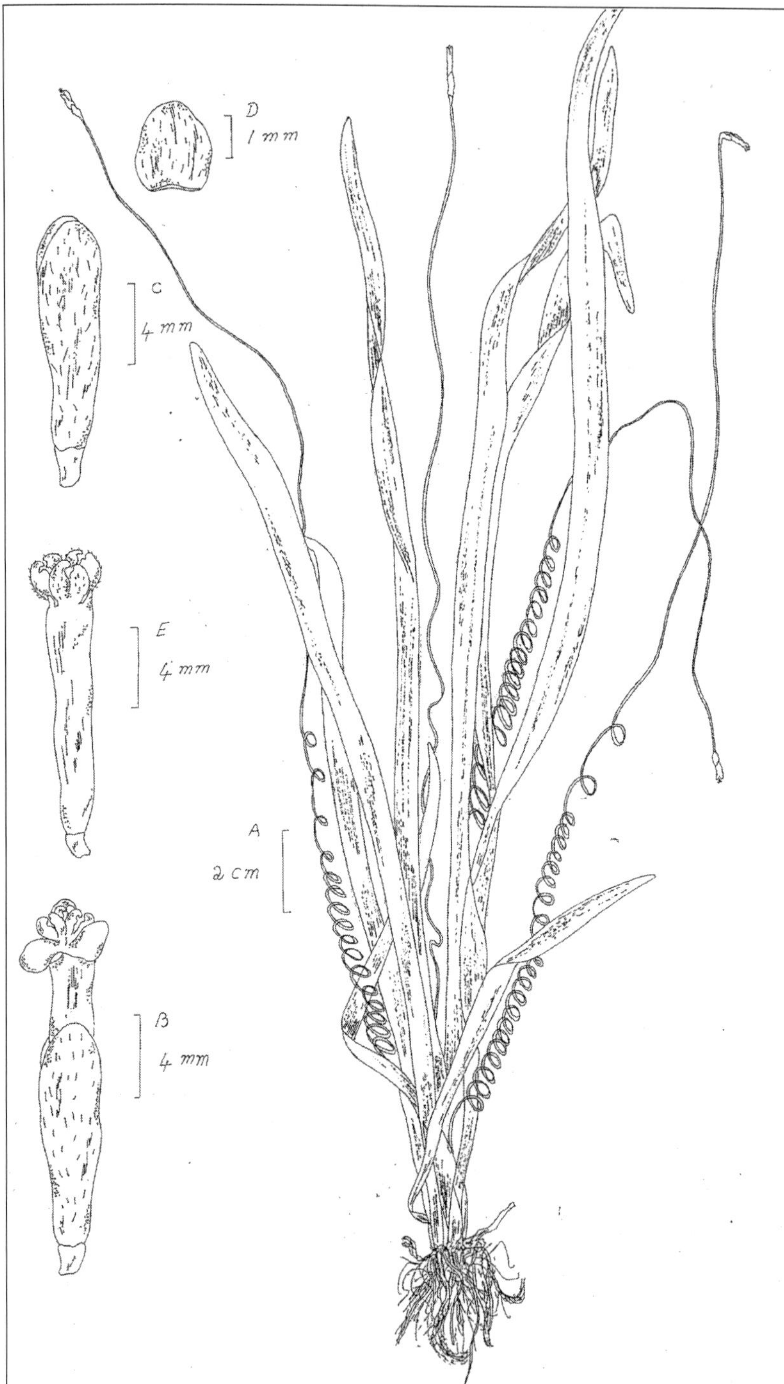

Figure 105: Vallisneria natans (Lour.) H. Hara, (*Vallisneria spiralis* auct non L.)
A. Twig, B. Flower, C. Spathe, D. Sepal, E. Pistil.

Sone (ADB), *TP* & *PVP* 4042; Kunthala (ADB), *GO* & *DAM* 5133; Janakampally (NZB), *TP* & *BR* 6322; Uppalwai (NZB), *BR* & *PSPB* 9565; Gandipeta (HYD), *KI* & *NVN* 18778; Kompalli hills (KMM), *MHR* & *MCK* 17045; Pakhal (WGL), *ANH* 15901 (MH); Near Dindi project (NLG), *BVRS* 677 (NU).

Note: Lowden (*loc. cit.*) revised the genus *Vallisneria* L. He followed Mikono, who erected var. *dense-serrulata* for *Vallisneria spiralis.* Accordingly var. *spiralis* is distinguished from the var. *dense-serrulata* in having the adnation of staminodia of the pistillate flowers to the shallowly cleft apex of the fused stigmatic lobes. In the latter case the staminodia are adnate to deeply cleft base of the stigmatic lobes. In the present study Hara (*loc. cit.*) concept is followed who treated *Vallisneria natans* as the correct and for the taxon, which is considered as synonym to *Vallisneria spiralis* var. *dense-serrulata* by Lowden.

ORCHIDACEAE

1. Terrestrial plants:
    2. Plants with distrinct corms or tuberoids:
        3. Subterranean stem a corm:
            4. Corm hypogeal or epigeal; inflorescence erect; flowers well spaced on a long rhachis, lip inferior .... **Eulophia**
            4. Corm always hypogeal; inflorescence decurved; flowers clustered on a short rhachis; lip superior .... **Geodorum**
        3. Subterranean stem a root-stem tuberoid .... **Habenaria**
    2. Plants without corm or tuberoid .... **Zeuxine**
1. Epiphytic plants:
    5. Leaves terete .... **Lusia**
    5. Leaves dorsiventrally flattened:
        6. Inflorescence simple .... **Vanda**
        6. Inflorescence branching .... **Acampe**

ACAMPE Lindl.

Acampe praemorsa (Roxb.) Blatter & McCann, J. Bombay Nat. Hist. Soc. 35: 495. 1932. *Epidendrum praemorsum* Roxb., Pl. Coromandel t. 43. 1795. *Saccolabium praemorsum* (Roxb.) Hook. f., Fl. Brit. India 6: 12. 1890. *S. wightianum* Hook. f., Fl. Brit. India 6: 12. 1890. *Acampe wightiana* Lindl., Fol. Orchid. Acampe 2. 1853; Fischer, Fl. Madras 3: 1447. 1928.

An epiphyte, stem non-psuedobulbous, 50-60 cm long. Leaves lorate, coriaceous, unequally bilobed at apex, 15-17 × 2-2.8 cm. Flowers in spiciform racemes, yellow, barred with red; peduncle short, stout, with sterile, scaly cupular bracts; floral bracts triangular, fleshy, with spotted pink; sepals and petals thick, fleshy, barred pink or red on the inside and spotted outside, dorsal sepal oblong, obtuse; lateral sepals broad, oblong, lip fleshy, saccate, lateral lobes small, blunt; midlip lobe thick, ovate, white, barred light pink; column short, thick, anther terminal, 2-loculed, pollinia 2, globose, attached by short caudicles to a slender strap and a small gland. Capsules fusiform, subsessile, to 4.5 × 1 cm.

Common in the interior hilly areas, in open and moist deciduous forests in Khammam district.

Fl. & Fr.: May – August.

Edrallapalli RF (KMM), *RCS* 102406 (BSID).

EULOPHIA R. Br. ex Lindl. *nom. cons*

1. Column not produced into a foot .... **E. graminea**
1. Column produced into a foot .... **E. spectabilis**

Eulophia graminea Lindl., Sp. Orchid. 182. 1833; Hook. f., Fl. Brit. India 6:2. 1890; Fischer, Fl. Madras 3: 1435. 1928. *E. decipiens* Kurz, J. Asiat. Soc. Bengal 45: 155, t. 13, f. 8-12. 1876; Hook. f., Fl. Brit. India 6:2. 1890.

A terrestrial, erect herb with conical pseudobulbous stems clothed in membranous sheaths; roots vermiform. Leaves 5-7, grass-like, linear, 15-24 × 0.4-0.7 cm, distichous, base sheathing, margin entire, tip acuminate, 5-veined. Panicles on 25-60 cm long peduncle, racemes lax-flowered, peduncle purplish-green, terete, with 3-5-sheaths. Flowers greenish with white lip; floral bracts small, lanceolate and persistent; sepals subequal, pale green, with purple veins, reflexed in distal halves; dorsal sepal linear to oblong-lanceolate, lateral sepals linear-oblong, falcate; petals tessellate, broader than sepals, elliptic-lanceolate, acute, tessellate, lip obovate, 1-1.2 × 0.6-0.8 cm, white, with maroon or purple veins, 3-lobed; lateral lobes smooth and embracing the column; mid lobe obovate, 0.5-0.6 × 0.5-0.7 cm, recurved, emarginated, crenulate, apex rounded; spur short, conical or cylindrical, to 0.3 × 0.2 cm, purple green, disc with five ridges with hooked processes; column long, not produced into a foot. Capsules clavate, to 3 × 0.5 cm.

Rare in open scrub forests in Mahabubnagar district (Sadasivaiah & *al.*, 2010).

Fl. & Fr.: March – May.

Eulophia spectabilis (Dennst.) Suresh in Nicolson & *al.*, Interpret. Hort. Malab. 300. 1988. *Wolfia spectabilis* Dennst., Schlussel sum Hort. Int. Malabar. 11, 25, 38. 1818. *Eulophia nuda* Lindl., Gen. Sp. Orchid. Pl. 180. 1833; Hook. f., Fl. Brit. India 6:5. 1890; Ramakrishna in Pullaiah, Fl. Andhra Pradesh 3: 940. 1997.

A terrestrial herb with large tuberous, slightly branched root-stock; roots long, vermiform. Leaves 2 at flowering stage, afterwards more, linear-lanceolate, 25-60 × 2-3.5 cm, base narrowing into a long sheath and attaching to the short stem, margin entire, apex acuminate. Racemes loosely flowered, arising from the base of stem; peduncle 40-50 cm long. Floral bracts lanceolate, membranous; sepals oblong, adnate to the foot of the column; petals oval-oblong, obtuse, lip ovate, recurved; lateral lobes erect, rounded, embracing the column, margin crisped; midlobe oblong, bifid, undulate, column continuing into the foot, spur dark purplish-green, conical; disc with many crested ridges, anthers 2-loculed, pollinia 2, connected to a large gland by a short, stout strap; ovary purplish. Capsules fusiform.

Occasional in moist deciduous open hill forests of Karimnagar district.

Fl. & Fr.: February – June.

Way to Mukunus (KRN), *PVS* 84027 (BSID).

GEODORUM G. Jackson

Geodorum densiflorum (Lam.) Schltr., Repert. Spec. Nov. Regni Veg. Beih. 4: 259. 1919; Fischer, Fl. Madras 3: 1437. 1928. *Limodorum densiflorum* Lam., Encyl. 3: 516. 1792. *Geodorum dilatatum* sensu Hook. f., Fl. Brit. India 6: 17. 1890, (non R. Br.) *G. purpureum* R. Br., Hort. Kew. 5: 207. 1813; Hook. f., Fl. Brit. India 6: 17. 1890.

A terrestrial herb with underground, tuberous, pseudobulbs; stem sheathed. Leaves bright green and shiny, 2 or 3, oblong-lanceolate, 12-28 × 6-10 cm, base narrow, margin wavy, apex acuminate. Flowers crowded in decurved racemes, white with yellow and pink lip; floral bracts lanceolate, membranous; sepals linear-oblong; petals about the same length and shape as sepals, obtuse, lip superior, sessile, cymbiform recurved, undulate with a broad ridge ending in a yellow warted callus; column short. Capsules pendulous, fusiform, 2 × 0.3 m.

Frequent in moist deciduous open forests of Karimnagar district.

Fl. & Fr.: August – November.

Aklaspur (KRN), *GVS* 20216 (MH).

HABENARIA Willd.

1. Petals bipartite .... **H. ramayyana**
1. Petals entire:
    2. Side lobes of lip broad:
        3. Leaves radical, adpressed to ground .... **H. roxburghii**
        3. Leaves not adpressed to ground, cauline .... **H. plantaginea**
    2. Side lobes of lip filiform or linear .... **H. marginata**

Habenaria marginata Coleb., Exot. Fl. t. 136. 1824; Hook. f., Fl. Brit. India 6: 150. 1890; Fischer, Fl. Madras 3: 1471. 1928.

Vern.: *Kusuma gadda.*

A terrestrial, tuberous herb, 10-25 cm long. Leaves few, radical, sessile, 5-10 × 1.2-2.5 cm, oblong, obtuse, with yellow or white margins, reduced upwards to lanceolate, finely acuminate sheaths. Flowers yellow, sessile, numerous, in 3-6 cm long, cylindric spikes, bracts ovate-lanceolate, lateral sepals broad, obliquely oblong, acute, dorsal sepal broadly ovate, acute, cucullate, petals oblong-lanceolate, acute, falcate from a broad base, closely appressed to the dorsal sepal, lip nearly 1.2 cm long with broad claw, 3-lobed, spur stout, curved, clavate at the tip, anther-tubes short, spreading, funnel-shaped, stigmatic processes elongate, adnate to the lip, rostellum triangular, acute. Capsules to 1 × 0.2 cm, fusiform, turgid, curved, obscurely beaked.

Occasional in moist deciduous open forests and hills in Nizamabad, Medak and Mahabubnagar districts.

Fl. & Fr.: September – November.

Narsapur (MDK), *KMS* 6809 (MH); Patharajampet (NZB), *BR & CPR* 7163; Narsapur (MDK), *KMS* 6809 (MH); Thupakulagudem (WGL), *PVS* & *NRR* 76988 (BSID).

Habenaria plantaginea Lindl., Gen. Sp. Orchid. Pl. 323. 1835; Hook. f., Fl. Brit. India 6: 141. 1890; Fischer, Fl. Madras 3: 1470. 1928.

A terrestrial herb with a very short stem, oblong tubers and fleshy radical fibres. Leaves radical, narrow oval, to 7.5 × 2.5 cm, margin entire, apex acute, membranous. Scape slender, 20-30 cm long with many, open lanceolate, acuminate sheaths. Racemose spikes 10-15 cm long. Flowers white, floral bracts lanceolate, much shorter than the ovary; sepals subequal, dorsal ovate-oblong, laterals falcately oblong; petals linear-lanceolate, lip more than twice as long as the sepals, flabelliform, puberulous, 3-lobed, lobes subacute, entire or toothed, lateral flabelliform, puberulous, 3-lobed, lobes subacute, entire or toothed, lateral half ovate, midlobe long, narrowly linear, spur very slender, column short; rostellum broad, triangular; stigmatic processes prominent, clavate; base of the column continuous with lip, anther lateral, fused to column, 2-loculed. Capsules to 1.5 × 0.4 cm, fusiform, curved; beak short, slender.

Frequent in moist deciduous forests, under shade, on moist forest floors in Mahabubnagar district.

Fl. & Fr.: September – October.

Habenaria ramayyana Ram. Chary & J.J. Wood, Kew Bull. 36: 235. 1981.

Tuberous herbs. Leaves narrowly elliptic to oblong-elliptic, margin minutely papillose. Flowers purplish white on a terminal raceme, dorsal sepal ovate-elliptic, subfalcate, petals bipartite; lip 3-lobed, sidelobes filiform, acute at apex; midlobe deflexed; spur narrowly cylindrical, column short.

Rare in moist deciduous forest in Mahabubnagar district.

Fl. & Fr.: July – September.

Amrabad forest reserve (MBNR), *STR* 642 (Holotype HH; Isotype CAL, K and MH).

Habenaria roxburghii Nicolson in Saldanha & Nicolson, Fl. Hassan Dist. 834. 1876. *Orchis roxburghii* Pers., Syn. Pl. 2:503. 1807 *nom. illeg.* (incl. *O. plantaginea* Roxb.) *Habenaria platyphylla* (Willd). Spreng. in L., Syst. Veg. (ed. 16) 690. 1826, *nom. illeg* (incl. *Orchis plantaginea* Roxb.) Hook. f., Fl. Brit. India 6: 140. 1890; Fischer, Fl. Madras 3: 1470. 1928.

A terrestrial orchid, to 40 cm long. Leaves 3-6, appressed to the earth, sessile, broadly elliptic or orbicular, 4-10 × 4-8 cm, apex rounded or acute. Scape 10-40 cm long, densely-flowered, bracts much shorter than the long beaked ovary. Flowers white; sepals broadly oblong, petals linear, lip side lobes broad, midlobes narrower, spur about 1.5 cm long, slender, clavate at tip, longer than the ovary, anther cells divergent; stigmatic processes subglobose, rostellum short, triangular.

A frequent herb in most of the districts, often sheltered inside bushes.

Fl. & Fr.: July – October.

Sirpur RF (ADB), *GO & PVP* 4908; Ghanapur RF (NZB), *BR & KH* 7101; Muttapally (MDK), *CP* 11779; Anantagiri (RR), *TP* & *MSM* 11080; Pakhal (WGL), *KMS* 13133 (CAL); Mukunur (KMM), *PVS* 84028 (BSID).

LUISIA Gaud.

Luisia tristis (G.Forst.) Hook.f.,Fl. Brit. India 6: 251. 1890. *Epidendrum triste* G. Forst., Fl. Ins. Austr. 60. 1786. *Luisia zeylanica* Lindl., Fol. Orchid. 4: 3. 1853. *L. teretifolia* sensu Hook. f., Fl. Brit. India 6: 22. 1890 (non Gaud. 1829). *L. truncata* Blatt. & McCann, J. Bombay Nat. Hist. Soc. 35: 491. 1932.

A tufted epiphytic herb, 15-30 cm long; stem terete, sheathed, roots vermiform. Leaves green, spotted with purple, 5-16 × 0.5-1 cm, terete. Flowers in short 2-3-flowered spikes, sepals greenish-yellow; lateral sepals oblong, subacute, with a winged keel at the back; dorsal sepal linear-oblong, petals yellow, as long as the sepals, linear-oblong, obtuse, lip 0.6 cm long, hypochite green spotted with purple, subsaccate, epichite dark purple, broadly cordate, rhomboid-ovate, obtuse, obscurely 3-lobed, deflexed, fleshy, column stout. Capsules oblong, to 3 × 0.8 cm, winged.

Rare, growing on the branches of trees in the moist deciduous forests of Mahabubnagar district

Fl. & Fr.: August – November.

VANDA W. Jones ex R. Br.

1. Flowers more than 2.5 cm long; sepals and petals brown tessellated; lip white; capsules 6-8 cm long .... **V. tessellata**
1. Flowers less than 2.5 cm long; sepals and petals not tessellated, lip pink; capsules 2.5-4 cm long .... **V. testacea**

Vanda tessellata (Roxb.) Hook. ex G. Don, Hort. Brit. 372. 1830; Fischer, Fl. Madras 3: 1445. 1928. *Epidendrum tessellatum* Roxb., Pl. Coromandel t. 42. 1795. *Vanda roxburghii* R. Br. in Bot. Reg. 6: t. 506. 1820; Hook. f., Fl. Brit. India 6: 52. 1890.

A stout epiphyte with non-pseudobulbous stem 30-60 cm long. Leaves linear-oblong, 15-20 × 1.6-2.5 cm, thick, coriaceous, distichous, strap-shaped, margin entire, apex obtusely keeled with two unequal lobes and a short pointed one in between. Racemes 4 to 10-flowered, 15-20 cm long; peduncle stout. Flowers large, 5 cm across, grey, greyish-blue, buff or red, bracts broadly triangular-ovate, clasping the rhachis, obtuse, 1-veined, dorsal sepal obovate, faintly trifid, margin wavy; lateral sepals obovate, obtuse; wavy; petals obovate, obtuse, lip funnel-shaped, 3-lobed; lateral lobes erect, oblong, acute; midlobes panduriform, constricted below the bifid apex;

disc swollen with fleshy curved ridges, spur conical, short, blunt and laterally compressed, column fleshy, cylindrical. Capsules 6-9 cm long, ribbed with a short pedicel.

Common in almost all districts, common on the trees in deciduous forests.

Fl.: July – August. Fr.: August onwards.

Ardinal beat (ADB), *TP & PVP* 4129; Ichampalli RF (ADB), *DAM & GO* 5157; Bhimnagar RF (NZB), *TP* & *BR* 6210; Gandhari (NZB), *BR* & *CPR* 9252; Dharmapur gutta (RR), *MSM* & *KH* 10290; Bollarum (NLG), *BVRS* 334 (NU).

Vanda testacea (Lindl.) Rchb. f., Gard. Chron. n.s., 8: 166. 1877. *Aerides testacea* Lindl., Gen. Sp. Orchid. Pl. 238. 1833 (*testaceum*). *Vanda parviflora* Lindl. in Edwards Bot. Reg. 30: Misc. 57. 1844; Hook. f., Fl. Brit. India 6: 50. 1890; Fischer, Fl. Madras 3: 1444. 1928.

An epiphyte with a non-pseudobulbous stem, 10-20 cm long; roots large, thick and vermiform. Leaves linear-lorate, distichous, 10-25 × 1-2 cm, coriaceous, margin entire, apex unequally lobed or toothed, with the midrib forming the third tooth. Raceme erect, many-flowered, 5-15 cm long; peduncle stout. Flowers cream-yellow with a white, purple or reddish-pink lip, bracts broadly ovate, acute, dorsal sepal obovate and incurved at the apex; lateral sepals obovate, incurved and twisted upwards laterally at the apex; petals spathulate, concave, narrow and twisted at the base downwards, lip 3-lobed, spur narrow, funnel-shaped; column short, truncate. Capsules 2-3 cm long, pedicelled, clavate.

Occasional in deciduous forests in hilly tracts in Mahabubnagar district.

Fl.: April – May, Fr.: June onwards.

Kudichintalabailu (MBNR), *TDCK* 15385

ZEUXINE Lindl. *nom. cons.*

Zeuxine strateumatica (L.) Schltr., Bot. Jahrb. Syst. 45.394. 1911; Fischer, Fl. Madras 3: 1456. 1928. *Orchis strateumatica* L., Sp. Pl. 943. 1753. *Zeuxine sulcata* Lindl., ex Wight, Cat. Ind. Pl. 123. 1836; Hook. f. Fl. Brit. India 6: 106. 1896.

A terrestrial glabrous herb, 6-15 cm long; stem erect but for a short, creeping portion in the ground below. Leaves erect, sessile, sheathing of different sizes; lower ones 1.5-8 × 0.3-0.8 cm, upper one smaller, linear-lanceolate, margin entire, recurved, apex acute, midrib prominent and stout. Flowers small, white with a yellow lip; floral bracts erect, lanceolate, sepals oblong, membranous and subequal; laterals free, dorsal concave cohering with petals forming a hood and enclosing the base of the lip; petals oblong; hyaline, lip saccate at base, clawed, emarginated or bilobed apically, yellow, hypochile at sac with 1-2 glands on each side inside, column without foot, rostellum short, deeply divided, pollinia 2, pyriform, granular, attached by a gland, with or without a caudicle. Capsules ovoid, 0.5 long.

Rare in moist places among grasses along the banks of rivers in Adilabad and Hyderabad districts.

Fl.: & Fr.: December – February.

Potchera stream (ADB), *GO & TP* 4701.

Spathoglottis plicata Bl. is cultivated as ornamental.

ZINGIBERACEAE

1. Leaves spiral .... **Cheilocostus**
1. Leaves distichous:
    2. Inflorescence terminal on a leafy shoot .... **Globba**
    2. Inflorescence on a leafless scape:
        3. Lateral staminodes large, petaloid; inflorescence not radical ..... **Curcuma**
        3. Lateral staminodes minute; inflorescence radical .... **Zingiber**

CHEILOCOSTUS C.Specht

Cheilocostus speciosus (J.König) C.Specht, Taxon 55: 159. 2006. *Banksia speciosa* J.König, Observ. Bot. 3: 75. 1783. *Costus speciosus* (J.König) Smith, Trans. Linn. Soc. London 1: 249. 1800; Baker in Hook. f., Fl. Brit. India 6: 249. 1892; Fischer, Fl. Madras 3: 1490. 1928. *Costus speciosus* var. *nipalensis* (Rosc.) Baker in Hook. f., Fl. Brit. India 6: 249. 1892.

An erect herb, up to 2.5 m long; root stock tuberous, insipid; stem subwoody at the base, spirally twisted so that the leaves appear spirally arranged. Leaves subsessile, oblong or lanceolate, 10-30 × 3-8 cm, base rounded, margin entire, apex acute or acuminate, often cuspidate, glabrous above, silky-pubescent beneath, ligule short, to 2 cm; sheath tubular. Flowers white, numerous in terminal spikes; bracts oblong-ovate, reddish, acuminate, often pungently mucronate; bracteole solitary below the calyx; calyx tubular, glabrous, 3-lobed; lobes triangular, corolla white, lobes elliptic-oblong, labellum white, with a yellow centre, broadly obovate with incurved margins, base densely hairy; staminodes medium on an oblong, petaloid process; ovary globose, ovules arranged on axile placentae. Capsules globose-ovoid, tardily-dehiscing; seeds black with white aril.

Occasional in Karimnagar, Warangal and Mahabubnagar districts.

Fl. & Fr.: July – October.

Farahabad (MBNR), *TDCK* 13659; Pakhal (WGL), *KMS* 25547 (MH).

CURCUMA L.

1. Rhizome without sessile tubers:
    2. Corolla deep purple; labellum purple or with a median yellow band .... **C. inodora**
    2. Corolla yellowish white; labellum bright yellow .... **C. pseudomontana**
1. Rhizome with sessile tubers:
    3. Rhizome light orange yellow or deep orange yellow inside:

4. Rhizome light orange yellow inside; labellum deep yellow ... **C. montana**
4. Rhizome deep orange yellow; labellum white with median dark yellow band .... **C. longa**
3. Rhizome cream or ivory inside .... **C. aromatica**

Curcuma aromatica Salisb., Parad. Lond. t. 96. 1808; Baker in Hook. f., Fl. Brit. India 6: 210. 1890; Fischer, Fl. Madras 3: 1483. 1928.

A plant with large root-stock, of palmately branched, sessile, annulate, biennial tubers, yellow and aromatic inside. Leaves elliptic or lanceolate-oblong, 30-60 cm long, caudate-acuminate, green, often variegated above, pubescent beneath, base deltoid; petioles as long as or longer than the blade. Spike 15-30 cm long; flowers pink, fragrant, shorter than the bracts, bracts ovate, recurved, cymbiform, obtuse, fertile, pale green, coma more or less tinged with red or pink; calyx irregularly 3-lobed; corolla tube 1.5 cm long, the upper half funnel-shaped, lateral segments oblong, upper longer, ovate, concave; lip yellow, obovate, deflexed, subentire or obscurely 3-lobed; staminodes obtuse, as long as the corolla-segments; ovary oblong; style filiform; stigma obscurely 2-lobed. Capsules globose, seeds ovoid.

Rare in Mahabubnagar district (Raghava Rao, 1990).

Fl. & Fr.: June – September.

Curcuma inodora Blatter, J. Proc. Asiat. Soc. Beng. 26: 357. 1931; Santapau, J. Bomaby Nat. Hist. Soc. 51: 135. 1952.

Vern.: *Adavi pasupu.*

Herbs up to 60 cm high; rhizome with many tuberiferous fibrous roots; tubers white inside, odourless. Leaves radical, 16-30 × 6-13 cm, broadly elliptic-lanceolate, plicate, with a red band on midnerve. Spikes 10-20 cm long; flowers purplish with yellow streak on the lip. Capsules 3-valved; seeds arillate.

Rare in shady, moist localities of forests in Warangal district (C.S. Reddy, 1999; C.S. Reddy & *al.*, 2001). Prof. M.Sabu of Calicut University is of the opinion that occurrence of this species in Telangana is doubtful (Pers. Commun.).

Pasra RF (WGL), *CSR* 1356 (KU); Tadvai RF (WGL), *CSR* & *VSR* 1336 (KU).

Curcuma longa L., Sp. Pl. 1: 2. 1753; Hook.f. Fl. Brit. India 6 : 214. 1890; C. Fischer, Fl. Madras, 3 : 1483. 1928; Leong-Skornickova & al., Taxon 59: 274. 2010. *C. domestica* Valeton, Bull. Jard. Bot. Buitenzorg (ser.2) 27 : 31. 1918; Holttum, Gard. Bull. Straits Settlem. (ser.4) 13 : 68. 1950; B.L. Burtt & R. M. Smith, Notes Roy. Bot. Gard. Edinburgh 31 : 185. 1972.

Vern.: *Pasupu.*

Rhizomatous herbs, up to 45 cm high; rhizomes cylindrical, orange yellow within, aromatic. Leaves oblong-elliptic or oblong lanceolate, acuminate, 25 x 12 cm; petioles, 25 cm long. Flowers in terminal, shortly exserted, dense, 15-20 cm long spikes; bracts pale green, ovate, 2.5-4 cm long, those of the coma pale pink; calyx scarsely 3-lobed;

corolla slightly longer with oblong lobes; lip yellow; fertile stamen 1, not crested, finely spined, 6 mm; ovary hairy.

Cultivated extensively in Nizamabad district and sporadically in other districts.

Fl. & Fr. : July-August.

Curcuma montana Roxb., Pl. Coromandel 2: 28. 1800; Baker in Hook. f., Fl. Brit. India 6: 214. 1890 *p.p.*; Fischer, Fl. Madras 3: 1483. 1928; A.S.Rao & Verma, Bull. Bot. Surv. India 14: 122. 1972.

Vern.: *Maccha pasupu.*

Herbs, up to 70 cm high; rhizome and sessile tubers yellowish inside and aromatic; root fibres with elliptic tubers, 4 cm long, white inside and not aromatic. Leaf tuft 50-70 cm long; leaves oblong, elliptic-oblong or oblong-lanceolate, up to 50 × 20 cm, acute, midrib purplish, base tapering. Inflorescence rising in the centre of leaf tuft, spike 10 × 4.5 cm, coma bracts pink or purplish. Flowers pale yellow, longer than bracts; labellum suborbicular, entire, margins undulate.

Rare in shady, moist localities of forests.

Fl. & Fr.: June – September.

Pakhal RF (WGL), *CSR* & *VSR* 1567 (KU).

Curcuma pseudomontana J. Graham, Cat. Pl. Bombay 210. 1839; Fischer, Fl. Madras 3: 1483. 1928. *C. montana* sensu Baker in Hook. f., Fl. Brit. India 6: 214. 1890 *p.p.*,non Roscoe 1807.

Vern.: *Adavi pasupu.*

An erect herb, 75-80 cm long with small rootstock bearing small almond-like or subglobose tubers at the ends of the fibres, tubers pure white inside and edible. Leaves uniformly green, reaching 30-60 cm or more long (including the petiole), lanceolate-oblong, margin entire, apex acuminate, tapering to the base. Flowers bright yellow, in racemose, oblong, spikes on 7-10 cm long peduncle, flowering bracts obovate-lanceolate, the lowest with purple edges only; the upper more or less uniformly mauve-purple, stamen 1 perfect, filament short; anthers not crested, with contiguous, spurred at the base; lateral staminodes oblong, petaloid, lip orbicular, with a deflexed tip, ovary 3-celled, ovules many on axile placentation, style filiform, stigma 2-lipped. Fruit a dehiscent, globose, 3-valved capsule; seeds ovoid or oblong, usually arillate.

In shady localities in most districts.

Fl. & Fr.: May – September.

Vempalli RF (ADB), *GO & PVP* 4780; Chedmal RF (NZB), *BR* & *CPR* 9255; Narsapur (MDK), *TP* & *CP* 14060; Edrallapalli (KMM), *RCS* 102410 (BSID); Way to Mukunur (KRN), *PVS* 84023 (BSID); Mannanur RF (MBNR), *VBH* 86622 (BSID); 8 km towards north from Pasra (WGL), *RKP* 108242 (BSID); way to Gangaram (WGL), *PVS* 84004 (BSID).

Figure 106: Curcuma pseudomontana J. Graham

A. Habit

GLOBBA L.

Globba marantina L., Mant. Pl. 2: 170. 1771; Baker in Hook. f., Fl. Brit. India 6: 206. 1890. *G. bulbifera* Roxb., Asiat. Res 11: 358. 1810; Hook.f., Fl. Brit. India 6: 206. 1890; Fischer, Fl. Madras 3: 1481. 1928.

A slender herb, 40-50 cm long, with creeping rhizome and erect stem; roots fleshy. Leaves distichous, oblong or oblong-lanceolate, 9-15 × 3-5 cm, base cuneate, margin entire, apex caudate; ligule broad, ciliate; sheath ciliate along the apical margin; petiole reduced or short. Spike terminal, 10-15 cm long; flowers orange-yellow, fragrant and buds sometimes replaced by bulbils, peduncle 6-8 cm long, exerted from leaf-sheaths; lower sterile bracts oblong, upper one persistent, calyx funnel-shaped, 3-lobed, ciliate; corolla yellow, tubular, slender, longer than the calyx, lobes ovate, subequal, spreading or reflexed; labellum small, 2-fid, base auricled, narrow; lateral staminodes petaloid; stamen with a long 2-appendaged filament; anther oblong, with spreading 2-fid wings, ovary globose, 1-celled; ovules arranged on 3 parietal placentae, style filiform, stigma turbinate. Capsules globose, tardily dehiscent, to 4 mm; seeds ovoid, often tomentose.

In shady localities of forests in Mahabubnagar district.

Fl. & Fr.: August – October.

Farahabad (MBNR), *GS* 20818.

ZINGIBER Boehmer

1. Spikes terminating the leafy stem .... **Z. capitatum**
1. Spikes radical, produced directly from the root stock:
    2. Flowers red or scarlet .... **Z. roseum**
    2. Flowers greenish or greenish yellow .... **Z. officinale**

Zingiber capitatum Roxb., Asiat. Res. 11: 348. 1810; Baker in Hook. f., Fl. Brit. India 6: 248. 1892.

A tuberous herb. Leaves linear, ascending, 20-40 × 3-5 cm, tapering gradually to the point, usually pubescent beneath. Racemose spike at the end of the leafy stem, densefid, erect, oblong-cylindrical, 7-15 × 3-5 cm; bracts closely imbricate, subcoriaceous, green with a narrow brown edge; corolla tube as long as the bract, pale yellow, unspotted, midlobes orbicular, emarginated, basal auricles large, oblong, obtuse, bright red; stamen pale yellow, nearly as long as the lip; ovary 3-celled, ovules many on axile placentae. Capsules bright red; seeds black, shining, aril large, lacerated white.

In Mahabubnagar and Warangal districts.

Fl. & Fr.: July – August.

Pakhal (WGL), *KMS* 11704 (MH & CAL).

Zingiber officinale Roscoe, Trans. Linn. Soc. London 8 : 348. 1807; Baker in Hook. f. Fl. Brit. India 6 : 247. 1892.

Vern.: *Allamu.*

Perennial herbs with short corm, to 60 m high; rhizome horizontal, branched, creeping with brown scale leaves. Leaves subsessile, oblong-lanceolate, acute, glabrous. Flowers greenish or greenish yellow with a small dark purple labellum in dense oblong-ellipsoid spikes, 15 x 4 cm; peduncle 40-50 cm high; calyx persistent, reddish; labellum 3-lobed, midlobe oblong-obovate, lateral lobes short, obtuse; stamens dark purple, as long as the labellum, rather shorter than the corolla. Capsule red; seeds black.

Cultivated occasionally for its rhizome which is used as spice.

Fl. & Fr. : July-August

Zingiber roseum (Roxb.) Roscoe, Trans. Linn. Soc. London 8: 348. 1807; Baker in Hook. f., Fl. Brit. India 6: 244. 1892; Fischer, Fl. Madras 3: 1489. 1928. *Amomum roseum* Roxb., Pl. Coromandel t. 126. 1800.

A tall herb. Leaves oblong or oblong-lanceolate, 30-40 × 6-10 cm, margin entire, apex acuminate, sessile, more or less pubescent below. Flowers red or scarlet in racemose spikes arising directly from tuberous root, bracts red, exterior ones broadly ovate, interior linear-lanceolate, more or less hairy; calyx cylindric, shortly 3-lobed; corolla red or scarlet, lip 3-lobed, midlobes oblong-cuneate, margins recurved, crisped, lateral lobes short, broad; stamens arching over the lip and equaling it in length, ovary 3-celled, ovules many on axile placentation, style filiform, stigma small. Capsules subglobose, dehiscent; seeds many, minute.

Occasional in forests in Mahabubnagar district.

Fl. & Fr.: August – September.

Farahabad (MBNR), *TDCK* 15378.

MUSACEAE

MUSA L.

Musa paradisiaca L., Sp. Pl. 1043. 1753; Backer & Bakhuizen, Fl. Java 3: 36. 1968. *M. sapientum* L., Syst. Nat. Ed. 10. 2: 1303. 1759; Hook.f., Fl. Brit. India 6: 262. 1888; Fischer, Fl. Madras 3: 1491. 1928.

Vern.: *Arati.* English: Banana

Stoloniferous plant, 1-1.5 m; pseudostem with leaf sheaths; leaves very large, spirally arranged, oblong. Flowers monoecious, on stout, drooping, elongate, bracteate spikes; male flowers above, female flowers below. Fruit oblong, yellowish green when ripe,.

Extensively cultivated for its edible fruits.

Fl. & Fr.: Round the year.

Ravenala madagascarensis Sonn. is cultivated in parks and gardens as an ornamental.

CANNACEAE

CANNA L.

Canna indica L., Sp. Pl. l. 1753. *C. orientalis* Roscoe, Scit. Pl. t. 12. 1828; Fischer, Fl. Madras 3 : 1496. 1928. *C. indica* var. *orientalis* (Roscoe) Baker in Hook. f., Fl. Brit. India 6: 260. 1892.

Perennial, erect herb, to 1.5 m high. Leaves elliptic-lanceolate, 12-50 × 8-20 cm, glabrous, base obtuse, margin entire, apex caudate-acuminate. Flowers in terminal spikes scarlet, orange or yellow; sepals 3; more or less tubular below; stamens partly adnate to corolla tube, usually 1-5, one bearing single anther cell on the margin of a petaloid stamen, staminodes petaloid, one opposite the fertile stamen secured, other erect; ovary 3-celled, style flattened, stigma terminal. Capsules globose or ellipsoid, echinate, to 2 cm.

Commonly grown as ornamental throughout the State. Also as an esape especially in moist places in villages, fields.

Fl. & Fr.: Throughout the year.

AMARYLLIDACEAE

1. Filaments basally united by a coronal membrane into a toothed cup .... **Pancratium**
1. Filaments not united by a coronal membrane .... **Crinum**

CRINUM L.

1. Perianth-tube straight, erect; lobes stellately patent; stamens spreading
   2. Leaves over 10 cm wide; umbels 10-30-flowered; neck of the bulb 10 cm or more across .... **C. asiaticum**
   2. Leaves under 5 cm wide; umbels 6-12-flowered; neck of the bulb 6 cm or less across:
      3. Perianth not tubular; lobes linear .... **C. viviparum**
      3. Perianth tubular, lobed, lobes as long as the tube or shorter than the tube .... **C. defixum**
1. Perianth-tube curved, declinate, lobes ascending; stamens declinate .... **C. latifolium**

Crinum asiaticum L., Sp. Pl. 292. 1753; Hook. f., Fl. Brit. India 6: 280. 1892; Fischer, Fl. Madras 3: 1504. 1928.

A stout herb, up to 90 cm high; stem with white bulbous root stock, 12 × 10 cm; neck 10-20 cm, clothed with old leaf-sheaths; leafy stem not dying down annually and plant flowering with the leaves. Leaves linear-lanceolate, to 150 × 10 cm, glabrous, base narrow, margin entire, apex gradually tapering. Scape 0.5 – 1 m long, 1-2 cm across. Flowers white, 15-50 in an umbel; bracts hooded, ovate-lanceolate; perianth-tube white, salver-shaped, lobes oblong-linear, glabrous, cuspidate; stamens reddish, filaments slender, 4 cm, shorter than the lobes of the perianth, ovary 3-celled, style filiform, stigma minute, subcapitate. Fruit sub-globose, 2-5 cm in diameter, 1-(rarely 2)-seeded, dehiscing irregularly.

Often grown as an ornamental.

Fl. & Fr.: July – December.

Crinum defixum Ker-Gawl., J. Sci. Arts (London) 3: 105. 1817; Hook. f., Fl. Brit. India 6: 281. 1892; Fischer, Fl. Madras 3: 1504. 1928.

A stout herb with bulbous rootstock; bulb ovoid, 8 × 6 cm, fusiform, stoloniferous base; neck 15 cm long, cylindric. Leaves 6-8 to a bulb, 60-90 cm long, deeply canaliculated, coriaceous, scaberulous, apex gradually tapering, scapes from the axils of the lowest leaves, 45-75 cm long, erect, cylindric. Umbels 14-20 cm long, 10 cm across. Flowers white, fragrant at night, 6-12-flowered, pedicels very short; bracts lanceolate, perianth salver-shaped; lobes white, oblong, glabrous, apex obtuse, cuspidate; stamens 6, bright-red; ovary 3-celled, oblong, style delicate, stigma simple. Capsule subglobose, seeds 1-2, rugose.

Common in all districts along streams and back waters.

Fl. & Fr.: November – January.

Nizamsagar (NZB), *BR & CPR* 9248; Mohammadabad (RR), *MSM* & *NV* 10425; Borapur (MBNR), *BSS* & *SKB* 32389; Godavari (KRN), *GVS* 22473 (MH); Dosalagunda lake,Rusthempet-Narsapur (MDK), *KMS* 6802 (MH).

Crinum latifolium L., Sp. Pl. 291. 1753; Hook.f. Fl. Brit. India 6 : 283. 1892; Fischer, Fl. Madras 3 : 1504. 1928.

Stout perennial herbs, of about 2 m high; rosette-like, arises from an underground bulb; bulbs 10-15 cm with short and stout neck. Leaves long, linear, coriaceous, acuminate with thin scabrid margins and ligulate. Scape 50-100 cm long; umbels 10-20 flowered; perianth white, funnel-shaped, streaked or tinged with red or red purple; stamens shorter than the tube; style longer than the stamens. Capsule 2-5 cm in diam.

Grown in gardens and natrualized elsewhere.

Fl. & Fr.: July.-August.

Kudichintalabailu Ravu area (MBNR), *TDCK* 15351.

Crinum viviparum (Lam.) R. Ansari & V. J. Nair, J. Econ. Taxon. Bot. 11 : 205. (1987) 1988. *Amaryllis viviparum* Lam., Encycl. 1 : 123. 1983.

Plants 0.5-1 m high; bulbs 5-8 cm, ovoid with 5-15 cm long cylindrical neck. Leaves 30-50 cm long, about 3 cm wide, dark green. Flowers arise in an umbel of 6-12 atop a leafless scape, 40-75 cm high; bracts lance-shaped,white; perianth tube 6-8 cm, slender; tepals narrow, spreading, 6-8 cm long; stamens, 4-6 cm long; filaments red.

Grown in gardens.

Fl. & Fr.: September.-October.

PANCRATIUM L.

Pancratium triflorum Roxb., Fl. Ind. 2: 126. 1832; Hook. f., Fl. Brit. India 6: 285. 1892; Fischer, Fl. Madras 3: 1604. 1928. *P. verecundum auct. non.* Ait. 1789; Hook. f., Fl. Brit. India 6: 286. 1892.

A herb with globose bulb, bulb 6 × 5 cm, neck 0 (or) short, 1 cm. Leaves linear-lanceolate, 12-28 × 1-2.5 cm, flat, thin coriaceous, glabrous, base narrow, apex subacute. Scape slender, 12-20 cm long. Flowers fragrant, 3-8 in an umbel, spathe ovate, acute; pedicel very short, perianth funnel-shaped, lobes 6, white, narrow, soon reflexed, stamens 6, filaments connate at base into a petaloid 2-dentate membranous cup; teeth ca 7, ovary oblong, 3-celled, ovules many per cell, 2-seriate, style filiform, stigma small, capitate. Capsule subglobose.

In hill areas of Visakhapatnam district up to 600 m.

Fl. & Fr.: July – Short lived.

AGAVACEAE

1. Ovary superior .... **Sansevieria**
1. Ovary inferior:
    2. Stamens exserted .... **Agave**
    2. Stamens inserted .... **Furcraea**

AGAVE L.

1. Leaves over 90 cm long, in lax rosette or tufted:
    2. Leaves narrow, linear-oblong, marginal spines hooked, curved upwards .... **A. cantula**
    2. Leaves distinctly broader at or above the middle, marginal spines not as above: .... **A. americana**
1. Leaves never exceeding 90 cm, in a close globose rosette .... **A. angustifolia**

Agave americana L., Sp. Pl. 323. 1753; Hook. f., Fl. Brit. India 6: 277. 1892; Fischer, Fl. Madras 3: 1505. 1928.

A perennial stemless, scapigerous herb. Leaves atleast in their upper halves, distinctly exsculptate between the marginal spines, not with horny margins, erecta patent, lower ones partly decumbent, larger ones very often with recurved, rarely incurved upper halves, leaves not combined into dense squarrose crown, lanceolate, with very strong widely patent or recurved marginal spines and 2-5 cm long apical one, thick, rigid, glaucous, very often with longitudinal white or yellow streaks or bands. Panicle (inclusive of peduncle) 4-8 m long; branches numerous, widely patent, sigmoid, apically repeatedly branched, flowers densely crowded, with a strong evil smell; perianth funnel-shaped, segments oblong, yellowish green, stamens exerted. Capsule oblong-clavate, beaked, ca 4 cm long; seeds shining, robust.

Cultivated, often as escape in wastelands.

Fl. & Fl.: September – January.

Yedapally – Bodhan Road (NZB), *BR & KH* 6476; Medak (MDK), *BR* & *CP* 11416; Zaheerabad (MDK), *CP* 11591; Parigi (RR), *MSM* & *PRSR* 15102.

Agave angustifolia Haw., Syn. Pl. Succ. 72. 1812; var. **angustifolia**. *A. wightii* Drumm, & Prain, Bengal Agric, Bull. 8: 15. 1906 & Agric. Ledger 12(7): 91. 1906; Fischer, Fl. Madras 3: 1505. 1928. *A. vivipara* Wight, Icon. Pl. Ind. Orient. t. 2024. 1853; Hook. f., Fl. Brit. India 6: 277. 1892.

A perennial, somewhat stout herb; stem 30-45 cm long; woody, simple, with a rosette of leaves. Leaves erect, narrowed at both ends, constricted at base with a spinescently dentate margin, and an aciculate tip with 1-2 cm long greyish green spine. Spines stout, upcurved. Scape stout, scaly, 2-4 m high. Flowers in terminal panicles; perianth 4-5 cm long, perianth segments oblong-lanceolate, green or brown, thin, with a hardened apex, apically with minute hairs; stamens 6, filaments subequl. Capsule oblong, shortly beaked; seeds numerous, dull. Bulbils often produced.

Common in waste lands, along road sides, forest clearings and often planted as hedge plant.

Fl. & Fr.: July – September.

Agave cantula Roxb., Fl. Ind. 2: 167. 1832; Fischer, Fl. Madras 3: 1505. 1928. *A. americana* sensu Hook. f., Fl. Brit. India 6: 277. 1892 (non L. 1753).

A perennial, somewhat stout, scapigerous herb. Stem very short, woody, simple with a crown of rosette leaves. Leaves patent, recurved in the upper part, remotely spinescently dentate with upwardly hooked spines, spines 1-2 cm long, sharp, dark brown; apical spine robust, 1-2.5 cm long. Scapes stout, scaly, up to 5 m high, patently branched at the top. Flowers greenish, in panicles; flowers shortly pedicelled, perianth 6-8 cm long, tube widened upwards, with 6 longitudinal grooves; segments oblong, on the innerside of the top with a tuft of hairs; filaments brown-blotched. Bulbils present.

Common in waste places road sides, often planted as hedge plant.

Fl. & Fr.: July – September.

FURCRAEA Ventenat

Furcraea foetida (L.) Haw., Syn. Pl. Succ. 78. 1812. *Agave foetida* L., Sp. Pl. 323. 1753. *Furcraea gigantea* Vent., Bull. Soc. Philom. Paris 1: 65. 1793; Fischer, Fl. Madras 3: 1505. 1928.

A perennial herb; superterraneous stem short or long, woody. Leaves densely crowded, squarrose, lanceolate or linear-lanceolate, acute, tough-fibrous, spinous-dentate or entire. Flowers in terminal, large, widely branched panicles; flowers along the rhachises of the inflorescence fascicled or solitary, shortly pedicelled, pendulous; perianth constricted above the ovary, chorophyllous, tepals 6, erecto-patent, subequal, the inner ones broader; stamens 6, inserted at the base of the tepals, erect, filaments in the middle with a spongy thickening, anther medifixed; ovary 3-celled, cells numerousovuled, style basally much thickened and triquetrous-trilobed, higher up filiform-subulate, stigma small. Capsule ellipsoid, trigonous, loculicidally 3-valved, seeds numerous, flat.

Occasional in dry tracts of some districts.

Fl. & Fr.: August – October.

Bakarapeta (CTR), *KRN* & *DAM* 9841; Anantagiri (VSKP) *GVS* 32823 (MH).

SANSEVIERIA Thunb. *nom. cons.*

Sansevieria roxburghiana Schult. & Schult. f., Syst. Veg. 7: 357. 1829; Hook. f., Fl. Brit. India 6: 271. 1892; Fischer, Fl. Madras 3: 1520. 1928. *S. zeylanica* Roxb., Pl. Coromandel t. 184. 1805,(non (L.) Willd. 1799).

A herb with very short stem or stemless; root stock creeping. Leaves about 8 or 9 in a tuft, 75-90 × 2-5 cm, towards the middle, subrect, rigid, pale green with transverse, dark green bands, thick, with spine-like tip. Flowers greenish-white tinged with violet, in fascicles of 3-6 on 30-60 cm long racemes, bracts membranous: perianth-tube slender, lobes 6, long, narrow; stamens 6, on the perianth tube, filaments filiform, anthers dorsifixed; ovary 3-celled, ovules solitary in each cell, erect, style filiform, stigma simple. Fruit memberanous, indehiscent; seeds 1-3, large, globose, fleshy, ripening outside the pericarp.

Occasional in all districts both in forests and waste lands.

Fl. & Fr.: June – August.

HYPOXIDACEAE

CURCULIGO Gaertn.

Curculigo orchiodes Gaertn., Fruct. Sem. Pl. 1: 63. 1788; Hook. f., Fl. Brit. India 6: 279. 1892; Fischer, Fl. Madras 3: 1502. 1928.

Vern.: *Neella thati.*

A plant with stout root stock, short or elongated (sometimes 30 cm long) with fleshy root fibres. Leaves lanceolate, 5-20 × 0.8-1.5 cm, membranous, plicate, glabrous or sparsely softly hairy, the tips sometimes rooting and reaching the ground, base sheathing, margin entire, apex acuminate; petiole usually short or often absent; scapes 3-5 or more, slender, hidden by leaf-sheaths. Flowers bright-yellow, in 8 cm long raceme, distichous, spathaceous, glabrous, densely imbricating; perianth 6-lobed to the base, lobes yellow, oblong-elliptic, outer ones pilose without, inner ones sparsely pilose along nerves; stamens 6, filaments filiform; ovary 3-celled, oblong, stigmas 3, lobes elongate. Berry oblong, glabrescent, pericarp membranous; seeds ca 8, globose, glossy, beaked.

Abundant on forest floor in all districts.

Fl. & Fr.: Throughout the year.

Rebbana (ADB), *GO & PVP* 4838; Dichpally (NZB), *BR & KH* 6441; Gnanapur (MDK), *CP* 11791; Rudraram RF (RR), *MSM* 12712; Farahabad (MBNR), *TDCK* 15363;

East of Rechapalli (KRN), *GVS* 20154 (MH); Pakhal RF (WGL), *KMS* 13152 (MH & CAL); Kannaigudem – Bhadrachalam (KMM), *AJR* & *PANReddy* 20467; Rathamhutta hills (KMM), *RCS* 104232 (BSID); Vatavarlapalli (MBNR), *VBH* 84956 (BSID); 8 km towards north of Pasra (WGL), *RKP* 108232 (BSID).

TACCACEAE

TACCA Forst. *nom. cons*

Tacca leontopetaloides (L.) Kuntze, Revis. Gen. Pl. 2: 704. 1891. *Leontice leontopetaloides* L., Sp. Pl. 313. 1753. *Tacca pinnatifida* J.R.Forst. & G.Forst., Char. Gen. Pl. 35. 1775; Hook. f., Fl. Brit. India 6: 287. 1892; Fischer, Fl. Madras 3: 1506. 1928.

A plant with globose root stock, up to 30 cm in diameter; rootlets superficial. Leaves 30-90 cm in diameter, circular in outline, 3-partite, the segments variously pinnatifid, margins undulate; petioles 30-90 cm long, terete, with pale and dark green stripes. Flowers pedicellate, drooping, green, tinged with purple, involucral bracts 6-12, oblong-lanceolate, acuminate, recurved, striped with purple, bracteoles filiform, longer than the bracts; perianth subglobose, greenish, lobes margined with purple, connivent; stamens 6, adnate to the perianth tube, filaments short; ovary inferior, 3-angular, 1-celled, ovules many, on 3 parietal placentas. Fruit globose, ovoid, or oblong, 3-6-ribbed, forming an indehiscent berry; seeds numerous, ovoid, compressed, longitudinally striate.

Occasional in forest in Adilabad, Nizamabad and Karimnagar districts.

Fl. & Fr.: June – November.

Sarvaipet (ADB), *GO* 4507; Nasrullabad beat (NZB), *TP & BR* 6145.

DIOSCOREACEAE

DIOSCOREA L.

1. Leaves compound:
    2. Fertile stamens 6 .... **D. hispida**
    2. Fertile stamens 3:
        3. Leaves glabrous .... **D. pentaphylla**
        3. Leaves tomentose .... **D. tomentosa**
1. Leaves simple:
    4. Stems twining to left .... **D. bulbifera**
    4. Stems twining to right:
        5. Bases of leaves acute-rounded .... **D. oppositifolia**
        5. Bases of leaves cordate .... **D. belophylla**

Dioscorea belophylla (Prain) Haines, For. Fl. Chota Nagpur 530. 1910. *D. nummularia* var. *belophylla* Prain, Bengal Pl. 2: 1065, 1067. 1903. *D. glabra* auct non Roxb.: Hook. f., Fl. Brit. India 6: 294. 1892; Fischer, Fl. Madras 3: 1522. 1928.

A vine; stem unarmed, enlarged into a small rhizome at base emitting long fleshy tuber-bearing fibres. Leaves variable, usually ovate or tapering to an acute point, 7-9-ribbed, veins rather regular, close and parallel, petioles 2-4 cm long. Male flowers in short spikes in the axils or on leafless axillary shoots; sepals ovate-oblong;, petals cuneately obovate; pistillode minute. Capsules sub-orbicular, to 5 × 4 cm, slightly broader than long.

Hills forests of Mahabubnagar district.

Fl. & Fr.: October – December.

Tekulapenta (MBNR), *BSS* & *SKB* 32628.

Dioscorea bulbifera L., Sp. Pl. 1033. 1753, var. **bulbifera**; Fischer, Fl. Madras 3: 1511. 1928. *D. sativa* Thunb., Fl. Jap. 151. 1784 (non L. 1753); Hook. f., Fl. Brit. India 6: 295. 1892.

Vern.: *Nallagadda.*

A vine, stems twining to the left; tubers solitary or 2-3, large, globose; stem terete, unarmed, glabrous, bulbiferous in leaf axils; bulbils warted. Leaves broadly ovate-sub-orbicular, 8-20 × 4-12 cm, dark green, base deeply cordate, margin entire, apex acuminate to shortly caudate. Male flowers variable in size, green or purplish in 3-10 cm long spikes, close or scattered on long, pendulous, axillary panicles; bracts ovate, acuminate; perianth sessile by a broad base; stamens 6, filaments short, anthers minute, didymous, pistillode 3-lobed. Female flowers sessile, axillary, solitary or fascicled, pendulous spikes, perianth sessile; staminodes 6; bracts below the ovary minute, ovate, acuminate, style short, conical, stigmas 3, very short. Capsules recurved, quadrately oblong, slightly widened upwards, seeds winged at the base only.

Common throughout the State.

Fl. & Fr.: September – December.

Neelwai RF (ADB), *TP & GO* 5430; Pathur (MDK), *TP* & *CP* 12121; Narsapur RF (MDK), *TP* & *CP* 14067; Anantasagar (RR), *MSM* 10511; Mamidipally RF (NZB), *BR* 6435; Pakhal (WGL), *KMS* 13117 (MH); Murugampadu (KMM), *RCS* 102451 (BSID); Way to Gangaram (WGL), *PVS* 84016 (BSID); Jakaram (WGL), *RKP* 105262 (BSID).

Dioscorea hispida Dennst., Schlüssel. Hort. Malab. 15. 1818; Fischer, Fl. Madras 3: 1511. 1928. *D. daemona* Roxb., Fl. Ind. 3: 805. 1832; Hook. f., Fl. Brit. India 6: 289. 1892.

A vine; tubers lobed, biennial. Leaves 3-foliolate, on 10-20 cm long, prickly petiole, leaflets broadly cuneate-oblovate, 10-20 × 5-7 cm, all petiolulate, cuspidately caudate-acuminate, villous when young, glabrous with age, sometimes reticulately veined, base tapering, lateral leaflets very oblique. Male flowers in dense cylindric pedunculate spikes, arranged in clusters along the more or less prickly, pubescent or villous rhachis of a 15-40 cm long raceme; bracts shorter than the flowers; sepals orbicular-ovate, membranous, shorter than the coriaceous incurved petals; stamens 6, all antheriferous; anthers sub-sessile, pistillode very low, broad. Female flowers in spikes, distant,

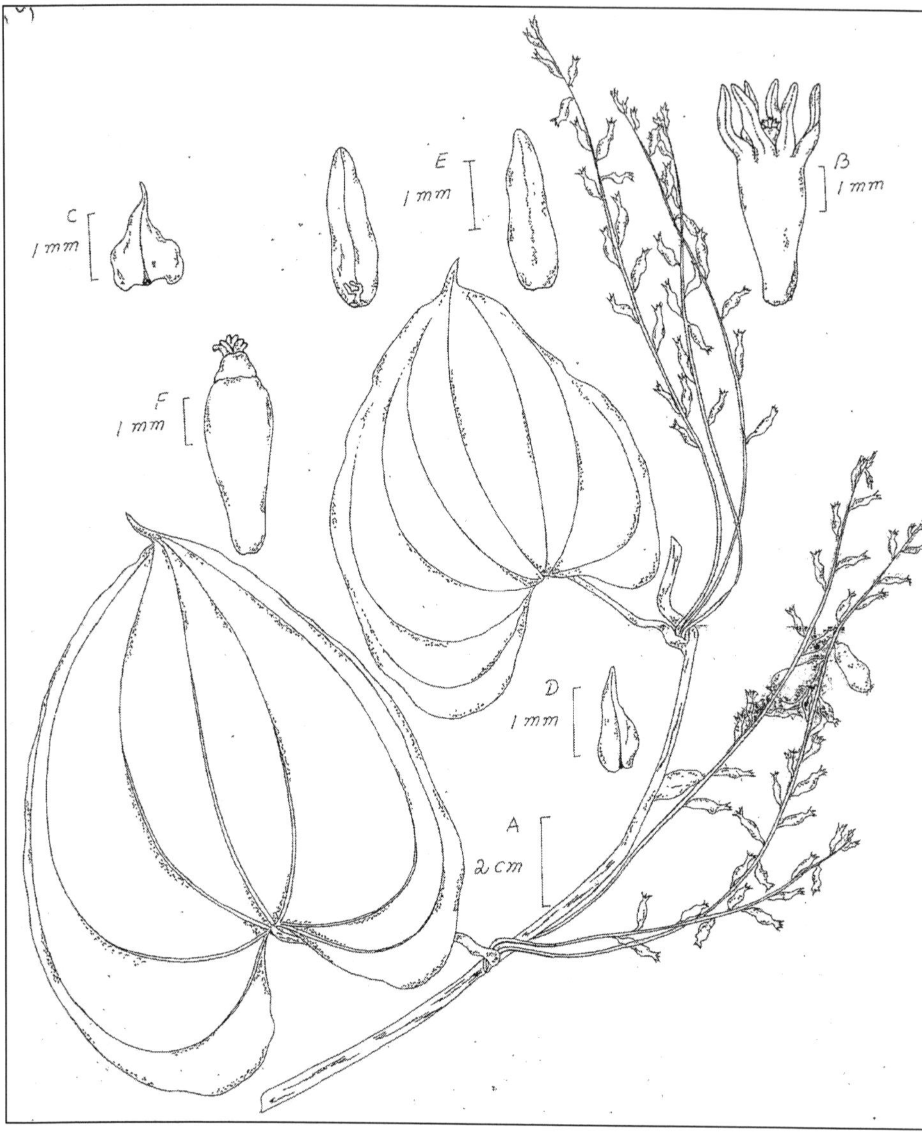

Figure 107: Dioscorea bulbifera L. –Female

A. Twig, B. Female flower, C. Bract, D. Bracteole, E. Perianth lobe (dorsal and ventral), F. Pistil.

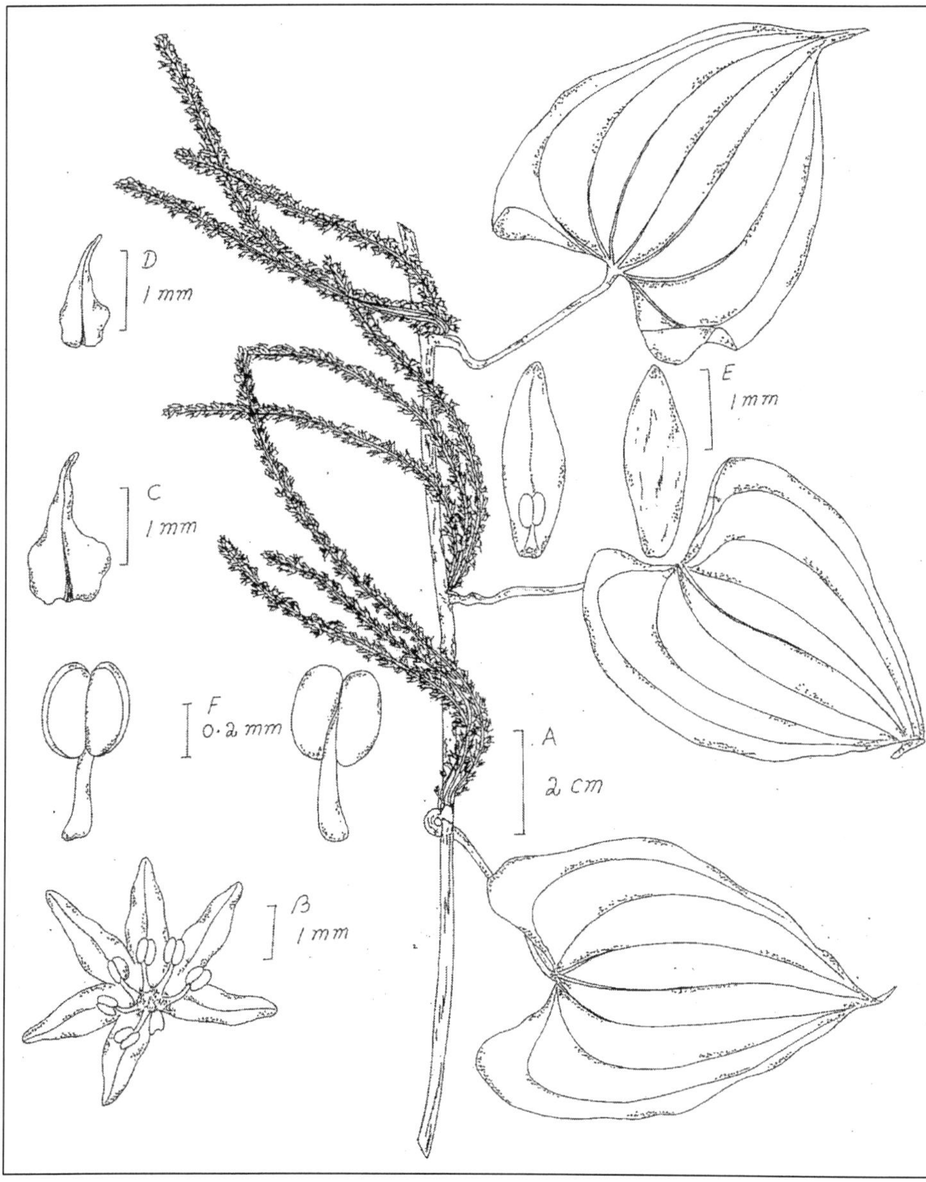

Figure 108: Dioscorea bulbifera L. – Male

A. Twig, B. Male flower, C. Bract, D. Bracteole, E. Perianth lobe (dorsal and ventral)., F. Stamens.

perianth like that of the male. Capsules quadrately oblong, ends truncately rounded, smooth; seeds winged at the base only.

More or less common throughout the State.

Fl. & Fr.: October – December.

Annaram RF (NZB), *BR* & *CP* 7185; Narsapur RF (MDK), *BR* & *CP* 14103; Aklaspur (KRN), *GVS* 25667 (MH); Pakhal (WGL), *KMS* 13102 (MH & CAL).

Dioscorea oppositifolia L., Sp. Pl. 1033. 1753; var. **oppositifolia**; Hook. f., Fl. Brit. India 6: 292. 1892; Fischer, Fl. Madras 3: 1512. 1928. *D. lanceolata* Heyne ex Hook. f., Fl. Brit. India 6: 292. 1892.

A vine, stem twining to the right; rhizomes branched with many long cylindric roots; stem slender, unarmed, glabrous. Leaves oblong or ovate-suborbicular, 3-12 × 1.5-9 cm, base obtuse-rotund, margin entire, apex sub-acute-obtuse, mucronate, glabrous. Male flowers yellowish-green, slighty scented, in dense shortly pedunculate spikes, which are fascicled in the leaf axils or along a slender axillary rhachis; bracts linear, bracteoles scarious, triangular, subequal; perianth-lobes oblanceolate; stamens 6. Female flowers distant, in solitary or fasciculate, axillary, 10-20 cm long spikes; flowers distant, perianth-lobes ovate. Capsules obovoid, stipitate; seeds orbicular, winged throughout.

Commong throughout the State.

Fl. & Fr.: August – November.

Jaipur RF (ADB), *GO & PVP* 4951; Annaram RF (NZB), *BR& CP* 7172; Narsapur RF (MDK), *TP* & *CP* 14056, *KMS* 6701 (MH); Anantasagar (RR), *MSM* 10399, 10509; Thattilanga RF (KMM), *RCS* 102418 (BSID); Siddharam (KMM), *PVS* 84079 (BSID); Nilagiri view point (MBNR), *SRS* 109724 (BSID); Mannanur RF (MBNR), *VBH* 86609 (BSID).

Dioscorea pentaphylla L., Sp. Pl. 1032. 1753 var. **pentaphylla**; Hook.f., Fl. Brit. India 6: 292. 1892; Fischer, Fl. Madras 3: 1511. 1928. *D. jacquemontii* Hook.f., Fl. Brit. India 6: 281. 1892.

A gregarious tuberous vine; tubers oblong, 50-100 cm long; stem slender, glabrous, prickly towards the base, often bulbiferous in the leaf axils. Leaves alternate, 3-5-foliolate, glabrous or sparsely pubescent beneath, terminal one elliptic-ovate, 4-8 × 1.5-3 cm, lateral ones inequilateral, 3-7 × 2-3 cm, thin coriaceous, 3-5-nerved at base, rusty-pubescent below, sparsely so above, base subacute-attenuate, margin entire, apex acute, mucronate; petiole grooved. Male flowers pale greenish, fragrant, in very slender shortly pedunculate 2-4 cm long racemes which are solitary or in fascicles along the hairy branches, bracteoles connate, orbicular-ovate, woolly; fertile stamens 3, staminodes 3, shorter than the outer perianth lobes. Female flowers in axillary pendulous pubescent spikes, flowers distant; bracts lanceolate, outer perianth lobes 2 mm; ovary densely tomentose. Capsules oblong, deeply angled, winged, glabrous; seeds 4-6, subquadrate, apically winged.

Common throughout the State.

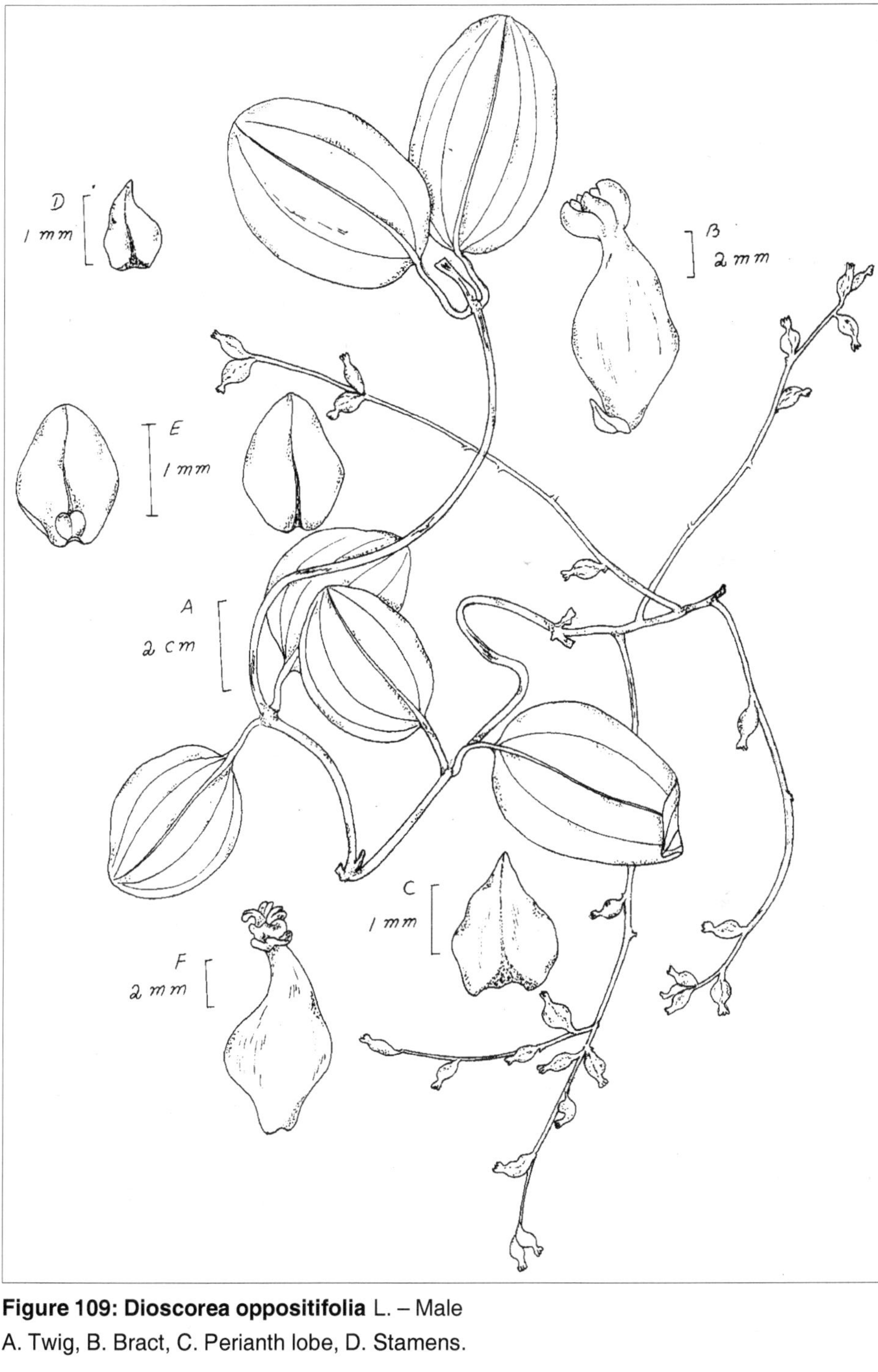

Figure 109: Dioscorea oppositifolia L. – Male
A. Twig, B. Bract, C. Perianth lobe, D. Stamens.

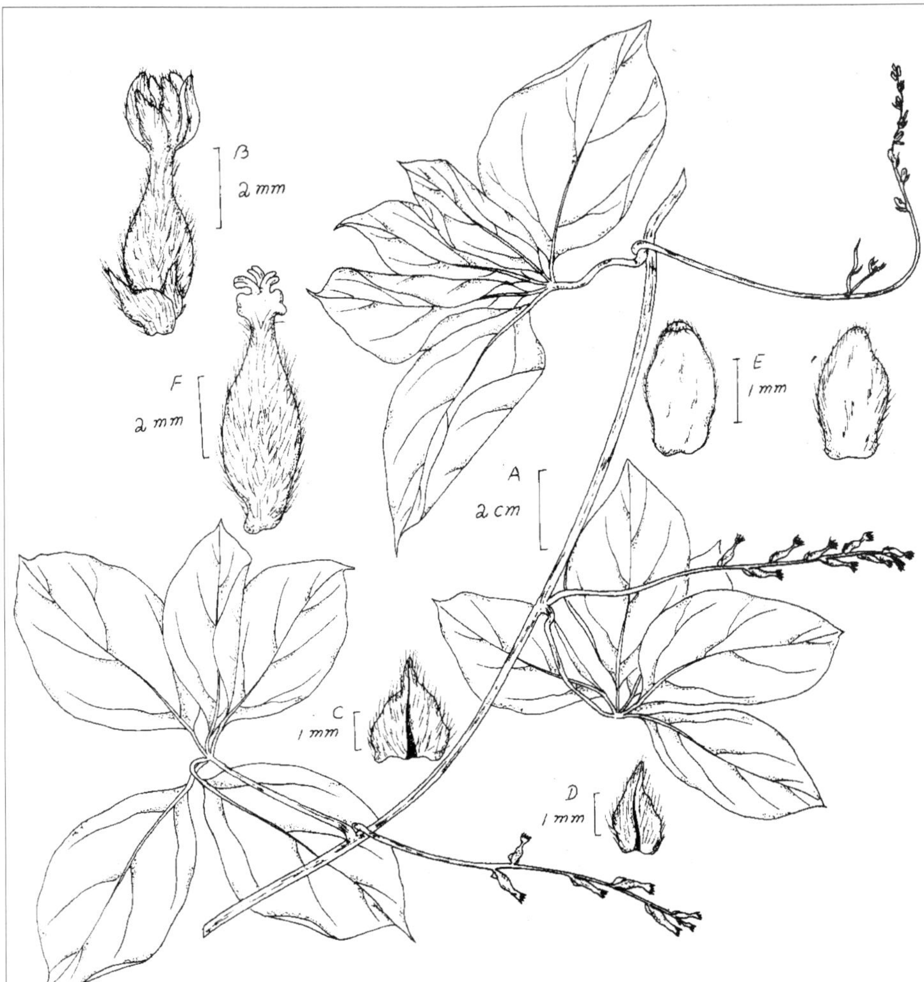

Figure 110: Dioscorea pentaphylla L. – Female

A. Twig, B. Female flower with bract and bracteole, C. Bract, D. Bracteole, E. Perianth lobe (Dorsal and Ventral), F. Pistil.

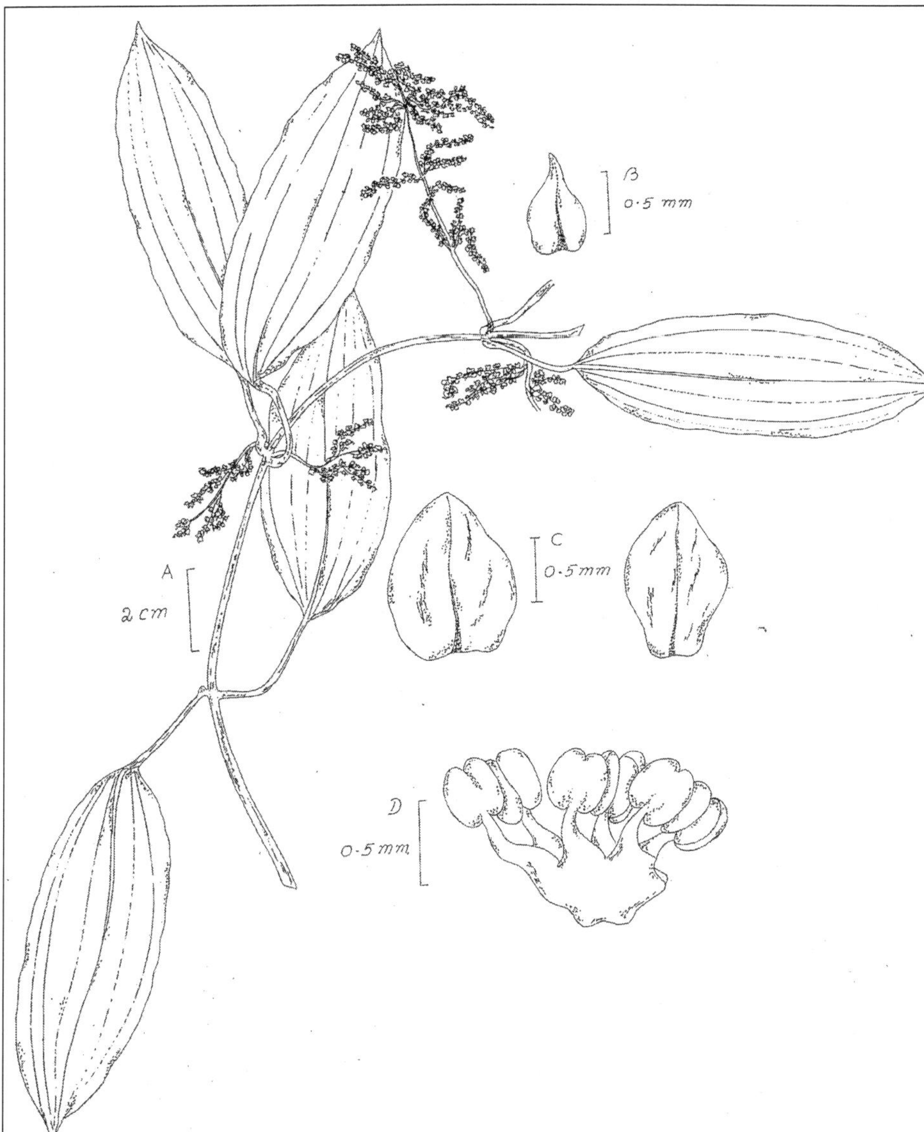

Figure 111: Dioscorea pentaphylla L. – Male

A. Twig, B. Bract, C. Bracteole, D. Perianth lobe (dorsal and ventral), E. Stamens.

Fl. & Fr.: February – September.

Mamidipalli RF (NZB), *BR* 6431; Nadipally (NZB), *BR* 6449; Pochammaralu (MDK), *CP* 11760; Narsapur (MDK), *KMS* 6677 (MH); Anantasagar (RR), *MSM* 10515; Mannanur (MBNR), *TDCK* 13645; Pakhal RF (WGL), *KMS* 11607 (MH); Thoniganla RF (MDK), *RG* 112485 (BSID); Lawal beat, Pasra (WGL), *RKP* 1086161 (BSID).

Dioscorea tomentosa J. König ex Spreng., Pl. Min. Cogn. Pug. 2: 92. 1815; Hook.f., Fl. Brit. India 6: 289. 1892; Fischer, Fl. Madras 3: 1511. 1928.

A vine; tubers elongate; branchlets tomentose, rarely sparsely prickled. Leaves 3-5-foliolate, ovate-broadly elliptic or lanceolate, laterals oblique at base; leaflets 4-12 × 1.5-7 cm, coriaceous, tomentose, 3-5-nerved, margin entire, apex round-acute, mucronate. Male flowers in axillary and/or terminal 10-20 cm long spikes; bracteoles ovate; perianth-lobes coriaceous, tomentose; stamens 3, staminodes 3; pistillode short. Female flowers axillary, paired, 5-12 cm long spikes, perianth-lobes ovate, ovary oblong, 3-celled, ovules 2 per cell, stigma capitate. Capsules oblong, apex broad, tomentose, glabrescent later; seeds obovoid, apically winged.

In hill forests of Warangal and Mahabubnagar districts.

Fl. & Fr.: March – November.

Pakhal RF (WGL), *KMS* 13139 (MH).

STEMONACEAE

STEMONA Lour.

Stemona tuberosa Lour., Pl. Cochinch. 2: 490. 1790; Hook.f., Fl. Brit. India 6: 298. 1892; Fischer, Fl. Madras 3: 1514. 1928.

A plant with twining stem, leafing and flowering at the same time. Leaves ovate, membranous, 4-7.5 × 3-5 cm, opposite, rarely alternate, 9-nerved, glabrous, base shallow-cordate, margin entire, apex acuminate-caudate. Flowers 1-3, perianth segments lanceolate, many-nerved, greenish with purple; stamens large, filaments red, stout, deeply grooved in from with crenulate margins, connective green; ovary free, compressed, stigma small, ovules 2 or more. Capsules ovoid-oblong, 5-8-seeded.

Rare in moist deciduous forests in Warangal district (C.S.Reddy, 2001).

Fl. & Fr.: February – April.

LILIACEAE

1. Leaves minute, often spinescent scales, bearing axillary tufts of needle like or slightly flattened cladodes .... **Asparagus**
1. Leaves well developed:
    2. Climbing plants .... **Gloriosa**
    2. Erect undershrubs or herbs:
        3. Leaves very thick, cartilaginous or fleshy .... **Aloe**

3. Leaves not very thick, neither cartilaginous nor fleshy:
    4. Root stock small with fleshy or tuberous roots:
        5. Leaves semi-terete, fistular .... **Asphodelus**
        5. Leaves flat, not fistular .... **Chlorophytum**
    4. Root stock a bulb or corm:
        6. Wild plants; inflorescence not an umbel, never enveloped by a scarious spathe:
            7. Flowers several-many, racemose on a simple naked scape:
                8. Flowers distant, usually appearing before leaves, dingy – brown .... **Drimia**
                8. Flowers close, appearing with leaves, greenish purple .... **Ledebouria**
            7. Flowers solitary or few corymbose; scape with a few leaves .... **Iphigenia**
        6. Cultigens; inflorescence an umbel, at first wholly enveloped by a scarious spathe .... **Allium**

ALLIUM L.

1. Leaves fistular, terete or semi terete; filaments not cuspidate, sometimes the inner with a tooth on either side of the broad base .... **A. cepa**
1. Leaves flat, sometimes keeled; filaments of inner whorl 3-cuspidate; central cups anther bearing .... **A. sativum**

Allium cepa L., Sp. Pl. 300.1753; Hook. f. Fl. Brit. India 6 : 337.1892; C. Fischer, Fl. Madras 3 : 1528. 1928.

Vern.: *Ullipayalu.* English: Onion.

Annual herbs; bulbs not rhizomatous, depressed-globose, 5–8 x 3–10 cm. Leaves subdistichous, 30-50 cm long, fistulose, acute, glaucous green. Scapes persistent, solitary, erect, fistulose, inflated below middle, 30-90 cm high. Flowers numerous in globose umbels, stellate to campanulate to urceolate, 3–7 mm; white to pink with greenish midveins. Capsule globose-trigonous; seeds angular, black shining.

Cultivated for its bulbs.

Fl. & Fr.: March-May.

Allium sativum L., Sp. Pl. 296.1753; Hook.f. Fl. Brit. India 6: 337.1892; C. Fischer, Fl. Madras 3 : 1528. 1928.

Vern.: *Thellagaddalu, Thellavayalu.* English: Garlic.

Annual herbs; bulbs ovoid with 6-10 bulblets; scales fleshy, white. Leaves linear, flattened, 20-40 cm long. Flowers in globose umbels. Scape solitary, slender, 30-60 cm long; tepals white or tinged with pink, lanceolate, acuminate; filaments shorter than the tepals. Capsule oblong-ovoid, many-seeded; seeds black angular.

Cultivated for its bulbs.

Fl. & Fr.: March-May.

ALOE L.

Aloe vera (L.) Burm. f., Fl. Indica 83. 1768; Fischer, Fl. Madras 3: 1520. 1928. *A. perfoliata* L. var. *vera* L., Sp. Pl. 320. 1753.

A stoloniferous herb. Leaves dense, aggregated, succulent, ensiform, 40-60 × 3-8 cm, narrowed from base to apex, pale green, with distant horny prickles on the margins. Scape 60-90 cm long, simple or branched. Raceme dense, 15-30 cm long, bracts lanceolate, flowers reddish-yellow, bisexual; perianth-tube terete, campanulate, somewhat curved, lobes 6, reddish-yellow or orange, oblong, 3-nerved; stamens 3 + 3, filaments inserted in a pit in the connective; ovary 3-celled, ovules numerous per cell on axile placentae, style elongated, stigma obscurely lobed. Capsule ellipsoid-oblong.

Common throughout the State in open wastelands and field hedges.

Fl. & Fr.: September – December.

Domakonda (NZB), *BR* 9694; Errapenta (MBNR), *TDCK* 15330.

ASPARAGUS L.

1. Cladodes triquetrous or slightly compressed:
   2. Cladodes 2-6-nate, falcate-divaricate, sometimes slightly compressed, 2.5-4 cm long; bracts minute .... **A. racemosus**
   2. Cladodes 3-6-nate, stout, acuminate, spreading, 0.6-0.8 cm long; bracts comparatively large .... **A. laevissimus**
1. Clododes flat, falcate-ensiform, 2-6-nate, acute or acuminate .... **A. gonoclados**

Asparagus gonoclados Baker, J. Linn.Soc. 14 : 627. 1874; Fischer, Fl. Madras 3 : 1517. 1928.

Much branched, subscandent, armed undershrub; main stems smooth, terete, branches firm, green, 3-quetrous. Leaves spured at base with hard spines. Cladodes 2-6 nate, flat, usually falcate, finely spinous pointed. Flowers white in racemes, pedicels jointed above the middle; bracts ovate, boat-shaped; perianth segments spreading, the outer linear to oblong, the inner sub spathulate. Berry globose, twice as broad.

Occasional in scrub jungles on shrubs in Warangal (C.S. Reddy, 2001) and Hyderabad districts (Ramana, 2010).

Fl. & Fr.: Mid summer.

Tadwai (WGL), *CSR* 1624 (KU).

Asparagus laevissimus Steud. ex Baker, J. Linn. Soc. Bot. 14: 623. 1875; Hook.f., Fl. Brit. India 6: 317. 1892; Fischer, Fl. Madras 3: 1517. 1928; Narasimha Rao, Indian J. Bot. 12: 26. 1989.

A subscandent under shrub; main stem terete, polished; branched conspicuously striate; branchlets ascending, angular; spines short, strong. Leaves with stout conical deflexed rigid spinous spurs; cladodes 3-6-nate, spreading or ascending, 3-quetrous. Flowers in racemes, pedicels short; bracts large, cymbiform; perianth segments linear-oblong, obtuse, stamens as long as the perianth, style very short and stout; stigmas 3, spreading. Fruit berry.

Rare in Medak district (Narasimha Rao, 1989).

Fl. & Fr.: June – September.

Asparagus racemosus Willd., Sp. Pl. 2: 152. 1799; Hook.f., Fl. Brit. India 6: 316. 1892; Fischer, Fl. Madras 3: 1517. 1928. var. **racemosus**

An extensively scandent, spinous, much-branched under shrub; root stock tuberous; branchlets angular. Leaves scaly, triangular, 4-6 mm, stiff-acuminate, cladodes 2-6-nate, triquetrous, some times slightly compressed, 1.25-4 × 0.5-0.7 cm, base narrow, margin entire, apex acuminate. Flowers bisexual white, fragrant, in simple (rarely branched) 5-15 cm long racemes; peduncle much reduced; bracts triangular; perianth-lobes 6, white, oblong-obovate, obtuse; stamens 6, adnate to the base of perianth; ovary obovoid, 3-celled, ovules 2 per cell on axile placentae, style terminal, apex 3-fid, stigmas 3, recurved. Berry globose; seeds 3-6, globose or angled.

Common throughout the State.

Fl. & fr.: September – October.

Sattenapalli (ADB), *TP & PVP* 4163; Varni Nursery (NZB), *TP & BR* 6175; Medak (MDK), *BR & CP* 11423; Nagulabanda (MDK), *CP* 11339; Lingampalli (RR), *MSM* 12738; Mannanur (MBNR), *TDCK* 13657; Borapur (MBNR), *BR & BSS* 32361; Edrallapalli RF (KMM), *RCS* 102404 (BSID); Jakaram (WGL), *RKP* 105278 (BSID).

ASPHODELUS L.

Asphodelus tenuifolius Cav., Anales Cl. Nat. 3: 46. 1801; Hook.f., Fl. Brit. India 6: 332. 1892; Fischer, Fl. Madras 3: 1524. 1928.

An annual herb with slender root-fibres. Leaves 15-30 cm long, terete, fistulous, acute, sheathing at the base, finely puberulous. Scapes several, 30-60 cm long, glabrous or sparsely minutely puberulous. Flowers white, distant, laxly racemose, solitary in each bract; jointed at or below the middle; bracts broadly ovate, acute, scarious, with a strong brownish keel, perianth segments oblong, obtuse, with a strong conspicuous brownish costa; ovary 3-celled, ovules 2 in each cell. Capsule globose, erect, the valves deeply wrinkled; seeds sharply trigonous, acute, black.

A weed of fields and gardens, rare.

Fl. & Fr.: December – March.

Kadthala (ADB), *GO & DAM* 5010.

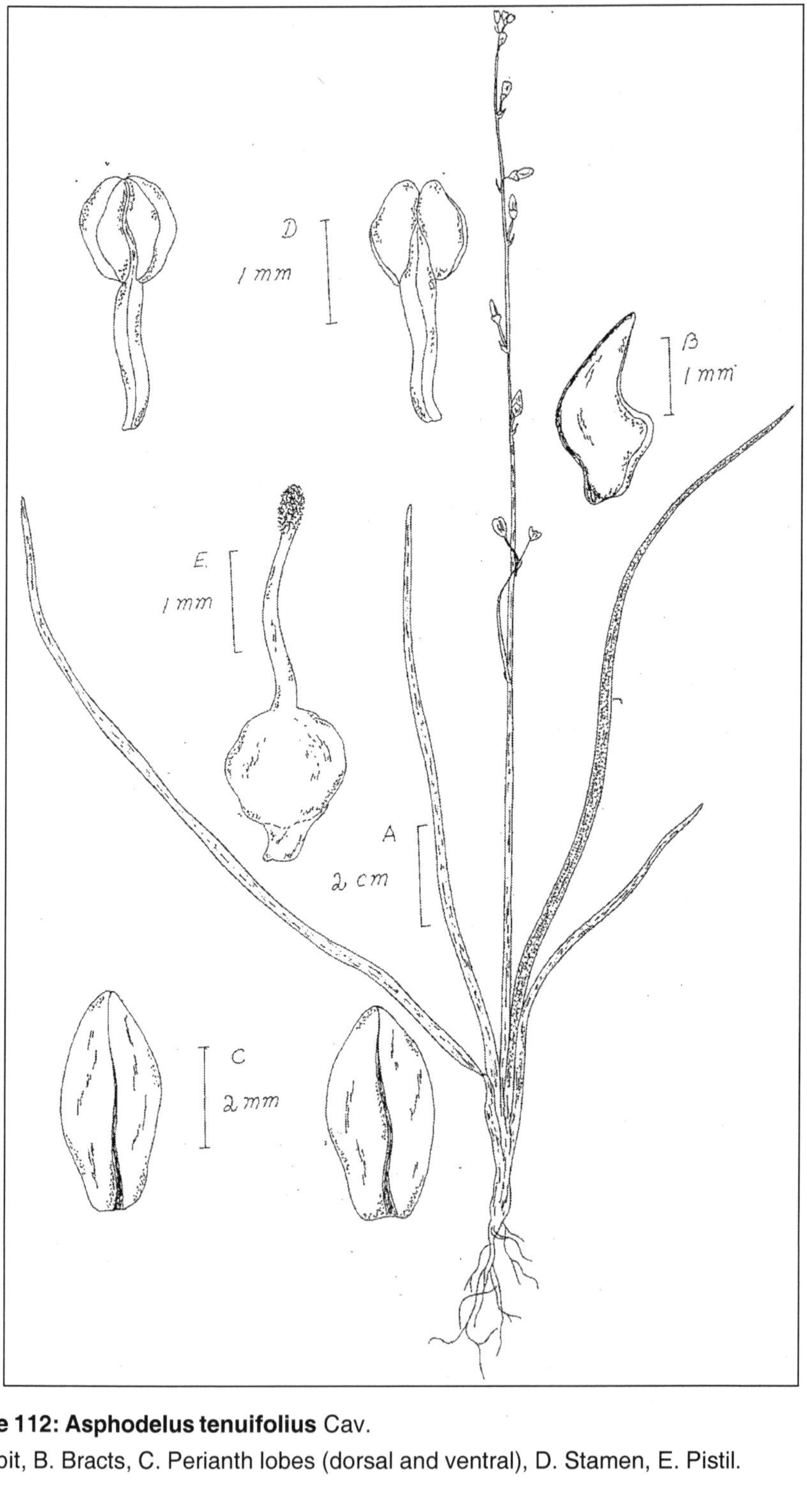

Figure 112: Asphodelus tenuifolius Cav.

A. Habit, B. Bracts, C. Perianth lobes (dorsal and ventral), D. Stamen, E. Pistil.

CHLOROPHYTUM Ker – Gawl.

1. Flowers remotely scattered on racemes, solitary, or in pairs; leaves < 1.2 cm wide .... **C. laxum**
1. Racemes dense-flowered; leaves 2.5 – 6 cm wide:
   2. Leaves narrow-lanceolate or oblanceolate, narrowed into a petiole .... **C. arundinaceum**
   2. Leaves linear or lorate, not or rarely narrowed into a petiole:
      3. Scape usually longer than the leaves; perianth-segments 7-9-nerved; capsules orbicular .... **C. tuberosum**
      3. Scape shorter than the leaves; perianth-segments 3-nerved; capsules broadly obcordate .... **C. malabaricum**

Chlorophytum arundinaceum Baker, J.Linn. Soc. Bot. 15 : 323. 1876; Hook., f. Fl. Brit. India 6 : 34. 1892; Fischer, Fl. Madras 3 : 1526. 1928.

Perennial herbs, 45 cm high; roots fascicled, tuberous. Leaves elliptic-lanceolate, to 40 × 5 cm. Flowers white, 6 mm long; scape 40 cm high; filaments 1 mm, anthers 4 mm long. Capsule trilobed, apically depressed; seeds suborbicular, black.

Found as under growth of moist places.

Fl. & Fr.: July-November.

Chlorophytum laxum R. Br., Prodr. Fl. Nov. Holl. 277. 1810; Hook.f., Fl. Brit. India 6: 336. 1892; Fischer, Fl. Madras 3: 1526. 1928.

A low plant rarely exceeding 30 cm high; root-fibres usually with small, oblong tubers hanging from them. Leaves 6-12, grass like, 15-30 × 0.3-1.2 cm, subdistichous, usually spreading and recurved, longer than the scape, narrowly linear, falcate, flat, acute, glabrous, with numerous conspicuous nerves. Scape very slender, flexuous, 2-6 cm long, naked. Flowers few, greenish-white, very distant, in lax very slender simple or forked racemes, 10-25 cm long; bracts ovate, acuminate, pedicels jointed, about the middle, short, but becoming longer and drooping in fruit; perianth lobes greenish-white, oblong, obtuse, obscurely 3-nerved; stamens alternately short and long. Capsule broadly obcordate, 3-winged, the cells 1-4-seeded; seeds irregularly angled, black, minutely papillose.

Rare in Nizamabad and Hyderabad districts.

Fl. & Fr.: May – June.

Ghanapur RF (NZB), *BR & KH* 7115; Sardhana (MDK), *CP* 11778.

Chlorophytum malabaricum Baker, J.Linn. Soc. Bot. 15 : 331. 1876; Hook.f. Fl. Brit. India 6 : 335.1892; C. Fischer, Fl. Madras 3 : 1526. 1928.

Annual herbs, roots tuberous, fibrous. Leaves linear, 5.6-20 x 0.4-0.6 cm, keeled, margins scaberulous, apex obtuse-subacute. Flowers white, distant, paired or solitary in simple or branched racemes. Capsule globose, 3-winged; seeds ovoid, warty.

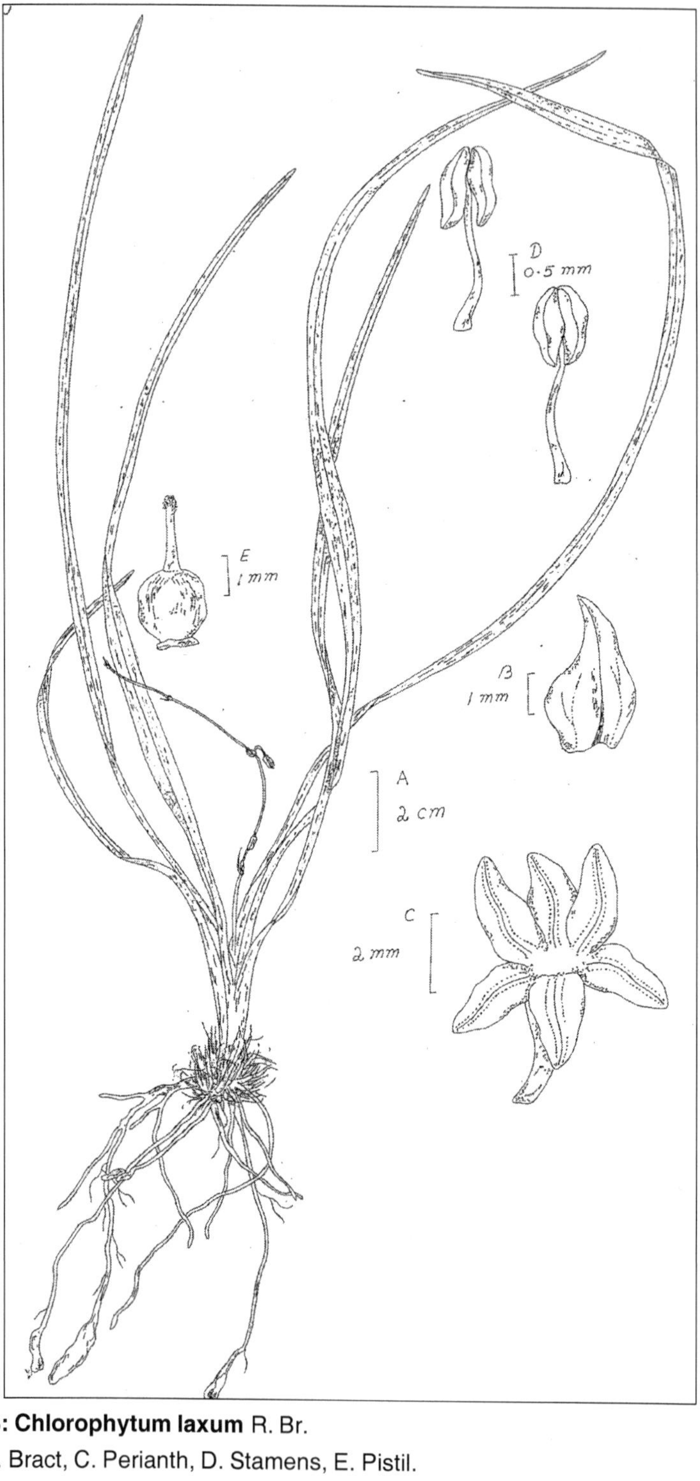

Figure 113: Chlorophytum laxum R. Br.

A. Habit, B. Bract, C. Perianth, D. Stamens, E. Pistil.

In moist shallow habitats in Hyderabad district (Ramana, 2010)

Fl. & Fr.: June-November.

Chlorophytum tuberosum (Roxb.) Baker, J. Linn. Soc. Bot. 15: 332. 1876; Hook.f., Fl. Brit. India 6: 334. 1892; Fischer, Fl. Madras 3: 1526. 1928. *Anthericum tuberosum* Roxb., Fl. Ind. 1: 149. 1832.

A herb with cylindric root-fibres and ellipsoid tubers hanging from them. Leaves 6-12, membranous, sessile, 15-30 × 1.5-2.5 cm, shorter than the scape, usually falcate, recurved, acuminate, the margins undulate. Scape terete, naked, 8-25 cm long. Flowers white, in simple or shortly branched recemes, peduncle stout, angular, bracts elongate, pedicel 1-2 cm, jointed at or below-middle; perianth lobes white, oblong-obovate; stamens 6; ovary 3-celled, ovules 4 or more per cell. Capsule oblong, triquetrous, strongly nerved, glabrous; seeds subquadrate, irregularly pitted.

Weed in fields and waste lands in Adilabad and Mahabubnagar districts.

Fl. & Fr.: June – November.

Vempalli RF (ADB), *TP & GO* 5499; Nizamsagar (NZB), *BR* & *CP* 9247; Bannaloddi (MBNR), *TDCK* 15361; Mannanur (MBNR), *BR* & *BSS* 30652.

DRIMIA Jacq. ex Willd.

Note: Jessop (1977) is of the opinion that the genera *Urginea* Steinhill, *Idotheae* Kunth, *Thuranthos* Wt., *Urginepsis* Compton and *Drimia* Jacq. ex Willd are closely allied to each other and preferred to treat them all under *Urginea* Jacq. ex Willd., the earliest valid name. The difference between *Urginea* and *Drimia* are essentially based on their perianth being reflexed or not and the nature of perianth tube and colour of the bulbs.

The degree of the fusion of the perianth tube considerably varies and overlaps in the both the genera and although a majority of species under *Urginea* from India have only white bulbs, a few (from South Africa) have red bulbs. Further in the Indian species of *Urginea*, the perianth segments are spreading and get reflexed with either partially or wholly when fully opened. M.Y. Ansari and Sundara Raghavan (1980) based on Jessop treatment provide combinations for Indian species of *Urginea* with other combinations effected by Jessop.

1. Flowers appearing before the leaves; bulb 5-10 cm long .... **D. indica**
1. Flowers appearing with the leaves; bulb 2-5 cm long .... **D. wightii**

Drimia indica (Roxb.) Jessop, J. South African Bot. 43: 272. 1977. *Scilla indica* Roxb., Fl. Ind. 2: 147. 1832. *Urginea coromandeliana* (Roxb.) Hook. f., Fl. Brit. India 6: 347. 1892; Fischer, Fl. Madras 3; 1527. 1928. *Scilla coromandeliana* Roxb., Fl. Ind. 2: 148. 1832. *Urginea wightiana* Hook. f., Fl. Brit. India 6: 347. 1892. *U. indica* (Roxb.) Kunth., Enum. Pl. 4: 333.1843.

A herb with tunicate bulb; bulb the size of an apple, pale, 5-10 cm long, ovoid, thick. Leaves appearing after the flowers, 15-45 × 1-2 cm, nearly flat, sub-bifarious, linear, acute. Scape erect, brittle, 30-45 × 0.4-0.6 cm, at the base. Flowers dingy brown, very distant, in slender, laxy-flowered racemes, 15-25 cm long; bracts minute, soon falling, slender, spreading or decurved; perianth campanulate, lobes oblong-lanceolate, obtuse; stamens 6, adnate at or near the base of the perianth lobes; ovary 3-celled, ovules numerous in each cell. Capsule ellipsoid, tapering to both ends, the cells 6-9-seeded; seeds elliptic, 6 mm, flattended, black.

Common throughout the State in shady localities of forest.

Fl. & Fr.: March – April.

Chilapathi RF (ADB), *GO* & *PVP* 4906 B; Annaram RF (NZB), *BR* 7179; Mannanur (MBNR), *TDCK* 15373.

Drimia wightii Lakshmin., Kew Bull. 58(2): 507. 2003. *Urginia congesta* Wight, Icon. Pl. Ind. Orient. t. 2004. 1853; Hook.f., Fl. Brit. India 6: 348; Fischer, Fl. Madras 3: 1527. 1928.

Annual herbs; bulbs elliopsid, 2.5 cm diam. Leaves linear, 8-15 cm long, filiform, grooved above, appearing with the flowers. Scape 15 cm long, flexuous. Flowers on 5-8 cm long racemes; bracts deltoid; perianth white to purpulish; stamens 6. Capsule subglobose, 8 mm broad.

In rocky situations and open grounds in Hyderabad (Ramana, 2010).

Fl. & Fr.: April-May.

GLORIOSA L.

Gloriosa superba L., Sp. Pl. 305. 1753; Hook. f., Fl. Brit. India 6: 358. 1892; Fischer, Fl. Madras 3: 1519. 1928.

A herbaceous, glabrous branching climber, root stock tuberous, solid, fleshy-white, cylindric, 15-25 × 2.5-4 cm, pointed at each end, V-shaped, producing a new joint at the end of each branch; roots fibrous; stems annual, 3-6 m long, given off from the angles of the young tubers, herbaceous. Leaves alternate, opposite or whorled, oblong-lanceolate, sessile, 5-12 × 1-3 cm, flat, glabrous, base cordate, margin entire, apex tendrillar. Flowers large, 8-10 cm across, bisexual, solitary or some what subcorymobose at the ends of branches; perianth lobes 6, oblong-lanceolate, with crisply waved margins, greenish at first, then yellow, passing through orange and scarlet to crimson, flower exhibits herkogamy; stamens 6, spreading; ovary oblong, ovules many per cell on axile placentae, stigmas 3. Capsule ellipsoid-oblong, torulose; seeds numerous, globose, dorsally compressed, warty.

In all districts, common, often climbing on hedge-row plants.

Fl. & Fr.: August – September.

Ankusapuram (ADB), *GO* 4333; Vempalli (ADB), *GO* & *PVP* 4783; Mamidipally RF (NZB), *BR* 6433; Nagulabanda (MDK), *CP* 11343; Anantasagar (RR), *MSM* 10513; Hajipur hills (MBNR), *TDCK* 13616; Aklaspur (KRN), *GVS* 25610 (MH); Edrallapalli (KMM), *RCS* 98759 (BSID); Pasra (WGL), *RKP* 108102 (BSID).

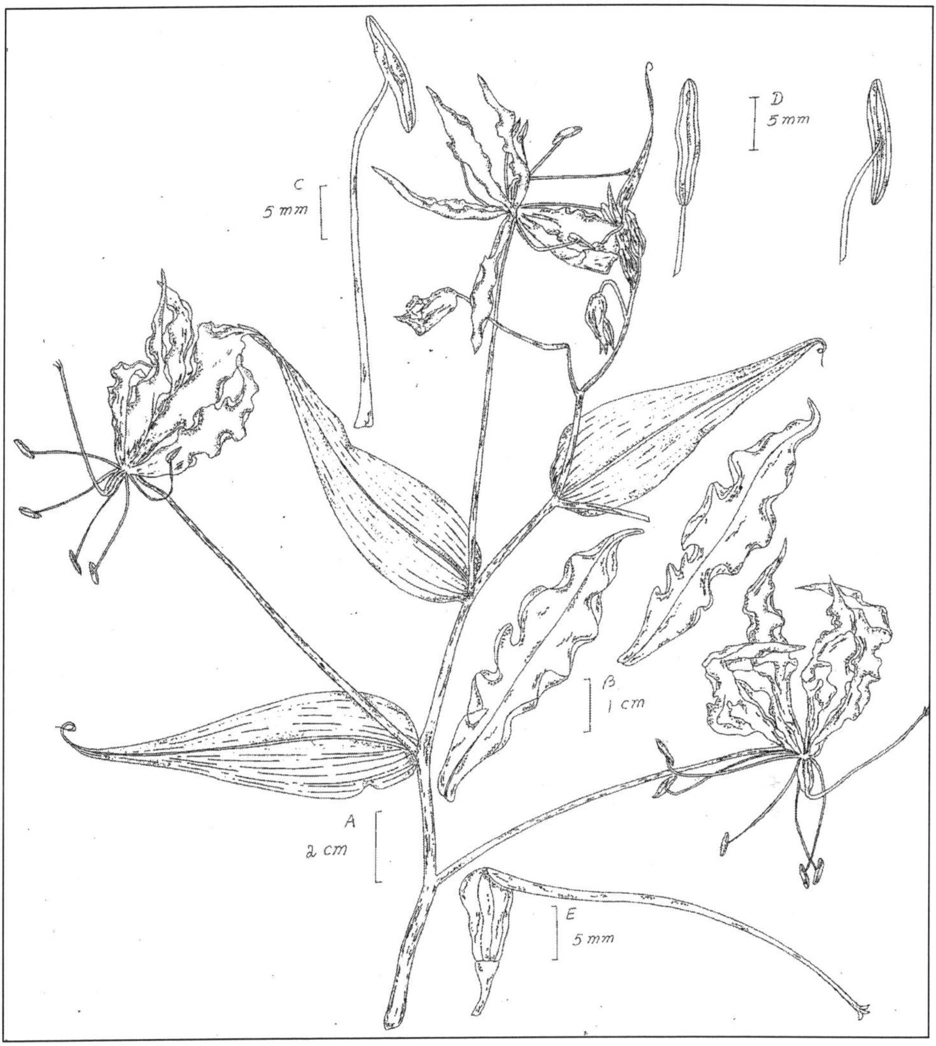

Figure 114: Gloriosa superba L.

A. Twig, B. Perianth lobe (dorsal and ventral), C. Stamen, D. Anthers, E. Pistil.

IPHIGENIA Kunth

1. Perianth purple; segments linear, subulate:
    2. Corms 1.5 cm across; internodes short; leaves linear-lanceolate .... **I. indica**
    2. Corms 0.8-1 cm across; internodes longer; leaves acicular .... **I. mysorensis**
1. Perianth whitish; segments oblanceolate .... **I. pallida**

Iphigenia indica (L.) A. Gray ex Kunth, Enum. Pl. 1: 213; 1833; Hook. f., Fl. Brit. India 6: 357. 1892; Fischer, Fl. Madras 3: 1528. 1928. *Melanthium indicum* L., Mant. Pl. 226. 1771.

A herb with small tunicate corm, corm subglobose, 0.5-1 cm long, 1.5 cm across, narrowed into short neck, tunicate with pale brown sheaths, stem 8-15 cm high, rigid or flexuous. Leaves alternate, few, sessile, the lower 8-15 × 0.2-0.4 cm, the upper gradually smaller, all narrowly linear or linear-lanceolate, glabrous, margin entire, apex acuminate; leaf sheath 4 cm, tubular. Flowers purple, few or many in a terminal erect raceme; bracts linear or subulate, the lower leafy; perianth lobes 6, deep purple, linear-oblong, spreading; stamens oblong; ovules many per cell on axile placentae. Capsule oblong, loculicidal; seeds ovoid, warty, brown.

Occasional in deciduous forests in Adilabad, Nizamabad, Hyderabad and Karimnagar districts.

Fl. & Fr.: July – September.

Jaipur RF (ADB), *PVP* 9436; Ghanapur RF (NZB), *BR* 7119; Sardhana (MDK), *CP* 11777; Kusumasamudram (RR), *MSM* & *NV* 10374, 10374; Saleswaram (MBNR), *BR* & *BSS* 30732; Rachepalli (KRN), *GVS* 20153 (MH).

Iphigenia mysorensis Arekal & S.N.Ramaswamy, Bot. Not. 125: 220. 1972.

A perennial slender herb, corm 0.8-1 cm across, bulb like with pale brown sheath and somewhat flattened on the side, associated with remains of the old corms, stem 15-20 cm long, slender. Leaves usually three (sometimes 2 or 4), acicular, 10-12 × 0.1-0.15 cm, generally thick. Flowers dark purple in terminal, 2-4-flowered raceme, occasionally a second axillary inflorescence is also produced. Flowers with longer ridged pedicels; perianth of 6 free filiform lobes; stamens 6; gynoecium with three free styles. Capsule small, obovoid, loculicidal; seeds several, trigonous, verrucose.

Rare in Nizamabad and Mahabubnagar districts (Raghava Rao & *al.*, 1990).

Fl. & Fr.: June – August.

Chedmal RF (NZB), *BR* & *CP* 9256.

Iphigenia pallida Baker, J. Linn. Soc. Bot. 17: 451. 1879; Hook. f., Fl. Brit. India 6: 357. 1892.

A herb with globose corm, 1.25 cm thick, tunicate with pale brown sheaths, neck hypogeal, stem above the ground 8-10 cm high, flexuous. Leaves usually 4, linear, grass-like, 7.5-10 × 0.3-0.6 cm, the upper smaller. Flowers 1-4, white or tinged with

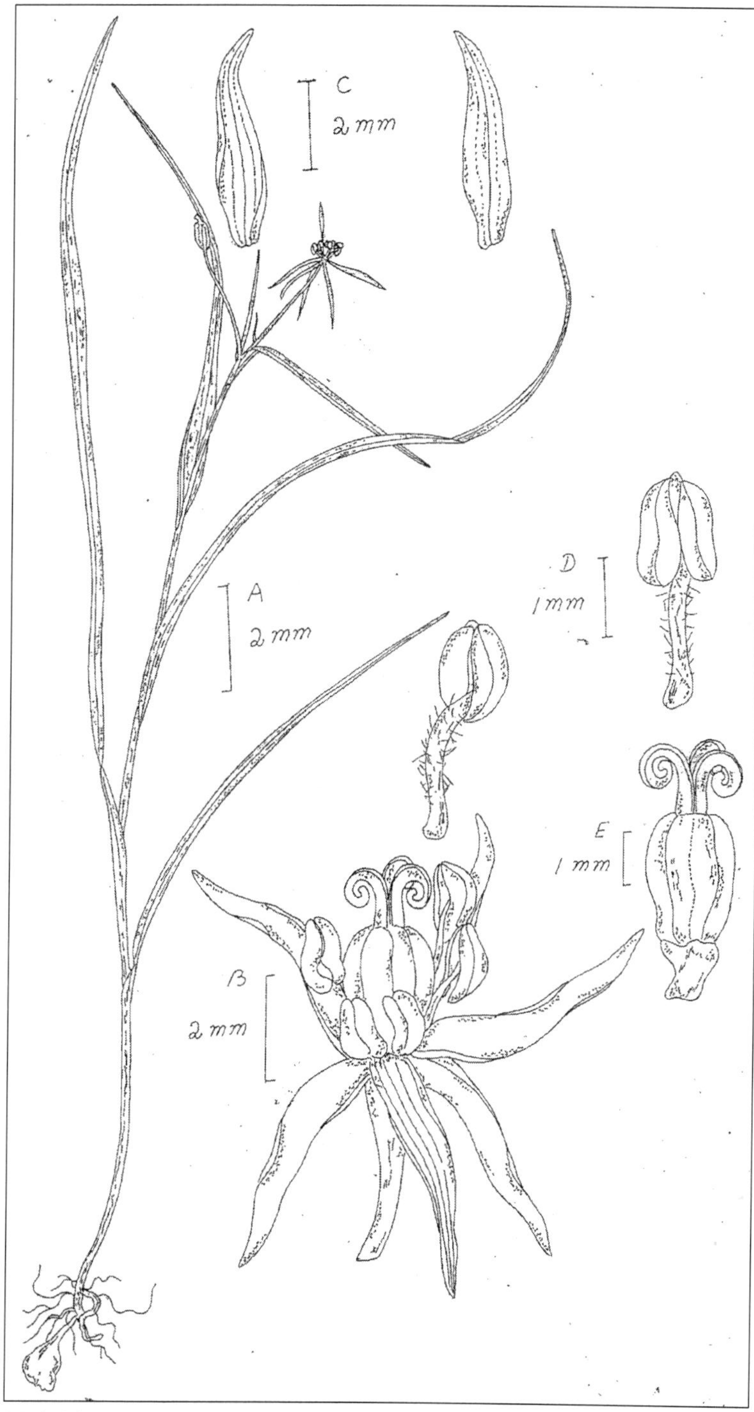

Figure 115: Iphigenia mysorensis Arekal & S.N.Ramaswamy,
A. Habit, B. Flower, C. Perianth lobe (dorsal and ventral), D. Stamen, E. Pistil.

purple; bracts linear, foliacous; perianth-segments oblanceolate, acute, many-nerved; stamens 6; ovary obovoid, styles 3, recurved. Capsule obovoid.

Reported from Mahabubnagar district only.

Fl. & Fr.: June – July.

Thimmapur (MBNR), *SRR* 1765 (HH).

LEDEBOURIA Roth

1. Capsule obovoid; scape 10-25-flowered .... **L. hyderabadensis**
1. Capsule globose; scape 30-60-flowered .... **L. revoluta**

Ledebouria hyderabadensis M.V.Ramana & *al.*, Kew Bull. 67: 561. 2012.

Bulbous herb, scapigerous, up to 15 cm high including scape; bulb ovoid, 1.5 × 1.5-2 cm, tunicated, tunic dark brown. Leaves radical, 2-6, bright green, fleshy, devoid of blotches on both surfaces, oblong, linear-lanceolate, 5-10 × 0.5-1.0 cm, base attenuate and sheathing, margin entire, apex acute, glabrous, devoid of bulbils. Scape slender, unbranched, longer than leaves, 8-14 cm long, 10-25-flowered, racemose; flowers pinkish green, campanulate, nodding, 5-8 mm long, rotate when fully opened, bracteate; bract 1, lunate with a semicircular notch above, ca. 2 mm, pedicels 3-4 mm long; perianth lobes divided near to the base; lobes 6, biseriate, oblong, ovate, acute,ca 6 × 2 mm, valvate, pinkish purple with green veins, reflexed from middle in mature flowers, persistent; stamens 6, filaments free, adnate at base to perianth, pinkish, 4 mm long, anthers oblong, 2-celled, dorsifixed and versatile, dehiscing longitudinally; ovary tricarpellary, stipitate, shallowly but distinctly 3-lobed, 3-celled, 1 × 1 mm, lobes more foliar, ovules 6, 2 per cell, attached at base, styles simple, erect, pinkish, ca 4 mm long, stigma triquetrous, penicillate. Capsule obovoid, membranous, 1 -2-lobed, rarely 3-lobed, ca 5 × 7 mm with persistent perianth and staminal filaments; seeds 1 per cell, attached at base, smooth, ovoid, 4 mm, brownish, albuminous.

Found in open fallow lands on gravelly soils and in between rock boulders in dense tufts; associated with *Striga aisatica*, *Kylling triceps*, *Cyperus meebodlii* and *Ophioglossum costatum* in Hyderabad city.

Fl. & Fr.: June – September.

Dammaiguda hills (HYD), *PVP* 1459 (holotype CAL; isotype BSID); Osmania University campus (HYD), *Ramana* 1923 (BSID).

Ledebouria revoluta (L.f.) Jessop, J. S. African Bot. 36: 255. 1970. *Hyacinthus revolutus* L.f., Suppl. Pl. 204. 1782. *Scilla hyacinthina* (Roth) J.F.Macbr., Contr. Gray. Herb. 56: 14. 1918; Rao in Pullaiah, Fl. Andhra Pradesh 3: 992. 1997. *Ledebouria hyacinthina* Roth, Nov. Pl. Sp. 195. 1821 (illegitimate). *Scilla indica* (Wight) Baker in Saunders, Refug. Bot. 3(App.): 12. 1870; (non Roxb. 1832); Hook. f., Fl. Brit. India 6: 348. 1892; Fischer, Fl. Madras 3: 1527. 1928. *Barnardia indica* Wight, Icon. Pl. Ind. Orient. t. 2041. 1853.

A herb with tunicate bulb; bulb ovoid or globose, 2-6 × 1.5-3 cm, apex narrowing into a short neck. Leaves radical, 4-15 × 2-3.5 cm, variable, oblong to lanceolate or oblanceolate, chartaceous, occasionally with purplish blotches, glabrous, base gradually attenuate and sheathing, apex obtuse-subacute, mucronate. Scape (s) 1-3, racemes 5-12 cm, 30-60-flowered; flowers greenish-purple, bisexual, bracts scarious, bracteoles 2, adnate to the base of pedicel; perianth deeply 6-lobed to the base, greenish-purple, oblong, subequal; stamens 6, adnate to the base of perianth; ovary 3-lobed, 3-celled, ovules 2 or more per cell. Capsule globose, loculicidal, thin-walled.

In all districts in forests and waste lands.

Fl. & Fr.: June – September.

Kadam dam field (ADB), *BR* 9416; Mudheli RF (NZB), *BR* & *KH* 7137; Kamareddy – Indelwai X roads (NZB), *BR* & *CP* 9229; Neelakandakurthi (NZB), *BR* 9220; Mohammadabad (RR), *MSM* & *KH* 10226; Kusumasamudram (RR), *MSM* 10374; Kodimial (KRN), *GVS* 24507 (MH).

URGINEA Steinh.

Urginea raogibikei Hemadri, Med. Flora Pragati Resorts 386. 2006.

Vern.: *Adavi ulligadda* (This name is common for all the Indian species of *Urginea*).

Scapigerous herb with tunicate, compressedly subglobose-ovoid, perennial herb, 4-5 cm long and 4-6 cm broad with a neck of 2-4 cm long; bulb sometimes longer than broad if grown amongst stones; outer scales dirty white, inner white. Leaves 3-7, appearing after flowering (i.e., during the months of July-August and retain till December – January), 20-30 x 1.1-2 cm, green, ashy both sides, linear-oblong with acute apex. Flowering scape solitary, (sometimes a second one at a later stage), appearing during April-May (-June), 15-20 cm high including 9-12 cm long raceme, 0.4-0.6 cm thick at base, cylindrical, greenish-brown. Inflorescence a raceme of 20-30 flowers, ending with a few ill-formed buds. Bracts ovate, 0.2-0.3 x 0.15 -0.2 cm, the lowest 2-3 bracts deltoid and spurred. Flowers 0.75-0.9 cm long, closely arranged; pedicel 0.3-0.4 cm long; perianth lobes 6 in two whorls, connate at base, non-reflexed (non revolute) when fully opened, 0.7-0.8 x 0.35-0.4 cm, elliptic-oblong, light yellow with scarious margins, brownish-buff coloured along the midrib; stamens 6, arising from the base of perianth segments; filaments about 0.2 cm long, gynoecium 0.6-0.7 cm long including 0.3 cm long style; ovary 0.3 cm thick, green. Fruit an ovoid, triquetrous capsule of 1-1.2 x 0.8-1 cm with 0.4-0.5 cm long stalk; seeds many, 0.08-0.1 x 0.5-0.7 cm, black and broadly winged.

Collected and grown in Pragati resorts near Hyderabad. Also in Central University Campus (Seshagiri Rao, 2012).

Fl. & Fr.: May – June.

SMILACACEAE

SMILAX L.

1. Leaf sheath not auricled; umbels 1-3 per inflorescence:
    2. Leaf base obtuse-rotund .... **S. zeylanica**
    2. Leaf base rounded or shortly wedge-shaped .... **S. ovalifolia**
1. Leaf sheath auricled; umbels many on an axillary peduncle .... **S. perfoliata**

S. ovalifolia Roxb. ex D.Don, Prodr. Fl. Nepal 49. 1825. *S. macrophylla* Roxb., Fl. Ind. 3: 793. 1832 (illegitimate); Hook. f., Fl. Brit. India 6: 310. 1892.

Vern.: *Darra*

Armed or unarmed climber. Leaves coriaceous, shining, 7-15 × 4-11 cm, broadly ovate to elliptic, base rounded or shortly wedge-shaped, 3-5-nerved, petiole 1.5 cm long, base sheathing with tendrils at the end. Flowers white, in dense umbels in leaf axils; bracts ovate; perianth recurved in mature flowers, outer 3 segments 4 mm long, oblong, inner narrower; stamens as long as the perianth.

Rare in Warangal district

Fl.: January - April

Gangaram (WGL), *PVS* 84019 (BSID).

Smilax perfoliata Lour., Fl. Cochinch. 2: 622. 1790. *S. prolifera* Wall. ex Roxb., Fl. Ind. 3: 795. 1832; Hook. f., Fl. Brit. India 6: 312. 1892; Fischer, Fl. Madras 3: 1519. 1928.

A vine of 8-10 m long; branchlets glabrous; prickles recurved. Leaves broadly ovate or elliptic-lanceolate, 7-16 × 4-12 cm, coriaceous, 5-nerved, glabrous, base rotund-subacute, margin undulate, entire, apex obtuse, sometimes retuse; abruptly cuspidate; leaf-sheath broad, auriculate, amplexicaul. Male flowers: umbels ca. 7, axillary, alternate or verticillate; bracts ovate, cuspidate, bracteoles scarious; perianth lobes greenish, oblong, outer ones broad, apiculate, falcate, inner ones narrow, obtuse; stamens 6. Female flower: umbels axillary, alternate or verticillate; ovary 3-celled, 3-gonous, ovules 1-2 in each cell. Berry globose.

Rare in forests in Khammam district.

Fl.: February – April. Fruits April onwards.

Bandirevu RF (KMM), *RCS* 98920 (BSID).

Smilax zeylanica L., Sp. Pl. 1029. 1753; Hook. f., Fl. Brit. India 6: 309. 1892; Fischer, Fl. Madras 3: 1519. 1928.

A large climber; stem smooth, striate, armed with a few small distant prickles or almost unarmed. Leaves alternate, oblong-broadly elliptic or ovate-lanceolate, 8-20 × 4-10 cm, coriaceous, glabrous, base obtuse-rotund, margin entire, apex obtuse, somewhat notched, cuspidate, leaf-sheath narrow, not auricled. Flowers in pedunculate many-flowered umbels. Male flowers: umbels axillary, 1-3, bracts oblong, bracteoles triangular; perianth lobes greenish, linear, obtuse, erect when young, afterwards reflexed, outer ones broad, inner ones narrow, obtuse; stamens 6. Female flowers: ovary 3-lobed, 3-celled, ovule(s) 1 or 2 per cell, staminodes 3-6, linear. Berry globose, smooth, remaining green for a long time, becoming red when ripe.

Occasional in deciduous forests in Mahabubnagar district.

Fl. & Fr.: June – September.

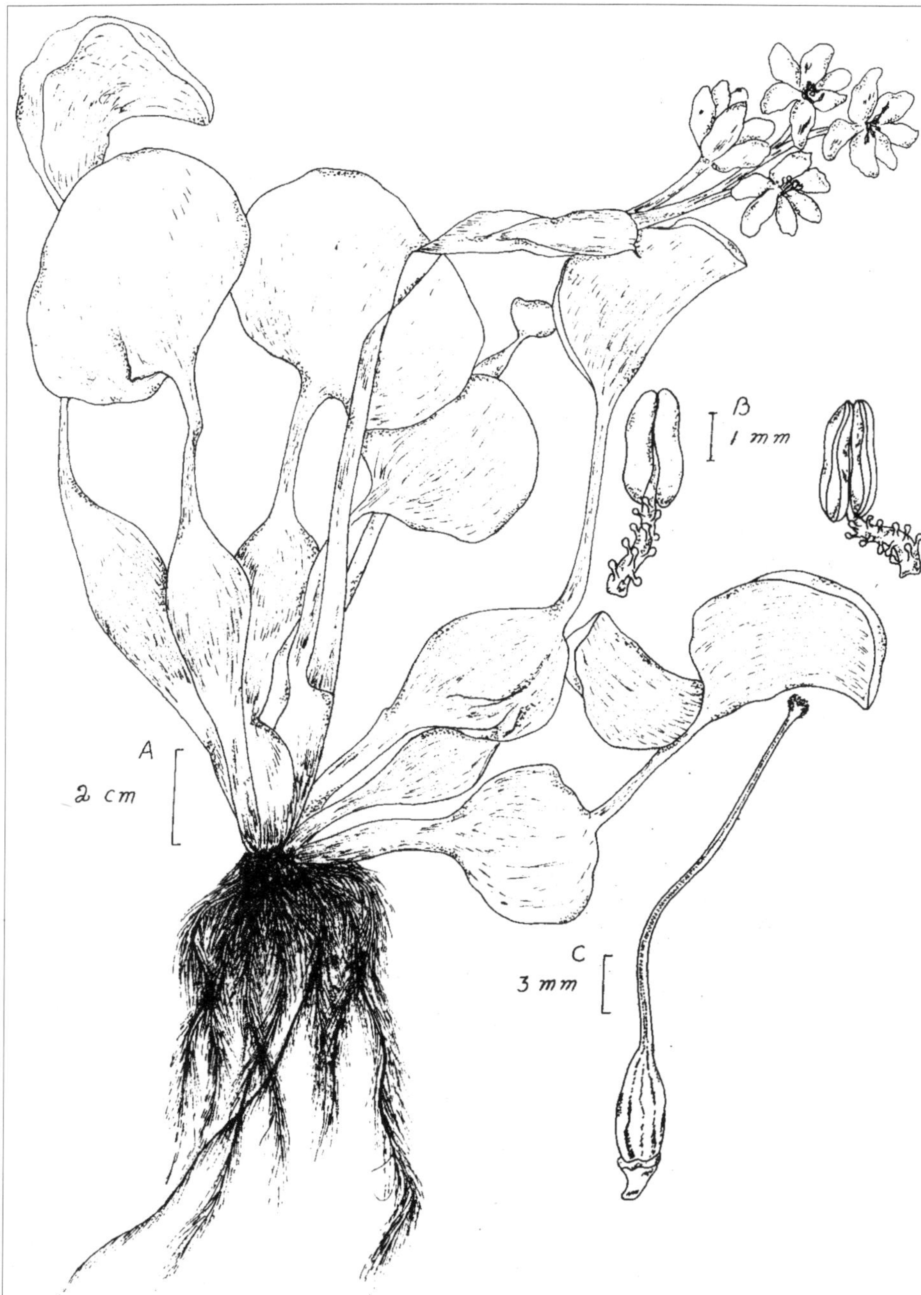

Figure 116: Eichhornia crassipes (Mart.) Solms.
A. Habit, B. Stamens, C. Pistil.

PONTEDERIACEAE

1. Plants with swollen petioles .... **Eichhornia**
1. Plants without swollen petioles .... **Monochoria**

EICHHORNIA Kunth *nom. cons*

Eichhornia crassipes (Mart.) Solms, Monogr. Phan. 4: 527. 1833; Fischer, Fl. Madras 3: 1530. 1928. *Pontederia crassipes* Mart., Nov. Gen. Sp. Pl. 1: 9. 1823.

A free floating or rooted herb, stolons slender, roots elongate, fibrous. Leaves in a rosette, spoon or paddle shaped, 5-10 × 5-10 cm, margin entire, apex rounded or obtuse; petiole elongate, 6-30 cm long, spongy with a fusiform bulbous portion to form floats. Scape erect from the centre of the rosette, 15-25 cm long; spike terminal, ca 15-flowered. Flowers violet-blue, perianth funnel-shaped, tube often 2-lipped, upper one lilac, with a blue, bearded, yellow median blotch; lower one gradually smaller; stamens 6, declinate, anthers subequal, dorsifixed; ovary 3-celled, oblong, glabrous, ovules numerous per cell on axile placentae. Capsule ovoid-oblong; seeds ovoid, many-ribbed.

In still or slow-flowing waters, often clogging irrigation canals in most districts.

Fl. & Fr.: September – February.

Salura (NZB), *BR* 7275; Pothangal (NZB), *BR* & *KH* 9647; Appapur (MBNR), *TDCK* 16841; Uppal (HYD), *KI* & *NVN* 18753; Along Moosi River (HYD), *KMS* 5951 (MH).

MONOCHORIA Presl.

Monochoria vaginalis (Burm. f.) Presl., Reliq. Haenk 1: 128. 1827; Hook. f., Fl. Brit. India 6: 363. 1892; Fischer, Fl. Madras 3: 1529. 1928. *Pontederia vaginalis* Burm. f., Fl. Indica 80. 1768.

A rooted aquatic herb, 20-60 cm long; root stock short, suberect, spongy. Leaves very variable, 5-10 × 1.5-5 cm, from linear to ovate or ovate-cordate, margin entire, apex gradually acuminate. The peduncle emerging from the channeled sheaths of the upper most leaves. Inflorescence centripetal. Flowers blue, usually spotted with red, regular; perianth campanulate, deeply lobed, lobes 3 + 3, oblong-obovate, inner ones broad, outer ones narrow; stamens 6, inserted at the base of the perianth; ovary 3-celled, globose, ovules many per cell on deeply forked, swollen axile placentae. Capsule oblong, 1 × 0.8 cm, glabrous; seeds oblong, ribbed.

In all districts in stagnant ponds, tanks and slow flowing water.

Fl. & Fr.: August – December.

Dabba (ADB), *GO* 4807; Ibrahimpet (NZB), *TP & BR* 6129; Mingaram (NZB), *TP* & *BR* 6306; Gangapur (MDK), *TP* & *CP* 12082; Indira park (HYD), *KI* & *NVN* 18743; Mohammadabad (RR), *MSM* & *NV* 10434; Malkalcheruvu (RR), *MSM* & *KH* 10541; Narsapur (MDK), *KMS* 6719 (MH & CAL); Pakhal (WGL), *ANH* 15979 (MH & CAL); Khammam to Dhaniyayigudem (KMM), *RR* 108516 (BSID).

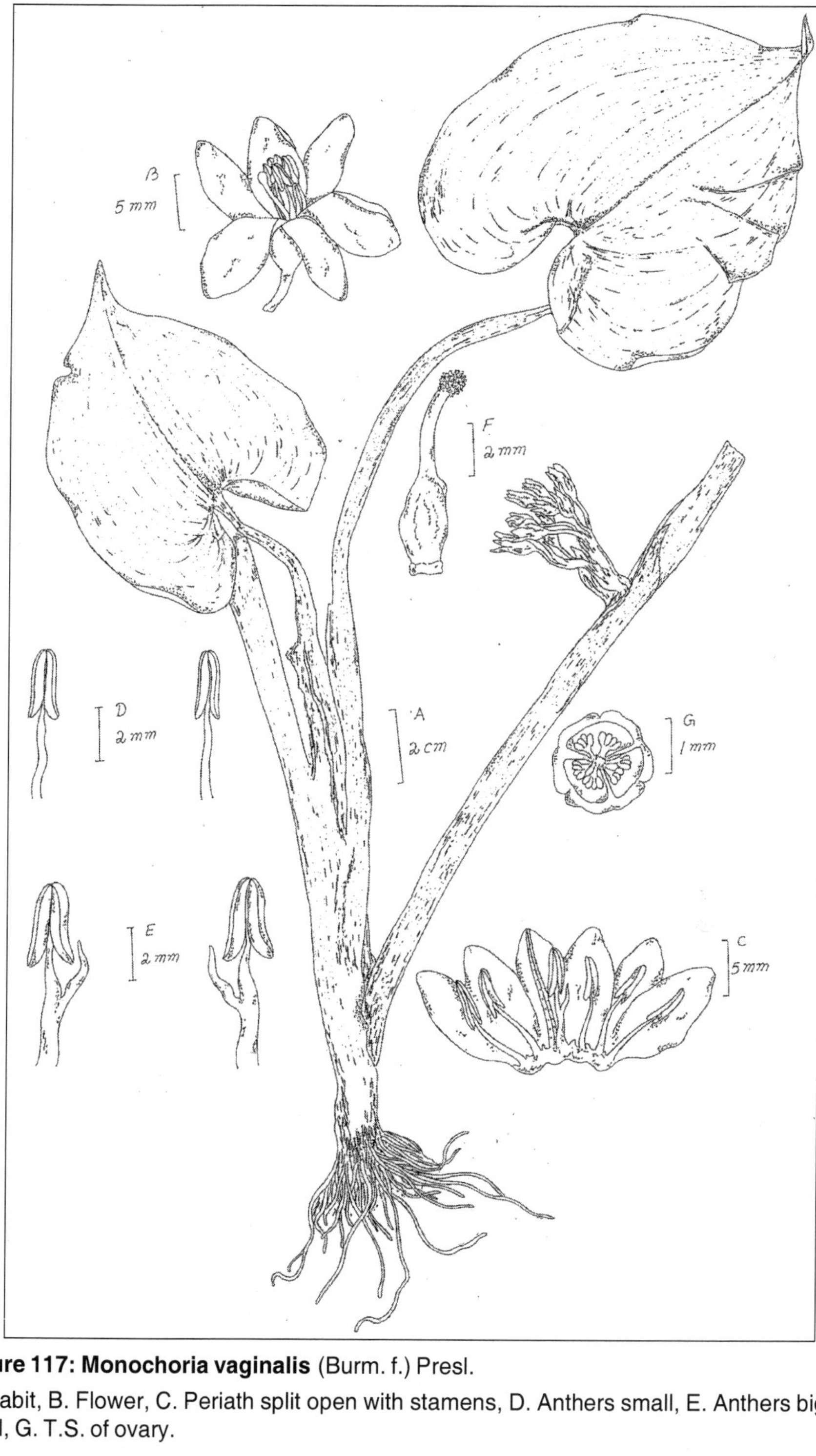

Figure 117: Monochoria vaginalis (Burm. f.) Presl.

A. Habit, B. Flower, C. Periath split open with stamens, D. Anthers small, E. Anthers big, F. Pistil, G. T.S. of ovary.

XYRIDACEAE

XYRIS L.

Xyris pauciflora Willd., Phytographia 1: 2. 1794; Hook. f., Fl. Brit. India 6: 365. 1892; Fischer, Fl. Madras 3: 1532. 1928.

An erect tufted, reed-like scapigerous herb. Leaves narrowly linear, 10 × 0.2 cm, papillose. Scape 15 cm long, slightly compressed, hardly winged, papillate, ribbed. Flower-head globose, bracts chestnut brown, apex dorsally keeled, median bract elliptic, mucronate, basal bract orbicular, margin membranous, apex triangular, papillate, spinulose; sepals 3, median one broad, laterals curved; petals 3, yellow, limbs obovate, emarginated; stamens 3; ovary oblong, 1-celled, ovules on parietal placentae. Fruit capsule.

In marshy localities in Medak and Hyderabad districts.

Fl. & Fr.: January – February. Fruit February onwards.

Medak to Bodhan (MDK), *RG* 113539 (BSID).

COMMELINACEAE

1. Plants cultivated .... **Tradescantia**
1. Plants wild:
    2. Inflorescence leaf-opposed (atleast some of them) .... **Commelina**
    2. Inflorescence terminal or axillary:
        3. Petals connate below into a tube .... **Cyanotis**
        3. Petals free .... **Murdannia**

COMMELINA L.

1. Spathes funnel-shaped or cucullate; margins fused at least basally:
    2. 2 anticous cells of the ovary 2-ovuled, the posticous 1-ovuled:
        3. Leaves ovate or ovate-lanceolate:
            4. Leaves glabrous .... **C. petersii**
            4. Leaves pubescent .... **C. benghalensis**
        3. Leaves linear or linear-oblong .... **C. forsskalaei**
    2. All the cells of the ovary 1-ovuled:
        5. Capsules subequally 3-valved .... **C. paludosa**
        5. Capsules 2-valved:
            6. Leaves glabrous above, puberulous below .... **C. ensifolia**
            6. Leaves appressed pubescent above, glaucous below .... **C. erecta**

1. Spathes complicate, margins free or connate only at the base:
    7. Capsules 2-celled, rarely posticous cell present usually empty or with an imperfect seed:
        8. Spathes base not auricled .... **C. clavata**
        8. Spathes base distinctly sagittate-auricled .... **C. attenuata**
    7. Capsules 3-celled:
        9. Spathes sessile .... **C. subulata**
        9. Spathes distinctly peduncled:
            10. Seeds tuberculate and reticulate .... **C. diffusa**
            10. Seeds smooth:
                11. Seeds appendaged at one end .... **C. undulata**
                11. Seeds not appendaged .... **C. caroliniana**

Commelina attenuata K.D. Koenig ex Vahl, Enum. Pl. 2: 168. 1805; Hook.f., Fl. Brit. India 6: 372. 1892; Fischer, Fl. Madras 3: 1539. 1931.

A much slender, diffuse, annual herb, stem and branches angled; stem 30-60 cm long, slender, much branched, glabrous or sparsely hairy. Leaves sessile, linear or linear-lanceolate, 4-8 × 0.5-1 cm, chartaceous, sparsely scattered-pubescent above, glabrous below, base obtuse, margin entire, apex tapering, acuminate; sheath 0.4 -1 cm. Flowers blue, axillary with slender peduncle, spathes narrowly ovate, 2-4 cm long, base auriculate; sepals 3, pale white – pale green, paired sepals united basally, odd sepal boat-shaped; petals 3, two petals whitish yellow, one rose-purple and bluish; fertile stamens 3, sterile stamens 3; ovary ovoid, 2-celled, the cells 2-ovulate, style long, white, apex coiled, bends to one side, stigma capitate, trifid. Capsules oblong, to 8 × 5 m, dorsal locule not developed; seeds elliptic, whitish-appendaged at the ends.

Common in Nizamabad, Medak, Ranga Reddi, Karimnagar and Mahabubnagar districts.

Fl. & Fr.: July – December.

Mudheli RF (NZB), *BR & KH* 7122; Patancheru (MDK), *BR* & *CP* 11303; Lingampalle (RR), *MSM* 12751; Aklaspur (KRN), *GVS* 25633 (MH).

Note: *C. attenuata* is allied to *C. wightii* in its densely branching habit, stout stem, attenuate spathe apex and smooth seeds. But *C. attenuata* has glabrous peduncles, blue flowers and appendaged seeds against pubescent peduncles, yellow flowers and non –appendaged seeds as in *C. wightii.* The pits on the appendages of seeds are a peculiar for this species.

Commelina benghalensis L., Sp. Pl. 41. 1753; Hook.f., Fl. Brit. India 6: 370. 1892; Fischer, Fl. Madras 3: 1539. 1931. var. **benghalensis.** *Commelina hirsuta* (Wight) C.B.Clarke in A. DC., Monogr. Phan. 3: 163. 1881; 6: 371. 1892; Fischer, Fl. Madras 3: 1538. 1931. *Heterocarpus hirsutus* Wight, Icon. Pl. Ind. Orient. t. 2067. 1853.

Figure 118: Commelina benghalensis L.

A. Twig, B. Flower, C. Bract, D. Calyx, E. Stamens with pistil, F. Anthers, G. Pistil.

An ascending herb; root stock with cleistogamous flowers; stem 30-80 cm long, slender, dichotomously branched from the base upwards; branches diffuse, glabrous or pubescent, creeping and rooting below. Leaves ovate or oblong, 2-5 × 2-3 cm, base subtruncate, margin ciliate, apex acute-obtuse; sessile or shortly with 1 cm long petiole, sheaths short or long, pubescent or villous, the margins ciliate or sometimes bearded with rufous hairs. Spathes clustered, funnel-shaped, pubescent; upper cymes 2-3-flowered; lower one 1 or 2-flowered. Flowers blue; sepals 3, outer one linear, inner ones orbicular; petals 3, two 6 × 5 mm, lanceolate-boat-shaped; fertile stamens 3, antisepalous, sterile stamens 3; ovary 3-celled, 2 cells 2-ovulate, 1 cell 1-ovulate, glabrous, style 3.5 mm long, stigma capitate. Capsule pyriform, membranous, trilocular, anterior locules 2-seeded, posterior 1-seeded; seeds oblong, closely pitted.

Very common in all districts, weed in agriculatural fields, road sides and waste lands.

Fl. & Fr.: July – November.

Surzapur (ADB), *TP & PVP* 4113; Mosra-Malkapur Road (NZB), *TP & BR* 6097; Lingampalle (RR), *MSM* 19219; Vempalli (KRN), *GVS* 20203 (MH); Kambalpally (NLG), *BVRS* 328 (NU); Aklaspur (KRN), *GVS* 20213 (MH); Tupakulagudem (WGL), *PVS & NRR* 76972 (BSID); Karepally (KMM), *RR* 105959 (BSID).

Commelina caroliniana Walter, Fl. Carol. 68. 1788. *C. hasskarlii* C.B. Clarke, Commelyn. Cyrtandr. Bengal t. 5. 1874; Hook.f., Fl. Brit. India 6: 370. 1892; Fischer, Fl. Madras 3: 1538. 1931; B.R.P. Rao in Pullaiah, Fl. Andhra Pradesh 3: 1002, 1997.

A herb with much branched stem, glabrous or pubescent, sometimes scaberulous. Leaves narrowly lanceolate, 2-6 × 1-1.8 cm, subacute; sheaths 1.2 cm long, base broad, ciliate. Spathes 1-3 cm long, ovate-lanceolate, cordate at the base with rounded lobes, glabrous, scabrid or hispid. Flowers in pubescent cymes, the upper branch 2-4, the lower 1-2-flowered; sepals 3; petals 3, paired petals reniform, blue; fertile stamens 3, sterile stamens 3. Capsules quadrate, subtruncate, membranous; seeds cylindric, truncate at one end, rounded at the other without an appendage.

Common in most districts in moist places both in fields and waste lands.

Fl. & Fr.: August – November.

Sattenapalli beat (ADB), *TP & PVP* 4099; Bijjur (ADB), *GO & PVP* 4862; Birsaipet (ADB), *GO & PVP* 4596; Laxmapur (NZB), *TP & BR* 6268; Mingaram (NZB), *TP & BR* 6304; Gadirayal RF (RR), *MSM* 10519; Rechapalli (KRN), *GVS* 22263 (MH & CAL); Bhadrachalam (KMM), *JSG* 16036 (MH & CAL); Kanapur (WGL), *KMS* 11580 (MH & CAL).

Commelina clavata C.B. Clarke, Commelyn. Cyrtandr. Bengal. t. 5. 1874; Hook.f., Fl. Brit. India 6: 371. 1892; Fischer, Fl. Madras 3: 1539. 1931.

A herb with 30-90 cm long stem, sparingly branched; branches diffuse, rooting at the nodes. Leaves ovate, elliptic-lanceolate or linear, 3-5 × 0.6-1 cm, glabrescent, base tapering, margin entire, apex acute-acuminate; sheath 2.5 cm, apically hirsute, peduncle 3.5 cm. Spathes ovate-lanceolate, 2.5 cm long, base cordate, apex attenuate; sepals with small, brown, linear spots; petals lilac. Capsules usually 2, oblong-quadrate, constricted in the middle; seeds 4, 2 per cell, black, two lower often imperfect.

Occasional in Medak and Hyderabad districts.

Fl. & Fr.: June – December.

Toopran (MDK), *BR* & *CP* 11283.

Commelina diffusa Burm. f., Fl. Indica 18. 1768; Panigr. & Kammathy, J. Indian Bot. Soc. 43: 299. 1964. *C. nudiflora* sensu Hook. f., Fl. Brit. India 6: 369. 1892; non L. 1753; Fischer, Fl. Madras 3: 1538. 1931.

A herb with creeping stem and rooting at the nodes; stems 60-90 cm long, branching from the base. Leaves sessile, ovate or lanceolate, 4-6 × 0.5-0.8 cm, chartaceous, base subcordate, margin entire, apex acuminate. Cymes usually 1-3-flowered. Spathes complicate, oblong-lanceolate, 3-5 cm long, glabrous, base cordate; sepals 3, petals 3 and two interior petals obovate with long claws, dark blue, the exterior sub-sessile, pale blue or nearly white; fertile stamens 3, sterile stamens 2, ovary globose, 3-celled of which two cells are 2-ovulate, the third 1-ovulate, style up to 5 mm long, coiled at apex, stigma capitate. Capsule broadly oblong, 6 mm long, acuminate, coriaceous, 5-seeded; seeds oblong-cylindric, tuberculate and brown.

Occasional along streams in Ranga Reddi district.

Fl. & Fr.: July – December.

Commelina ensifolia R.Br., Prodr.l. Nov. Holl. 69. 1810; Hook.f., Fl. Birt. India 6: 374. 1892; Fischer, Fl. Madras 3: 1540. 1931. *C. undulata* var. *setosa* C.B.Clarke in DC. Monogr. Phan. 179. 1881; Hook.f., Fl. Birt. India 6: 373. 1892; Fischer, Fl. Madras 3: 1539. 1931.

Profusely branched, perennial herbs; root fibres thick. Leaves linear-lanceolate, 5-12 × 0.5- 1.2cm, attenuate, undulate, apex acute-obtuse, glabrous above, puberulous below; sheath 2.5 cm. Flowers blue to violet-purple; spathe solitary or 3-5 together, cucullate. Capsules 2-3 per spathe, 2-celled, ovoid, 5 mm across, 3-seeded; seeds smooth, brown.

Common along streams and in marshes in Karimnagar district (Naqvi, 2001).

Fl. & Fr.: August – October.

Commelina erecta L., Sp. Pl. 41. 1753. *C. undulata* R. Br., Prodr. 270. 1810; Hook.f., Fl. Brit. India 6: 373. 1892 *p.p.*

A slender herb; stem 30-60 cm long, hairy or glabrate. Leaves lanceolate or oblong-lanceolate, 5-13 × 1.5-3 cm, chartaceous, appressed-pubescent above, glaucous below, base oblique, obtuse, margin entire, apex gradually tapering, acuminate; pubescent, margin ciliate. Spathes 3-5 in a terminal cluster, cucullate, broadly cordate, glabrous, apex acute-acuminate, subsessile. Flowers 3-6, pale violet in racemes. Capsule 3-celled, 2-valved; seeds 3, ashy, adnate to capsule wall.

A weed of cultivated fields and wastelands in Medak, Ranga Reddi, Warangal, Khammam and Hyderabad districts.

Fl. & Fr.: August – September.

Narsapur (MDK), *BR* & *CP* 14109, *KMS* 6711 (MH); Pakhal (WGL), *KMS* 11628 (MH); Parnasala (KMM), *RCS* 104242 (BSID); Bhadrachalam RF (KMM), *RCS* 10430 (BSID);

Commelina forskalaei Vahl, Enum. Pl. 2: 172. 1805; Hook.f., Fl. Brit. India 6: 371. 1893; Fischer, Fl. Madras 3: 1539. 1928.

A herb with long, branched, slender, diffuse stem, rooting at the nodes. Leaves linear or linear-oblong, 2.5-5 × 0.6-1.2 cm, base narrowed, margins usually undulate or crenulate, apex obtuse; sheaths 0.8-1.2 cm long, finely striate, pubescent, ciliate. Spathe 1.2 cm long, broadly ovate, hooded, sparsely hairy with long slender hairs; upper cyme-branch 1-3-flowered; sepals 3, pale white, odd sepal boat-shaped, paired sepals ovate, free; petals 3, blue, two obovate with long slender claw and one small, lanceolate with short claw; stamens 3, unequal, median stamen with short filament and large yellow anther, staminodes yellow; ovary globose, style long, stigma capitate. Capsule ovoid, trilocular, ventral locules 2-seeded, dorsal locule 1-seeded; seeds subglobose, smooth, without appendages, orange-brown.

In dry tracts of Mahabubnagar district (Raghava Rao, 1989).

Fl. & Fr.: June - November.

Commelina paludosa Blume, Enum. Pl. Jav. 1: 2. 1827; Rolla Rao & Kammathy, J. Bombay Nat. Hist. Soc. 59: 60. 1962; Fischer, Fl. Madras 3: 1539. 1931. *C. obliqua* Buch.-Ham ex D. Don, Prodr. Fl. Nep. 45. 1825 (non Vahl 1805); Hook.f., Fl. Brit. India 6: 372. 1892.

A stout subscandent herb; stem 60-90 cm long, branched, glabrous. Leaves lanceolate or elliptic-lanceolate, 10-20 × 2-5 cm, sessile or petiolate, acute or caudate-acuminate, membranous, glabrous, scabrous or villous; sheaths reaching 3 cm long, the mouth bearded with long hairs. Spathes sessile, solitary or crowded in terminal heads, funnel-shaped, acute, glabrous, subscabrid, usually filled with a clear glutinous liquid. Flowers blue, in simple raceme, large petals clawed, orbicular, stamens whitish yellow, ovary 3-celled, the cells 1-ovulate. Capsules trigonous-obovoid, subequally 3-valved, 3-celled, 3-seeded; seeds oblong or ellipsoid, smooth, lead-coloured.

Common in Adilabad, Nizamabad, Medak and Ranga Reddi districts.

Fl. & Fr.: June – October.

Vempalli (ADB), *GO* 4750; Ibrahimpet (NZB), *TP* & *BR* 6133; Mudheli RF (NZB), *BR* 7126; Kusumasamudram (RR), *MSM* & *NV* 10363; way to Siddharam (KMM), *PVS* 84075 (BSID); way to Gangaram (WGL), *PVS* 84035 (BSID).

Commelina petersii Hassk., Flora 46: 385. 1863; Faden, in Dassanayake (ed.), Rev. Handb. Fl. Ceylon 14: 191. 2000. *C. persicariifolia* Wight ex C.B.Clarke in DC. Mon. Phan. 3: 171. 1881. *nom. illegit.* non Dwile, 1815; Hook.f., Fl. Brit. India 6:

372. 1892; Fischer, Fl. Madras 3: 1536. 1931. *C. imberbis* sensu Rolla Rao, Blumea 14: 352. 1966; B.R.P.Rao in Pullaiah, Fl. Andhra Pradesh 3: 1002. 1997. *C. alisagarensis* Kumar & Deodikar, Proc. Indian Acad. Sci. 13: 168. 1941.

A herb with diffuse stem; leaves ovate-lanceolate, acuminate, base rounded and then narrowed into short petiole, 5-12 × 2-4 cm long, sheaths 1-2.5 cm long, mouth usually minutely pubescent. Spathes ovate, base truncately rounded. Capsule oblong, obtuse, seeds oblong, terete, smooth or obscurely rugose.

Rare in Nizamabad district.

Fl. & Fr.: December – February.

Note: Kumar and Deodikar (1941) reported a new species, *C. alisagarensis,* which was collected from Alisagar, Nizamabad district. Rolla Rao (*loc. cit*) considered the species as closely related to *C. imberbis* and reduced it as a synonym to the latter. *C. imberbis* is an African species and does not occur in India. *C. petersii* is closely allied to *C. imberbis,* an African species in its general appearance and tuberculate seeds (Nampy, 2005).

Commelina subulata Roth, Nov. Pl. Sp. 23. 1821; Hook.f., Fl. Brit. India 6: 369. 1892; Fischer, Fl. Madras 3: 1538. 1928.

A herb with slender erect stem, 20-40 cm long, simple or branched. Leaves narrow lanceolate or elliptic, 4-9 × 0.2-0.5 cm, glabrescent, base obtuse, margin entire, apex acute; sheath 1.2 cm long. Spathe ovate, 6-8 cm long, base cordate, margin cilate, apex acute. Flowers small, orange-purple, or chrome-yellow, when dry violet, sepals 3, unequal, broadly ovate-orbicular, paired sepals laterally fused, pale white, odd sepal boat-shaped; petals 3, paired petals reniform, golden yellow, free, odd petal ovate, pale white, margin pink; fertile stamens 3, staminodes 3; ovary oblong, 3-lobed, style 1.5 mm, dark purple, stigma capitate, pale purple. Capsules oblong, 3-celled, ventral locules 2-seeded, dorsal locule 1-seeded; seeds ovoid, dark, 4-grooved.

Reported from Hyderabad district.

Fl. & Fr.: August – October.

Osmania University campus (HYD), *Kammathy*, 73965 (BSI, Pune).

Commelina undulata R.Br., Prodr. Fl. Nov. Holl. 270. 1810. *Commelina kurzii* C.B. Clarke, J. Linn. Soc., Bot. 11: 444. 1871; Hook.f., Fl. Brit. India 6: 373. 1892; Fischer, Fl. Madras 3: 1539. 1931. *C. longifolia* Lam., Tabl. Encyl. 1: 129. 1791; Rolla Rao, Taxon 10: 254. 1961.

A slender, decumbent herb; stem slender, sometimes rooting, glabrous, with long internodes. Leaves linear or linear-lanceolate, 3-8 × 0.5-1 cm, chartaceous, glabrous, base obtuse, margin entire, apex acute-acuminate; sheath 2 cm, margin ciliate. Spathes ovate-lanceolate, axillary, glabrous, base rounded. Flowers small, polygamous; branches of the cyme equal, usually 1-2-flowered; sepals free, ovate, obtuse, the 2 inner connate below, larger than the outer; petals dark blue, the 2 larger ovate, entire, subsessile or with a very short claw; stamens 3, fertile, one anther large, lunate, the other two smaller, ellipsoid, staminodes 3, clavate. Capsules oblong, 3-celled; seeds 2-5, ovoid-subglobose, truncate at one end, black.

Occasional in Warangal, Hyderabad, Nalgonda and Mahabubnagar districts.

Fl. & Fr.: July – October.

Daravagu (MBNR), *SRR* 2631; Pakhal (WGL), *ANH* 15974 (MH & CAL); Ramanagutta (KMM), *RR* 107971 (BSID); Kambalpally (NLG), *BVRS* 528 (NU).

CYANOTIS D. Don *nom. cons.*

1. Cymes enclosed in leaf sheaths:
   2. Filaments bearded .... **C. axillaris**
   2. Filaments naked .... **C. cucullata**
1. Cymes not enclosed in the leaf-sheaths:
   3. Plants cobwebby-woolly throughout:
      4. Cymes densely aggregate on elongate flowering shoots; basal leaves prominently dissimilar to upper ones .... **C. arachnoidea**
      4. Cymes lax, terminal and axillary on stems; leaves similar .... **C. fasciculata**
   3. Plants not cob-webby-woolly:
      5. Roots tuberous; plants pubescent; leaves 10-30 cm long .... **C. tuberosa**
      5. Roots not tuberous; plants almost glabrous; leaves < 5 cm long .... **C. cristata**

Cyanotis arachnoidea C.B. Clarke, Monogr. Phan. 3: 250. 1881; Hook.f., Fl. Brit. India 6: 386. 1892; Fischer, Fl. Madras 3: 1550. 1931.

A herb, prostrate or creeping with cottony or cobwebby woolly stem; stems 15-60 cm long, often decumbent and rooting below. Leaves very variable, basal ones oblong-elliptic or lanceolate, 6-15 × 0.4-0.8 cm, chartaceous, cobwebby, base obtuse, margin entire, apex rounded, sheath 2 cm. Cymes terminal and/or axillary; bracts ovate-lanceolate, 1 cm, exceeding cymes; bracteoles 5-8 pairs, falcate. Flowers blue; sepals 3, green, lanceolate, cobwebby, bearded; petals 3, violet; stamens 6, filaments purple-violet, bearded hairs blue-violet or purple, with sub apical bulging, anther lobes orange yellow; ovary 3-celled, style pilose. Capsules to 3 mm; seeds terete, pitted.

Common in Ranga Reddi and Karimnagar districts.

Fl. & Fr.: August – October.

Cyanotis axillaris (L.) D. Don ex Sweet, Hort. Brit. 430. 1826; Hook.f., Fl. Brit. India 6: 388. 1892; Fischer, Fl. Madras 3: 1550. 1931. *Commelina axillaris* L., Sp. Pl. 42. 1753. *Tonningia axillaris* (L.) Kuntze, Revis. Gen. Pl. 2: 722. 1891. *Amischophacelus axillaris* (L.) R.S.Rao & Kammathy, J. Linn. Soc. Bot. 59; 306. 1963

An annual glabrous herb, with variable, prostate or erect branches, long and spreading, rooting at nodes, stem 10-40 cm long, glabrous. Leaves narrow-lanceolate,

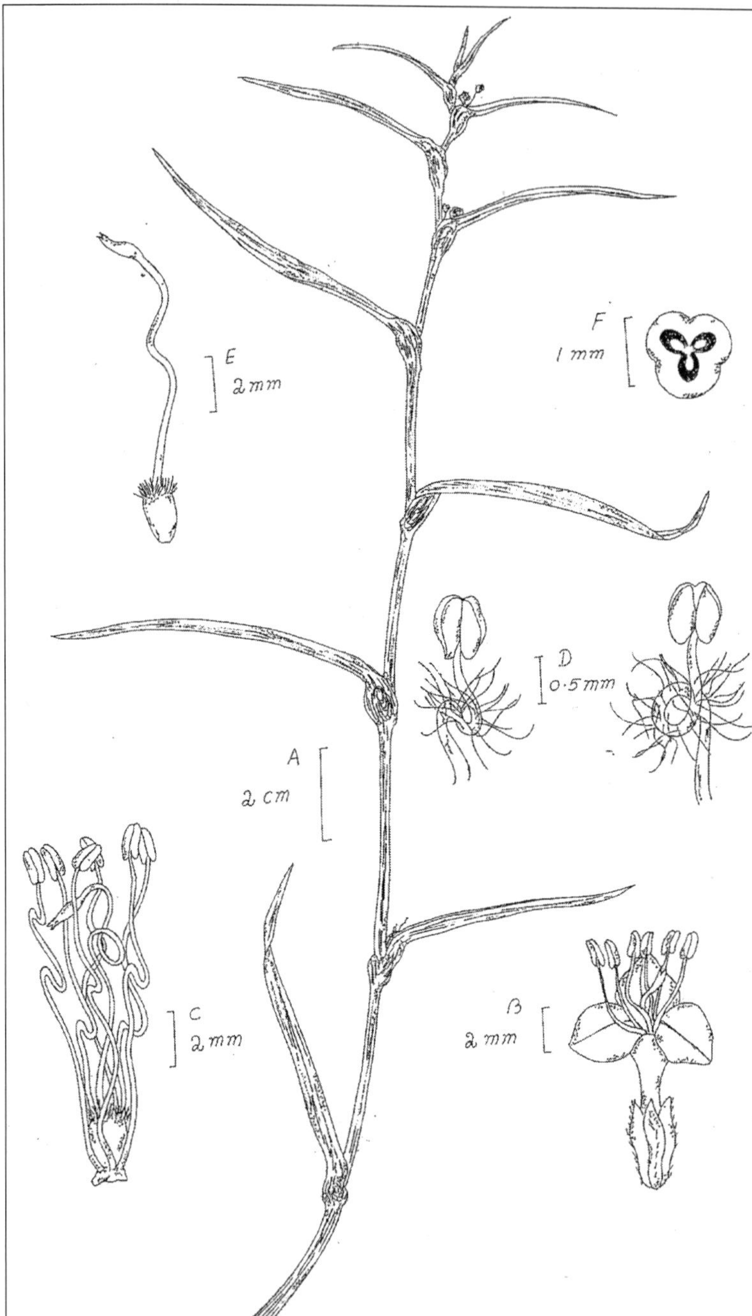

Figure 119: Cyanotis axillaris (L.) D. Don ex Sweet [Syn.: *Tonningia axillaris* (L.) Kuntze] A. Habit, B. Flower, C. Perianth split open with stamens, D. Anthers, E. Pistil, F. T.S. of ovary.

4-10 × 0.5-1 cm, succulent, plicate or flat, base obtuse, margin entire, apex acuminate; sheaths 6 mm, pilose, pouched. Flowers violet, blue and white clustered in the axils of leaf sheaths; bracteoles linear. Flowers 6 mm across; calyx lobes lanceolate; petals ovate; stamens 6, filaments blue with pilose hairs, anthers oblong, ovary woolly, 3-celled, ovules 2 per cell on axile placentae, pilose, stigma shortly 3-fid. Capsules ellipsoid, apex beaked; seeds oblong, pitted.

Common in moist places and cultivated fields in all districts.

Fl. & Fr.: July – November.

Gupalapatnam (ADB), *GO* 4395; Manichippa RF (NZB), *TP & BR* 6028; Sivanoor (MDK), *BR* & *CP* 11545; Narsapur RF (MDK), *TP* & *CP* 14029, *BR* & *CP* 14111; Gandipet (HYD), *KI* & *NVN* 18757; Sagar left canal (KMM), *MHR* & *MCK* 17009; Mohammadabad RF (RR), *MSM* 10495; Pogila (NLG), *BVRS* 298 (NU); Narsapur (MDK), *KMS* 6678 (MH); Pakhal (WGL), *KMS* 11659 (MH); Rechapalli (KRN), *GVS* 22260 (MH).

Cyanotis cristata (L.) D. Don, Prodr. Fl. Nepal 46. 1825; Hook.f., Fl. Brit. India 6: 385, 1892; Fischer, Fl. Madras 3: 1549. 1931. *Commelina cristata* L., Sp. Pl. 42. 1753.

A herb, prostate or creeping; stem branched from the base, the branches 15-45 cm long, slender, creeping and rooting below. Leaves sessile, ovate or oblong-lanceolate to 2.5 × 1 cm, chartaceous, sparsely pubescent, base cordate, margin ciliate, apex obtuse; sheath loose, pubescent. Flowers in scorpioidly recurved cymes, bracts ovate-lanceolate, exceeding cymes; bracteoles 5-8 pairs; sepals 3, pale white; petals 3, rose to purple, united up to middle; stamens 6, filaments blue to violet, bearded, hairs blue, anther lobes yellow; ovary globose, style glabrous, stigma capitate. Capsules oblong, 2 mm long, trigonous, seeds trigonous, with 2 large pits on 2 sides.

Common in Adilabad, Medak, Karimnagar, Warangal, Nalgonda, Hyderabad and Mahabubnagar districts in exposed to partially shaded areas, moist crevices, shallow road side ditches and banks, near streams.

Fl. & Fr.: July – November.

Ankusapuram (ADB), *GO* 4301; Nagulabanda (MDK), *CP* 11350; Kodimial (KRN), *GVS* 25671 (MH & CAL); Rekularam (NLG), *BVRS* 504 (NU); Kanapur RF (WGL), *KMS* 11585 (MH & CAL); Parnasala RF (KMM), *RCS* 102465 (BSID); Krishnarajasagar (KMM), *RCS* 102460 (BSID).

Cyanotis cucullata (Roth) Kunth, Enum. Pl. 4: 107. 1843; Hook.f., Fl. Brit. India 6: 389. 1892; Fischer, Fl. Madras 3: 1550. 1931. *Tonningia cucullata* (Roth) Kuntze, Revis. Gen. Pl. 2: 722. 1891. *Amischophacelus cucullata* (Roth) Rao & Kamm., J. Linn. Soc. Bot. 59: 306. 1966. *Tradescantia cucullata* Roth, Nov. Pl. Sp. 189. 1821.

An annual prostrate slender herb. Leaves oblong-lanceolate, succulent, obtuse, sheath prominently inflated. Flowers lilac, in axillary fascicles. Capsule depressed at apex with 3 projections; seeds obscurely pitted.

A weed of cultivated fields in Adilabad, Medak and Nalgonda districts.

Fl. & Fr.: July – December.

Kadam dam (ADB), *TP* & *PVP* 4147; Gummadidala (MDK), *BR* & *CP* 14115; Nagarjunasagar (NLG), *KMS* 9731 (MH & CAL).

Cyanotis fasciculata (B.Heyne ex Roth) Schult. & Schult. f., Syst. Veg. 7: 1152. 1830; Hook.f., Fl. Brit. India 6: 387. 1892; Fischer, Fl. Madras 3: 1550. 1931. *Tradescantia fasciculata* B. Heyne ex Roth, Nov. Pl. Sp. 189. 1821.

A decumbent herb with slender stems, 5-18 cm long, branchlets cottony or cobwebby. Leaves linear or oblong-lanceolate, 1-3 × 0.4-0.8 cm, chartaceous, silky-cobwebby or floccose, base cordate, margin entire, apex subacute; sheath 5 mm. Cymes axillary and/or terminal, solitary or 2 or 3 in a cluster; pink or green, cobwebby; peduncles slender, bracts narrowly lanceolate, recurved, bracteoles narrowly ovate, 3-5 pairs, apex acuminate. Flowers 5 mm across; calyx cottony, lobes lanceolate; petals 3, rose; stamens 6, filaments blue-purple; ovary ovate-oblong, style blue, bearded, hairs blue-purple. Capsules oblong; seeds 2-6, oblong, faintly rugose.

Common throughout along marshy patches.

Fl. & Fr.: June – December.

Nadipally (NZB), *BR* 6451; Gandhari (NZB), *BR* & *CPR* 7240; Sangareddy (MDK), *BR* & *CP* 11286; Sardhana (MDK), *CP* 11773; Dharmapur (RR), *MSM* 10489; Kodimial (KRN), *GVS* 25678 (MH & CAL); Narsapur (MDK), *KMS* 6688 (MH).

Cyanotis tuberosa (Roxb.) Schult. & Schult. f., Syst. Veg. 7: 1153. 1830; Hook.f., Fl. Brit. India 6: 386. 1892; Fischer, Fl. Madras 3: 1549. 1931. *Tradescantia tuberosa* Roxb., Pl. Coromandel 2:5. 1799. *Cyanotis adscendens* Dalzell in Hook. Kew J. Bot. 343.1852; Rolla Rao, Blumea 14(2) : 348.1966.

A perennial herb with roots of fusiform tubers; stem 15-75 cm long, swollen, hirsute at the base, suberect or prostrate below, almost glabrous. Basal leaves oblong, 10-30 × 1-3 cm, upper ones linear-ensiform, chartaceous, glabrous or villose above, downy-sericeous below, base obtuse, margin ciliate, apex subacute; sheath 1-2 cm. Cymes terminal and/or axillary, corymbose; bracts ovate, equal or slightly exceeding cyme; bracteoles 10-15 pairs, in 2 series; calyx-tube villous; corolla bluish-purple, lobes ovate, subacute; stamens 6, blue, bearded, hairs blue to violet; style violet-blue, stigma simple. Capsules ellipsoid, the upper half hairy, the lower half glabrous; seeds 4-6, pyramidal, obscurely rugose, brown.

Common throughout the State as forest undergrowth.

Fl. & Fr.: August – September.

Bhimnagar (NZB), *TP* & *BR* 6211; Nagulabanda (MDK), *CP* 11345; Narsapur RF (MDK), *TP* & *CP* 14041; Kusumasamudram RF (RR), *MSM* & *NV* 10341; Central University campus (RR), *MSM* 15180; Borapur (MBNR), *BSS* & *SKB* 32397; Kodimial (KRN), *GVS* 20066 (MH).

MURDANNIA Royle *nom. cons.*

1. Roots tuberous .... **M. edulis**
1. Roots fibrous:
   2. Leaves linear, 10 times longer than broad:
      3. Leaves 1-2 mm broad; bracts ochreate; staminal filaments naked .... **M. semiteres**
      3. Leaves 3-8 mm broad; bracts ovate-oblong; staminal filaments bearded:
         4. Leaves plicate, recurved .... **M. esculenta**
         4. Leaves flat, not recurved:
            5. Capsule subglobose; seeds 2 per cell .... **M. nudiflora**
            5. Capsule globose, seeds 1 per cell .... **M. vaginata**
   2. Leaves ovate-lanceolate, to 3 times longer than broad .... **M. spirata**

Murdannia edulis (Stokes) Faden, Taxon 29: 77. 1980. *Commelina edulis* Stokes, Bot. Mat. Med. 1: 184. 1812. *Aneilema scapiflorum* (Roxb.) Kostel. in Allg. Med.-Pharm. Fl. 1: 127. 1831; Hook.f., Fl. Brit. India 6: 375. 1892; Fischer, Fl. Madras 3: 1544. 1928. *Commelina scapiflora* Roxb., Fl. Ind. 1: 178. 1820. *Murdannia scapiflora* (Roxb.) Royle, Ill. Bot. Himal. Mts. 403. 1840.

A tufted herb with elongate pisiform tubers. Leaves radical, erect, 10-20 × 1-1.5 cm, linear-ensiform, finely acuminate, glabrous, slightly narrowed at the base. Flowers blue-mauve in panicles on terminal leafless scape; scape together with the panicles 20-45 cm long; upper bracts amplexicaul, ovate, acuminate or truncate, membranous, often spotted with small spots; sepals elliptic-oblong, purple-green; petals blue, obovate; stamens 3 perfect and 3 staminodes, filaments all bearded with blue hairs, anthers of fertile stamens blue and those of staminodes yellow. Capsules obovoid; seeds 5 or 6 in a cell, superposed, sharply 3-gonous.

In plain fields of Mahabubnagar district.

Fl. & Fr.: May – June.

Appapur (MBNR), *BSS* & *SKB* 30703 (BSID); Kattaribandalu (MBNR), *SRR* 752 (HH).

Murdannia esculenta (Wall. ex C.B. Clarke) R.S.Rao & Kammathy, Bull. Bot. Surv. India 3: 394. 1961. *Aneilema esculentum* Wall. ex. C.B. Clarke, Monogr. Phan. 3: 206. 1881; Hook.f., Fl. Brit. India 6: 377. 1892; Fischer, Fl. Madras 3: 1545. 1928.

A herb, subsolitary stem, roots tuberous. Leaves linear, 4-7.5 × 0.4-0.8 cm, glabrous, often recurved, base obtuse, margin entire, apex subacute; sheath 1.5 – 2.5 cm, oblique at the mouth. Flowers in panicles on 4-6 cm long peduncle; bracts oblong, persistent; sepals 3, oblong, subequal; petals 3, pink to violet, obovate; fertile stamens 3, staminodes 3, filaments bearded; ovary linear, style pale white, stigma capitate. Capsule 3-angled; seeds subcubical, dark brown, obscurely scabrid or pitted.

In marsh and moist rocky places of Warangal district.

Fl. & Fr.: September – November.

Subedari – North West (WGL), *PSR* 292 (HH).

Murdannia nudiflora (L.) Brenan, Kew Bull. 7: 189. 1952. *Commelina nudiflora* L., Mant. 177. 1771. (non Sp. Pl. 41. 1753). *Aneilema nudiflorum* (L.) R. Br., Prodr. 271. 1810; Hook.f., Fl. Brit. India 6: 378. 1892; Fischer, Fl. Madras 3: 1545. 1931.

An erect or decumbent, slender branching herb, rooting at nodes; stem simple or branched from the base, branches 15-30 cm long. Leaves linear-lanceolate, 3-8 × 0.4 cm, chartaceous, glabrous, base obtuse, margin entire, apex tapering, acuminate; sheath 0.6 – 1 cm, the margins and mouth strongly ciliate. Flowers in subglobose cymes on terminal (rarely axillary) panicles; sepals 3, oblong, obtuse; petals 3, blue or purple suborbicular; stamens 2 perfect, 2-4 sterile, filaments of the fertile only or of all bearded; ovary ovate, style white, stigma capitate. Capsules trigonously subglobose, membranous, 3-celled, with 2 seeds in each cell; seeds cuboid, deeply pitted, dark brown.

Common throughout the State on river banks, near waterlogged paddy fields etc.

Fl. & Fr.: June – October.

Mingaram (NZB), *TP* & *BR* 6310; Gandhari (NZB), *BR* & *CP* 7225; Thornal (MDK), *CP* 11704; Mohammadabad RF (RR), *MSM* 10443; Pakhal lake (WGL), *MHR* & *MCK* 17064; Pakhal (WGL), *KMS* 13151 (MH); Aklaspur (KRN), *GVS* 25639 (MH); Nellippakka RF (KMM), *RCS* 99080 (BSID).

Murdannia semiteres (Dalzell) Santapau, Poona Agric. Coll. Mag. 41(4): 15. 1951. *Aneilema semiteres* Dalzell, Hooker's J. Bot. Kew Gard. Misc. 3: 138. 1851. *A. paniculatum* Wall. ex. C.B. Clarke, Monogr. Phan. 3: 215. 1881, non Wight, 1853; Hook. f., Fl. Brit. India 6: 351. 1892. *p.p.* Fischer, Fl. Madras 3: 1546. 1931. *p.p.* *Commelina nimmoniana* Grah., Cat. Bombay Pl. 224. 1839; Hook.f., Fl. Brit. India 6: 374. 1892. *Murdannia juncoides* (Wight) R.S.Rao & Kammathy, Bull. Surv. India 6: 3. 1965.

Erect tufted herbs, to 15 cm high. Leaves 7-10 × 0.1-0.3 xm, linear, base acute, apex acute or acuminate, glabrous, bracts minute. Flowers blue in 5-7 cm long panicles; pedicles 3-4 mm long; sepals 3, elliptic-obovate; petals 3, obovate; stamens 3, connate at base, staminodes 3; ovary 3-locular, ovules 2 in each locule. Capsule brown, elliptic; seeds 6 or 8 per locule, dark brown, smooth with minute striations.

Rare in Mahabubnagar, Hyderabad and Medak districts.

Fl. & Fr.: July – December.

Thummalakunta (HYD), *TR* 1404 (HH); Narsapur (MDK), *KMS* 6767 (MH); Bonthapalli (MDK), *VNR* 4451 (HH); Shadnagar scrub jungle (MBNR), *SR* 2796 (HH).

Murdannia spirata (L.) Brückn., Nat. Pflanzenfam. (ed. 2) 15 a: 173. 1930. *Commelina spirata* L., Mant. Pl. 2: 176. 1771. *Aneilema spiratum* (L.) R. Br., Prodr. 271. 1810; Hook.f., Fl. Brit. India 6: 377. 1892; Fischer, Fl. Madras 3: 1545. 1931.

A small, erect or diffuse, annual herb, often rooting at nodes; branches 15-25 cm long, decumbent. Leaves sessile, the lower ones broadly linear-lanceolate, the upper ones oblong, cordate, all amplexicaul, glabrous, 1.5-3.5 × 0.6 cm, margin entire, apex acute. Flowers in dichotomously branched panicles; sepals 3, oblong; petals 3, blue, broadly oblong, obtuse; stamens 3 perfect, the anthers apiculate, blue, the filaments bearded, staminodes 3, shorter; ovary 3-celled, cells 3-many-ovulate. Capsules oblong, smooth, shining, trigonous, the cells 3-7-seeded; seeds 1-seriate, minute, cubical, white or straw-coloured.

Common in Adilabad, Nizamabad, Medak, Hyderabad, Karimnagar and Warangal districts.

Fl. & Fr.: August – November.

Bijjur (ADB), *GO* & *PVP* 4851, 4855; Mosra (NZB), *TP* & *BR* 6112; Sagar left canal (KMM), *MHR* & *MCK* 17021; Narsapur (MDK), *KMS* 6690 (MH); Pakhal (WGL), *KMS* 11673 (MH & CAL).

Murdannia vaginata (L.) Brückn., Pflanzenfam. (ed. 2.) 15 a: 173. 1930. *Commelina vaginata* L., Mant. Pl. 2: 177. 1771. *Aneilema vaginatum* (L.) R. Br., Prodr. 271. 1810; Hook.f., Fl. Brit. India 6: 381. 1892; Fischer, Fl. Madras 3: 1546. 1931.

A slender grass like herb with erect and prostrate branches from near the root; roots fleshy; stem 15-40 cm long, very slender, flaccid, rooting at the lower nodes; branches sub erect, ending in filiform peduncles with distant flowering bracts. Leaves linear, 5-13 × 0.6-1 cm, apex acuminate, glabrous or sparsely hairy; sheaths short, open ciliate. Flowers small, 1-3 together, fascicled in distant bracts; bracts narrowly oblong, obtuse, strongly ribbed, slender, usually twice jointed; sepals glabrous or pubescent; petals orbicular, blue; stamens 2 perfect, 3-4 sterile, filaments of the perfect stamens bearded, ovary 2-celled, cells 1-ovulate. Capsules globose, cuspidate, shining, 3-celled; seeds trigonously conical, obtuse, slightly rugose, black.

Rare in Mahabubnagar and Khammam districts.

Fl. & Fr.: July – October.

Jatprole (MBNR), *SRR* 1977 (HH); Rathamhutta hills (KMM), *RCS* 104231 (BSID).

TRADESCANTIA L.

1. Plants erect; flowers in axillary cymes .... **T. spathacea**
1. Plants decumbent; flowers developed on top of the stems (terminal) .... **T. zebrina**

Tradescantia spathacea Sw. Prodr.57.1788. *T. discolor* L'Herit, Sert. Angl. 8, t. 12.1788. *Rhoeo discolor* (L' Herit.) Hance in Walp. Ann. 3:659.1853; Cooke, Fl. Pres. Bombay 3:306 (BSI reprint 1958). *R. spathacea* (Swartz) Stearn, Baileya. 5: 198. 1957; Brenan in Hutch. & Dalz., Fl. W. Trop. Afr. 3(1): 23. 1968.

Perennial, robust herbs with short stem arising from the underground rhizome, 30-60 cm high. Leaves crowded on congested nodes, imbricately arranged, erect, lanceolate- oblanceolate, sheathing at base, 20-30 x 3-5 cm, purplish red beneath. Flowers white, in condensed, axillary cymes, enclosed by short, boat-shaped involucral bracts, 2.5-4 cm long, on short peduncles with few, smaller, sterile bractlets; sepals ovate-lanceolate; petals ovate; stamens all perfect; filaments barbellate. Capsules ovoid, 3-valved.

Cultivated in gardens as a border ornamental and also grown in pots.

Fl. & Fr. : Almost round the year.

Tradescantia zebrina Hort ex Bosse, Vollstandiges Handb. Blumengart. 4: 655.1849. *Zebrina pendula* Schnizl., Bot. Zeit. 3:870.1899; Bailey, Man. Cult. Pl. 198; 1949.

Decumbent herbs. Leaves lanceolate-elliptic to ovate-elliptic, 3-9 x 1.5-3 cm, variegated, lower surface reddish purple, upper surface striped green and white, base oblique, cuneate, apex acute to acuminate. Flowers in terminal, pairs of sessile cymes enclosed in sheaths of spathaceous bracts, pedunculate; spathaceous bracts foliaceous, reduced; sepals basally connate, 4-5 mm; petals pink, clawed, claws basally connate forming tube; filaments bearded. Capsules 3-locular; locules 2-seeded.

Grown in pots and in borders as an ornmental.

Fl. & Fr. : September-February.

ARECACEAE

1. Leaves simple, palmately fan-shaped .... **Borassus**
1. Leaves compound; pinnate or 2-pinnate:
    2. Leaves pinnate:
        3. Leaves armed; fruit drupe .... **Phoenix**
        3. Leaves unarmed; fruit nut .... **Cocos**
    2. Leaves bipinnate .... **Caryota**

BORASSUS L.

Borassus flabellifer L., Sp. Pl. 1187. 1753; Becc. in Hook.f., Fl. Brit. India 6: 482. 1893; Fischer, Fl. Madras 3: 1562. 1931. *B. flabelliformis* Murr. in L. Syst. Veg. (ed. 13). 827. 1774.

A tree with tall stem, stout, 40 m high and 0.5 m diameter near the ground. Leaves simple, palmately fan-shaped, plicate, the margin multifid; leaflets 60-80, induplicate, apex acuminate; petiole stout, spinous, base broad, split. Spadices interfoliar, large, dioecious, branched; peduncles sheathed with open spathes. Male flowers small, clustered, mixed with scaly bracts, second in 2 series in a small spikelet; sepals 3, oblong; petals 3, obovate-spathulate; staments 6; pistillode of 3 bristles. Female flowers larger, globose, perianth fleshy, accrescent; sepals reniform, imbricate; petals smaller, convolute; staminodes 6-9; ovary globose, subtrigonous, entire or 3-4-partite, 3-4-celled, ovules basal, erect, stigmas 3, sessile, recurved. Fruit a large subglobose drupe with 1-3 fibrous pyrenes, yellow when ripe; pericarp thinly fleshy; seed oblong, top 3-lobed, testa adhering to the pyrene.

In most districts, wild and cultivated.

Fl. & Fr.: February – April.

Dharmaram (NZB), *BR* & *KH* 6462; Narsapur (MDK), *BR* & *CP* 14132; Lingampalle (RR), *MSM* 19222.

CALAMUS L.

Calamus rotang L., Sp. Pl. 325. 1753; Becc. in Hook.f., Fl. Brit. India 6: 447. 1892; Fischer, Fl. Madras 3: 1568. 1931.

An armed, scandent shrub; stem ringed at nodes. Leaves pinnatisect, 30-60 × 0.6-1 cm, base cuneate, margins bristly ciliate, apex gradually acuminate, setose, nerves prominent, midnerve with scattered, straight spines below; petiole very short, with straight or recurved spines; rachis apically 3-gonous, with recurved spines; leaf-sheath glabrous; ochrea short, truncate. Spadices axillary, elongate, branched, 60 cm long, apical flagellum with recurved spines; spathes tubular, sparingly armed. Flowers small, polygamodioecious. Male flowers in 3 or 4 series; calyx cupular, 3-lobed; petals 3, yellow, free; stamens 6, shortly connate at base. Female flower: ovary imcompletely 3-celled, closed with retrorse scales, ovules 3, basal erect, style terminal. Fruit subglobose, mucronate, with scales in vertical series; seed solitary.

Occasional clumps in Ramappa lake margins (C.S. Reddy, 2001).

Fl. & Fr.: September – February.

Ramappa lake (WGL), *CSR* 850 (KU).

CARYOTA L.

Caryota urens L., Sp. Pl. 1189. 1753; Becc. in Hook.f., Fl. Brit. India 6: 422. 1892; Fischer, Fl. Madras 3: 1560. 1931.

A tree up to 20 m high; stem annulate, cylindric, smooth, grey, shining, covered with long shallow cracks with corky edges. Leaves bipinnate, 5 m; pinnae ca 10 pairs; subopposite, curved. Leaflets induplicate, clustered or alternate, 15-20 × 7-10 cm, base cuneate, oblique, margin praemorse (ragged), irregularly serrate, upper serration caudate, nerves flabellate; petiole stout. Spadices interfoliar, 3 m long; branches simple, forming a dense tassel drooping from the stout short peduncle, all reaching the same level. Spathes 0.5 m long, closely embracing the peduncle of the spadix. Flowers monoecious, solitary and/or in threes with the central one pistillate between the 2 staminate. Male flowers: sepals 3, rounded, imbricate; petals linear-oblong, valvate; stamens ca. 40, filaments short; pistillode 0. Female flowers subglobose, sepals broader; petals rounded, valvate; staminodes 3; ovary trigonous, 3-celled, ovules 1 per cell, erect, stigma sessile, 3-lobed. Fruit globose; seed(s) 1 or 2.

Grown as ornamental.

Fl. & Fr.: February – August.

COCOS L.

Cocos nucifera L. Sp. Pl. 1188. 1753; Beccari & Hook.f. Fl. Brit. India 6 : 482. 1893; Blatter, Palms Brit. India 499. 1926.

Solitary palm; trunk 6-12 m high, annulate, thickend at the base with a mass of rootlets. Leaves pinnatisect, 2-4 m or more long; segments linear, inserted on rachis in 2 ranks, 60-120 cm long; petioles stout, 1-2 m long, sheathing at base. Plants monoecious; spadices stout, 1-2 m long, androgynous, panicled; spathes 60-90 cm long, hard, splitting length wise, persistent; staminate flowers creamy yellow, unsymmetric; perianth accrescent in fruit. Fruits green, yellow, or bronzy red when immature, brown when mature; mesocarp dry, fibrous; endocarp brown, bearing 3 germination pores.

Planted in gardens, farms etc.

Fl. & Fr.: Round the year.

PHOENIX L.

1. Stem more than 15 m high .... **P. sylvestris**
1. Stem less than 15 m high:
   3. Leaflets still .... **P. pusilla**
   3. Leaflets pliable .... **P. loureirii**

Phoenix loureiri Kunth., Enum. Pl. 3: 257. 1841. *P. humilis* Royle var. *pedunculata* Becc., Malesia 3: 373. 1890; Becc. in Hook.f., Fl. Brit. India 6: 427. 1892; Fischer, Fl. Madras 3: 1560. 1931.

A plant with usually well developed stem, up to 2 m high, densely covered with the bases of the fallen petioles, more or less spirally arranged; root suckers developing when the primary stem has been burnt or injured. Leaves 2.5 m long; leaflets 15-30 × 0.8-1 cm, pliable, base narrow, apex acuminate, spinous. Spadices interfoliar. Male: spadix 40 cm; spikes 10 cm, in clusters; spathes 15-20 cm; calyx lobes triangular; petals oblong; stamens 6. Female: Spadix 1 m; spikes 8 cm. Drupe oblong, 1.5 cm, at first orange, black when ripe.

Common in waste lands, scrub jungles and hedges in Nizamabad, Medak and Mahabubnagar districts.

Fl. & Fr.: March – May.

Indalwai (NZB), *TP* & *BR* 6411; NSF fence (MDK), *BR* & *CP* 11649; Farahabad (MBNR), *TDCK* 15347.

Phoenix pusilla Gaertn., Fruct. 1: 24. 1788; Becc. in Hook. f. Fl. Brit. India 6: 426. 1892. *Phoenix farinifera* Roxb., Pl. Coromandel 1: 55, t. 74. 1796 & Fl. Ind. 3: 783. 1832; Becc. in Hook.f., Fl. Brit. India 6: 426. 1892; Fischer, Fl. Madras 3: 1559. 1931.

Vern.: *Chitti-ita, Churuta-ita.*

A small shrub, 15 cm – 3 m tall, stem thickly clothed and hidden by the old leaf-sheaths. Leaves 1-1.5 m long, leaflets fascicled more or less 4-farious, rigid, shining, usually with an orange-red pulvinus at the junction with the rachis, 7-20 × 1-1.5 cm, several basal forming stout flat spines of 5-7.5 cm long. Fruit 1.2-2 cm long, first green, black when ripe.

Rare in Mahabubnagar district.

Fl. & Fr.: January – April.

Nilagiri view point (MBNR), *SRS* 109720 (BSID).

Phoenix sylvestris (L.) Roxb., Fl. Ind. 3: 787. 1832; Becc. in Hook.f., Fl. Brit. India 6: 425. 1892; Fischer, Fl. Madras 3: 1559. 1931. *Elate sylvestris* L., Sp. Pl. 1189. 1753.

A tall graceful palm, up to 15 m high with a large thick hemispherical crown; trunk clothed with the persistent bases of the petioles. Leaves greyish green, 2-4 m long; leaflets 10-30 × 1-2.5 cm, glaucous, rigid, base narrow, apex acuminate, spinous; lower leaflets spiny. Spadices 1 m; spathes as long as spadices, coriaceous; spikes numerous, in clusters, 10-30 cm long. Male flowers white, scented, angular, oblique; calyx cup-shaped, with 3 short rounded teeth; petals 3 or 4 times as long as the calyx, concave, warted on the outside; stamens 6, anthers linear, filaments very short, free. Female flowers distant, roundish; calyx cup-shaped, obsoletely 3-toothed; petals 3, very broad, convolutedly imbricate; staminodes 3 or 4. Fruit oblong – ellipsoid, orange-yellow, edible; seeds rounded at the ends, grooved on the face, pale brown.

Common in waste lands, scrub jungles and field hedges in all districts.

Fl. & Fr.: January – April.

Indampalli (ADB), *GO* & *DAM* 5410; Lingampalle (RR), *MSM* 19221; Amrabad (MBNR), *SKB* & *BSS* 32547; Farahabad (MBNR), *TDCK* 15347; Hamsanpalli (MDK), *RG* 104045 (BSID).

PANDANACEAE

PANDANUS L. f.

Pandanus fascicularis Lam., Encycl. 1: 372. 1785; Hook.f., Fl. Brit. India 6: 485. 1893. *P. odoratissimus* L. f., Suppl. Pl. 424. 1781. *nom. illeg. P. tectorius* sensu Fischer, Fl. Madras 3: 1570. 1931.

A much branched shrub or tree, 6 m high; rarely erect, stem supported by aerial roots. Leaves linear-ensiform, stoutly armed with white prickles on margins and dorsal midrib, under-surface of leaf glaucous; leaves usually less than 2 m long and 8 cm wide. Male inflorescence of several spikes with white or cream axillant bracts. Spikes 4-12 cm long, composed of numerous phalanges, these racemose of numerous stamens; anthers elongate, briefly apiculate, white. Female inflorescence a solitary terminal cephalium, pendulous in fruit. Phalanges mostly 5-15-celled, the carpels concentrically arranged. Fruit an oblong syncarpium, 15-25 cm long, yellow or red.

Rare in all districts, in marshy localities.

Fl. & Fr.: July – December.

Pocharam RF (MDK), *BR* & *CP* 11601; Amrabad (MBNR), *TDCK* 15342.

TYPHACEAE

TYPHA L.

1. Leaf sheath angular, keeled .... **T. elephantina**
1. Leaf sheath not angular, not keeled .... **T. domingensis**

Typha domingensis Pers., Syn. Pl. 2: 532. 1807; Napper in Redhead & Polhill, Fl. Trop. E. Afr. 1. 1971; Cook in Tutin & *al.*, Fl. Eur. 276. 1980. *T. angustata* Bory & Chaub., Nouv. Fl. Pelop 4. 1838; Hook.f., Fl. Brit. India 6: 489. 1893; Fischer, Fl. Madras 3: 1571. 1931.

A robust herb, 2 m high, stem terete. Leaves distichous, linear, erect, 2 m long and 3 cm broad, exceeding the flowering stem, apex tapering, acute, base sheathing. Spikes cylindric, the male and female spikes often separated by a considerable interval. Male spike 10-20 cm long; rhachis (separating male from female) 3 cm long; stamens 2 or 3; anthers linear, connections shortly produced, apex rounded, hairs around stamens linear, spathulate. Female spike 20 cm long, 1-1.5 cm broad, cushion-like; bracteoles numerous, gynophores white, ovary linear, stigma flattened. Fruit fusiform, 2 mm.

Common throughtout the State in stagnant water and slow flowing streams.

Fl. & Fr.: May – July.

Sone (ADB), *TP* & *PVP* 4027; Mandora (NZB), *TP & BR* 6227; Medak (MDK), *CP* 11735; Lingampalle (RR), *MSM* 19223; Molachintapally (MBNR), *TDCK* 16824; Kodimial (KRN), *GVS* 217878 (MH); Rusthumpet (MDK), *KMS* 800 (MH);

Typha elephantina Roxb., Fl. Ind. 3: 566. 1832; Hook.f., Fl. Brit. India 4: 489. 1894; Sharma & Gopal, Aquatic Bot. 9: 381-387. 1980.

Perennial herb, 1.5-4 m high. Leaves linear or broadly linear; trigonous above the sheath, angularly keeled dorsally, 3-angled; lamina 25-40 mm broad. Upper part flat, about as long as inflorescence. Male and female parts separate; axis of male spike coverd with hairs; female spike cylindric, blackish brown or brown; pistillodes present; female flowers bracteate; stigma lanceolate; bracts spathulate and longer than the hairs.

Occasional in stagnant water, ponds and slow flowing streams.

Fl. & Fr.: May – June.

ARACEAE

1. Floating herbs; leaves in close spirals .... **Pistia**
1. Terrestial or marsh or epiphytic; leaves not in close spirals:
   2. Leaves simple, not pedatisect:
      3. Ephiphytic or in pockets of bare rock .... **Remusatia**
      3. Terrestrial:

4. Leaves variegated .... **Caladium**
4. Leaves not variegated:
   5. Spadix covered with sterile flowers to apex and the sterile appendix is clelarly roughened .... **Colocasia**
   5. Spadix with smooth and naked terminal appendix which is green or purplish .... **Theriophonum**

2. Leaves compound, pedatisect or radiatesect:
   6. Inclorescence appearing with the leaves; leaf blade pedatisect or radiatisect .... **Arisaema**
   6. Inflorescence appearing without leaves; leaf blade trisect, each division again variously cut .... **Amorphophallus**

AMORPHOPHALLUS Blume ex Decne. *nom. cons.*

Amorphophallus sylvaticus (Roxb.) Kunth, Enum. Pl. 3: 34. 1841; Fischer, Fl. Madras 3: 1587. 1931. *Arum sylvaticum* Roxb., Fl. Ind. 3: 511. 1832. *Synantherias sylvatica* (Roxb.) Schott., Gen. Aroid. t. 28. 1858; Hook.f., Fl. Brit. India 6: 518. 1893.

A cormous herb; corms to 3-5 cm. Leaves 20-100 × 10-60 cm, segments lanceolate; petioles purplish with dark greenish-black spots; leaf segments bases strongly decurrent on the main petiolules to the junction with the petiole. Inflorescence appearing without leaves. Spathes convolute with open top, 3-5 cm long, cream-colured outside, maroon within, 10-50 cm long; spadix 20-30 cm long, very shortly stipitate, appendage linear-subulate, about 20 cm long. Male inflorescence 9-13 cm long, female inflorescence 3-4 cm long.

Occasional in Medak, Mahabubnagar and Nizamabad districts in moist localities.

Fl. & Fr.: June – September.

Annaram (NZB), *BR* 7181; Narsapur RF (MDK), *TP* & *CP* 14042; Farahabad (MBNR), *TDCK* 13685.

ARISAEMA Mart.

1. Leaves usually 2, rarely 3, shortly pedatisect; spadix sigmoid, long exserted .... **A. tortuosum**
1. Leaf solitary, radiatisect; spadix cylindrical, barely exserted..... **A. leschenaultii**

Arisaema leschenaultii Blume, Rumphia 1: 93. 1836; Hook.f., Fl. Brit. India 6: 504. 1893; Fischer, Fl. Madras 3: 1585. 1931; Malhotra & Balodi, J. Econ. Taxon. Bot. 7: 585. 1985. *A. pulchrum* N.E. Br., J. Linn. Soc., Bot. 18: 252. 1880.

A cormous herb; corm subglobose. Leaf solitary, radiatisect; petiole to 1 m; pseudostem to 75 cm, mottled; leaflets 5-13, obovate-lanceolate, to 30 × 10 cm, glossy above and light green below. Spathe to 15 cm long, with longitudinal white stripes, pale green but sometimes purplish; tube to 9 cm, cylindric, shortly outrolled at top. Spadix usually dioecious. Berries with 1-few globose seeds.

Rare in Ranga Reddi district.

Fl. & Fr.: August – December.

Rudraram RF (RR), *MSM* 12716.

Arisaema tortuosum (Wall.) Schott & Endl., Melet. Bot. 17: 1832; Hook.f., Fl. Brit. India 6: 502.1893; Fischer, Fl. Madras 3: 1584. 1931. *Arum tortuosum* Wall., Pl. Asiat. Rar. 2: 10. t. 114. 1830.

A cormous herb, to 1 m long; corms globose, 4-6 cm across. Leaves 2-3, pedatisect; petioles 30-90 cm long; sheaths often mottled with purple; leaflets ovate-lanceolate or linear-lanceolate, subcaudately acuminate, distant or crowded; peduncle 60-120 cm long; spathe 10-15 cm long, green outside, gradually dilated into ovate or oblong acuminate incurved limb. Spadix usually monoecious, rarely androecious, to 17 cm long, with a basal female inflorescence portion about 1 cm long followed by the male inflorescence about 2.5 cm and the apical, naked appendix with a few or no neuters at the base. Male flowers stalked; appendix very long, sigmoid, much exserted, tapering; ovary ovoid, attenuated into a short style. Berries orange-red, 4-5-seeded.

Occasional in the forests of Warangal district.

Fl. & Fr.: June – September.

Way to Gangaram (WGL), *PVS* 84013 (BSID).

CALADIUM Vent.

Caladium bicolor (Aiton) Vent. Mag. Encycl. 4(16) : 464. [22 Dec 1800-21 Jan] 1801 (Descr. Pl. nouv. ad t. 30. 1801 Mar). *Arum bicolor* Aiton, Hort. Kew. 3: 316. 1789.

Herbaceous perennials, arising from tuberous rhizomes. Leaves radical, ovate – sagittate, or ovate-triangular, variegated above, glaucous beneath; upper lobe semiovate, narrowing gradually to a cuspidate point; petioles smooth much longer than the blades.

Grown as pot herbs in green houses for its foliage.

COLOCASIA Schott *nom. cons.*

Colocasia esculenta (L.) Schott, Melet. Bot. 18. 1832. *Arum esculentum* L., Sp. Pl. 965. 1753. *Colocasia antiquorum* Schott, Melet. Bot. 18. 1832; Hook.f., Fl. Brit. India 6: 523. 1893; Fischer, Fl. Madras 3: 1580. 1931.

A robust tuberous herb; tubers to 15 cm across; leaves ovate or suborbicular, 15-50 × 8-30 cm, base cordate, margin undulate, apex rounded and usually apiculate, basal sinus triangular, dark-green, petiole 75-130 cm long, green or violet. Spathe 35 cm long with basal convolute, elliptic, green, tubular portion, 3-5 cm long, upper lanceolate, yellow, caducous limb portion about 25 cm long. Spadix with female portion about 3 cm long, neuteriflorous portion about 2 cm, male portion about 6 cm and the apical sterile appendix about 5 cm long. Female flowers many, arranged in close spirals; with few white tipped sterile flowers, stigma sessile, inconspicuously 3-5-lobed, white. Neuter flowers occur at the constriction of the spathe. Male flowers and appendix cream coloured, appendix tapering towards the tip. Berries ovoid, many-seeded.

Occasional in most of the districts along canals, irrigation channels, also cultivated.

Fl. & Fr.: August – November.

Mannanur (MBNR), *TDCK* 16804.

PISTIA L.

Pistia stratiotes L., Sp. Pl. 963. 1753; Hook.f., Fl. Brit. India 6: 497. 1893; Fischer, Fl. Madras 3: 1573. 1931.

A floating, stemless, stoloniferous aquatic herb with a peculiar muriatic odour; roots of tufted simple white fibres clothed with fibrillae. Leaves sessile, obovate-cuneate, 3-10 × 1-5 cm, tufted, rounded or retuse at the apex, densely and closely pubescent on both surfaces; nerves few or many flabellately arranged, converging within the margin, Spathe about 1.2 cm long, obliquely campanulate, white, gibbous and closed below, contracted about the middle, dilated and nearly orbicular above. Spadix adnate to the back of the tube of the spathe, free above. Male flowers in a whorl of a few connate stamens beneath the apex of the spadix. Neuters few, minute, confluent in a ring below the male. Female flowers solitary, ovary 1-celled, obliquely adnate to the spadix, the apex free and forming a conical style. Berries ovoid; seeds few.

Common in stagnant water in all districts.

Fl. & Fr.: December – February. Fruits throughout the year.

Ibrahimpet road (NZB), *TP* & *BR* 6136; Near Dindi project (NLG), *BVRS* 676 (NU).

REMUSATIA Schott

Remusatia vivipara (Roxb.) Schott, Melet. Bot. 18. 1832; Hook.f., Fl. Brit. India 6: 521. 1893; Fischer, Fl. Madras 3: 1583. 1931. *Arum viviparum* Roxb., Fl. Ind. 3: 496. 1832 (*viviparium*). *Caladium viviparum* (Roxb.) Lodd., Bot. Cab. 3: t. 281. 1818.

A tuberous herb, emitting 15-40 cm long, leafless bulbiliferous shoots from the crown of the tuber 2.5-4 cm across; bulbils with recurved tipped scales. Leaf usually solitary, sometimes 2-3, broadly ovate, 15-35 × 10-25 cm, base subcordate, acuminate, peltate, sometimes purplish tinge between the primary veins on the lower surface of the lamina; petiole 20-40 cm long. Spathe about 17 cm long, basal portion oblong, green, convolute and upper portion deflexed, yellow, broadly-obovate limb 5-13 × 9 cm. Spadix sessile, 5-7 cm long. Male and female flowers separated by neuters, stamens 2-3; anthers sessile, connate into a 4-6-angled and subclavate, flat-topped synandrum; synandria densely packed with neuters, ovaries crowded, ovoid, 1-celled or 2-4-celled upwards; stigma sessile, disciform, ovules many in 2 series on 4-6 parietal placentae. Berries small, obovoid, seeds small, ovoid.

Occasional in forests in Mahabubnagar district.

Fl. & Fr.: February – April.

Farahabad (MBNR), *TDCK* 13661; Mannanur (MBNR), *TDCK* 16804.

THERIOPHONUM Blume

1. Spadix with sterile flowers above and below the staminate flowers .... **T. minutum**
1. Spadix with sterile flowers only below the staminate flowers:
    2. Spadix appendix stipitate .... **T. infaustum**
    2. Spadix appendix sessile .... **T. dalzellii**

Theriophonum dalzellii Schott, Aroideae 1: 15. 1853; Hook.f., Fl. Brit. India 6: 513. 1893. *Tapinocarpus dalzellii* (Schott) Schott, Oesterr. Bot. Wochenbl. 7: 262. 1857 (*dalzelli*). *T. indicum* Engl., Pflanzenr. 73: 107. 1920; Fischer, Fl. Madras 3: 1579. 1928. *Theriophonum uniseriatum* Blatt. & McCann, J. Bombay Nat. Hist. Soc. 35: 25. 1931.

An acaulescent herb with a corm of a size of a walnut. Leaves appearing with the flowers, 10-20 × 4-8 cm, elliptic, oblong or linear, base rounded, hastate or sagittate, apex acute or subobtuse. Peduncle stout, 0.5 – 15 cm long, in fruit twisted and curved to the ground. Spathe 10-16 cm long, oblong-lanceolate, tube green when young, soon fading to white, limb flat, deep purple, rapidly fading, appendate or spadix terete, elongate, purple at the base, greenish-yellow above, anthers rosy-purple, opening by pore. Neuter flowers greenish-yellow; ovary green, ovules 4-6, pendulous. Fruit angular, 4-5-seeded.

In moist grasslands of Mahabubnagar district.

Fl. & Fr.: July – August.

Shadnagar (MBNR), *SRR* 1880 (HH).

Theriophonum infaustum N.E. Br., J. Linn. Soc., Bot. 18: 260. 1880 (excl. syn. *Arum minutum*); Hook.f., Fl. Brit. India 6: 513. 1893; Fischer, Fl. Madras 3: 1579. 1928.

A cormous acaulescent herb. Leaves cordately linear to ovate-hastate, not or indistinctly 3-lobed, acute, basal lobes rounded, sinus narrow; petiole slender, as long as or up to 3 times longer than the blade. Spadix abuts half the length of the spathe. Peduncle very slender, short; spathe white or purple with a long cylindric tube, limb narrowly lanceolate; appendage stipitate, slender, subulate.

Rare in Mahabubnagar district (Raghava Rao, 1989).

Fl. & Fr.: June – August.

Theriophonum minutum (Willd.) Baill., Hist. Pl. 13: 457. 1895; Fischer, Fl. Madras 3: 1579. 1931. *Arum minutum* Willd., Sp. Pl. 4: 484. 1805. *A. crenatum* Wight, Hook. J. Bot. Kew Gard. Misc. 2: 100 (Suppl.) t. 3. 1831. *Theriophonum crenatum* (Wight) Blume, Rumphia 1: 128. 1837; Hook.f., Fl. Brit. India 6: 512. 1893. *T. wightii* Schott. in Oesterr. Bot. 8: 3. 1858; Hook.f., Fl. Brit. India 6: 512. 1893; Gamble, Fl. Madras 3: 1579. 1931.

A cormous acaulescent herb; corm 1.5-2 cm across. Leaves with slender petioles, 8-15 cm long, basally sheathing, vaginate, 4 cm from base, lamina of various shapes, from broadly ovate to hastate, typically sub-3-lobed; the median lobe 4-7 × 3.5-6 cm, ovate, acuminate, mucronate; posterior lobes 1.5-4 × 1-2 cm. Inflorescence with peduncle 3-4.5 cm long and about 2 mm diameter. Spathe 5-12 cm long with the basal convolute portion 1-2.5 cm long and 0.6-1 cm in diameter and the upper erect, ovate-oblong, acuminate, greenish to purplish green, limb portion 4-10 × 2-3 cm, margins purplish and wavy or sometimes entire, the colour later fades to light greenish to white cream; spadix shorter than the spathe, 4-6 cm long, with neuter flowers below and above the male portion; neuters below the male flowers small, subulate. Female ones: few, in a single whorl, ovate. Male ones scattered and few, sessile, ovary unilocular; ovules 3-6, basal and apical. Berries 7-10 mm long, subconical with 3 seeds; seeds broadly ovoid.

Occasional in forests in most districts.

Fl. & Fr.: June – December.

Kotapalli RF (ADB), *TP* & *GO* 5427; Madharam RF (ADB), *TP* & *GO* 5476; Mudheli RF (NZB), *BR* 9296; Aklaspur (KRN), *GVS* 20214 (MH).

LEMNACEAE

1. Fronds with 1 or more roots:
   2. Fronds with single root .... **Lemna**
   2. Fronds with tuft of roots .... **Spirodela**
1. Fronds without any roots:
   3. Fronds minute like caryopsis of sand, rootless, proliferous, bearing the flowers on the upper surface; anthers sessile, 1-celled ..... **Wolffia**
   3. Fronds thin, orbicular, green towards both sides, rootless, the flowers situated in a depression; stamen 1, minute .... **Pseudowolffia**

LEMNA L.

Lemna perpusilla Torr., Fl. New York 2: 245. 1843; van der Plas in Steenis, Fl. Males I. 7: 231. 1971. *L. paucicostata* Hegelm., Lemnac. 139. t. 8. 1868; Hook.f., Fl. Brit. India 6: 556. 1893; Fischer, Fl. Madras 3: 1593. 1931.

A small or minute floating herb. Fronds in groups of 2 or 3, sometimes solitary, ovate-obovate or oblong, 1-4.5 × 1.25 mm; dorsal side flat, with a hook-shaped papilla at the tip; ventral side somewhat flat or convex, base obtuse, margin entire, apex obtuse-rounded, 3-nerved. Root 1, root-sheath with 2 lateral wings, root cap acute. Budding pouches 2, basal, lateral, one on either side of the axis, triangular, opening by a transverse slit, slit in line with the margin of the frond. Inflorescence 1 in the budding pouch, with 1 male and 2 female flowers, anthers transversely dehiscent. Fruit ellipsoid; seed with prominent longitudinal ribs.

In still waters in all districts.

Fl. & Fr.: September – January.

Ibrahimpet – Banswada road (NZB), *TP* & *BR* 6139; Patancheru (MDK), *BR* & *CP* 11301; Narsapur (MDK), *BR* & *CP* 14101, *KMS* 6788 (MH); Rachakonda (NLG), *BVRS* 540 (NU).

PSEUDOWOLFFIA Hartog & Plas

Pseudowolffia hyalina (Delile) Hartog & Plas, Blumea 18: 366. 1970; Ramayya & Rajagopal, Aquatic Bot. 1: 71. 1975. *Lemna hyalina* Delile, Descr. Egypte Hist. Nat. 4: 27. 1813.

A minute floating plant, rootless, differentiated into a frond and a stipe. Fronds thin, symmetric, orbicular to ovoid or broadly oblong, entire, about 1-1.5 × 1.5-2 mm, green towards both sides. Budding pouch situated on the upper surface, located at the junction of the frond and the stipe, appearing as a curved transverse slit. Daughter fronds two or three in budding pouch, piled up one over the other, the lower being the oldest. Flowers unisexual; male flower solitary, represented by a single minute stamen situated in a depression, lateral to and near the budding pouch; female flowers not seen.

Note: The plant occurs in putrid waters and it is invariably found in association with *Wolffia arrhiza* and frequently also with *Lema perpusilla*. It was reported by Ramayya and Rajagopal (1975) as a new record for Asia. Their report is without any Accession number as well as flowering and fruiting periods.

SPIRODELA Schleid.

Spirodela polyrrhiza (L.) Schleid., Linnaea 13: 392. 1839. *Lemna polyrhiza* L., Sp. Pl. 970. 1753; Hook.f., Fl. Brit. India 6: 557. 1893; Fischer, Fl. Madras 3: 1593. 1931.

A small or minute floating herb. Fronds herbaceous, not tailed, opaque, thick, flat above, slightly convex below, 6 mm diameter, dark green above, usually purplish beneath, 7-veined; young fronds sessile, each frond giving rise to a tuft of root-fibres. Spathe 2-lipped, stamens 2, ovules 1-2.

In still waters in all districts.

Fl. & Fr.: September – January.

Near Kadam (ADB), *GO* & *DAM* 5407, 5408; Dharmaram (NZB), *BR* & *KH* 6463; Patancheru (MDK), *BR* & *CP* 11302; Moosi River (HYD), *KMS* 5949 (MH); Rachakonda (NLG), *BVRS* 541 (NU).

WOLFFIA Horkel *nom. cons*

Wolffia globosa (Roxb.) Hartog & Plas, Blumea 18: 367. 1970. *Lemna globosa* Roxb., Fl. Ind. 3: 565. 1832. *Wolffia arrhiza* auct non Horkel ex Wimmer 1857; Hook.f., Fl. Brit. India 6: 557. 1893 *p.p.*; Fischer, Fl. Madras 3: 1593. 1931.

A small or minute rootless herb with very minute fronds, round or ovate, 0.2 – 0.8 × 0.1 – 0.3 mm, base and apex obtuse, margin with a few papillose cells or entire. Flowers in a groove on the upper surface of the frond, naked; male solitary, stamen solitary, anther 1-celled, sessile; ovay solitary, style short, stigma depressed, ovule solitary. Utricle spherical.

In still waters in all districts.

Fl. & Fr.: June – October.

Ibrahimpet – Banswada Road (NZB), *TP* & *BR* 6139; Rachakonda (NLG), *BVRS* 542 (NU).

ALISMATACEAE

1. Pistils inserted on a large, globose receptacle; stamens many .... **Sagittaria**
1. Pistils on a small receptacle; stamens 6; achenes round in section:
    2. Leaves sagittate; flowers polygamous; achenes with lateral air chambers .... **Limnophyton**
    2. Leaves cordate; flowers bisexual; achenes without air chambers .... **Caldesia**

CALDESIA Parl.

Caldesia parnassifolia (L.) Parl., Fl. Ital. 3: 599. 1858. *Alisma parnassifolium* L., Syst. Nat. ed. 12. 2: 230. 1768. *A. reniforme* D. Don, Prodr. Fl. Nepal 22. 1825; Hook.f., Fl. Brit. India 6: 560. 1863; Fischer, Fl. Madras 3: 1595. 1928.

A scapigerous herb. Leaves broadly ovate-suborbicular, 6-8 × 4-8 cm, chartaceous, glabrous, base cordate, margin entire, apex obtuse. Panicle 30-60 cm long, flowers white, bract solitary, lanceolate-oblong; sepals 3, elliptic, slightly concave; petals 3, white, sub-orbicular-ovate; stamens 6; ovaries ca. 15, obovoid, compressed, ovule 1. Achenes ellipsoid, laterally compressed, 3-5-ribbed.

Reported from Mahabubnagar district.

Fl. & Fr.: February – April & July – October.

Mahabubnagar, *SRR* 905 (HH).

LIMNOPHYTON Miq.

Limnophyton obtusifolium (L.) Miq., Fl. Ned. Ind. 3: 243. 1856; Hook.f., Fl. Brit. India 6: 560. 1893; Fischer, Fl. Madras 2: 1595. 1931. *Sagittaria obtusifolia* L., Sp.Pl. 993.1753.

A scapigerous, stemless, palustrine, perennial herb; root stock short, stout, with numerous long root-fibres. Leaves broadly sagittate, 4-20 × 3-15 cm, chartaceous, 7-20-nerved, stellate-pubescent, base broadly cordate; basal lobes lanceolate, margin entire, apex obtuse-rounded; petiole 6-60 cm. Panicles to 1.5 m long, branches in whorls of 2-6, 10-15-flowered per whorl; bracts 3, lanceolate, pedicel 4 cm. Flowers white, the upper whorls mostly male, the lower bisexual. Male flowers: sepals 3, oblong-elliptic, hyaline along margin; petals 3, white, obovate; stamens 6; ovaries 12-20, free, obovoid, ovules on basal placentae. Achene cluster globose, attenuate at base, reticulately ribbed, shortly beaked, turgid with air-chambers between exocarp and endocarp; seed oblong, 4 mm.

Common in Adilabad, Nizamabad, Ranga Reddi, Hyderabad and Mahabubnagar districts in stagnant waters.

Fl. & Fr.: January – March and July – September.

Salugupalli (ADB), *GO* 4837; Kammarapally (NZB), *TP* & *BR* 6229; Adluryellareddy (NZB), *BR* & *KH* 9059; Mohammadabad RF (RR), *MSM* & *KH* 10542; Rollapadu (KMM), *RR* 108038 (BSID).

Figure 120: Limnophyton obtusifolium (L.) Miq.

A. Habit, B. Perianth lobes (outer), C. Perianth lobes (inner), D. Stamens, E. Pistil.

SAGITTARIA L.

1. Leaves broadly ovate; wings of seeds prominently toothed; sepals appressed on mature pistillate flowers .... **S. guayanensis**

1. Leaves sagittate; wings of seeds entire; sepals not appressed on mature pistillate flowers .... **S. sagittifolia**

Sagittaria guayanensis Kunth, Nov. Gen. Sp. Pl. 1: 250. 1816 subsp. **lappula** (D. Don) Bogin, Mem. New York Bot. Gard. 9: 192. 1955. *S. lappula* D. Don, Prodr. Fl. Nepal 22. 1825. *S. guayanensis* sensu Hook. f. Fl. Brit. India 6: 561. 1893 *p.p*; Fischer, Fl. Madras 3: 1596. 1928.

An aquatic, floating, fibrous-rooted herb with small root stock. Leaves broadly ovate; membranous, indistinctly nerved, deeply cordate at base, lobes and blade obtuse or rounded. Flowers white, sepals appressed on the mature pistillate flowers; stamens 6-15. Achenes plump or compressed with shallowly or deeply crested, rarely smooth; seed prominently toothed and winged.

Rare in ponds and lakes in Medak, Khammam and Warangal districts (V.S. Raju, 1986).

Fl. & Fr.: October – December.

Pathur (MDK), *TP* & *CP* 12127.

Sagittaria sagittifolia L., Sp. Pl. 993. 1753; Hook.f., Fl. Brit. India 6: 561. 1893; Fischer, Fl. Madras 3: 1596. 1928.

A perennial, stoloniferous, deep-rooted aquatic herb, 30-100 cm high. Leaves radical, sagittate. Flowers white often with a purple claw. Scape with 3-5 flowers in each whorl, the lower whorls female, upper whorl male. Achenes obliquely obovate, flattened; wings of seeds entire.

Rare in Medak district.

Fl. & Fr.: October –December.

Medak (MDK), *RG* 104101 (BSID).

BUTOMACEAE

BUTOMOPSIS Kunth

Butomopsis latifolia Kunth, Enum. Pl. 3: 165. 1841; Hook.f., Fl. Brit. India 6: 562; Fischer, Fl. Madras 3: 1597. 1931. *Butomus latifolius* D. Don, Prodr. Fl. Nepal, 22. 1825. *Tenagocharis latifolia* (D. Don) Buchenau, Nachr. Königl. Ges. Wiss. Goorg-Augustus Univ. 1869. 238. 1869.

A marsh plant with milky juice. Leaves elliptic-lanceolate, acute at both ends, 5-15 × 1.5-5 cm; petioles dilated at the base. Scapes stout, usually longer than the leaves, bracts scarious, ovate, acute. Flowers 2 cm in diameter, 3-20 at the apex, sometimes with an additional whorl below; sepals herbaceous, persistent; petals white, membranous, deciduous; stamens 8-12; carpels 6-9, whorled, sessile on a flat receptacle, style short, ovules many. Follicles 6 mm long, shortly beaked, connate below; seeds numerous, minute, smooth.

Rare in Nizamabad and Warangal districts.

Fl. & Fr.: October – December.

Mudheli (NZB), *BR* & *CPR* 7216; Pakhal lake (WGL), *KMS* 11610 (MH).

NAJADACEAE

NAJAS L.

1. Leaf-sheaths apically produced into a linear-oblong auricle; leaf margin with more than 50 spiny teeth .... **N. graminea**
1. Leaf-sheaths apically obtuse, rarely with obscure auricle; leaf margins with less than 20 spiny teeth .... **N. minor**

Najas graminea Delile, Descr. Egypte Hist. Nat. 282. 1813; Hook.f., Fl. Brit. India 6: 569. 1893; Fischer, Fl. Madras 3: 1604. 1931.

A submerged, aquatic herb, rooting from the base and lower nodes; stems or shoots terete, up to 50 cm long. Leaves in pseudowhorls, linear-subulate, obtuse or acute with a spinulosely dentate margin, 1-2.5 × 0.005-0.1 cm, sheaths spinulosely denticulate with triangular, acute or obtuse, entire or rarely shallowly notched, 0.2 – 0.5 cm long, auricles 0.4 – 0.8 cm long. Flowers solitary or 2 – 4 together. Male: Spathe absent, perianth segements rounded, anther ellipsoid-ovoid or oblong. Female: Spathe absent; flowers solitary, axillary, subsessile, perianth lobes closely appressed to ovary, ovary elliptic. Fruit ellipsoid-oblong, attenuate at the apex, 0.15-0.2 cm long, areoles minute, subquadrate or polygonal.

Common in stagnant waters in Adilabad, Nizamabad, Hyderabad and Mahabubnagar districts.

Fl. & Fr.: August – November.

Nirmal (ADB), *TP* & *PVP* 4069; Gandivet (NZB), *BR* 7201.

Najas minor All., Auct. Syn. Stirp. Horti Taur. 3. 1773; Hook.f., Fl. Brit. India 6: 569. 1893; Fischer, Fl. Madras 3: 1604. 1931. (incl. var. *spinosa* Rendle).

A submerged aquatic herb; shoots 4-25 cm long. Leaves linear, 0.5-1.5 × 0.1-0.2 cm, base triquetrous, apex flat, dorsal surface smooth, septate, margins with 10-15 spinous teeth oblong. Male: spathe 2-3 mm long, 1 mm broad, beak 1.5 mm, with linear spinous teeth. Flowers solitary, sessile, perianth lobes obscure, anthers 1-celled. Female: spathe 0, flower solitary, ovary terete. Fruit oblong, trigonous, 2 × 0.8 mm, areoles subquadrate, in longitudinal rows.

In ponds of Adilabad district.

Fl. & Fr.: July – November.

Sirpur (ADB), *GO* 4419; Kadam dam (ADB), *GO* & *BR* 9420.

APONOGETONACEAE

APONOGETON L. f. *nom. cons.*

1. Tepals attenuate at the base; ovules 8 per ovary .... **A. natans**
1. Tepals broad at base; ovules 2 per ovary .... **A. crispus**

Aponogeton crispus Thunb., Nov. Gen. Pl. 73. 1784. *A. echinatus* Roxb., Fl. Ind. 2: 210. 1832; Hook.f., Fl. Brit. India 6: 564. 1893; H. Bruggen, Blumea 18: 483. t. 1970.

A submerged or floating, aquatic herb; tuber ovoid, 3 × 2 cm. Leaves oblong-lanceolate, submerged leaves oblong-elliptic, 15 × 5 cm, floating leaves oblong, 4-16 × 1-5 cm, glabrous, base subcordate, margin entire, apex subacute, obtuse. Spikes oblong, 15 cm; peduncle nearly as long as petiole. Flowers patent, dense at base, apically lax; tepals 2, pinkish, obovate sub-orbicular as long as broad; stamens 6; ovaries 3, globose, ovules 2, style short. Fruit ovoid, 4-6 × 2 mm; seeds oblong, flat on back, turged on the upper surface.

Common in ponds, tanks and ditches in Adilabad and Mahabubnagar districts.

Fl. & Fr.: January – April.

Mavala (ADB), *TP* & *GO* 5503.

Aponogeton natans (L.) Engl. & K.Krause, Pflanzenr. IV. 13: 11. 1906; Fischer, Fl. Madras 3: 1598. 1931. *Saururus natans* L., Mant. Pl. 227. 1771. *Aponogeton monostachyon* L. f., Suppl. Pl. 214. 1781; Hook.f., Fl. Brit. India 6: 564. 1893.

A submerged or floating, glabrous, aquatic herb, tuber linear-oblong, 2.5-3.5 × 0.8-1.5 cm. Leaves mostly floating, oblong, 2-10 × 0.8-2 cm, glabrous, base subcordate-obtuse, margin entire, apex obtuse subacute. Spikes 2.5-15 cm long, usually dense-flowered; peduncle 10-45 cm, usually overtopping leaves, tepals 2, violetish, obovate-suborbicular, 2.5 × 1-5 mm, nearly twice as long as broad; stamens 6, anthers bluish-purple; ovaries 3, oblong. Fruit globose, 4 × 2 mm, smooth, sharply beaked.

Common in ponds, tanks, ditches in all districts.

Fl. & Fr.: December – February.

Kunthala (ADB), *GO* & *DAM* 5139; Arsapally (NZB), *BR* & *KH* 6485; Annaram (NZB), *BR* & CPR 7166; Narsapur tank (MDK), *BR* & *CP* 11242, *TP* & *CP* 12008; Malkalcheruvu (RR), *MSM* & *KH* 10539; Pakhal (WGL), *ANH* 15945 (MH).

POTAMOGETONACEAE

1. Submerged leaves with stipules free from leaf base, or if adnate then adnate portion less than ½ length of stipule; leaves submerged and floating or all submerged; submerged leaf blades transluscent, not channeled, flattened; peduncle stiff, submerged or projecting above water .... **Potamogeton**
1. Submerged leaves with stipules adnate to leaf blade for 2/3 or more length of stipule; leaves all submerged, blades opaque, channeled; peduncle lax, not projecting inflorescence above water surface .... **Stuckenia**

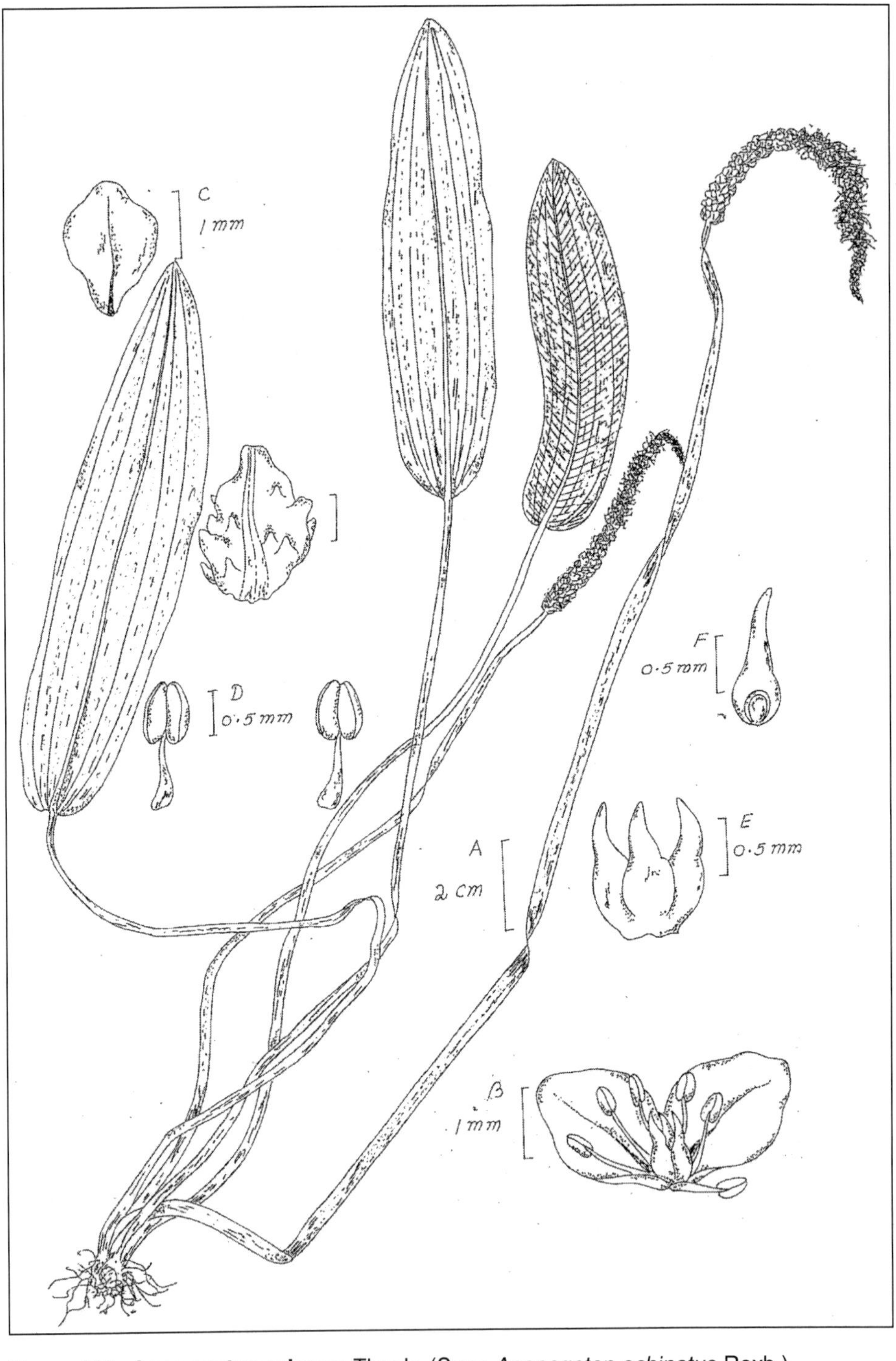

Figure 121: Aponogeton crispus Thunb. (Syn.: *Aponogeton echinatus* Roxb.)
A. Habit, B. Flowr, C. Perianth lobe, D. Stamens, E. Carpels, F. Carpel, G. Follicle.

POTAMOGETON L.

1. Floating and submerged leaves present .... **P. nodosus**
1. Floating leaves absent; leaves all submerged
   2. Leaves crisped, finely serrulate, 3-serrulate, 3-nerved .... **P. crispus**
   2. Leaves undulate, entire, 5-7-nerved .... **P. perfoliatus**

Potamogeton crispus L., Sp. Pl. 126. 1753; Hook.f., Fl. Brit. India 6: 566. 1893; Burkill, Rec. Bot. Surv. Ind. 4: 136. 1910.

An aquatic submerged herb with creeping root stock; stem branched, compressed. Leaves submerged, oblong-elliptic or oblanceolate, 1-5 × 0.3-0.8 cm, membranous, transluscent, 3-nerved, glabrous, base slightly amplexicaul, subcordate, margin serrulate, apex rotund, sessile. Peduncle long, often curved, tapering upwards. Spikes few-flowered, lax, up to 1 cm long; tepals clawed, sub-orbicular. Drupelets obliquely ovoid, tipped with a compressed, curved beak.

Rare in ponds and tanks.

Fl. & Fr.: September – February.

Belal tank (NZB), *BR* 7269.

Potamogeton nodosus Poir., Encycl. Suppl. 4: 535. 1816. *P. indicus* Roxb., Fl. Ind. 1: 471. 1820 *nom. illegitimate;* non Roth ex Roem. & Schult. 1818; Hook.f., Fl. Brit. India 6: 565. 1893; Fischer, Fl. Madras 3: 1600. 1831.

An aquatic submerged herb; stem terete, branched, smooth. Leaves heterophyllous, 4-12 × 2-3.5 cm, upper ones floating, broadly ovate-elliptic, sub-acute-obtuse at the ends, lower ones submerged, lanceolate, 7-9-nerved, base attenuate-cuneate, margin entire, apex acute-acuminate; stipules free, membranous, keeled, apex tapering. Spikes axillary, leaf-opposed, 2-5 cm long; peduncle 6 cm. Flowers sessile; tepals 4, suborbicular-ovate; stamens 4; ovaries 3, free, ovule 1. Drupelets obovoid, shortly beaked, 1-seeded.

Common in ponds and ditches in Adilabad, Medak and Hyderabad districts.

Fl. & Fr.: July – December.

Kadthala (ADB), *GO* & *DAM* 4991; Medak tank (MDK), *BR* & *CP* 11404.

Potamogeton perfoliatus L., Sp. 126. 1753; Hook.f., Fl. Brit. India 6: 566, 1893; Fischer, Fl. Madras 3: 1600. 1931.

An aquatic herb; stem stout, terete, slightly branched. Leaves 2.5-6 × 0.9-1.5 cm, sessile, ovate-lanceolate, obtuse, membranous, transluscent, amplexicaul, base cordate, margin wavy, 5-9-nerved, stipules small, caducous. Peduncle axillary, rather stout, spike dense-flowered, 1.2-2.5 cm long, tepals elliptic-obovate, clawed. Drupelets compressed-globose with short curved beak, hardly keeled, smooth.

Common in ponds and ditches in Adilabad and Hyderabad districts.

Fl. & Fr.: December – February.

Sone (ADB), *TP* & *PVP* 4043; Gandipet (HYD), *TR* 508 (HH).

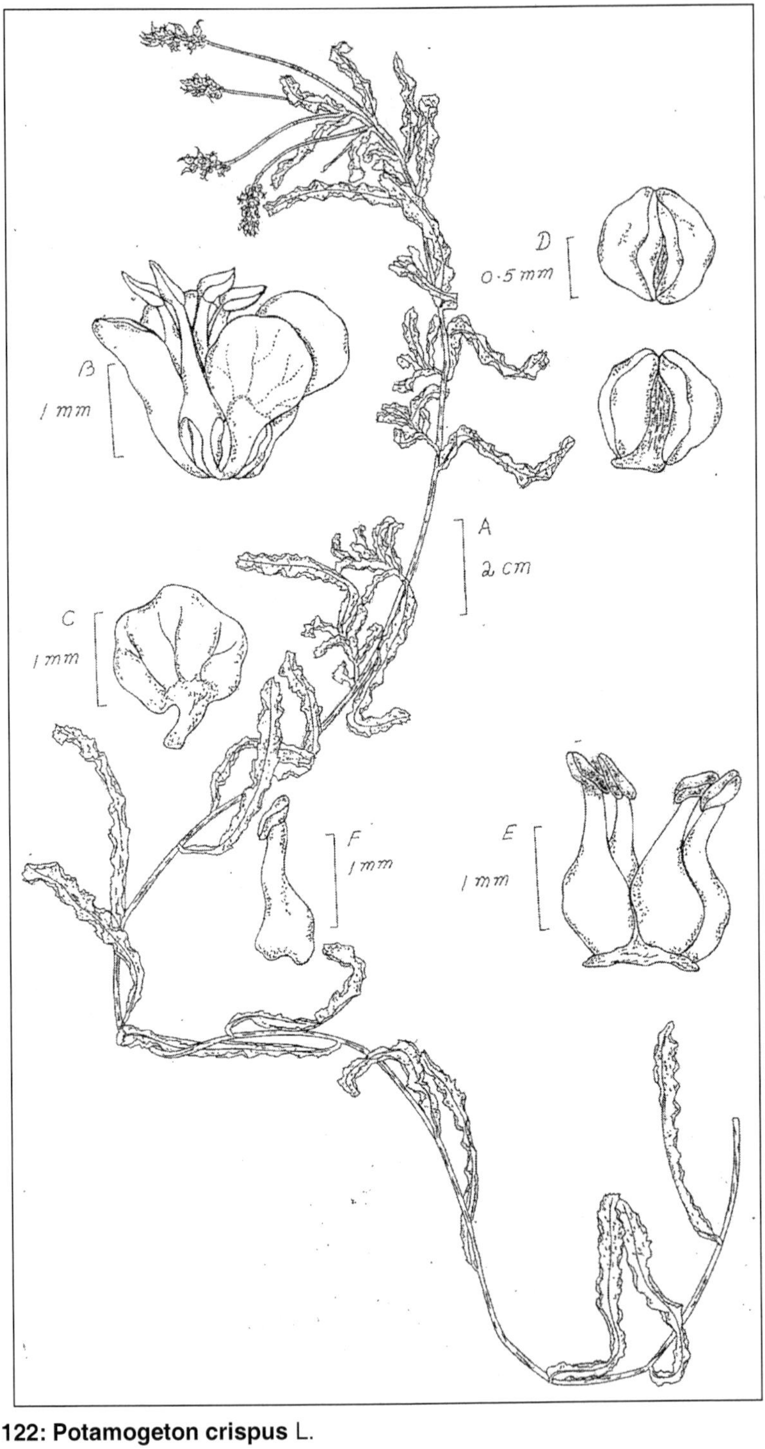

Figure 122: Potamogeton crispus L.

A. Twig, B. Flower, C. Perianth lobe, D. Anthers, E. Carpels, F. Carpel.

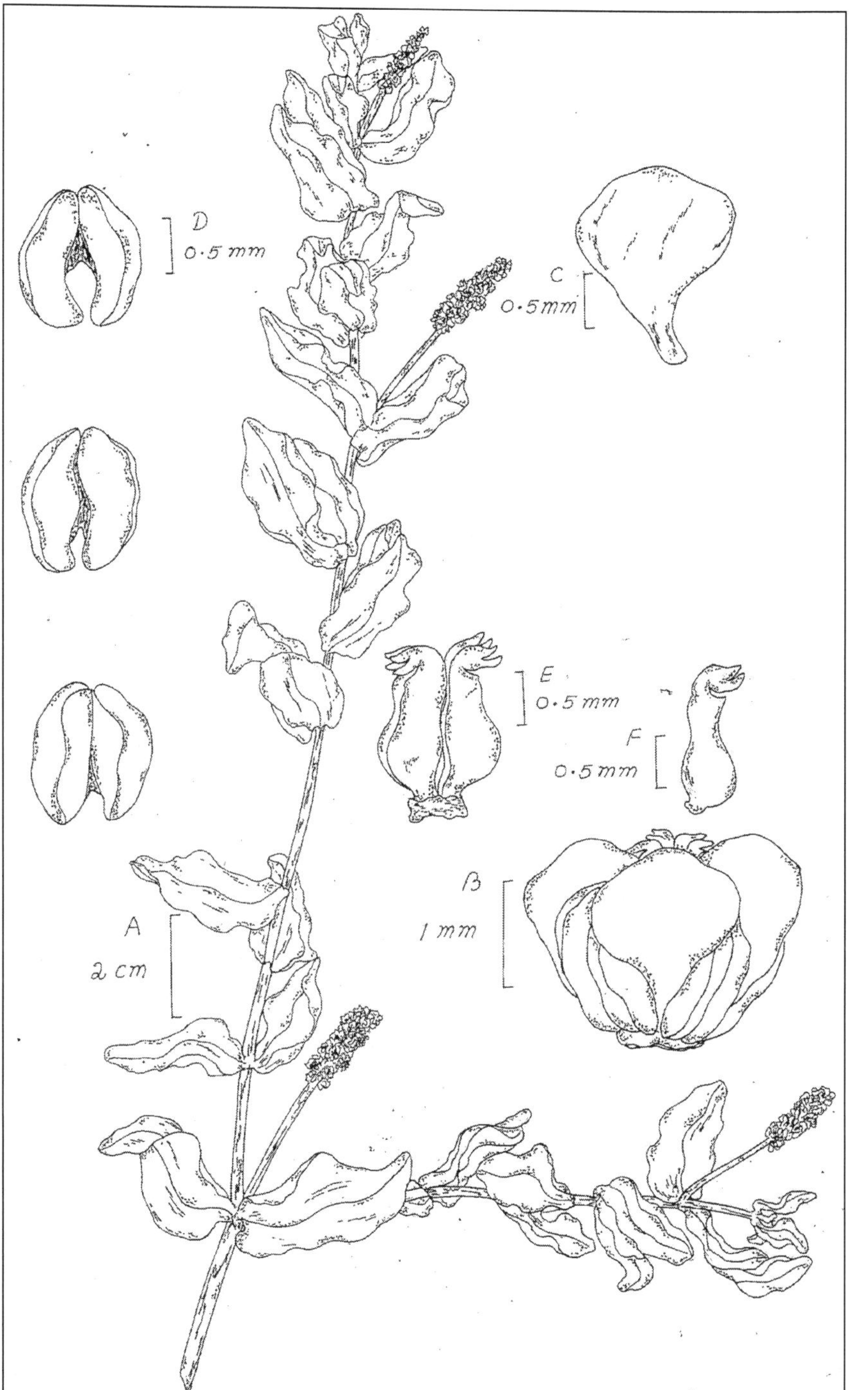

Figure 123: Potamogeton perfoliatus L.

A. Twig, B. Flower, C. Perianth, D. Anthers, E. Carpels, F. Carpel.

STUCKENIA Börner

Stuckenia pectinata (L.) Börner, Fl. Deut. Volk. 713. 1912. *Potamogeton pectinatus* L., Sp. Pl. 127. 1753; Hook.f., Fl. Brit. India 6: 567. 1893; Fischer, Fl. Madras 3: 1600. 1931.

An aquatic submerged herb; stem filiform, copiously distichously branched. Leaves all submerged, alternate, acicular, 2-8 × 0.2-0.5 cm, puberulous, apex acute, stipules adnate to the leaf-sheaths, the tips free. Spikes axillary, 2.5 cm. Flowers a few in a few distant whorls, minute, green; tepals 4, obovate-oblong, base clawed, cuneate, margin entire, apex rotund; stamens 4, filaments equal to anther lobes; ovaries 4, free, linear, stigma subsessile. Drupelets obovate, 3 × 2 mm, obscurely beaked.

Occasional in Hyderabad district.

Fl. & Fr.: December – March.

Hussainsagar (HYD), *TR* 1412 (HH).

ERIOCAULACEAE

ERIOCAULON L.

1. Stems elongate; aquatic plants .... **E. setaceum**
1. Stems short or none; marsh plants:
    2. Involucral bracts woolly or hairy, receptacles villous .... **E. wightianum**
    2. Involucral bracts glabrous; receptacles glabrous or sparsely hairy:
        3. Floral bracts glabrous .... **E. cinereum**
        3. Floral bracts bearded villous or pubescent:
            4. Receptacles glabrous .... **E. cuspidatum**
            4. Receptacles villous:
                5. Male flower calyx spathaceous, split down one side, 3-lobed or partite:
                    6. Sepals white .... **E. quinquangulare**
                    6. Sepals dark:
                        7. Plants with a rootstock of about 1.5 cm long .... **E. richardianum**
                        7. Plants without any rootstock .... **E. santapaui**
                5. Male flower sepals free or two sepals connate and one free .... **E. thwaitesii**

Eriocaulon cinereum R. Br., Prodr. Fl. Nov. Holl. 254. 1810; Ansari & N.P.Balakr., Eriocaula. in India 172. 2009. *E. sieboldianum* Siebold & Zucc. ex Steud., Syn. Pl. Glumac. 2: 272. 1855; Hook.f., Fl. Brit. India 6: 577. 1893; Fischer, Fl. Madras 3: 1619. 1931.

A tufted, annual marsh herb; leaves capillary or very narrowly linear, finely acuminate, 1.5 × 0.15 cm, 5-nerved. Head spherical, whitish or purplish; peduncles numerous, slender, faintly ribbed, sheaths shorter than the leaves, acute; involucral bract obovate, floral bract obovate-cuneate, grey, receptacle columnar, sparsely pilose. Male floret: calyx lobes spathaceous, 3-partite, glabrous; petals short, glabrous, with apical black gland; stamens 3+3, white. Female floret: sepals 3, free, subequal, spathulate, glabrous; petals 3, free, linear – lanceolate, ciliate, ovary 3-lobed; style branches 3, filiform. Seed ellipsoid, transversly striate.

Occasional in marshy localities in Medak and Warangal districts.

Fl. & Fr.: September – December.

Ooracheru lake, Narsapur (MDK), *KMS* 6757 (MH & CAL); Pakhal (WGL), *KMS* 11696 (MH & CAL), *ANH* 15910.

Eriocaulon cuspidatum Dalzell, Hooker's J. Bot. Kew Gard. Misc. 3: 281. 1851; Hook.f., Fl. Brit. India 6: 581. 1893; Fischer, Fl. Madras 3: 1618. 1931; Ansari & N.P.Balakr., Eriocaula. in India 88. 2009.

A terrestrial or marsh herb with very short stem. Leaves linear-oblong, 2.5-9 × 0.4-0.6 cm, rounded and cuspidate at the apex, glabrous, translucent, fenestrately 7-9-nerved. Head 0.4-0.8 cm in diameter, hemispheric or subglobose, white-puberulous; sheaths reaching sometimes 10 cm long, obliquely split, glabrous, involucres of bracts broadly obovate-cuneate, obtuse, glabrous; floral bracts slightly narrower; receptacle glabrous. Male flower: sepals 2, free, obovate-lanceolate, hyaline, slightly winged and puberulous near the apex; petals 3, unequal, narrowly linear-lanceolate; anthers black. Female flower: sepals 2, elliptic-lanceolate, acute, with a pubescent, rounded with wing on the back; petals 3, unequal, narrowly linear-lanceolate, apex villous and with a conspicuous black gland; ovary 3-lobed, style-branches 3, filiform. Seeds oblong and quite smooth.

Reported from Mahabubnagar district only (Raghava Rao, 1990).

Fl. & Fr.: August – November.

Note: Endemic to Peninsular India.

Eriocaulon leucomelas Stued., Nomencl. Bot. ed. 2. 1: 585. 1840; Ansari & N.P.Balakr., Eriocaula. in India 166. 2009. *Eriocaulon melaleucum* Mart., Pl. Asiat. Rar. 3: 29. 1832.

Acaulescent herbs; rootstock absent. Leaves rosulate, linear-ensiform, acuminate, up to 4 x 0.2 cm, glabrous. Peduncles solitary or few, up to 15 cm long, rigid glabrous; sheaths up to 3 cm long, glabrous; limb incised; heads spherical, ca 1 cm across, grey or black; receptacles flat or convex, sparsely pilose. Involucral bracts reflexed, oblong-ovate, acute, ca 2.5 x 1.5 mm, chartaceous, glabrous, black; floral bracts oblanceolate, acuminate, ca 3 x 0.75 mm, chartaceous, hoary towards apex, black. Male flowers: pedicels ca 0.5 mm long; sepals obovate, connate into a spathe of 1.75 x 2 mm, 3-lobed with acute lateral lobes and obtuse middle lobe, hoary towards apex, black; stipe of corolla ca 1.5 mm long; petals 3, subequal oblong or lanceolate, obtuse or acute, ca 0.5

mm long, hoary with a black gland; anthers 6, oblong-globose, pale yellow. Female flowers: pedicels ca 0.75 mm long; sepals 3, free, oblanceolate, conduplicate, acuminate, ca 2.5 mm long, hoary towards apex, black; petals 3, free, linear-spathulate, not clawed, obtuse, ca 2 mm long, hyaline, hoary towards apex, with a black gland; stipitate between sepals and petals; ovary stalked, ovoid-globose; style 3-fid; seeds oblong-obovoid, obtuse, ca 0.6 x 0.48 mm, purple; cells of seeds coat transversely elongated, aligned in vertical rows; appendages absent.

On laterite soil, occasional in Karimnagar and Medak districts.

Fl.: September.

Aklaspur (KRN), *GVS* 25646; Ramancheru, Perugunda (MDK), *RG* 96293 (BSID).

Eriocaulon quinquangulare L., Sp. Pl. 87. 1753; Hook.f., Fl. Brit. India 6 : 582. 1893; Fischer, Fl. Madras 3 : 1620. 1931; Ansari & N.P.Balakr., Eriocaula. in India 97. 2009.

A tufted annual marsh herb. Leaves lanceolate, purplish red, 3-5 × 0.2-0.3 cm, 7-nerved. Head globose, snow white, 6 mm across; peduncles numerous, twisted, 7-ribbed, puberulous; sheaths lax, shorter than the leaves, often purplish, obtuse; involucral bract obovate, obtuse, glabrous; floral bract hyaline, obovate, cuneate, ciliate, cuspidate; receptacles globose, sparsely villous. Male floret: calyx lobes oblong, apex white papillose hoary, petals subequal, ciliate, glandular or not; stamens 3 + 3, dark. Female floret: sepals 3, free, oblanceolate, densely ciliate; petals 3, subequal, pilose, apex with a black gland; ovary stipitate or subsessile, 3-lobed, style branches 3, filiform. Seed oblong, ellipsoid, pale brown, smooth or striate.

Occasional in marshy places in all districts.

Fl. & Fr.: August – December.

Way to Rusthempet (MDK), *KMS* 6806; Gadirayal RF (RR), *MSM* 10522; Kusumasamudram (RR), *MSM* 11032; Aklaspur (KRN), *GVS* 25646 (MH); Pakhal (WGL), *KMS* 11675 (MH); Nellippakka RF (KMM), *RCS* 99084 (BSID); Jakaram (WGL), *RKP* 112707 (BSID).

Eriocaulon richardianaum (Fyson) Ansari & N.P.Balakr., Eriocaula. India 51. 1994. *Eriocaulon dianae* Fyson var. *richardiana* Fyson, J. Indian Bot. 2: 259. 1921; Fischer, Fl. Madras 3: 1620. 1931.

A tufted annual marsh herb; leaves oblong-lanceolate, 12 × 0.7 cm, 7-nerved, apex acute. Head subglobose, 7 mm across, grey; peduncles many, striate, 6-20 cm long; sheaths lax, half as long as the leaves, apex obtuse; involucral bract lanceolate or ovate, glabrous; floral bract oblanceolate – cuneate, apically ciliate; receptacle conical, villous. Male floret: 1.8 mm long, calyx spathaceous, split down one side, lobes 3, small, rounded, apex white papillose-hairy, petals 3, ciliate, apically glandular; stamens 3 + 3. Female floret: 2 mm long; sepals 3, free, dark, subequal, 1.5-2 mm, petals 3, free, oblanceolate, keeled, 1.8 mm; eglandular; ovary 3-lobed, style branches 3, filiform. Seeds transversely ellipsoid.

Occasional in wet places, often in marshy areas in Nizamabad district.

Fl. & fr.: August – November.

Bhimnagar RF (NZB), *TP* & *BR* 6205.

Eriocaulon santapaui Mold., Phytologia 3: 166. 1949; Ansari & N.P.Balakr., Eriocaulac. in India 172. 2009.

Acaulescent herbs; rootstock absent. Leaves rosulate, linear-lanceolate, acute or obtuse, up to 4 × 0.3cm, glabrous. Peduncles many, up to 20 cm long; sheaths up to 3 cm long, glabrous. Heads ovoid or obovoid, ca 3 cm across, grey; receptacle cylindric, villous; involucral bracts erect or spreading, oblong-lanceolate, chartaceous, glabrous; floral bracts oblanceolate-cuneate, chartaceous, black. Male flowers: Pedicels 0.25 cm long; sepals obovate, connate into a spathe, 3-lobed with obtuse lobes, hoary at apex, black; petals 3, subequal, linear, sparsely hoary, with a black gland; anthers 6, globose, black. Female flowers: pedicels minute; sepals 3, free, unequal, the two lateral ones oblanceolate, conduplicate, keeled; petals 3, subequal, spathulate, not clawed, glabrous, without any black gland; ovary sessile, globose, style 3-fid. Seeds ovoid-globose, pale purple.

This taxon was repoted by V.S. Raju (1983) as a new record for Eastern Ghats. He collected this species from Vazeedu of Khammam district (*V.S. Raju* 685a MH).

Fl. & Fr.: November – January.

Eriocaulon setaceum L., Sp. Pl. 87. 1753; Hook.f., Fl. Brit. India 6: 572. 1893; Gamble, Fl. Madras 3: 1618. 1931; Ansari & N.P.Balakr., Eriocaula. in India 67. 2009.

Submerged herbs, to 25 cm high. Leaves filiform, 3-4 × 0.2 cm, 1-veined. Scapes 5-15, to 11 cm, 6-ribbed, sheath 2-2.5 cm, heads dusky, subglobose; involucral bracts reflexed, straw-coloured, ovate; floral bracts orbicular. Male flowers: sepals spathe like, 3-lobed; petals 3, elliptic to subulate. Female flowers: sepals 3, free; petals 3, free, oblanceolate-linear; ovary 3-loculed, style 3-cleft. Seeds narrowly ovoid.

Murthy & *al.* (2008) reported from Medak district.

Fl. & Fr.: July – October.

Eriocaulon thwaitesii Körn., Linnaea 27: 627. 1856; Hook.f., Fl. Brit. India 6: 583. 1893; Ansari & N.P.Balakr., Eriocaula. in India 60. 2009. *E. mariae* Fyson, Kew Bull. 1914: 331. 1914.

An annual tufted herb. Leaves linear or ensiform, 1-8 × 0.2-1 cm, sheaths lax, widened upwards. Peduncles many, 3-12 cm long, 5-ribbed; involucral bracts oblong-obovae, apex rounded, yellowish brown; floral bracts obovate-spathulate, shortly cuspidate, dark. Male florets: sepals rarely 3, glabrous, hairy at apex, corolla lobes with or without a black or red gland. Female florets: sepals 2, boat-shaped; petals 3, filiform, very short, white pubescent near apex, eglandular. Seeds with longitudinal rows of white papillae.

Occasional in wet places in Adilabad and Nizamabad districts.

Fl. & Fr.: August – October.

Vempalli (ADB), *GO* 4436; Gandhari RF (NZB), *BR* & *CPR* 7223.

Eriocaulon wightianum Mart., Pl. Asiat. Rar. 3: 28. 1832; Pearson, J. Linn. Soc. Bot. 34: 320. 357. 1899. *E. sericans* Mart. in Wall., Pl. Asiat. Rar. 3: 29. 1832; Hook.f., Fl. Brit. India 6: 576. 1893.

A tufted annual marsh herb; leaves linear, obtuse, glabrous, many-nerved, scarcely fenestrate. Head globose, 5 mm across, densely white-villous; peduncles stiff, aggregated, 15-30 cm, 7- ribbed, glabrous or sparsely hairy, twisted; involucral bract cuneate-obovate, hairy at the back in the upper part, ultimately reflexed; floral bracts obovate, acuminate, densely clothed with white hairs in the upper part; receptacle pilose. Male floret: sepals 3, obovate-oblong, cuneate at the base, free, hyaline, the tips pubescent; petals 3, densely hairy at the apex and without a black gland, one petal slightly larger than the other 2; stamens 3 + 3, anthers black. Female floret: sepals 3, free, slightly unequal, obovate, bearded towards the tip; petals 3, linear-spathulate, villose with long white hairs and with a minute black gland; ovary sessile, 3-lobed, style branches 3, filiform. Seeds ellipsoid, dark reddish brown, smooth, minutely papillose.

Reported from Khammam district.

Fl. & Fr.: October – November.

Godavari river bank – Vazeedu (KMM), *VSR* 675 A (MH).

CYPERACEAE

1. Florets unisexual:
    2. Nut enclosed in a utricle .... **Carex**
    2. Nut not enclosed in a utricle:
        3. Nut enveloped by two glumes and falling together with them .... **Diplacrum**
        3. Nut not enveloped by two glumes .... **Scleria**
1. Florets bisexual:
    4. Glumes on spikelets distichous:
        5. Stigmas 2; nutlets 2-sided with 1 margin facing spikelet axis:
            6. Spikelets with more than 2 glumes; spikelet axis and glumes persistent .... **Pycreus**
            6. Spikelets with 1 or 2 glumes; spikelet axis deciduous, spikelets falling whole:
                7. Spike sessile; lower glumes bisexual, upper male/sterile .... **Kyllinga**
                7. Spike stalked, all flowers bisexual .... **Queenslandiella**
        5. Stigmas 3, rarely 2; nutlets trigonous, rarely biconvex with 1 side facing spikelet axis:
            8. Annual or perennial; glumes not winged .... **Cyperus**
            8. Annual; glumes prominently winged .... **Courtoisina**

4. Glumes on spikelets spiral:
   9. Hypogynous scales present .... **Lipocarpha**
   9. Hypogynous scales absent:
      10. Style base dilated and constricted or articulated above the nut:
         11. Hypogynous bristles present, nut beaked:
            12. Leaves absent, style persistent .... **Eleocharis**
            12. Leaves present, style not persistent .... **Rhynchospora**
         11. Hypogynous bristles absent, but not beaked:
            13. Style base persistent; if falling not leaving a tumour on the nut .... **Fimbristylis**
            13. Style deciduous leaving a tumour on the nut .... **Bulbostylis**
      10. Style base not dilated, continuous with the nut:
         14. Leaves reduced to sheaths:
            15. One involucral bract over 1.5 cm, longest involucral bract erect, culmlike and apparently continuous with culm, inflorescence appearing to be lateral:
               16. Plants colonial from elongate rhizomes, perennial, usually more than 0.8 m high; hypogynous bristles present .... **Schoenoplectus**
               16. Plants caespitose, annual, less than 0.8 high; hypogynous bristles absent .... **Schoenoplectiella**
            15. At least 2 involucral bracts over 1.5 cm, longest involucral bract leaflike, erect or spreading .... **Bolboschoenus**
         14. Leaves well developed:
            17. Leaves with a collar around the stem .... **Fuirena**
            17. Leaves well developed .... **Actinoscirpus**

ACTINOSCIRPUS (Ohwi) Haines & Lye

Actinoscirpus grossus (L.f.) Goetgh. & D.A. Simpson, Kew Bull. 46: 171. 1991; Kukkonen in Rech.f., Fl. Iran 173: 15. 1998. *Scirpus grossus* L. f., Suppl. Pl. 104. 1782; C.B.Clarke in Hook. f., Fl. Brit. India 6: 659. 1893; Fischer, Fl. Madras 3: 1666. 1931. *Schoenoplectus grossus* (L.f.) Palla, Allg. Bot. Z. Syst. 17: Beibl. 3: 1911; Pullaiah & Hanumanthappa in Pullaiah, Fl. Andhra Pradesh 3: 1107. 1997.

Perennial, stoloniferous herb; stolons producing tubers. Culms to 200 cm high, sharply 3-angled, spongy, faces concave, septate-nodose, with a corm like hard enlargement at base. Leaves equaling the culm, 1-ribbed, flattish-plicate, septate-nodose, sheaths spongy, prominently septate-nodose, tightly surrounding the culm base. Inflorescence a large terminal corymb with primary, secondary and tertiary rays; rather dense, to 12 cm long. Involucral bracts 4, foliaceous, exceeding the inflorescence, the longest to 40 cm long. Spikelets solitary, ellipsoid, to 10 mm long, densely many-flowered, light brownish, rust-colured. Achenes broadly obovate, subcompressed-triangular, stramineous brown.

A rare species found in marshy fields.

Fl. & Fr.: August – December.

Rusthempet (MDK), *KMS* 6799 (MH & CAL).

BOLBOSCHOENUS (Ascherson) Palla

Bolboschoenus maritimus (L.) Palla in Koch., Syn. Deutsch. Fl. ed. 3. 2532. 1904. subsp. **maritimus**. *Scirpus maritimus* L., Sp. Pl. 75. 1753; C.B.Clarke in Hook. f., Fl. Brit. India 6 : 658. 1893; Fischer, Fl. Madras 3 : 1666. 1931. *Schoenoplectus maritimus* (L.) Lye, Blyttia 29: 145. 1971; Pullaiah & Hanumanthappa in Pullaiah, Fl. Andhra Pradesh 3: 1109. 1997. *Scirpus tuberosus* Desf., Fl. Atlant. 1: 50. 1798; Verma & Veenachandra, Rec. Bot. Surv. India 21: 263. 1981; Karthikeyan & *al*., Florae Indicae Enumeratio Monocotyledonae 71. 1989.

Perennial, erect, rhizomatous creeping herb with tubers, to 75 cm high; culms triquetrous. Leaves flat, linear, to 30 cm long, acuminate, ligulate, sheaths reaching to mid culm. Inflorescence terminal umbel, rayed, unequal. Involucral bracts 3-5, leaflike; spikelets ovoid to cylindric, acute. Achenes obovate, trigonous, smooth, to 4 mm long.

Occasaional in moist waste lands, marshes, in beds of drying ponds and canals.

Fl. & Fr.: October – March.

Penganga River (ADB), *GO* & *PVP* 4290, 4291; Swarna project (ADB), *MHR* 14558; Nasrullabad (NZB), *TP* & *BR* 6181; Mohammadabad RF (RR), *KH* & *MSM* 5576; Kampallicheruvu (KMM), *MHR* & *MCK* 17050; Lingalacheruvu, (KMM), *MHR* & *MCK* 17038; Aklaspur (KRN), *GVS* 22531 (MH & CAL); Rusthempet (MDK), *KMS* 8009 (MH & CAL); Rathamhutta hills (KMM), *RCS* 99054 (BSID).

Note: *Bolboschoenus maritimus* belongs to a difficult, worldwide complex, in which the delimitation of specific and intraspecific taxa is still unclear.

BULBOSTYLIS Kunth *nom. cons.*

Bulbostylis barbata (Rottb.) Kunth ex C.B. Clarke in Hook. f., Fl. Brit. India 6: 651. 1893. subsp. **barbata.** *Scirpus barbata* Rottb., Progr. 27. 1772. *Stenophyllus barbata* (Rottb.) Cooke, Fl. Pres. Bombay 2: 887. 1908; Fischer, Fl. Madras 3: 1662. 1931.

Annual, densely tufted erect herb, to 30 cm high, culms slender, striate, leaved at base. Leaf blade filiform, 4 to 8 cm long, sheaths membranous, brownish. Inflorescence a terminal head of 2 -20 sessile spikelets, 5-15 mm across, brownish. Involucral bracts 3, spreading, to 2 cm long, setaceous. Spikelets sessile, lanceolate, 3-8 mm long, angular, subdensely 7 to 15-flowered, pale brownish; glumes ovate, to 2 mm long with recurved short awn, membranous, yellowish-green and becoming rusty brown, prominently keeled, ciliolate; stamen 1; style base depressed-globular, style filiform, stigmas 3. Achenes obovate-orbicular, trigonous, pale rounded at apex.

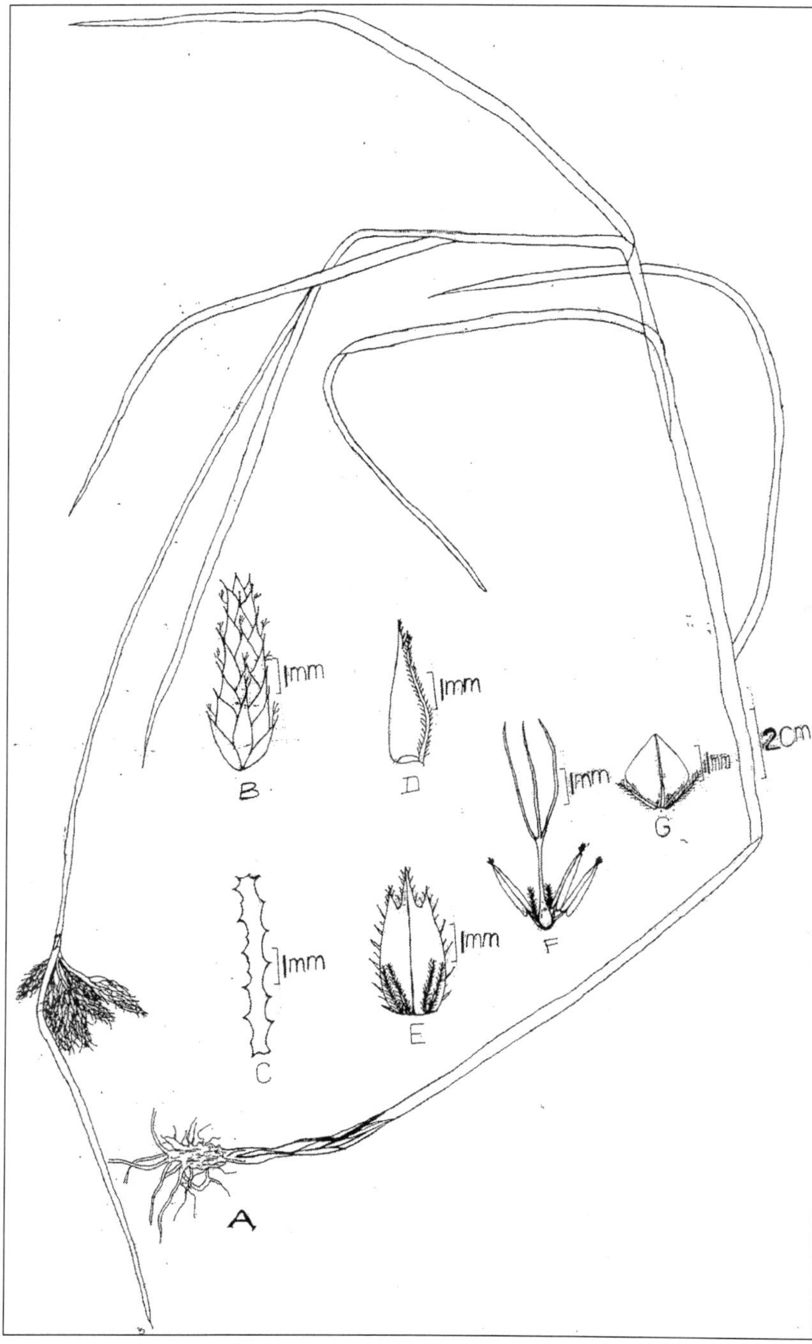

Figure 124: Bolboschoenus maritimus (L.) Palla (Syn.: *Schoenoplectus maritimus* Lye). A. Habit, B. Spikelet, C. Rhachilla, D. Closed glume, E. Opened glume, F. Pistil with stamens, G. Achene.

This is a very common species found everywhere. Usually seen in open dry, especially sandy areas, rocky surfaces, waste lands, road sides, and as a weed in cultivated lands.

Fl. & Fr.: August - December.

Ankusapuram (ADB), *GO* 4330; Borgaun (NZB), *BR* 9281; Choutkoor (MDK), *TP* & *CP* 11947; Medak (MDK), *RG* 104151 (BSID), Kodimial (KRN), *GVS* 20049 (MH).

CAREX L.

Carex hebecarpa C. A. Mey., Mém. Acad. St.-Petersbourg. 1: 223. 1831. *C. hebecarpa* var. *ligulata* (Nees) Kük. in Engler, Pflanzenr. IV.20, 38: 745. 1909; Fischer, Fl. Madras 3 : 1686. 1931. *C. ligulata* Nees in Wight, Contr. Bot. Ind.: 127. 1834; C.B.Clarke in Hook.f. Fl. Brit. India 6: 747.1894.

Tufted perennials, to 75 cm high; stems trigonous, scabrous along edges. Leaves 5-10 mm wide, scabrous towards apex, longer than stem; basal sheaths reddish; ligule conspicuous. Inflorescence of terminal male and 2-6 peduncled female spikes; involucral bracts longer than spikes; sheaths long, pilose, with conspicuous, reddish-hairy auricles. Female spikes cylindrical, 2-4 cm, glumes 1.5 mm, mucronate. Utricles 3.5 x 1.5 mm, ellipsoid, trigonous, hispid, beak bi-dentate, 0.7 mm; style base bulbous. Nut ellipsoid, trigonous.

In marshes of open grass lands (Ramana, 2010)

Fl. & Fr. : August-December.

COURTOISINA Soják

Courtoisina cyperoides (Roxb.) Soják, Cass. Nar. Mus. Odd. Prir. 148: 193. 1979 publ. 1980. *Mariscus cyperoides* (Roxb.) A. Dietr. S. Pl. 2: 348. 1832. *Cyperus pseudokyllingioides* Kuk., in Engler, Pflanzenreich IV. 20, (Ht.101): 501. 1936. *Indocourtoisia cyperoides* (Roxb.) Bennet & Raizada, Indian Forester 107: 432. 1981. *Kyllinga cyperoides* Roxb., Fl. Ind. 1: 187. 1820. *Courtoisia cyperoides* (Roxb.) Nees, Linnaea, 9: 286. 1834; C.B.Clarke in Hook. f., Fl. Brit. India 6: 625. 1893; Fischer, Fl. Madras 3: 1645. 1931.

Annual, erect, tufted, non rhizomatous herb, to 30 cm high, culms trigonous, glabrous. Leaves few, basal, as long as the stem, gradually acuminate; sheaths 2.5 cm; yellow, membranous. Inflorescence compound, branches ending in spikes. Involucral bracts 6, lower one much exceeding the inflorescence, spreading. Spike globose, 6 mm long; spikelets many, 4 mm long, clustered, compressed, 1-flowered, broadly elliptic, golden-brown-coloured, shining, rhachilla slender, narrowly winged. Achenes trigonous, linear-oblong, 2.8 mm long, apiculate, golden brown, strongly puncticulate, about ¾ of the length of the glume.

Occasionally seen in rice fields as a weed and also found in waste lands.

Fl. & Fr.: November – January.

Pocharam RF (MDK), *TP* & *CP* 12061; Nellippakka RF (KMM), *RCS* 99079 (BSID); Vempalli (ADB), GO 4446; Akkannapet to Laxmipuram (MDK), *RG* 106725 (BSID); Ramancheru, Perugunda (MDK), *RG* 96297 (BSID).

CYPERUS L.

1. Florets bisexual
    2. Rhachilla not articulated, persistent:
        3. Stigmas 3, nut trigonous:
            4. Rhachilla not winged; stamen 1 .... **C. leucocephalus**
            4. Rhachilla winged; stamens 1 to 3:
                5. Spikelets spicately disposed on conspicuously elongated rhachis thus forming spikes:
                    6. Rhachilla of spikelets winged with base of glumes, which is decurrent along the rhachilla internode:
                        7. Spikes cylindrical, rhachis 3-7 cm long; spikelets numerous:
                            8. Rhizome short; spikes umbelled or corymbose; spikelets distinctly compressed: involucral bracts 3-6 .... **C. exaltatus**
                            8. Rhizome woody, creeping; spikes digitate; spikelets terete; involucral bracts 3-12 .... **C. digitatus**
                        7. Spikes broadly elliptic, rhachis 1-4 cm long; spikelets occasionally up to 30:
                            9. Plants with creeping rhizomes or stolons (except in *C. bulbosus* which bears tunicate bulb at base of the culm).
                                10. Culms thick, more than 5 cm wide and 60 cm high, clothed at base with bladeless sheaths:
                                    11. Culms obtusely trigonous to subterete; involucral bracts much shorter than the corymb:
                                        12. Culms not septate .... **C. corymbosus**
                                        12. Culms transversely septate .. **C. articulatus**
                                    11. Culms acutely trignous; involucral bracts much surpassing the corymb .... **C. pangorei**
                                10. Culms slender, 1-2 mm wide and less than 60 cm high, leaves with elongated blade:
                                    13. Stolons filiform, soon disappearing; culms arising from bulb like tuber; .... **C. bulbosus**

13. Stolons conspicuous and persistent, culm base forming corm like enlargement:

14. Glumes 2-2.5 mm long; leaves as long as the culm .... **C. stoloniferus**

14. Glumes 3-4 mm long; leaves much shorter than the culm:

15. Culms 10 to 30 cm high; longest involucral bracts to 15 cm long .... **C. rotundus**

15. Culms 30-60 cm high, longest involucral bract to 10 cm long .... **C. bifax**

9. Plants without creeping rhizome or stolons:

16. Spikelets linear to lanceolate, 1.5-2 mm wide, weakly flattened with two acute edges; glumes close together .... **C. tenuiculmis**

16. Spikelets filiform, subterete, less than 1 mm wide; glumes spaced:

17. Spikelets spreading; glumes rounded, obtuse at apex bearing narrow hyaline margins towards apex .... **C. distans**

17. Spikelets erect patent; glumes shallowly emarginated at apex, bearing very broad hyaline margin towards apex .... **C. nutans**

6. Rhachilla of spikelets without conspicuous wings, base of glumes not decurrent along the rhachilla internode:

18. Plants perennial with stolons:

19. Rhachis of spikes glabrous; glumes obtuse at white hyaline apex .... **C. procerus**

19. Rhachis of spikes densely hispid; glumes acute at apex with broad hyaline apex .... **C. pilosus**

18. Plants annuals with fibrous roots only:

20. Spikelets 1-2.5 cm long, 2-3 mm wide; glumes 2-3.5 mm long, acute at apex .... **C. compressus**

20. Spikelets 1 cm long, 1-1.5 mm wide; glumes 1-1.5 mm long.

21. Rhachilla of the spike glabrous; keel 3-5-nerved, smooth, not winged .... **C. iria**

21. Rhachilla of the spike scabrid on the angles, keel 7-nerved, winged with serrate margins at apex .... **C. alulatus**

5. Spikelets digitately disposed or capitately congested at apices of inflorescence rays, thus forming heads without elongated rhachis:

22. Inflorescence with open elongated rays:

23. Large sized perennials with conspicuous rhizomes; atleast some leaves and involucral bracts more than 10 mm wide; culms 30-90 cm high .... **C. platystylis**

23. Small to medium sized annuals, leaves and bracts almost 5 mm wide; culms up to 30 cm high:

24. Leaves flattish; glumes muticous; achenes obovate:

25. Spikelets digitate in clusters of 3-15; glumes acute at apex:

26. Glumes sublaxly imbricate and slightly spaced; thus achenes exposed between glumes; plants tufted with fibrous roots only .... **C. tenuispica**

26. Glumes densely imbricate and close together so that the achenes not shown between the glumes; rhizomes frequently elongated .... **C. halpan**

25. Spikelets many, congested on globose heads .... **C. difformis**

24. Leaves canaliculated; glumes awned or mucronate at apex; achenes oblong:

27. Glumes darker in colour, long with short erect mucro at apex .... **C. amabilis**

27. Glumes awned at apex:

28. Spikelets chestnut black; glumes with straight awn .... **C. castaneus**

28. Spikelets ferrugineus; glumes with recurved awn:

29. Involucral bracts 5; style shorter than the achenes .... **C. cuspidatus**

29. Involucral bracts 3; style longer than the achens .... **C. rubicundus**

22. Inflorescence congested in a singled head:
    30. Annual, tufted herb .... **C. meeboldii**
    30. Perennial rhizomatous herbs:
        31. Spikelets 3- or 4-flowered; nutlets with spongy base of style extending downward along angles of nutlet .... **C. cephalotes**
        31. Spikelets 8-flowered or more; nutlets without enlarged spongy apex:
            32 Culms usually tall; spikelets several to more than 20, laxly fascicled; glumes conspicuously distichous, ca 4 mm, apex obtuse and muticous .... **C. niveus**
            32. Culms dwarf; spikelets numerous, very densely arranged; glumes spirally imbricate or obscurely distichous, ca 2 mm, apex curved mucronate .. **C. michelianus**

3. Stigmas 2; nutlets biconvex:
    33. Inflorescence umbelled, stems robust up to 120 cm tall .... **C. alopecuroides**
    33. Inflorescence capitate, stems densely tufted, up to 25 cm high .... **C. laevigatus**

2. Rhachilla articulate, deciduous, leaving a knob above the two lowest glumes:
    34. Spikelets flattened, 3 to many-flowered, glumes folded on a conspicuous keel:
        35. Glumes awned .... **C. squarrosus**
        35. Glumes not awned:
            36. Plants 40-60 cm tall .... **C. compactus**
            36. Plants 15-30 cm tall .... **C. dubius**
    34. Spikelets terete, 1 or 2-flowered, glumes not or scarcely keeled:
        37. Base of the stem hardly enlarged:
            38. Spikelets shortly cylindric, oblong or ovoid, 1.2-1.8 × 0.8-1.2 cm, with sides narrowed toward base; anthers 1.5 mm; nutlets 0.6-0.9 mm wide; leaf blade margin scabrous:
                39. Spikelets 1 or 2-fruited, greenish straw-coloured .... **C. cyperinus**
                39. Spikelets 1-fruited, whitish-green .... **C. paniceus**

38. Spikes cylindric, 2-3.5 × 0.6-1 cm, with parallel sides; anthers 0.8-1 mm; nutlets 0.5mm wide; leaf blade margin smooth .... **C. cyperoides**

37. Base of the stem bulbous .... **C. clarkei**

1. Florets unisexual .... **C. mollipes**

Cyperus alopecuroides Rottb., Descr. Pl. Rar. Progr. 20. 1772. *Juncellus alopecuroides* (Rottb.) C.B.Clarke in Hook. f., Fl. Brit. India 6: 595. 1893; Fischer, Fl. Madras 3: 1629. 1931.

Perennial, erect, stout, rhizomatous herb, to 120 cm high, culms rounded below, trigonous above. Leaves many, flat, as long as the stem, coriaceous, acute; sheaths dark. Inflorescence compound umbel with 4-8 primary rays, to 15 cm long; secondary rays 3-7, to 2 cm long, bearing clusters of oblong spikes, densely covered with small spikelets. Spikelets spicate, to 1 cm long, 15-30-flowered, persistent, elliptic-lanceolate. Involucral bracts 5, far overlapping, to 40 cm long, leaf like. Achenes broadly elliptic, light brown, stipitate, apiculate.

Usually found in marshy low lands, along the banks of canals, ponds, lake margins and in swamps.

Fl. & Fr.: September – February.

Chandoor (NZB), *BR* & *GO* 9209; Along Moosi River bank (HYD), *KMS* 5932 (MH & CAL); Medak (MDK), *RG* 104018 (BSID).

Cyperus alulatus Kern, Reinwardtia 1: 464. f. 1. 1952. *C. iria* var. *rectangularis* Kuckenth. in Engl., Pflanzenr. 101: 152. 1935. *C. rectangularis* (Kuekenth.) Bennet, Indian Forester 95: 692. 1969.

Annual, tufted herb, to 40 cm high, culms triquetrous. Leaves as long as the culms, linear, serrulate margin near apex. Inflorescence terminal, lax, compound umbel. Involucral bracts 8, unequal, the longest one up to 30 cm long. Spikes to 3 cm long; rhachis flexuous, scabrid on the angles; rays 8, unequal, to 15 cm long. Spikelets greenish yellow, patent. Achenes obovoid, triquetrous, reddish-brown, apiculate.

Occasional in dry rice fields, ditches and agricultural fields.

Fl. & Fr.: August – November.

Nirmal (ADB), *VSR* 2104 (KU); Vidyaranyapuri (WGL), *VSR* 1990 (KU).

Cyperus amabilis Vahl, Enum. Pl. 2: 318. 1806; Hook.f., Fl. Brit. India 6: 598. 1893.

Annual, slender, tufted herb, to 30 cm high, culms triquetrous. Leaves 2, sub-basal, half as long as the culms. Umbel compound; rays 9 to 10 cm long; involucral bracts 2, leaflike, longest up to 15 long; spikelets linear, golden brown. Achenes obovoid, brown, faintly rugose.

Occasional weed in cultivated fields, grass lands and marshy localities.

Fl. & Fr.: August – December.

Raghava Rao (1989) reported it from Mahabubnagar District.

Cyperus articulatus L., Sp. Pl. 44. 1753; C.B.Clarke in Hook.f., Fl. Brit. India 6: 611. 1893; Fischer, Fl. Madras 3: 1641. 1931.

Perenial, erect, glabrous, stout herb, culms to 1.2 cm wide, terete, transversely septate. Leaves reduced to 7-20 cm long, brown to grey purple-coloured sheaths and covers the base of the stem. Inflorescence compound umbel; involucral bracts 5, much shorter than the corymb, to 1 cm long, primary rays 5-8, to 9.5 cm. Spikes shortly racemose, to 4 cm long, 2-15-spikeletted, spikelets oblong, narrow, linear, compressed, acute, ca. 35-flowered; rhachilla 2-winged. Achenes oblong, trigonous, maturing brown.

A water loving sedge, usually found in standing waters of ponds, canals and along river banks in large colonies.

Fl. & Fr.: March – September.

Mosra Road (NZB), *TP* & *BR* 6092; Along the Moosi River bank (HYD), *KMS* 5929 (MH & CAL).

Cyperus bifax C.B.Clarke, Kew Bull. 8: 13. 1908. *C. retzii* Nees in Wight, Contr. Bot. Ind. 82. 1834. *C. tuberosus* sensu C.B.Clarke in Hook. f., Fl. Brit. India 6: 616. 1893 ex parte.

Perennial, rhizomatous, stoloniferous herb; stolons covered with brown scales; tubers ellipsoidal, 0.3 cm long. Culms to 70 cm high, trigonous. Leaves few, blade shorter than the culm, to 6 mm wide, flattish, herbaceous; sheaths subloosely surrounding the culm, pale and tinged with light reddish-purplish brown. Anthela subcompound, to 14 cm long. Involucral bracts 4, foliaceous, patent, two longer surpassing the anthela, the longest to 10 cm long, slender, erect patent. Spikes broadly ovoidal, bearing 15 spikelets on short rhachis. Spikelets spreading to erect, linear, to 20 mm long, gradually narrowed distally to acuminate apex, moderately flattened, pale stramineous and tinged with light reddish brown, 20-flowered, rhachilla straight, internodes broadly winged, wings hyaline, persistent. Achenes obovoid, trigonous, greyish brown, puncticulate.

Occasional in open wet ground, the margins of ditches, tanks, irrigation canals and rice fields etc.

Fl. & Fr.: July – December.

Anantasagar beat (RR), *KH* & *MSM* 10928; Parvathagiri (WGL), *AR* 2003 (KU).

Cyperus bulbosus Vahl, Enum. Pl. 2 : 342. 1806; C.B. Clarke in Hook. f., Fl. Brit. India 6 : 611. 1893; Fischer, Fl. Madras 3 : 1641. 1931.

Perennial, stoloniferous herb, to 30 cm high; tuber ovoid-globose, clothed with coriaceous black-brown straite coat splitting into several segments; stolons filiform, to 10 cm long, terminated by tuber. Culms triquetrous arising from bulblike tuber. Leaves to 20 to a culm, slightly over-topping the culm; blade to 20 cm long, incurved, plicate, scabrid on upper margins; sheaths to 6 cm long, broader than the blade, membranous, cinnamon-coloured, the inner ones pale white. Inflorescence simple, to 3 cm long, spicately bearing 2 subdigitate clusters of spikelets on very short axis.

Involucral bracts 3, patent, the lowest surpassing the inflorescence, the longest to 10 cm long. Spikelets subdigitate-spicate on very short rhachis, patent linear, to 2 cm long, to 20-flowered, sanguineous brown; rhachilla flexuous, the internodes winged. Achenes ellipsoid, trigonous, maturing black, apiculate at obtuse apex.

Occasional in sandy places, on river sides and marshy areas in Warangal and Hyderabad districts (Ramana, 2010).

Fl. & Fr.: September – December.

Hanumakonda (WGL), *VSR* 1984 (KU).

Note: Fischer (*loc. cit*) reported it from Deccan and Carnatic; Beypore in South Malabar. Obviously its distribution in Deccan was not specifically mentioned. This species was added to the sedge flora of Andhra Pradesh State by Raju and Ragan (1989) from Warangal district as new record.

C. bulbosus can easily be distinguished by the black-coated bulbils which give rise to new plants. In the absence of the bulbils *C. bulbosus* may easily be confused with *C. rotundus*. But the former can be distinguished by spaced lower bracts, and more distinctly nerved glumes.

Cyperus castaneus Willd., Sp. Pl. 1: 278. 1797; C.B.Clarke in Hook. f., Fl. Brit. India 6: 598. 1893; Fischer, Fl. Madras 3: 1639. 1931.

Annual, tufted, glabrous herb, to 20 cm high, culms slender, triquetrous, 10-leaved at base. Leaves 9 cm long; blade setaceous, weakly scabrid, reddish brown. Anthela simple, to 8 cm long, frequently congested in a single head. Involucral bracts 4, surpassing the inflorescence; rays 5, to 5 cm long. Spikelets stellately clustered in groups of 5 to 10, slenderly linear, to 10 mm long, flattened, 20-40-flowered, chestnut black; rhachilla striaghtish, wingless. Achenes oblong with parallel sides to oblong-obovoid, brown, puncticulate.

Common in open wet lands, grassy place, rice fields, and in swamps etc.

Fl. & Fr.: August – December.

Rebbana (ADB), *GO & PVP* 4844; Gadirayal RF (RR), *MSM* 10525.

Cyperus cephalotes Vahl, Enum. Pl. 2: 311. 1805; C.B.Clarke in Hook. f., Fl. Brit. India 6: 597. 1893; Fishcer, Fl. Madras 3: 1639. 1931. *Anosporum cephalotes* (Vahl) Kurz, J. Asiat. Soc. Beng. Pt. 2, Nat. Hist. 39(2): 84. 1870.

Perennial, rhizomatous, stoloniferous herb; stolons to 15 cm long, brownish, rooting at nodes; culms to 40 cm high, triquetrous at least distally. Leaves 2 to several, half as long as the culm, blade narrowly linear, to 50 cm long, 1-costate, stiffish, scabrid, sheaths to 70 cm long, the dorsal hyaline part eventually brownish, the lower ones cinnamon-coloured. Inflorescence on ovoidal, conical, lobed head, bearing 4 glomerules of 8 spikelets. Involucral bracts foliaceous. Spikelets ovoid, to 8 mm long, densely to 20-flowered, brown to fuscous with yellowish margins. Achenes ovoidal, trigonous and dorsally flattened, brown, minutely puncticulate, the wings developing into yellowish corky ridges.

Occasional in wet or flooded rice fields, in tanks etc. (V.S. Raju & *al.*, 2008).

Fl. & Fr.: July – December.

Cyperus clarkei Cooke, Fl. Bombay 2: 873. 1908. *Mariscus clarkei* (Cooke) T. Koyama, J. Jap. Bot. 51: 313. 1976. *M. bulbosus* C.B.Clarke in Hook. f., Fl. Brit. India 6: 620. 1839. non Steud.; Fischer, Fl. Madras 3: 1644. 1931.

Perennial, rhizomatous, stoloniferous herb; stolons to 6 cm long, the internodes to 5 mm long, clothed with pale brownish, membranous scales, culms solitary, to 22 cm high, triquetrous, the base thickened, to form a lanceolate enlargement, to 4 cm long, clothed with dusky brown, membranous sheaths. Leaves 6 to a culm, shorter than the culm; blade 18 cm long, herbaceous, 1-costate, plicate with incurved margins, glaucous green; sheaths 4 cm long, membranous. Dusky brown inflorescence simple, the longest 10 cm long; spikes cylindrical, subsessile, 2 cm long, obtuse. Achenes oblong-elliptic, trigonous, 1.5 mm long, maturing dark brown.

Occasional in marshy places and along streams in Adilabad, Karimnagar, Hyderabad and Mahabubnagar districts.

Fl. & Fr.: September – December.

Siddipet – Hyderabad Road (MDK), *CP* 11379; Kusumasamudram beat (RR), *MSM* & *KH* 10380; Mohammadabad (RR), *MSM* & *NV* 10430; Pakhal lake (WGL), *MHR* & *MCK* 17072.

Cyperus compactus Retz., Observ. Bot. 5: 10. 1788. *C. spinulosus* Roxb., Fl. Ind. 1: 203. 1832. *Mariscus microcephalius* C. Presl, Rel. Haenk. 1: 182. 1830; C.B.Clarke in Hook. f., Fl. Brit. India 6: 624. 1893. *M. compactus* (Retz.) Bold., Zakfl. Java 77. 1916; Fischer, Fl. Madras 3: 1645. 1931.

Perennial, rhizomatous herb, rhizome corm like; culms solitary, erect, to 50 cm high, obtusely trigonous, septate nodulose. Leaves 3, longer than the culm; blade elongated, linear, thickish, canaliculated, glaucous green, markedly septate nodose; sheaths elongated, cylindrical, septate nodose, purple brown. Anthela compound, ample, 7 cm long and as wide. Involucral bracts 8, patent, the lower ones much surpassing the anthela, the longest 60 cm long; primary rays 6, patent, the longer one up to 7 cm long; secondary corymbs with 5 raylets; spikes nearly globose, bearing many stellately arranged spikelets; rhachis very short. Spikelets linear, red brown, jointed at base, congested in a globsoe head, 12 mm long, pale green and tinged with sanguineous brown; rhachilla flexuose with pale white hyaline wings. Achenes linear-oblong, trigonous, to 2 mm long, yellow brown.

This is a rare taxon, this species can be seen occasionally along ravines, flowing water streams in Warangal district.

Fl. & Fr.: March – September.

Pakhal lake (WGL), *MHR* & *MCK* 17084.

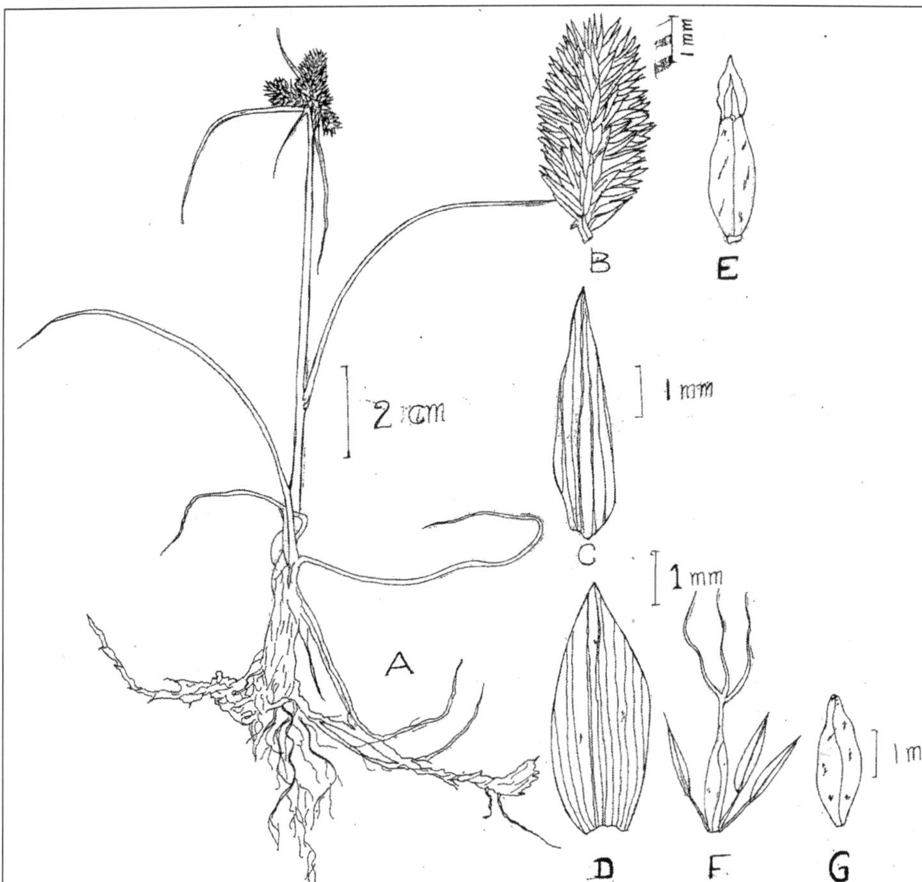

Figure 125: Cyperus clarkei Cooke [Syn.: *Mariscus clarkei* (Cooke) Koyama]

A. Habit, B. Spikelet, C. Closed glume, D. Opened glume, E. Rhachilla with a reduced glume, F. Pistil with stamens, G. Achene.

Cyperus compressus L., Sp. Pl. 46. 1753; C.B. Clarke in Hook. f., Fl. Brit. India 6 : 605. 1893; Ficher, Fl. Madras 3 : 1640. *C. pectiniformis* Roem. & Schult., Mant. 2 : 128. 1824.

Annual, tufted herb. Culms patent, to 30 cm high, 3-sided. Leaves 4 to a culm; basal; blade linear, shorter than the culms, light green; sheaths membranus, pale brownish, striate. Inflorescence anthelate, to 10 cm long; involucral bracts 7, foliaceous, unequal, the longest one to 20 cm long; rays 8, patent, to 10 cm long. Spikes bearing to 10 spikelets on abbreviated axis. Spikelets lance-oblong, 1-2.5 × 0.2-0.3 long, ca. 20-flowered, compressd, greenish and stramineous at maturity; rhachilla flexuose. Achenes 1.8 mm long, broadly obovate, 3-sided, dark brown, shiny.

Common weed of cultivated fields, waste lands, in open grass lands, along the sides of water courses, along paths on moist sandy soils.

Fl. & Fr.: September – February.

Ankusapuram (ADB), *GO* 4295; Bellampalli (ADB), *PVP* & *KH* 9448; Nirmal tank (ADB), *MHR* 14431; Manchippa (NZB), *TP & BR* 6029; Choutkoor (MDK), *TP* & *CP* 11922; Siddipet (MDK), *CP* 11376; Mohammadabad RF (RR), *MSM* 10478; Bhadrachalam – Kannaigudem (KMM), *AJR* & *PAN Reddy* 20466; Pakhal lake (WGL), *MHR* & *MCK* 17095; Katasintha (HYD), *KMS* 6011 (MH & CAL); Aklaspur (KRN), *GVS* 25617 (MH); Narsapur (MDK), *KMS* 6664 (MH & CAL); Mamillagudem (KMM), *RR* 105916 (BSID); Koriguttalu (KMM), *RR* 10605 (BSID); Kajipet (WGL), *RKP* 105226 (BSID).

Cyperus corymbosus Rottb., Descr. Icon. Rar. 42. t. 4. 1773; C.B.Clarke in Hook. f., Fl. Brit. India 6: 612. 1893; Fischer, Fl. Madras 3: 1641. 1931.

Perennial, glabrous, glaucous-green, rush-like herb; rhizome creeping, clothed with dark brown scales; culms to 90 cm long, 6 mm wide, terete below, trigonous above. Leaves reduced to sheaths. Umbel compound, to 14 cm long; involucral bracts 3, longest to 13 cm long, green; lowest involucral bract much shorter than the corymb; primary rays 6-9, very unequal, the longest reaching 10 cm long; secondary rays terminated by spikes or corymbs of 4-8 very slender spikelets. Spikes ca. 10, 6-9-spikeletted. Spikelets spicate, very variable in length, 2.5 cm long, rusty brown, rhachilla winged, 15-20-flowred. Achenes obovoid, tapering towards the base, trigonous.

Common in marshy areas near river banks, canals, along the sides of the ponds, in ditches etc.

Fl. & Fr.: August – December.

Sone (ADB), *TP* & *PVP* 4033; Chandoor (NZB), *BR* & GO 9209; Medak (MDK), *BR* & *CP* 11455, 11454; Edupayalu RF (MDK), *RG* 113432 (BSID); Rudraram RF (RR), *MSM* 12730; Mohammadabad RF (RR), *KH* & *MSM* 10922; Molachintapally (MBNR), *TDCK* 16825; Kampallicheruvu (KMM), *MHR* & *MCK* 17055.

Cyperus cuspidatus Kunth, Nov. Gen. Pl. 1: 204. 1815; C.B. Clarke in Hook. f., Fl. Brit. India 6: 598. 1893. *C. uncinatus* auct. non Poiret: Fischer, Fl. Madras 3: 1639. 1931.

Annual, erect, tufted herb, 9 cm high. Leaves setaceous, as long or half as long as culms, filiform, sheaths purple. Umbel simple, consisting of a central sessile head and few or many filiform spreading unequal rays, each terminated by a head of stellately spreading spikelets; involucral bracts 5, longest to 5 cm long, filiform. Spikes obovoid, digitately and stellately spikeletted. Spikelets 5-15 in a cluster, compressed, 12-24-flowered, 1 cm long, reddish-brown, rhachilla persistent. Achenes obovoid, triquetrous, winged, dark-brown.

Common in moist or damp sandy soil or cultivated and fallow fields.

Fl. & Fr.: July – December.

Ankusapuram (ADB), *GO* 4299; Mingaram (NZB), *BR* 9289; Mohammadabad (RR), *MSM* & *KH* 10530; Thadvai (WGL), *VSR* & *AR* 2010 (KU).

Cyperus cyperinus (Retz.) Suringar, Cyperus 154. 1898. *Kyllinga cyperina* Retz., Observ. Bot. 6: 21. 1791. *Mariscus cyperinus* (Retz.) Vahl, Enum. Pl. 2: 377. 1806; C.B.Clarke in Hook.f., Fl. Brit. India 6: 621. 1893; Fischer, Fl. Madras 3: 1644. 1931.

Perennial, rhizomatous herb, rhizome woody, clothed with brown fibres, culms few, erect, to 40 cm high, triquetrous, leaves at thickened base. Leaves 7, nearly equaling the culm; blade narrowly linear, long attenuate, 2.5 mm wide, herbaceous, plicate; sheaths pale green and stained with purplish pink, the basal sheaths dark red brown. Umbel simple, terminal, open with short rays, rays 9, to 2 cm long, each terminated by a spike; spikes cylindrical, narrowed to base, 2 cm long, greenish, densely bearing up to 60 spikelets. Involucral bracts 8, patent, the lower few surpassing the anthela, the longest up to 16 cm long. Spikelets patent, greenish straw-coloured, linear-lanceolate, 5 mm long, bearing 4 glumes and 2 achenes; rhachilla with lanceolate wings; stamens 3; stigmas 3. Achenes oblong, trigonous, 2.5 mm long, brownish.

Occasional in marshy areas, banks of streams, and in open grassy places in hilly areas.

Fl. & Fr.: July – October.

Ankusapuram (ADB), *GO* 4300; Jaipur RF (ADB), *PVP* & *KH* 9437; Gandhrari RF (NZB), *BR & CPR*, 7241; Mallipet (NZB), *BR* & *CP* 9234; Gnanapur (MDK), *CP* 11785; Nagulabanda (MDK), *CP* 11331,11340; Mohammadabad RF (RR), *MSM* & *KH* 10430; Bhadrachalam RF (KMM), *RCS* 104313 (BSID).

Cyperus cyperinus (Retz.) Suringer forma **curvata** Suringer. *Mariscus tenuifolius* Schrader & Nees, Fl. Brasil. 2(1) 46. 1842; C.B.Clarke in Hook. f., Fl. Brit. India 6: 622. 1893; Fischer, Fl. Madras 3: 1644. 1931.

Perennial tufted herb, culms about 50 cm high with small basal tubers. Leaves shorter than culm, sheath purplish red; leaf blade 5-7 mm wide, slightly flaccid or stiff. Involucral bracts 6-10, leaf like, longer or slightly shorter than inflorescence, rays 6-10, spikes oblong-obovoid; spikelets narrowly linear-ovoid to linear-oblong, each 2-4-flowered, crowded, rhachilla wings lanceolate, glumes crowded, brownish to purplish red on both the surfaces; stamens 3; stigmas 3. Nutlets trigonal, reddish-brown, narrowly oblong, 3-sided.

Rare in open grassy places in hilly areas in Nalgonda district.

Fl. & Fr.: September – December.

Nagarjunakonda (NLG), *BVRS* 80 (SKU).

Cyperus cyperoides (L.) Kuntze, Revis Gen. Pl. 3(3): 333. 1898. *Mariscus sumatrensis* (Retz.) Raynal, Adansonia 15: 110. 1975; T.Koyama, Gard. Bull. Singapore 30: 154. 1977. *Kyllinga sumatrensis* Retz., Observ. Bot. 4: 13. 1786. *Mariscus sieberianus* Nees (Linnaea 9: 286. 1835, nom. nud.) ex C.B.Clarke in Hook.f., Fl. Brit. India 6: 122. 1893.

Annual herbs, culms up to 80 cm high. Leaves shorter than culms, herbaceous; bracts 3-10. Inflorescence open, simple, rays 3-15; spikes cylindrical, spikelets linear to linear-lanceolate; glumes lanceolate-oblong to ovate-oblong. Nuts linear-oblong, straw-coloured.

Rare in marshy areas along streams in Adilabad district.

Fl. & Fr.: August – November.

Jaipur RF (ADB), *PVP* 9437.

Cyperus difformis L., Cent. Pl. II. 6. 1756; C.B.Clarke in Hook. f., Fl. Brit. India 6 : 598. 1893; Fischer, Fl. Madras 3 : 1640. 1931.

Annual, glabrous, tufted herb with many reddish fibrous roots; culms to 40 cm high, weak, triquetrous towards the top. Leaves few, shorter than the stem, acuminate, flaccid, to 13 cm long; sheaths stramineous or brown. Spikes in a simple or compound or contracted into a head, to 1.5 cm across; involucral bracts 3, overtopping, longest to 20 cm long, leaflike; rays of the umbel 3-6, to 3 cm long. Spikelets many in dense heads, dusty or brown, oblong, 10-20-flowered, compressed, rhachilla not winged. Achenes ellipsoid, trigonous, yellow or brown.

Very common species found as a weed in rice fields, and cultivated lands, often associated with *C. haspan*, *C. iria* and *Pycreus sanguinolentus*.

Fl. & Fr.: Throughout the year.

Alampalli (ADB), *TP* & *PVP* 4175; Nirmal (ADB), *MHR* 14413; Mingaram (NZB), *TP* & *BR* 6307; Mombojipally (MDK), *BR* & *CP* 11647; Pocharam sanctuary (MDK), *BR* & *CP* 11610; Mohammadabad RF (RR), *KH* & *MSM* 5584; Pakhal lake (WGL), *MHR* & *MCK* 17076; Godavari river (KRN), *GVS* 22474 (MH & CAL); Narsapur (MDK), *KMS* 62346 (MH & CAL); Lingalacheruvu (KMM), *MHR* & *MCK* 17037; Sagar left canal (KMM), *MHR* & *MCK* 16700; Gandipet (HYD), *KI* & *NVN* 18777; Dhaniyayigudem (KMM), *RR* 108508 (BSID); Lakshmigudem (KMM), *RCS* 104266 (BSID).

Cyperus digitatus Roxb., Fl. Ind. 1: 209. 1820; C.B. Clarke in Hook. f., Fl. Brit. India 6: 618. 1893; Fischer, Fl. Madras 3: 1642. 1931. *C. digitatus* Roxb. var . *hookeri* (Boeck.) C.B.Clarke in Hook. f., Fl. Brit. India 6: 618. 1893.

Perennial, erect, stout, robust herb with woody creeping, rhizome, culms triquetrous, to 100 cm high. Leaves as long as the stem, coriaceous, multistriate, flat, with scaberulous margins; sheaths purplish. Umbel very large, broad, to 15 cm long; involucral bracts 3-12, to 40 cm long, leaflike; primary rays 3-6, rigid, trigonous, to 9 cm long, terminated by stellately spreading sessile or cylindric spikes of unequal length, loosely set all round with innumerable horizontally spreading spikelets; secondary rays 1-3. Spikes digitate, terete, 7-15 in a cluster, to 2.5 cm long, sessile, 10-20-flowered. Spikelets terete, 3 × 0.5 cm, spicate, at right angles to rhachis, acute, yellowish-brown, rhachilla winged. Achene ellipsoid, trigonous, yellowish.

Occasional near water logged places, at the sides of ponds, channels ditches, and in marsh areas in Nizamabad district.

Fl. & Fr.: July – December.

Kotgir (NZB), *BR* & *KH* 9644.

Cyperus distans L. f., Suppl. 103. 1781; C.B.Clarke in Hook.f., Fl. Brit. India 6: 607. 1893; Fischer, Fl. Madras 3: 1640. 1931. var. **distans**

Perennial, erect, robust, leafy herb; culms not tufted, to 100 cm high, trigonous. Leaves flat, as long as stem, scaberulous; sheaths purplish. Inflorescence brown, decompound, to 25 cm long. Involucral bracts 5-7, leaflike, very much elongated, primary rays 7-9, secondary rays 1-5. Spikes ovate, with 15-20 patent spikelets, at right angles to rhachis. Spikelets distant, spreading, narrowly linear, 2 cm long, purplish, 13-15-flowered; rhachilla persistent. Achenes ellispoid, triqetrous, grey apiculate.

Occasional on the margins of ponds, ditches, rivers, rice fields and in wet places.

Fl. & Fr.: September – December.

Ramagiri (KRN), *VSR* 4003 (KU).

Cyperus dubius Rottb. Descr. Icon. Rar. 20. t. 4. f. 5. 1773. *Mariscus dubius* (Rottb.) Kukenth. ex Fischer, Fl. Madras 3: 1644. 1931. *M. dregeanus* Kunth, Enum. Pl. 2: 120. 1837; C.B.Clarke in Hook. f., Fl. Brit. India 6: 620. 1893.

Perennial, tufted herb with short rhizome; culms thickish, stiffly erect, to 20 cm high, triquetrous, with a narrowly ovoidal bulbose enlargement with thickened basal sheaths, 30 mm long, clothed with brown outer sheaths and their dusky brown fibrous remnants. Leaves 7 to a culm; blade linear surpassing the culm, soft, herbaceous, scabrid on upper margins; sheaths to 5 cm long, scarious, whitish. Inflorescence a dense, hemispherical, ovoid, deltoid head, to 12 mm long and as wide. Involucral bracts 4, spreading, the longest 13 cm long; spikes broadly ovoidal, densely bearing 20 spikelets, light greenish; rhachis subulate, 2 mm long. Spikelets patent, ovate, 5 mm long, bearing 6 glumes, 6-flowered, rhachilla straightish; wings lanceolate, white-hyaline. Achenes elliptical, 1.5 mm long, trigonous, apiculate, light brown.

In marshy places in Karimnagar district (Naqvi, 2001).

Fl. & F.: August – December.

Cyperus exaltatus Retz., Observ. Bot. 5: 11. 1789; C.B. Clarke in Hook. f., Fl. Brit. India 6: 617. 1893; Fischer, Fl. Madras 3: 1642. 1931.

Perennial, stout, glabrous, tufted erect herb with a short rhizome; culms to 110 cm high, rigid, trigonous. Leaves folded, to 75 cm long, thick, to 8 mm wide, nerves prominent above, multistriate; sheaths brown to purple, spongy. Umbel compound, to 20 cm long; involucral bracts 3-6, overtopping, longest to 55 cm long, leaf like; primary rays 7-10, unequal, to 20 cm long, spreading; secondary rays 3-5, to 5 cm long, bearing spikes. Spikes umbelled or corymbose, terete, loose, to 4 × 1 cm long, 30-40-spikeletted, rhachis visible, flexuous. Spikelets spicate, at right angles to rhachis, oblong, ascending chestnut brown, much compressed, 1 cm long, acute, 10-20-flowered; rhachilla winged. Achenes ellipsoid, tapering to both ends, trigonous, brown.

Common in wet fields, swamps, marshy areas, along streams and canals.

Fl. & Fr.: August – December.

Sone (ADB), *TP* & *PVP* 4051; Kunthala (ADB), *MHR* 13266; Kalmankalva (RR), *MSM* 10470; Mohammadabad RF (RR), *KH* & *MSM* 10914; Rechapalli (KRN), *GVS* 22271 (MH);

Cyperus halpan L., Sp. Pl. 45. 1753 (*haspan*); C.B.Clarke in Hook. f., Fl. Brit. India 6: 600. 1893; Fischer, Fl. Madras 3: 1640. 1931.

Key to Subspecies

1. Rhizome long, creeping, stem spaced; glumes 1.5 – 1.8 mm; stamens 3 .... subsp. **juncoides**
1. Rhizome short, stem tufted; glumes 1 mm long; stamens 1 .... subsp. **halpan**

Cyperus halpan L., subsp. **halpan**

Perennial, tufted, rhizomatous herb, culms to 50 cm high, acutely triquetrous, leaved at base. Leaves 2 to a culm, blade linear, to 15 cm long, herbaceous, light green; sheaths to 4 cm long, lower ones membranous, red purple. Umbel compound, dense to subloose, to 10 cm long. Involucral bracts 2, the lowest one surpassing the anthela, the longest one to 8 cm long; primary rays to 10, unequal, patent, up to 7 cm long; secondary rays to 2 cm long bearing 3-5 stellately spreading spikelets. Spikelets linear-oblong, to 10 mm long, subdensely to 28-flowered, chestnut brown, the wingless rhachilla hidden by densely disposd glumes. Achenes broadly obovate, trigonous, cream-yellow.

Common weed in rice fields, abundant in wet and marshy areas.

Fl. & Fr.: August – December.

Thornal (MDK), *CP* 11701; Madikonakutta (MDK), *KMS* 6723 (MH).

Cyperus halpan L., subsp. **juncoides** (Lam.) Kuekenth., Feddes Rep. 23: 184. 1926. *C. juncoides* Lam. Tabl. Encycl. 1: 147. 1791. *C. halpan* var. *evoluta* Valck., Sur. Gesl. Cyp. Mal. Arch. 92. 1898. *C. halpan* var. *flaccidissimus* Kuekenth. in Engl., Pflanzenr. 101: 248. 1936.

Perennial herb with long creeping rhizome, along with the culms are disposed in a row in a spaced manner.

Occasionally seen in marshy areas, in wet grass lands.

Fl. & Fr.: September – November.

Nirmal (ADB), *MHR* 14410; Pakhal (WGL), *VSR* & *AR* 1972 (KU); Eturunagaram (WGL), *AR* 1255 (KU); Nellippakka RF (KMM), *RCS* 99090 (BSID).

Cyperus iria L., Sp. Pl. 45. 1753; C.B.Clarke in Hook. f., Fl. Brit. India 6: 606; Fischer, Fl. Madras 3: 1640. 1931. *C. parviflorus* Nees in Wight, Contr. Bot. India 87. 1834 excl. Syn. non Vahl, 1806. *C. iria* var. *parviflorus* (Nees) Miov. in Fl. Ind. Bot. 3: 270. 1859. *C. iria* var *paniciformis* (Franch. & Sav.) C.B.Clarke in Hook. f., Fl. Brit. India 6: 607. 1893.

Annual, glabrous, erect, tufted, marshy herb; culms to 50 cm long, triquetrous, striate, rigid. Leaves flat or channeled, to 30 cm long, 6 mm wide, scaberulous, multistriate, acuminate, flaccid; sheaths purplish. Umbel decompound, 5-15 cm long; primary rays 5-7, to 11 cm long; secondary rays 3-5, to 8 cm long, bearing irregularly fascicled umbellules, formed of narrow interrupted spikes of 15-20, small, few-flowered spikelet. Rhachilla of the spike glabrous. Spikelets spicate, oblong, 8 × 1 mm, to 15-flowered. Achenes obovoid, triquetrous, black or brown.

Common weed in wet and flooded paddy fields.

Fl. & Fr.: July – December.

Vempalli (ADB), *GO* 4446; Ankusapuram (ADB), *GO* 4342; Mingaram (NZB), *TP* & *BR* 6313; Manchippa RF (NZB), *TP* & *BR* 6030; Venkatapuram (NDK), *TP* & *CP* 14005; Mombojipally (MDK), *BR* & *CP* 11663; Mohammadabad RF (RR), *KH* & *MSM* 5572, 5579; Kusumasamudram (RR), *MSM* & *NV* 10377; Lakkavarapucheruvu (KMM), *MHR* & *MCK* 16694; Parnasala RF (KMM), *RCS* 99053 (BSID); Narsapur (MDK), *KMS* 6638 (MH & CAL); Pakhal lake (WGL), *KMS* 13138 (MH & CAL); Jakaram (WGL), RKP 105268 (BSID).

Cyperus laevigatus L., Mant. Alt. 179. 1771. *Juncellus laevigatus* (L.) C.B.Clarke in Hook. f., Fl .Brit. India 6: 596. 1893; Fisher, Fl. Madras 3: 1629. 1931.

Perennial, tufted, rhizomatous herb, rhizome creeping horizontally with solitary distant stems, their bases enclosed by imbricate, shining, chestnut red scales. Culms to 30 cm high, round, flelshy. Leaf blade undeveloped. Spikelets 30 in one appearently lateral head, straw-coloured. Involucral bracts 2, the lower as though a continuation of stem, to 7 cm long, the other much shorter. Spikelets ovate, compressed but thick, straw-coloured dotted with brown spots. Achenes obovoid, brown, stipitate.

Occasional in moist localities, sandy beds of rives and muddy soil.

Fl. & Fr.: September – March.

Puranapul (HYD), *KMS* 9761 (MH & CAL); Yadagirigutta (NLG), *KI* & *NVN* 18744.

Cyperus leucocephalus Retz., Observ. Bot. 5: 11. 1789; C.B.Clarke in Hook. f., Fl. Brit. India 6: 602. 1893; Fischer, Fl. Madras 3; 1640. 1931. *C. coronarius* Kunth, Enum. Pl. 2: 44. 1837; C.B.Clarke in Hook. f., Fl. Brit. India 6: 619. 1893. *Sorostachys leucocephalus* (Retz.) Lye, Nord. J. Bot. 1: 190. 1981.

Perennial, tufted, rhizomatous herb; rhizome very short, woody; culms to 30 cm high, very slender, nodose at the base, each carrying one head. Leaves shorter than the stem. Heads dense, globose, white, containing 40 spikelets; involucral bracts 3, the longest reaching to 6 cm long, long acuminate. Spikelets compressed, elliptic, to 18-flowered. Achenes oblong-ellipsoid, trigonous, shortly apiculate, black.

Occasional in marshy and swampy areas.

Fl. & Fr.: July – October.

Kodimial (KRN), *GVS* 20071 (MH).

Cyperus maderaspatanus Willd., Sp. Pl. 1: 278. 1797. *Mariscus maderaspatensis* (Willd.) Napper, J.E.Africa Nat. Hist. Soc. Natl. Mus. 28(124): 1971.

Perennial erect, tufted herb, 3-16 cm high. Leaves shorter than culms, flat, to 0.25 cm wide, apex acuminate. Involucral bracts 3-8, unequal, over-topping, to 8 cm long; primary rays 3-5, to 1.5 cm long. Spikelets terete, narrowly linear, digitate, compressed, rhachilla persistent, flexuous, winged; glumes ovate-oblong, membranaceous, retuse, aristate, sides nerveless; arista erect, keel brownish, 3-nerved; stamen 1. Nut oblong-ellipsoid, trigonous, apiculate, stalked, stramineous.

Occasional in wastelands in Adilabad and Medak districts (Murthy & *al.*, 2008)

Fl. & Fr.: August – October

Cyperus meeboldii Kuekenth., Feddes Rep. 18. 345. 1922; Hepper in Hutchin. & Dalz. Fl. West. Trop. Afr. 3(2): 393. 1922.

Caespitose annual herbs, to 10 cm high; rhizome tuberous, covered with black fibrous hair-like roots; stems trigonous. Leaves in a whorl, equal or sub-equal to stem; sheaths scarious. Inflorescence spicate with about 20 spikelets clustered into a head, 0.5-1.0 cm across; involucral bracts 3-many, deflexed; rays absent; spikelets oblong to linear, 5 x 1.75 mm; glumes 1.5 mm; keel lip curved; nuts obovoid, trigonous, 0.5 mm.

Occasional in moist gravelly soil, along the banks of rivers and ravines.

Fl. & Fr.: August – December.

It was reported from Hyderabad district as a new record by Rajagopal (1973).

Cyperus michelianus (L.) Delile, Egypte, Hist. Nat. 3: 50. 1813 subsp. **pygmaeus** (Rottb.) Asch. & Graebn., Syn. Fl. Mitteleur 2(2): 312. 1936. *Scirpus michelianus* L., Sp. Pl. 52. 1753; C.B.Clarke in Hook.f., Fl. Brit. India 6: 662. 1893. *Juncellus pygmaeus*

(Rottb.) C.B. Clarke in Hook. f., Fl. Brit. India 6: 596. 1893; Fischer, Fl. Madras 3: 1629. 1931. *Cyperus pygmaeus* Rottb., Descr. Icon. 20. t. 14, f. 5. 1773.

Annual, dwarf, tufted herb; culms to 5 cm high, triquetrous. Leaves several, to 3 cm long, flaccid, scaberulous, green, acuminate; sheaths reddish. Spikelets very many, sessile, densely packed in a terminal head; involucral bracts 6, unequal, to 4 cm long, base dilated, acuminate, leaflike. Spikelets broadly ovate, to 3 mm, pale-brown, 5-10-flowered, compressed. Achenes ellipsoid, trigonous, orange-brown.

Frequent along the sandy or silty beds of rivers, in rice fields as tufted cushions.

Fl. & Fr.: September – March.

Potchera (ADB), *GO* & *DAM* 5082; Laxmapur (NZB), *TP* & *BR* 6280; Venkatammacheruvu-Gadirayal RF (RR), *KH* & *MSM* 5575; Narsapur (MDK) *KMS* 7968 (MH & CAL); Krishna river bank (NLG), *KMS* 9811 (MH & CAL); Kanapur (WGL), *KMS* 11581 (MH & CAL); Pocharam Wild Life Sanctuary (MDK), *RG* 116403 (BSID).

Cyperus mollipes (C.B.Clarke) K. Schum., Pflanzenw. Ost-Afrikas, C: 122. 1895. *Ascopholis gamblei* Fischer, Bull. Misc. Inform. Kew 1931: 105. 1931 & Fl. Madras 3: 1679. 1931. *Mariscus mollipes* C.B.Clarke, Consp. Fl. Afric. 5: 590. 1894.

Perennial, erect, glabrous herb, to 15 cm high; culms bulbously thickened at base, covered with papery, brown leaf sheaths. Spikes sessile, oblong, aggregated in a globose terminal bracteate head; involucral bracts 4. Spikelets spirally imbricate, 1-flowered. Rhachilla disarticulating above the 2 empty glumes, leaving a knob. Achene enclosed in the scale, oblong, dark-brown, sessile, with a beak continous with style, obscurely trigonous, 2.25 × 0.75 mm.

Rare in dry grass lands in Mahabubnagar district.

Fl. & Fr.: August – December.

Shadnagar (MBNR), *SSR* 1858 (HH).

Note: Govaerts (1995), Govaerts and Simpson (2007) and Hoenselaar & *al.* (2010) treat *Ascopholis gamblei* as synonym of *Cyperus mollipes* (C.B.Clarke) K.Schum.

Cyperus niveus Retz., Observ. 5. 12. 1789; C.B.Clarke in Hook. f., Fl. Brit. India 6: 601. 1893; Fischer, Fl. Madras 3: 1640. 1931.

Perennial, glaucescent, rhizomatous herb, 20-70 cm high; rootlets wiry; culms nodose at the base. Leaves much shorter than the stem, to 25 cm long, narrowly linear, setaceo-acuminate. Spikelets 6-16 in one head, 2 cm long, oblong-elliptic, compressed, pale or sometimes almost white, 20-30-flowered, rhachilla not winged. Involucral bracts 2, longest reaching 5 cm long, finely acuminate, leaflike. Achenes obovoid, triquetrous, dark brown.

This is a rare species usually seen in sandy localities of river beds, in gravelly soils and in dry grass lands.

Fl. & Fr.: May – November.

Manchippa RF (NZB), *BR* & *KH* 9035; Nagulabanda (MDK), *CP* 11333; Kodimial (KRN), *GVS* 20049 (MH).

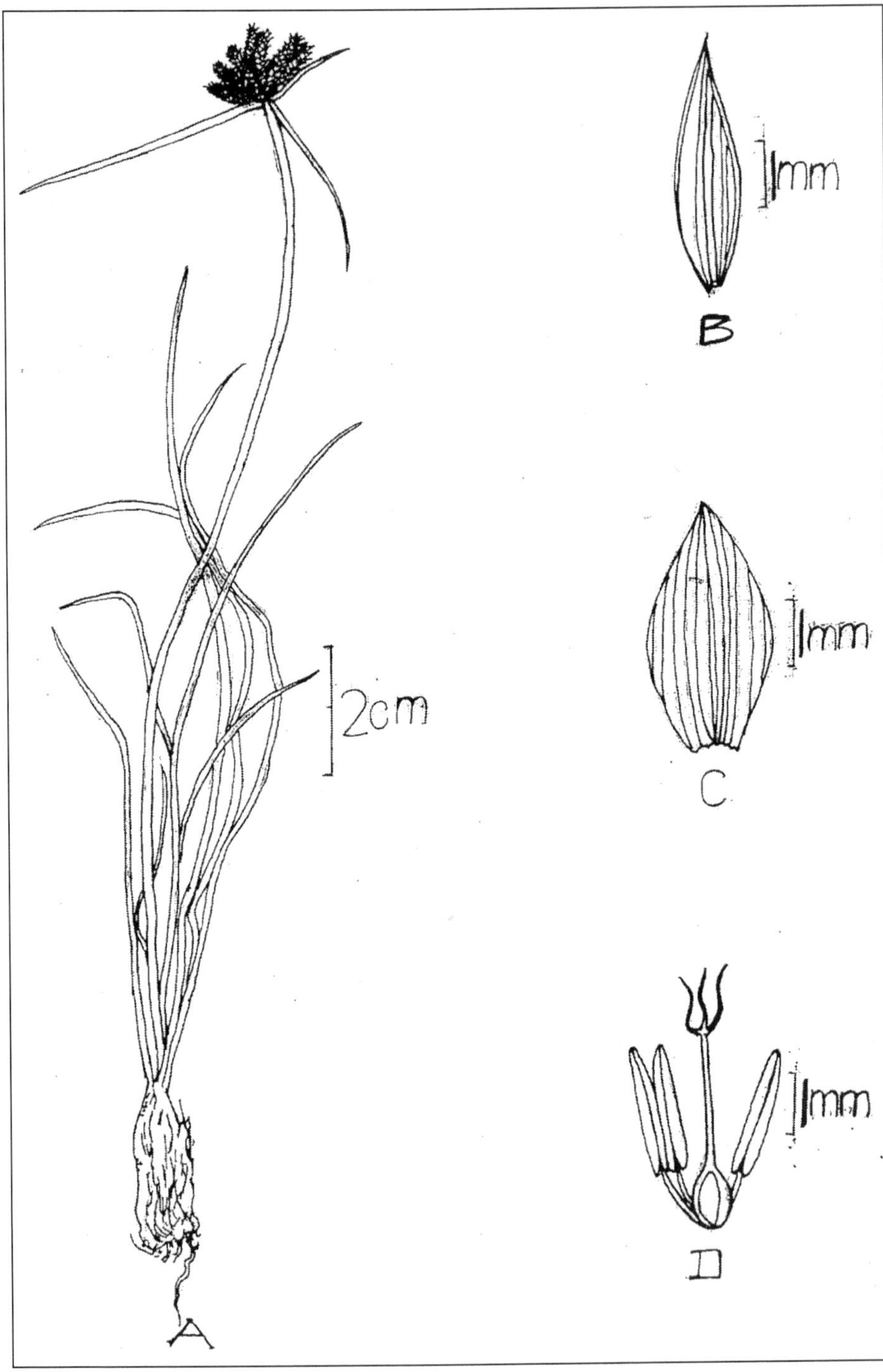

Figure 126: Cyperus niveus Retz.

A. Habit, B. Closed glume, C. Opened glume, D. Pistil with stamens.

Cyperus nutans Vahl, Enum. Pl. 2: 363. 1806; C.B.Clarke in Hook. f., Fl. Brit. India 6: 697. 1893; Fischer, Fl. Madras 3: 1640. 1931. *Cyperus distans* L. var. *major* Thwaites, Enum. Pl. Zeyl. 432. 1864.

subsp. **eleusinoides** (Kunth) T. Koyama, Gard. Bull. Singapore 30: 136. 1977. *Cyperus eleusinoides* Kunth, Enum. Pl. 2: 39. 1937; C.B.Clarke in Hook. f., Fl. Brit. India 6: 608. 1893; Fischer, Fl. Madras 3: 1640. 1931.

Culms to 90 cm high. Leaves as long as the culms, thick. Inflorescence decompound, 15 cm long. Involucral bracts 7, longest to 30 cm long; primary rays 7. Spikes erect, oblong, close, densely spikeletted, bearing spikelets nearly to the base. Achenes obovoid, trigonous 1.2 mm long, more suddenly narrowed to apiculate tip.

In swamps or wet areas, along banks of streams and rivers.

Fl. & Fr.: September – December.

Swarna project (ADB), *MHR* 14570; Nirmal (ADB), *MHR* 14407; Mohammadabad RF (RR), *KH* & *MSM* 10921; Gandipet (HYD), *KI* & *NVN* 18763; Nagarjunasagar (NLG), *KMS* 9757 (MH & CAL); Pagarikutta (MDK), *KMS* 6628 (MH & CAL).

Cyperus pangorei Rottb., Descr. Icon. Rar. 31. 7. f. 3. 1773; Fischer, Fl. Madras 3: 1641. 1931. *C. tegetum* Roxb., Fl. Ind. 1: 211. 1820; C.B.Clarke in Hook. f., Fl. Brit. India 6: 613. 1893. *C. corymbosus* Rottb. var. *pangorei* (Rottb.) C.B.Clarke, J. Linn. Soc. 21: 292. 1884 & in Hook.f., Fl. Brit. India 6: 613. 1893.

Perennial, erect, glaucous green, rhizomatous, robust, marshy herb; culms to 130 cm high, 7 mm wide, rigid, trigonous, ribbed, sulcate. Leaves flat, to 25 cm × 1 cm, acuminate, nerves prominent; sheath long, purplish, lax. Umbel compound, to 10 cm long; involucral bracts 3, unequal, overtopping, longest 15 cm long, erect-patent, with strong midrib and scaberulous margins which are usually recurved in dried specimens; primary rays 5-7, to 7.5 cm long; secondary rays 5, to 2 cm long. Spikes corymbose, 3-10-spikeletted. Spikelets spicate, narrow-oblong, to 3 cm long, purplish brown, acute, compressed; rhachilla with large, dark-brown, deciduous wings. Achenes ellipsoid, trigonous, tapering towards the base, yellowish-brown.

Occasional along the streams in marshy and swampy areas.

Fl. & Fr.: September – December.

Penganga river (ADB), *GO* & *PVP* 4282; way to Kunthala (ADB), *MHR* 13268; Lingalacheruvu (KMM), *MHR* & *MCK* 17042; Parnasala RF (KMM), *RCS* 99051 (BSID); Godavari (KRN), *GVS* 22475 (CAL); Pakhal (WGL), *RKP* 108267 (BSID); Pasra (WGL), *RKP* 105391 (BSID).

Cyperus paniceus (Rottb.) Boeckeler, Linnaea 36: 381. 1870. *Mariscus paniceus* (Rottb.) Vahl, Enum. Pl. 2: 373. 1806; C.B.Clarke in Hook. f., Fl. Brit. India 6: 620. 1893; Fischer, Fl. Madras 3: 1644. 1931. *Kyllinga panicea* Rottb., Descr Ic. Rar. 15, t. 4, f. 1. 1773.

Perennial, rhizomatous, stoloniferous herb; rhizome decumbent, knotty, 2 cm long, clothed with dark red brown fibres, emitting stolons; stolons covered with reddish-brown scales, soon disintegrating into brown fibres, culms to 40 cm high, trigonous. Leaves 4 to a culm; blade nearly equaling the culm, flattish-plicate; sheaths to 5 cm long, reddish-brown. Anthela simple. Umbel of spikes or digitate-contracted, to 3 cm long, light glaucous-greenish. Involucral bracts 7, patent, surpassing the anthela, the longest to 25 cm long. Spikes cylindrical, to 1.5 cm long, densely bearing up to 30 spikelets. Spikelets horizontally divergent, lanceolate, whitish green, slightly recurved, 3 mm long, bearing 4 glumes, and 1-flowered. Achenes elliptic-oblong, trigonous, yellow brown.

Though rare it is sometimes seen growing along forest margins, in damp places preferably under shades.

Fl. & Fr.: July – December.

Kodimal (KRN), *GVS* 20130 (MH); way to Gangaram (WGL), *PVS* 84032 (BSID).

Cyperus pilosus Vahl, Enum. Pl. 2: 354. 1806; C.B. Clarke in Hook. f., Fl. Brit. India 6: 609. 1893; Fischer, Fl. Madras 3 : 1641. 1931.

Perennial, slender, erect, rhizomatous herb; culms to 90 cm high, rigid, triquetorus. Leaves as long as stem, to 35 × 0.8 cm wide, flat or canaliculated, nerves prominent; sheaths pale to purplish. Umbel compound, to 17 cm long; involucral bracts 3-7, overtopping the inflorescence, longest to 20 cm long, primary rays 5-7, to 15 cm long, trigonous; secondary rays 5, 1-5 cm long. Spikes 15-20-spikeletted, at right angles to the rhachis, rhachis densely hispid. Spikelet spicate, oblong, narrow, 7 mm long, compressed, pale to grey-purple, 5-10-flowered; rhachilla hispid. Achenes ellipsoid, triquetrous, dark grey.

A common weed, usually seen in rice fields, river banks, and open wet places.

Fl. & Fr.: May – December.

Cyperus platystylis R. Br., Prodr. 214. 1810; C.B.Clarke in Hook. f., Fl. Brit. India 6: 598. 1893; Fischer, Fl. Madras 3: 1639. 1931.

Perennial, rhizomatous, aquatic herb, to 90 cm high, culms solitary, triquetrous, stiff, scabrid on angles. Leaves few, equaling the culm; blade linear, to 12 mm wide, coriaceous, 1 costate, flattish-plicate, septate, nodulose, scabrous on margins; sheaths light brown. Anthela compound, depressed hemispherical, to 10 cm long, loose. Involucral bracts 10, slightly spaced, spreading, the longest to 70 cm long, to 12 mm wide, primary rays 15, spreading, to 10 cm long, slender and rigid; secondary rays slightly spaced, hence anthelules truly corymbose, to 4 cm long. Spikelets digitate in groups of 8 at apices of raylets, each lance-oblong, flattened, to 2 cm long, densely to 50-flowered, straw-coloured and stained with brown. Achenes oblong-elliptic, trigonous with concave sides, glaucous brownish, the angles markedly spongy thickened, pale brownish.

Usually seen in marshy areas, swamps, and in the margins of ponds in large communities.

Fl. & Fr.: September – December.

Pakhal lake (WGL), *VSR* & *AR* 1190 (KU).

Cyperus procerus Rottb., Descr. Icon. Rar. 29. t. 5. f. 3. 1773; C.B.Clarke in Hook. f., Fl. Brit. India 6: 610. 1893; Fischer, Fl. Madras 3: 1641. 1931.

Perennial, rhizomatous, stoloniferous herb; stolons slender, long, distantly covered with scales; culms to 100 cm high, triquetrous. Leaves several, shorter than the culms; blade to 10 mm wide, sub coriaceous, scabrid margined distally; sheaths long, pale greenish, the basal sheaths stained with dusty brown. Anthela simple, subloose, to 15 cm long. Involucral bracts 4, erect-patent, the lowest one surpassing the anthela, the longest to 50 cm long; primary rays 7, very unequal, erect – patent, to 15 cm long; secondary rays to 2 cm long. Spikes broadly ovoid, to 4 cm long, loose, bearing 8 spikelets, rhachilla glabrous; spikelets oblong, often weakly curved, to 8 cm long, patent subdensely to 40-flowered, pale and tinged with red brown. Achenes elliptical trigonous, brownish.

Usually seen in marshy areas, rice fields, grows with *C. pilosus*.

Fl. & Fr.: August – January.

Pakhal lake (WGL), *KMS* 11680 (MH); Ramancheru, Perugunda (MDK), *RG* 96295 (BSID).

Cyperus rotundus L., Sp. Pl. 45. 1753; C.B.Clarke in Hook. f., Fl. Brit. India 6: 614. 1893; Fischer, Fl. Madras 3: 1641. 1931.

Perennial, glabrous, erect, stoloniferous herb; stolons bearing ovoid, tunicate, black, fragrant tubers; culms to 30 cm high, triquetrous at the top, tuberous at the base. Leaves several, narrowly linear, flat, 10-30 cm long, nerves prominent; sheaths brown. Umbel compound, 4-9 cm long. Involucral bracts 3-5, unequal, longest to 15 cm long; primary rays 4-6, to 6 cm long; secondary rays 3, bearing, short spikes of 4-10 slender, spreading, red-brown spikelets. Spikelets variable in length, linear, red-brown, compressed, spicate, to 3 cm long, pale or purplish, 15-30-flowered; rhachilla winged, pale. Achenes oblong, trigonous, greyish-black.

Very common weed in rice fields and cultivated lands in all districts.

Fl. & Fr.: October – December.

Mallipet (NZB), *BR* & *CPR* 9233; Mombojipally (MDK), *BR* & *CP* 11659; Narsingi (MDK), *BR* & *CP* 11531; Mohammadabad (RR), *KH* & *MSM* 5571; Moosi River Bank (HYD), *KMS* 5926 (MH); Bhupatipur (KRN), *GVS* 20192 (CAL); Amangal to Shadnagar (MBNR), *SRS* 104577 (BSID); Subedari (WGL), *RKP* 105233 (BSID).

Cyperus rubicundus Vahl, Enum. Pl. 2 : 308. 1805; Raynal, Adansonia (ser.2). 17: 45. 1976. *C. teneriffae* Poiret in Lam., Encycl. 7 : 245. 1806; Hook.f., Fl. Brit. India 6 : 601. 1893; C. Fischer, Fl, Madras 3: 1639. 1931.

Small caespitose annual sedge; stems tufted, reddish-brown 3-10 cm high, trigonous. Leaves, closely distichous, half or as long as stem. Inflorescence simple, capitate to 1.5 cm; involucral bracts 3, unequal. Spikelets 5-7, digitately arranged, 2 cm arcoss, stellate, purple, 15-40-flowered; rachilla persistent, corky, wingles; glumes broadly ovate, 7-nerved, margins hyaline, 3 x 2 mm. nut stipitate, obovoid, trigonous, chestnut coloured, 0.8 mm.

A common weed of road sides, old walls and waste places Karimnagar and Hyderabad districts (Naqvi, 2001; Ramana, 2010).

Fl. & Fr.: August-November.

Cyperus squarrosus L., Cent. Pl. 2: 6. 1756. *Mariscus squarrosus* (L.) C.B.Clarke in Hook. f., Fl. Brit. India 6: 623. 1893; Fischer, Fl. Madras 3: 1645. 1931.

Annual, densely tufted herb with reddish purplish fibrous roots, culms to 15 cm high, triquetrous with wing like angles. Leaves 3 to a culm, nearly equaling the culm; blade thinly herbaceous; sheaths to 2 cm long, membranous, pale-greenish and stained with reddish-purple. Anthela simple, capitate with numerous spikelets crowded into globose heads. Involucral bracts 3-5, patent, the lowest two much longer than the anthela, the lowest to 7 cm long. Spikes subglobose, light yellow-green. Spikelets densely spicately disposed on a short rhachis, oblong. Achenes broadly oblong, trigonous, greyish-green.

Common species usually seen in wet sandy soils and open grass lands.

Fl. & Fr.: July – November.

Bellampalli (ADB), *PVP* & *KH* 9446; Nirmal (ADB), *MHR* 14581; way to Potchera (ADB), *MHR* 14488; Narsapur town (MDK), *BR* 112456; Gandipet (HYD), *KI* & *NVN* 18765; Kusumasamudram (RR), *MSM* & *NV* 10379; Sagar left canal (KMM), *MHR* & *KI* 17002; Aklaspur (KRN), *GVS* 24640 (MH); Jakaram (WGL), *RKP* 105315 (BSID).

Cyperus stoloniferus Retz., Observ. Bot. 4: 10. 1786; C.B.Clarke in Hook. f., Fl. Brit. India 6: 615. 1893; Fischer, Fl. Madras 3: 1641. 1931.

Perennial, rhizomatous, stoloniferous herb; rhizome woody, creeping, forming ellipsoid or ovoid enlargement at base of short, clothed with dusky brown parallel fibres. Culms to 30 cm high, obtusely trignous, glaucous green, leaved at base. Leaves rigid, filiform, shorter than the culm; blades linear, to 4 mm wide, folded, recurved; sheaths pale, split into parallel brown fibres. Umbels simple and compact, to 5 cm long; Involucral bracts 3, foliaceous, lowest to 30 cm long, second briefly surpassing the anthela; primary rays 4, unequal, rigid, to 5 cm long. Spikes ovate, to 2 cm long with 3-6 spikelets in short axis. Spikelets linear-oblong, 10 mm long, to 18-flowered, slightly thickened with rather blunt edges, glaucous yellow and stained with sanguineous brown; rhachilla widely winged. Achenes ovate, oval, 3-sided, 1.5 mm long, maturing dark-brown.

Frequent in sandy river banks.

Fl. & Fr.: May – December.

Swarna Project (ADB), *MHR* 14557; Kambalpally (KMM), *MHR* & *MCK* 17052, 17054.

Cyperus tenuiculmis Boeck., Linnaea 36: 286. 1870. *C. zollingeri* Steud.: sensu C.B.Clarke in Hook. f., Fl. Brit. India 6: 613. 1893; Fischer, Fl. Madras 3: 1641. 1931.

Perennial herb with short, corm like, decumbent rhizome, emitting slender stolons, clothed with dark brown scales and their fibrous remnants; culms solitary, stiffy erect, 25 mm high, triquetrous. Leaves few, much shorter than the culm; blade narrowly linear, folded; sheaths light brown. Anthela simple. Involucral bracts 3, leafy, the lowest surpassing the anthela, the second one equaling the anthela; rays erect-patent, slender, unequal. Spikes ovoidal, up to 4 cm long, bearing ca. 12 spikelets. Spikelets linear, nearly quadrangular, straw-coloured, 1.5-2 mm wide; weakly flattened with acute edges, rhachilla dark brown, internodes winged, the wings hyaline, caducous, pale. Achenes elliptic, triquetrous with concave sides, 2 mm long, maturing black.

A very rare taxon seen only in swampy areas.

Fl. & Fr.: September – December.

Bellampalli (ADB), *PVP* & *KH* 9467; Bijjur RF (ADB), *GO* & *PVP* 4878; Mohammadabad RF (RR), *KH* & *MSM* 5582; Thadvai (WGL), *VSR* & *AR* 2043 (KU).

Cyperus tenuispica Steud., Synops. Pl. Glumac. 2 : 11. 1855; Fischer, Fl. Madras 3 : 1640. 1931. *C. flavidus* Retz : sensu C.B.Clarke in Hook. f., Fl. Br. India 6: 600. 1893.

Annual, slender, erect, tufted herb; culms to 20 cm high, triquetrous, finely ribbed. Leaves flat, linear, acute, to 15 cm long, Umbel compound, to 7 cm long; involucral bracts 3, longest to 13 cm long, overtopping; primary rays 7-10, to 6 cm long; secondary rays 5-7, to 1 cm long, bearing heads of 5-10 stellately spreading minute spikelets. Spikelets narrowly linear, to 5 mm long, pale brown, ca. 20-flowered; rhachilla flexuous. Achenes globosely obovoid, trigonous, at first pale-yellow becoming white when ripe.

A very common weed in rice fields, and cultivated low lands. Common in marshy fields and margins of ponds.

Fl. & Fr.: September – February.

Ankusapuram (ADB), *GO* 4299; Gadirayal (RR), *MSM* 10532; Kalmankalva (RR), *MSM* 10453; Sirnapalli (NZB), *BR* & *PSPB* 9578; Appapur (MBNR), *BR* & *BSS* 32296; Pakhal lake (WGL), *MHR* & *MCK* 17063; Nagarjunagar (NLG), *KMS* 9766 (MH & CAL); Way to Madikonakuta (MDK), *KMS* 6723 (MH); Pakhal lake (WGL), *MHR* & *MCK* 17063, *ANH* 15964 (MH & CAL); Dornacheruvu (MBNR), *SRS* 107460 (BSID).

DIPLACRUM R. Br.

Diplacram caricinum R. Br., Prodr. Fl. Nov. Holl. 1. 241. 1810; Fischer, Fl. Madras 3: 1678. 1931. *Scleria caricina* (R. Br.) Benth., Fl. Austral. 7: 426. 1878; C.B. Clarke in Hook. f., Fl. Brit. India 6: 688. 1893.

Annual, small, slender, caespitose herb, to 25 cm high; culms loosely tufted from purplish-reddish fibrous roots, sharply 3-angled, several-noded. Leaves basal and on the culm, to 5 cm long, to 4 mm wide; sheaths to 9 mm long, acutely 3-angled without wings, slightly dilated upwards. Inflorescence of about 20 small globular clusters, one or two at leaf axils of much of the culms, scarcely exerted from the sheath of subtending leaf like bract. Staminate partial spikelet one to a bisexual spikelet, borne at the lowest axil of glumes, to 2 mm long. Achenes tightly enveloped in 2 subtending glumes and falling off with them, globular with about 3 prominent longitudinal ribs, white, irregularly ribbed, slightly hispidulous at apex; disc adnate to the base of achene, obtusely triangular in outline.

This is a rare species collected from hilly areas, moist grassy localities in the forest of Khammam district.

Fl. & Fr.: August – January.

Vazeedu (KMM), *VSR* 927 (MH).

ELEOCHARIS R. Br.

1. Spikelets cylindrical, as wide as the thick culms that range from 3-6 mm in width:
   2. Culms solid, not septate .... **E. acutangula**
   2. Culms hollow, transversely septate .... **E. dulcis**
1. Spikelets ovoidal to ellipsoidal, much wider than slender culms that range from 0.3 – 2.5 mm in width;
   3. Floral glumes spirally imbricate:
      4. Culms 25-60 cm high; hypogynous bristles present:
         5. Stigmas 3; achenes olive-coloured to yellow green at maturity, trigonous .... **E. congesta**
         5. Stigmas 2; achenes black when mature, biconvex:
            6. Hypogynous bristles 4-6, equallling or shorter than achene; stamens 2 .... **E. atropurpurea**
            6. Hypogynous bristles 5-7, longer than the achene; stamens 3 .... **E. geniculata**
      4. Culms 5-25 cm high; hypogynous bristles absent .... **E. setifolia**
   3. Floral glumes 2-ranked .... **E. retroflexa**

Eleocharis acutangula (Roxb.) Schult. in Roem. & Schult., Mant. 2: 91. 1824. *Scirpus acutangulus* Roxb., Fl. Ind. 1 : 216. 1820. *Eleocharis fistulosa* (Poir.) Schult. in Roem. & Schult., Mant. 2: 89. 1824; C.B.Clarke in Hook. f., Fl. Brit. India 6: 626. 1893; Fischer, Fl. Madras 3: 1648. 1931. *Scirpus fistulosus* Poir., Encycl 6: 749. 1804.

Perennial, tufted, aquatic, partly submerged, rhizomatous and stoloniferous herb, to 70 cm high. Culms sharply triquetrous, light green, solid, not septate. Basal sheaths 4, the lower ones scale like, brownish, the uppermost ones to 10 cm long, pale green, obliquely truncate at orifice. Spikelets cylindrical, to 4 cm long, 4 mm wide, subacute at apex, obtusely angular, loosely many-flowered, pale green. Achenes broadly obovate, 2 mm long, compressed-triangular, the apex contracted to an annular neck, the sides yellowish brown, cancelloid with 15 rows of transversely oblong epidermal cells.

Rare in marshy areas near river sides, shallow stagnant waters in ponds.

Fl. & Fr.: June – December.

Pakhal (WGL), *VSR* & *AR* 110 (KU).

Eleocharis atropurpurea (Retz.) Presl., Rel. Haenk. 1: 196. 1828; C.B.Clarke in Hook. f., Fl. Brit. India 6: 627. 1893; Fischer, Fl. Madras 3: 1648. 1931. *Scirpus atropurpurea* Retz., Observ. Bot. 5: 14. 1788.

Annual, erect, tufted, non rhizomatous herb, to 10 cm high; culms terete, capillary, smooth, sulcate. Leaves reduced to bladeless sheaths; sheaths membranous, purple-coloured. Inflorescence with a single spikelet; spikelets globose or ovoid, 3-5 × 2 mm, angular, densely many-flowered. Achenes obovoid, biconvex, to 1 mm long, glossy.

Common in wet fields, marshy areas etc.

Fl. & Fr.: November – December.

Kadam dam (ADB), *TP* & *PVP* 4144; Nirmal (ADB), *MHR* 14580; Potchera (ADB), *GO* & *DAM* 5052; *MHR* 13276; Mohammadabad RF (RR), *KH* & *MSM* 10927; Narsapur (MDK), *KMS* 6718 (MH & CAL); Pakhal RF (WGL), *VSR* & *AR* 115 (KU); Pocharam Wild Life Sanctuary (MDK), *RG* 116401 (BSID).

Eleocharis congesta D. Don, Prodr. Fl. Nepal 41. 1825; Clarke in Hook. f., Fl. Brit. India 6: 630, 1893; Fischer, Fl. Madras 3: 1648. 1931. subsp. **japonica** (Miq.) T. Koyama in Fl. Taiwan 5: 220. t. 1324. 1978 & in Dassan. & Fosb., Rev. Fl. Ceylon 5: 264. 1985. *E. japanica* Miq., Ann. Mus. Bot. Lugduno-Batavum 2: 142. 1865. *E. atropurpurea* Trimen, Handb. Fl. Ceylon 5: 71. 1900. non Kunth; C.B.Clarke in Hook. f., Fl. Brit. India 6: 627; Fischer, Fl. Madras 3: 1648. 1931.

A perennial, rhizomatous herb, culms slenderly filiform, up to 40 cm tall. Leaves purplish. Spikelets ovate, densely many-flowered, 2 mm long, 1.8 mm wide; glumes 1.5 mm long, basal most glume empty, amplexicaul; perianth bristles 6, longer than nut; stamen 1; stigmas 3. Achenes obovoid, trigonous, smooth, black, 1 mm long.

Common in ponds and ditches in Ranga Reddi district.

Fl. & Fr.: October – January.

Mohammadabad RF (RR), *KH* & *MSM* 10920.

Eleocharis dulcis (Burm. f.) Trin. ex Hensch. in Vita Rumph. 186. 1833. *Andropogon dulce* Burm. f., Fl. Ind. 219. 1768. *Eleocharis plantaginea* (Retz.) Roem. & Schult., Syst. Veg. 2: 150. 1817; C.B.Clarke in Hook. f., Fl. Brit. India 6: 625. 1893; Fischer, Fl. Madras 3: 1647. 1931. *Scirpus plantagineus* Retz., Observ. Bot. 5, 14. 1799.

Perennial, tufted, partly submerged, stoloniferous herb; stolons terminated by a small tuber; culms to 90 cm high, 7 mm thick, hollow, transversely septate, the inter septa to 12 mm long, empty, deeply green, somewhat shiny. Basal sheaths to 20 cm long, coloured with red-brown, truncate at orifice. Spikelet cylindrical, to 5 cm long, 4 mm wide, subobtuse at apex, terete, whitish to pale stramineous. Achenes obovate, to 3 mm long, biconvexed with obtuse edges, yellow, shiny.

Occasional in marshy areas, stagnant or slowly moving water.

Fl. & Fr.: September – December.

Pakhal (WGL), *KMS* 13111 (MH & CAL);

Eleocharis geniculata (L.) Roem. & Schult., Syst. Veg. 2: 150. 1817. *Scirpus geniculatus* L., Sp. Pl. 48. 1753. *Eleocharis capitata* R., Br., Prodr. 225. 1810; C.B.Clarke in Hook. f., Fl. Brit. India 6 : 627. 1893; Fischer, Fl. Madras 3 : 1648. 1931. non *Scirpus capitatus* L. 1753.

Annual, erect, caespitose herb, to 30 cm high; culms sleder, angular, sulcate, slightly glaucous green. Sheaths herbaceous, pale green, often tinged with red brown on lower part, the mouth oblique, acute. Spikelets ovoid-globose to ovoid-ellipsoidal, 5 × 3 mm, rounded at base, very obtuse at apex, reddish-ferrugineous, densely many-flowered. Achenes globosely obovoid, brown, shining, apiculate with broad style base.

A very common species usually on wet fields, marshy areas, including rice fields.

Fl. & Fr.: June – October.

Kadam Dam (ADB), *TP* & *PVP* 4144; Nirmal-Adilabad 20kms (ADB), *MHR* 14497; Sirnapalli river (NZB), *BR* & *GO* 9081; Pochammaralu (MDK), *BR* & *CP* 11473; Pocharam tank (MDK), *TP* & *CP* 12046; Anantasagar RF (RR), *KH* & *MSM* 10929; Gandipet (HYD), *KI* & *NVN* 18776; Kampally cheruvu (KMM), *MHR* & *MCK* 17047; Avatavarlapalli (MBNR), *BSS* & *SKB* 33119; Aklaspur (KRN), *GVS* 22538 (MH); Ramavaram (KMM), *RR* 112575 (BSID);

Eleocharis retroflexa (Poir.) Urban, Symb. Antil. 2: 165. 1900. subsp. **chaetaria** (Roem. & Schult.) T. Koyama, Bull. Natl. Sci. Mus. 17: 68, f. 2, 1974. *Scripus retroflexus* Poir. in Lam., Encycl. Meth. Bot. 6 : 753. 1804. *Eleocharis chaetaria* Roem. & Schult. Syst. Veg. 2 : 154. 1817; C.B.Clarke in Hook. f., Fl. Brit. India 6 : 629. 1893; Fischer, Fl. Madras 3 : 1648. 1931. *Scirpus chaetaria* Thwaites, Enum. Pl. Zeyl. 435. 1864.

Annual, densely tufted, erect herb, to 15 cm high, culms filiform, recurved, angular, light green, clothed at base with 2 sheaths. Sheaths pale, the orifice hyaline, obliquely truncate, lanceolate. Spikelets broadly ovate at maturity, to 5 mm long, bearing 8 distichously disposed flowers, sanguineous-tinged. Achenes broadly obovate, contracted to apex, rounded-truncate at apex, trigonous with ribbed edges, the sides lightly yellow, trabeculate with 9 rows of isodiametrical thick-walled cells, rather deeply pitted.

Common species found in marshy areas, near river sides margins of rice fields, wet sandy edges of ponds in Khammam district.

Fl .& F.: October – December.

Vazeedu (KMM), *VSR* 1010 (KU).

Eleocharis setifolia (A. Rich.) Raynal, Adansonia II, 7: 318. 1967. *Isolepis setifolia* A. Rich., Tent. Fl. Abyss. 2: 498. 1851. *Eleocharis minutiflora* Boeck. in Engl. Bot. Jahrb. 7: 274. 1886. *E. nigrescens* var. *minutiflora* (Boeck.) Sevenson, Rhodora 39: 226: 1937.

Annual, tufted, slender, glabrous herb, to 25 cm high. Culms filiform, quadrangular, sulcate. Sheaths 2, tubular, whitish to light brown, apex acute. Spikelets ovate, 3 mm long. Achenes obovate, obscurely trigonous, light brown.

Though rare, this species is found along the margins of the lakes, marshy places, in shallow puddles etc.

Fl. & Fr.: August – December.

Pakhal lake (WGL), *AR* 2061 (MH & KU).

Eleocharis zatei W.Khan & L'narasimhan, Bioinfolet 8: 5. 2011.

Leafless, tufted annual, 10-20 cm high, glabrous all over, stems terete, 0.5-1 mm thick, lowest sheaths often purplish, upper sheath-mouth oblique with ca 1 mm long limb. Inflorescence of single, terminal spikelet; involucral bracts 2-3, glume-like (empty glumes). Spikelets ovoid or ovoid-oblong (rarely longer than broad), densely many flowered, usually whitish, sometimes tinged with brown, 3-6 × 2-3 mm, wider than stems; rhachilla wingless, rather thick. Glumes spiral, tightly imbricated, thinly membranous, oblong-ovate, ca 1.5-1.7 × 0.8 mm, acutish, white scarious with fine, close, many visible nerves on sides, rarely with purplish vertical band along keel, broadly hyaline margined, keeled; keels 3-nerved (apparently uninerved) reaching the apex; stamens 2 or 1, anthers linear, 0.3-0.4 mm long, minutely appendiculate; styles 2-fid, glabrous, 0.4-0.5 mm long, almost as long as stigmas, stigmas glabrous or glabrescent, filiform, blackish above the persistent base. Nuts biconvex, ellipsoid to narrowly obovoid, 0.7-0.8 × 0.3-0.4 mm, olive-green or pale-greenish-brown, smooth, shining (never maturing black); hypogynous bristles 6-8, glistening white, transluscent, sparsely barbellate, much shorter than nuts, connate at base, the persistent style-base at nut-apex often apiculate.

Common in rice fields, swamps, ponds, ditches along road sides, often standing in water in Andhra Pradesh (Wadoodkhan & *al.*, 2011).

Fl. & Fr.: November – March.

Andhra Pradesh, *Solanke* 1375 (HCMCM).

FIMBRISTYLIS Vahl

1. Glumes on spikelets distichous:
   2. Inflorescence reduced to a single spikelet; involucral bracts glume lime .... **F. ovata**
   2. Inflorescence with 2-several spikelets; involucral bracts setaceous .... **F. fusca**
1. Glumes on spikelets spiral:
   3. Achenes sub-orbicular, obovate; stigmas 3 or 2:

4. Stigmas 3; achenes trigonous:
   5. Leaves surrounding culm bases bladed:
      6. Leaves with a ligule of a series of pubescence:
         7. Perennial rhizomatous herb .... **F. complanata**
         7. Annual tufted herb without rhizome
            8. Style conspicuously protruded .... **F. arnottiana**
            8. Style not protruding .... **F. microcarya**
      6. Leaves without ligule:
         9. Glumes atleast lower ones distichous; spiklelets strongly compressed:
            10. Inflorescence of 1-3 spikelets; nuts 2-3 mm long, coarsely trabeculate.............................. **F. woodrowii**
            10. Inflorescence of several to numerous spikeleted; nuts 1 mm long, smooth or verrucose .... **F. tenera**
         9. All glumes spirally imbricate; spikelets terete or angular, margins of glumes hyaline or scarious .... **F. falcata**
   5. All or upper few leaves surrounding culm bases blade less:
      11. Leaf blade dorsiventral; culms pentagonous . **F. quinquangularis**
      11. Leaf blade laterally flattened, culms compressed, tetragonal .... **F. littoralis**
4. Stigmas 2; achenes biconvex: (occasionally 3 in *F. cymosa*)
   12. Inflorescence corymbose with more than 2 rays and 5 spikelets:
      13. Styles hardly flattened, glabrous .... **F. cymosa**
      13. Styles dorsiventrally flattened, fimbriate or long ciliate atleast on upper part:
         14. Leaves with a ligule, usually a fringe of dense pubescence:
            15. Glumes pubescent on upper half:
               16. Lowest involucral bract shorter than inflorescence; leaves less than 10 cm long .... **F. ferruginea**
               16. Lowest involucral bract exeeding the inflorescence, leaves 15-40 cm long:
                  17. Perennials with decumbent rhizome; involucral bracts 3-5; stamens 2 .... **F. sieberiana**
                  17. Annuals without rhizome; involucral bracts 1-5; stamens 3 .... **F. pubisquama**

15. Glumes glabrous:
    18. Spikelets 1-1.8 mm wide, oblong .... **F. bisumbellata**
    18. Spikelets 2-4 mm wide, ovate:
        19. Leaves half as long as the culms; involucral bracts 2; spikelets whitish green .... **F. alboviridis**
        19. Leaves as long as the culms; involucral bracts up to 6; spikelets brown:
            20. Nuts with 5-10(12) vertical ridges or rows of thick walled transversely oblong (bar-like) epidermal cells, *i.e.*, coarsely trabeculate, 0.8-1.2 mm long; glumes nerveless on sides .... **F. dichotoma**
            20. Nuts with 15-24 fine close vertical striations or rows of thin walled epidermal cells, smoothly reticulate or subtrabeculate, usually large, 1.5-2.2 mm long; glumes distinctly nerved on sides .... **F. naikii**

14. Leaves without a ligule .... **F. aestivalis**

12. Inflorescence head like or consisting of 1-4 solitary spikes only:
    21. Inflorescence head like with many spikelets; style glabrous .... **F. argentea**
    21. Inflorescence with 1-4 solitary spikes; styles flimbriate (except in *F. acuminata*):
        22. Inflorescence with 1-3 spikelets; leaf blade well elongated .... **F. schoenoides**
        22. Inflorescence with a single spikelet; leaves bladeless:
            23. Spikelets obtuse at apex; laf blades occasionally remaining as a setaceous appendages of sheath .... **F. polytrichoides**
            23. Spikelets acute at apex; leaf blades always none .... **F. acuminata**

3. Achenes oblong-cylindrical; stigmas 2:
    24. Inflorescence corymbose with few to several small spikelets; spikelets 6 mm long .... **F. dipsacea**
    24. Inflorescence of a single terminal spikelet; spikelets 12 mm long .... **F. tetragona**

Fimbristylis acuminata Vahl, Enum. Pl. 2: 285. 1806; Fischer, Fl. Madras 3: 1658. 1931. *F. setacea* C.B.Clarke in Hook. f., Fl. Brit. India 6: 631. 1893; *p.p.,* non Benth.

Annual, densely tufted, rhizomatous herb; culms slender, to 35 cm high, obtusely trigonous, glaucous green, clothed at base with 3 bladeless sheaths only. Basal sheaths 1 cm long, cylindrical, membranous, pale green, the upper most sheath to 4 cm. Inflorescence a single terminal spikelet with a scale like bract. Spikelet ovate, terete, acute, to 12 mm long, greenish white, tightly ca. 15-flowered. Achenes obovate-orbicular, biconvexed, maturing whitish or cream-coloured, transversely coarsely wrinkled with 8 ridges.

Occasional in marshy palces in Adilabad district.

Fl. & Fr.: October – November.

Jaipur RF (ADB), *PVP* 9428.

Fimbristylis aestivalis (Retz.) Vahl, Enum. Pl. 2: 288. 1806; C.B.Clarke in Hook. f., Fl. Brit. India 6: 637. 1893. *Scirpus aestivalis* Retz., Observ. Bot. 4. 12. 1786.

Annual, densely tufted, small erect herb, to 10 cm high; culms trigonous, light green, few leaves at base. Leaf blade much shorter than the culm, flattish with incurved margins, light green, subdensely to sparsely pilose on both surfaces with spreading hairs; sheaths to 15 mm long, pale greenish to yellow-brownish, subdensely pubescent with spreading hairs, the ventral side hyaline, light brown; orifice obliquely truncate. Corymbs compound, to 4 cm long; rays 7, filiform, patent; involucral bracts 3, filiform. Spikelets solitary, lance-ovate, subterete, to 6 mm long, densely to 40-flowered. Achenes obovate, thickly biconvexed, rounded at apex, short-stipitate cuneate base, yellowish at maturity and smooth.

Occasional in wet or moist sandy low lands and as a weed in rice fields.

Fl. & Fr.: January – April.

Penganga river bed (ADB), *GO* & *PVP* 4275; Potchera (ADB), *GO* & *DAM* 5055, 5057; Medak (MDK), *BR* & *CP* 11401, 11459; Ramayampet (MDK), *BR* & *CP* 11509.

Fimbristylis alboviridis C.B.Clarke in Hook. f., Fl. Brit. India 6: 638. 1893. *F. annua* Roem. & Schult. var. *pluristriata* Back. & Bakh., Fl. Java (em. ed.) 101: 121. 1949. fam. 246. p. 21. (non *F. diphylla* Vahl var. *pluristriata* C.B.Clarke. 1893).

Annual, erect, tufted herb, to 20 cm high; culms slender, compressed, obtusangular. Leaves half as long as the culms, glabrous, scabrid on the margins in the upper part, greyish green; ligule a fringe of short white hairs. Inflorescence simple, loose, with several spikelets, 2 cm long; involucral bracts 2, suberect, the lowest somewhat shorter than the inflorescence, spikelets solitary, ovoid, terete, acute, densely many-flowered, whitish-green, often brownish variegated; rhachilla narrowly winged. Achenes biconvex, obovate, shortly stipitate, obsoletely reticulate by the transversely elliptic not impressed epidermal cells in 15 vertical rows on either face, stramineous.

Rarely seen in dry grass lands and grassy slopes in Adilabad and Ranga Reddi districts.

Fl. & Fr.: October – December.

Gadirayal (RR), *MSM* & *KH* 10529; Kunthala way (ADB), *MHR* & *KI* 13278 .

Fimbristylis argentea (Rottb.) Vahl, Enum. Pl. 2: 294. 1806; C.B.Clarke in Hook. f., Fl. Brit. India 6: 640. 1893; Fischer, Fl. Madras 3: 1659. 1931. *Scirpus argenteus* Rottb., Descr. Pl. Rar. 27. 1777 & Descr. Icon. Rar. 51. t. 17. f. 6. 1773.

Annual, caespitose, glaucous herb, to 10 cm high; culms filiform, trigonous, striate. Leaves filiform, to 5 cm long; ligule 0; sheaths stramineous. Inflorescence capitate, 5-12 in a cluster. Involucral bracts 4, overtopping, longest to 3 cm long. Spikelets sessile, cylindric-oblong, terete, obtuse, many-flowered, 3 mm long. Achenes obovoid, compressed, biconvex, fulvous.

Occasional on wet sandy grounds of open grasslands and rice fields.

Fl. & Fr.: September – December.

Mamidipalli (NZB), *BR* & *PSPB* 9600; Mohammadabad RF (RR), *KH* & *MSM* 5574; Kusumasamudram (RR), *MSM* & *NV* 10378; Pakhal lake (WGL), *MHR* & *MCK* 17071, 17081.

Fimbristylis arnottiana Boeck., Linnaea 37: 28. 1871; C.B.Clarke in Hook. f., Fl. Brit. India 6: 643. 1893; Fischer, Fl. Madras 3: 1659. 1931.

Annual, tufted herb, to 30 cm high; culms 4-5 angular under umbel. Umbel decompound, glabrous with 30 spikelets. Involucral bracts 4; spikelets obtuse, 10-fid; styles conspicuously protruded, cuboid-ellipsoid, one-coloured. Achenes small, obovoid, yellow-brown, transversely lineate, obscurely tuberculate.

Rarely seen in wet grassy lopes, open grass lands etc. in Mahabubnagar district (Raghava Rao, 1989).

Fl. & Fr.: October – November.

Fimbristylis bisumbellata (Forssk.) Bubani, Dodecanthea 30. 1850; Fischer, Fl. Madras 3: 1658. 1931. *Scirpus bisumbellata* Forssk., Fl. Aegypt. – Arab. 1: 15. 1775. *Fimbristylis dichotoma* C.B.Clarke in Hook. f., Fl. Brit. India 6: 636. 1893.

Annual, erect, tufted herb; culms to 30 cm high, trigonous. Leaves flat, to 12 cm long, pubescent; ligule hairy; sheaths membranous. Umbels compound, terminal, 5 × 5 cm. Involucral bracts 3, longest to 5 cm long; primary rays 3-5, to 3 cm long; secondary rays 3-5. Spikelets solitary, oblong, 7 × 1.8 mm, many – flowered. Achenes obovoid, biconvex, creamy-yellow.

This is common species usually found in rice fields, wet places, marshy areas.

Fl. & Fr.: September – December.

Alampalli (ADB), *TP* & *PVP* 4173; Sone (ADB), *TP* & *PVP* 4030; Potchera (ADB), *GO* & *DAM* 5057; Narsapur (MDK), *KMS* 7970 (MH & CAL); Akkannapet (MDK), *BR*

& *CP* 11560; Mohammadabad RF (RR), *KH* & *MSM* 5578, 5580; Lambaditanda (RR), *KH* & *MSM* 5586; Sagar left canal (KMM), *MHR* & *MCK* 17029; Krishna river bank (NLG), *KMS*, 9807 (MH & CAL); Puranapul (HYD), *KMS* 5996 (CAL); Eturnagaram (WGL), *RKP* 110839 (BSID); Bhadrachalam RF (KMM), *RCS* 99034 (BSID); Karkalpadu, Amangal RF (MBNR), *SRS* 104547 (BSID).

Fimbristylis complanata (Retz.) Link, Hort. Berol. 1: 292. 1827; C.B.Clarke in Hook. f., Fl. Brit. India 6: 646. 1893; Fischer, Fl. Madras 3: 1659. 1931. *Scirpus complanatus* Retz., Observ. Bot. 5; 14. 1791.

Perennial, tufted, rhizomatous herb, to 40 cm high; culms densely caespitose, apex compressed, base trigonous. Leaves crowded at the base, 20 cm long, coriaceous, glaucous, ligule shortly hairy; sheaths compressed, keeled. Umbels decompound, effuse, 6 cm long; involucral bracts 2, not overtopping, to 6 cm long, largest leaf like; primary rays 3-5, to 3 cm long; secondary rays 3-4. Spikelet solitary, ovoid-oblong, to 8 mm long, acute, ca 12-flowered, brown with distichously arranged glumes. Achenes obovoid, trigonous, stipitate and stramineous.

This species is usually found in wet or moist places, on grassy slopes, river banks, open grass lands at low altitudes, in cultivated lands and in forest margins.

Fl .& Fr.: August – April.

Ankusapuram (ADB), *GO* 4343; Swarna project (ADB), *MHR* 14564; Gangapur RF (MDK), *TP* & *CP* 12096; Tandur (RR), *MSM* 12704; Pakhal lake (WGL), *MHR* & *MCK* 17083; Rayagiri cheruvu (NLG), *KI* & *NVN* 18748; Katasintha (HYD), *KMS* 601& (MH & CAL); Kodimial (KRN), *GVS* 20110 (MH).

Fimbristylis cymosa R. Br., Prodr. Pl. Nov. Holl. 228. 1810. subsp. **spathacea** (Roth) T. Koyama, Micronesica 1: 83. 1964 & in Dassan. & Fosb., Rov. Handb. Fl. Ceylon 5: 301. 1985. *F. spathacea* Roth, Nov. Sp. Pl. 24. 1821; C.B.Clarke in Hook.f., Fl. Brit. India 6: 640. 1893; Fischer, Fl. Madras 3: 1659. 1931.

Perennial, erect, tufted, rhizomatous herb, to 40 cm high; culms densely leafy below, obtusely trigonous. Leaves densely crowded on the root stock, spreading and recurved, narrowly linear, to 15 cm long, rigidly coriaceous, margins recurved and scaberulous, sheaths short, coriaceous. Umbels compound, to 6 cm long; involucral bracts 3, filiform, to 2.5 cm long, leaf like, with broadly dilated base; primary rays 3-5, to 4 cm long; secondary rays 3, spikelet globose, to 3 mm long, pale brown, rhachilla winged. Achenes obovoid, biconvex, turbinate, compressed.

Widely distributed species. Usually seen in sandy areas near sea shores, river beds and marshy low lands.

Fl .& Fr.: March – October.

Swarna project (ADB), *MHR* 14561; Rangammagudem beat (RR), *KH* & *MSM* 5585; Gacchibowli (RR), *MSM* 15196; Kampalli cheruvu (KMM), *MHR* & *MCK* 17057.

Fimbristylis dichotoma (L.) Vahl, Enum. Pl. 2: 287. 1806; C.B.Clarke in Hook. f., Fl. Brit. India 6: 635. 1893; Fischer, Fl. Madras 3: 1658. 1931. *Scirpus dichotomus* L., Sp. Pl. 50. 1753.

Key to Subspecies

1. Leaves and bracts densely pubescent .... subsp. **podocarpa**
1. Leaves and bracts glabrous .... subsp. **dichotoma**

Fimbristylis dichotoma (L.) Vahl, subsp. **dichotoma**

Annual, tufted herb in a small clump; culms to 40 cm high, trigonous, leaves at base. Leaves to 10, as long as the culms; blade erect, patent, linear, to 30 cm high, herbaceous, glaucous green beneath, whitish green on upper surface, several veined; ligule of a fringe of dense white hairs; sheaths cylindrical with rounded back, the dorsal side pale greenish, the hyaline ventral side lightly brown, Umbels compressed, bearing ca 6 spikelets, lax to subdense, to 7 cm long. Involucral bracts to 6, the lowest one foliaceous, remaining setaceous; rays to 5, unequal, to 3 cm long, compressed, patent, terminated by 3 spikelets. Spikelets solitary, ovoid, to 7 mm long, terete, chestnut brown, densely many-flowered. Achenes obovate, biconvex, 1.8 mm long, excluding gynophore, rounded-truncate at mucronate apex, each side cancellated with 7-13 rows of transversely oblong cells.

Common along the sides of water courses, open grassy places, sandy localities and as a weed in cultivated lands.

Fl. & Fr.: March – December.

Way to Kunthala (ADB), *MHR* 13272, 13278; Pakhal lake (WGL), *MHR* & *MCK* 17080; Pullayapalli (MBNR), *BSS* & *SKB* 33116; Sagar left canal (KMM), *MHR* & *MCK* 17005, 17018; Narsapur (MDK), *KMS* 6634 (MH & CAL); Rathamhutta hills (KMM), *RCS* 104229 (BSID); Lawal beat, Pasra (WGL), *RKP* 108167 (BSID).

Fimbristylis dichotoma (L.) Vahl var. **dichotoma** forma **apiculata** W. Khan, Bioinfolet 6: 90. 2009.

Very similar to forma *dichotoma* in many respects and distinguished from the latter in having compound inflorescence, the single stamen, shortly but distinctly beaked nuts and very slender styles.

Rare in Adilabad district

Mahadev temple (ADB), *S.N.Solanke* 568 (HCNCM).

Fimbristylis dichotoma (L.) Vahl subsp. **podocarpa** (Nees & Meyen) T. Koyama, Micronesica 1: 87. 1964 & in Dassan & Fosb., Rev. Handb. Fl. Ceylon 5: 310. 1985. *F. podocarpa* Nees & Meyen ex Nees in Wight, Contr. Bot. Inda 98. 1834. *F. diphylla* Vahl var. *pluristriata* C.B.Clarke in Hook. f., Fl. Brit. India 6: 637. 1893. *F. annua* Roem. & Schult. var. *podocarpa* (Nees & Meyen) Kukenth., Bot. Jahrb. Syst. 59: 48. 1924. *F. diphylla* Vahl var. *podocarpa* (Nees & Meyen) Kukenth., Bot. Jahrb. Syst. 69: 257. 1938.

Leaves and involucral bracts densely pubescent. Achenes orbicular – obovate, thickly biconvexed, 1.5 mm long excluding gynophore, 1 mm wide, mucronate at apex, contracted at base; tuberculate on shoulder, cream-coloured, each side cancellated by 25 series of trasversely oblong hexagonal cells, gynophore obdeltoid.

Wet places, moist fields and in grass lands.

Fl .& Fr.: March – November.

Way to Kunthala (ADB), *MHR* 13285; Kanmankalva (RR), *MSM* 10455.

Fimbristylis dipsacea (Rottb.) C.B.Clarke in Hook. f., Fl. Brit. India 6: 635. 1893; Fischer, Fl. Madras 3: 1568. 1931. *Scirpus dipsaceus* Rottb., Descr. Ic. Rar. 56. t. 12. f. 1. 1773.

Annual, densely tufted herb, to 15 cm high, culms slender, trigonous, clothed at base with about 3 sheaths. Leaves reduced to short-bladed sheaths, 4 to a culm; blade 5 cm long, flattish with incurved margins, herbaceous, lightly green; sheaths to 3 cm long, cylindrical, pale green on dorsal side, the hyaline ventral side lightly brown, obliquely truncate at pale hyaline orifice, the lowest 2 sheaths scale like, brown. Corymbs simple, of a cluster of 3 spikelets, to 4 cm long, bearing ca 20 spikelets. Involucral bracts 7, filiform, 3 longer than the inflorescence; rays 12, subequal, to 3 cm long, each bearing one spikelet. Spikelets globose, to 6 mm long, squarrose, wholly greenish, densely many-flowered. Achenes linear-oblong, thickly biconvexed, cream-coloured, with clavate out growths, stipitate.

Common in sandy soils, in the sea shore, river beds and in sandy fields forming cushions.

Fl. & Fr.: April – August.

Narsapur (MDK), *KMS* 7974 (MH & CAL).

Fimbristylis falcata (Vahl) Kunth, Enum. Pl. 2: 239. 1837. *Scirpus falcatus* Vahl, Enum. Pl. 2: 275. 1806. *Trichelostylis junciformis* Nees in Wight, Contrib. Bot. India 106. 1834. *Fimbristylis junciformis* (Nees) Kunth, Enum. Pl. 2: 239. 1837; C.B.Clarke in Hook. f., Fl. Brit. India 6: 647. 1893; Fischer, Fl. Madras 1660. 1931.

Perennial, erect, rhizomatous herb; rhizome oblique, to 7 cm long, thickly clothed with brown leaf sheaths, culms solitary or sparsely tufted, rigid, 4-angled, to 40 cm high, deeply striate. Leaves crowded round the base of the stems, shorter than stems, spreading and recurved with triangular acute tip, to 20 cm long; sheaths short, ligule 0. Umbel compound, at times capitate, to 5 cm long, bearing distant or clustered small chestnut-brown spikelets. Involucral bracts 4, shorter than umbel, to 3.5 cm long; primary rays 3, slender, to 2 cm long; secondary rays 2. Spikelets ellipsoid, solitary, 5 mm long, purplish, rhachilla winged. Achenes obovoid, trigonous, stramineous.

Occasional in marshy localities.

Fl. & Fr.: June – November.

Vempalli (ADB), *GO* & *PVP* 4795; Vatavarlapalli (MBNR), *VBH* 84967 (BSID).

Fimbristylis ferruginea (L.) Vahl, Enum. Pl. 2: 291. 1806; C.B.Clarke in Hook. f., Fl. Brit. India 6: 638, 1893; Fischer, Fl. Madras 3: 1658. 1931. *Scirpus ferrugineus* L., Sp. Pl. 50. 1753.

Perennial, tufted, erect, rhizomatous herb; to 140 cm high; culms sub trigonous, rigid, to 7 mm wide. Leaves less than 10 cm long, scabrid, apex acuminate; ligule a dense fringe of short hairs; sheaths fuscous, margin hyaline. Umbel sub-compound of few spikelets, to 5 cm long. Involucral bracts 2, lowest involucral bract shorter than inflorescence. Spikelets solitary, ovoid or oblong, pale reddish brown, pubescent, to 2 cm long, lowest spikelet bractiform; rhachilla angular and winged. Achenes obovoid, biconvex, stipitate, stramineous.

A common species usually found in wet places, open swamps and marshy areas.

Fl. & Fr.: August – December.

Sone (ADB), *TP* & *PVP* 4048; Rushtempet (MDK), *KMS* 7976, (MH & CAL); Gacchibowli (RR), *MSM* 15191; Upparapalli cheruvu (WGL), *MHR* & *MCK* 17062; Katasintha (HYD), *KMS* 6014 (CAL).

Fimbristylis fusca (Nees) C.B.Clarke in Hook. f., Fl. Brit. India 6: 649. 1893; Wadoodkhan & Solanke, Bioinfolet 5: 104. 2008. *Abildgaardia fusca* Nees, Contr. Bot. India 95. 1834.

Perennial with short woody rhizome, clothed with fibrous remains of old sheaths; stems tufted, 30-50 cm high, slender but rigid, sulcate, angular, glabrous or scaberulous below inflorescence. Leaves eligulate, basal, crowded, exactly linear, ca ¼ as long as stems, 5-15 × 0.1-0.3 cm, flat, abruptly pointed, pubescent or glabrous, scaberulous on margins, greenish to purplish. Inflorescence compound or subparacompound, with many spikelets, ca 7 × 8 cm; rays slender, 7-8, very unequal; involucral bracts setiform, 3-5, much shorter than inflorescence, shortly pubescent at dilated base, 1-2 cm long; spikelets solitary, lanceolate, 5- 10 × 2-2.5 mm, acute, compressed; rhachilla broadly winged; glumes distichous, obliquely erect, lanceolate, 4-6.5 × 2.5 mm, acuminate with 1-nerved keel, yellowish, (fulvous), scaberulous-pubescent from short stiff hairs, broadly hyaline margined; stamens 3, anther linear, 1.5-2 mm long; styles trifid, very slender, glabrous, 4-6 mm long, distinctly pyramidally dilated as base; stigmas shorter than styles. Nuts trigonous obovoid, 0.8-1 × 0.3-0.5 mm, with convex faces, cuneate at base, umbonulate, densely verruculose.

Rare on wet gravel hills (low altitude ca 400 m) in Khammam district (Wadoodkhan and Solanke, 2008).

Fl. & Fr.: September – November.

Bhadrachalam (KMM), *S.N.Solanke* 681 (HCMCM).

Fimbristylis littoralis Gaud. in Frey. Voy. Bot. 413. 1826.

Culms tufted, erect portion arising from the axils of basal leaves, 15-60 cm high, 2-3 mm thick below, flattened, 4-angled with 2 sharp edges just below the inflorescence, smooth. Leaves reduced to bladeless sheaths, 2-10 cm long, clasping the stem, the ventral side splitting as the culm grows, laterally flattened with a sharp dorsal edge, brownish, hyaline on the margins, the lowest scale like. Involucral bracts 2-4, rigid, the lowest 1.8 cm, not overtopping the inflorescence. Inflorescence compound, 3-7 × 2-3 cm. Spikelets ovoid, globular, 2-3 × 2-2.5 mm, rusty brown, densely many-flowered; glumes broadly ovate, 1-1.5 × 0.5-0.7 mm, boat-shaped, obtuse at the apex, membranous, reddish brown, the margin with white hyaline margins, keel 1-nerved ending below the apex; stamens 1, 1.5 × 0.5 mm, anthers 0.5 × 0.1 mm; pedicel 1 mm; style flattened, 3-cleft at the apex, 1.8 mm. Nut trigonous, obovoid, 0.5-0.8 × 0.4 mm, creamish yellow, surface warty particularly at the ridges.

Rare in open, moist to swampy places in dry deciduous forests.

Fl. & Fr.: August – December.

Hamsanpally RF (MDK), *RG* 116438 (BSID); Mallurgutta (WGL), *RKP* 110959 (BSID).

Fimbristylis microcarya F.V.Muell., Fragm. 1: 200. 1859.

Annual, tufted, erect herb, 10-30 cm tall, culm flatly 3-angled, smooth. Leaves basal, few; sheath light straw-colored to pale brown; ligule ciliate; leaf blade linear, to 18 cm × 1.5-5 mm, flat, apical margin scabrid. Involucral bracts 3 or 4, basalmost leaflike and to 4 cm. Inflorescence a decompound anthela, lax; primary rays 2-5, 1-5 cm, unequal, apically scabrous; spikelets numerous per inflorescence, solitary, rust-colored, narrowly ovoid, slightly angular, sublaxly 5-10-flowered; glumes numerous, spirally imbricate, light rust-colored with pale brownish margins, ovate, 1-2 mm, membranous, 3-veined forming an abaxial keel, apex mucronate; stamens 1 or 2; style 3-sided, stigmas 3. Nutlet subsessile, light straw-colored, obovoid, smooth or with scattered verrucae and slightly cancellate with transversely oblong cells.

Along the margins of lakes, streams, in waterholes and moist places in Hyderabad, Medak and Adilabad districts.

Fl. & Fr.: August –October.

Nirmal (ADB), *MHR* & *KI* 14472; Medak (MDK), *RG* 104153 (BSID); Gandipet (HYD), *KI* & *NVN* 18766.

Fimbristylis naikii Wad. Khan & Lakshmin., J. Bot. Res. Inst. Texas 2(1): 381. 2008.

Annuals, 10-35 cm high; stems slender, trigonous to compressed trigonous, 0.3-1 mm thick, ribbed, glabrous, rather rigid. Leaves ligulate, subbasal, 1-2 mm wide, ½ to ¾ or more as long as the stems, flat with often thick rib-like margins, hairy with long white cilia or glabrescent, acute; sheaths densely hairy, glabrous or glabrescent, pale ferrugineus or dark brownish. Inflorescence small, simple to subcompound, 1.5-3.5 cm long and broad with usually 3-9 spikelets; rays if present 1-3, each one bearing usually paired (rarely 3) spikelets; involucral bracts up to 6, 1-2-foliaceous, the other ones setiform, the longest one as long as to much overtopping the inflorescence, up to 6 cm long, long ciliate at least on margins and base. Spikelets all solitary, ovoid or ovoid-oblong, 3-7 × 2-3.5 mm, acute, brown-white variegated, many-flowered;

peduncles 2-5 mm long; rhachilla winged. Glumes spiral, some what loosely imbricated, ovate to broadly ovate, 2-2.8 × 1.8 – 2.2 mm, almost as long as to slightly longer than broad, multistriate on sides, obtuse or acute, distinctly apiculate, often enrolled in fruiting, reddish-brown with whitish keel portion on the back, broadly hyaline margined; keel 3-5-nerved with strong midnerve; stamens 1 or 2, anther elliptic-oblong, 0.4-0.5 mm long, apiculate; styles flat, shorter than the nuts, 1-1.2 mm long, conspicuously broader than the staminal filaments, fimbriate throughout, not hyaline margined, pyramidally dilated at base or somewhat gradually narrowed upwards, stigmas shorter than the styles, shortly hairy. Nuts biconvex, broadly obovoid, 1.5-1.6 × 0.8 – 1.2 mm, turgid in the centre, subtrabeculate with 15-18 vertical striations in between transversely oblong thin walled epidermal cells, distinct in the centre and on the shoulder, nearly quadrate near base and apex, and on shoulder, apex rounded to subtruncate, muticous; gynophores discoid or cuboid, small, 0.1-0.2 mm long and broad, not or obscurely lobed, often reddish-brown (very rarely dull whitish), its thin rim often appressed to the nut base.

Occasional in wet open grassland and eutrophicated marshes in Medak district.

Fl. & Fr.: October – November.

Patancheru (MDK), *Solanke* 824 (HCMCM).

Fimbristylis ovata (Burm. f.) Kern, Blumea 15: 126. 1967. & in Steenis, Fl. Males. 1. 7: 595. 1974. *Carex ovata* Burm. f., Fl. Indica 194. 1768. *Abildgaardia ovata* (Burm. f.) Kral, Sida 4: 72. 1971. *Cyperus monostachyos* L., Mant. 2: 180. 1771. *Abildgaardia monostachyos* (L.) Vahl, Enum. 2: 296. 1805. *Fimbristylis monostachyos* (L.) Hassk., Fl. Jav. Rav. 61. 1848, (*monostachya*); C.B.Clarke in Hook. f., Fl. Brit. India 6: 649. 1893; Fischer, Fl. Madras 3: 1660. 1931

Perennial, erect, caespitose and leafy herb, to 40 cm high; culms filiform, trigonous. Leaves crowded on the swollen base of the stem, filiform, to 20 cm long, scabrid; ligule 0; sheaths stramineous. Spikelets solitary on the stem. Involucral bracts 2, not overtopping, to 1 cm long. Spikelets compressed, ovate, acute, shining, 2 lower glumes empty and longer than the others, cuspidate, to 1 cm long; rhachilla winged. Achenes obovoid, trigonous, stipitate, stramenious.

Occasional on wet grass lands, stream banks, in bunds of rice fields and in marshy areas in most districts.

Fl. & Fr.: December – March.

Rebbana (ADB), *GO* & *PVP* 4849; Jaipur RF (ADB), *PVP* & *KH* 9428; Ghanapur (NZB), *BR* & *KH* 7102; Chedmal (NZB), *BR* & *CPR* 9260; Pocharam tank (MDK), *TP* & *CP* 12046; Marrikunta (NZB), *CP* 11395; Anantasagar (RR), *MSM* & *KH* 10505; Rathamhutta hills (KMM), *RCS* 104228 (BSID).

Fimbristylis polytrichoides (Retz.) R. Br., Prodr. 226. 1810; C.B.Clarke in Hook. f., Fl. Brit. India 6 : 632. 1893; Fischer, Fl. Madras 3 : 1658. 1931. *Scirpus polytrichoides* Retz., Observ. Bot. 4 : 11. 1786/87.

Perennial, densely tufted herb, culms to 25 cm high, trigonous, shallowly striate, glaucous-green, clothed at base with few sheaths; sheaths to 3 cm long, herbaceous, yellowish-brown. Leaf blades to 10 cm long, canaliculated, remaining as setaceous appendages of sheath. Inflorescence a single, terminal spikelet. Involucral bracts 2, the longest up to 3 cm long, overtopping the spikelet. Spikelet ellipsoidal, to 6 mm long, terete, obtuse, brownish grey, densely many-flowered. Achenes obovoid, biconvex to 2 mm long, grey purple at maturity, minutely cancellated with transversely oblong cells.

Open wet grounds, sandy shores, rock crevices near the sea.

Fl. & Fr.: September – March.

Upparapalli cheruvu (WGL), *MHR* & *MCK* 17061; Rudraram RF (RR), *TP* & *MSM* 12726.

Fimbristylis pubisquama Kern, Blumea 8: 131. 1955. *F. compressa* Boeck., Linnaea 38: 1874; C.B.Clarke in Hook. f., Fl. Brit. India 6: 639. 1893; Fischer, Fl. Madras 3: 1658. 1931.

Annual densely tufted herb, to 40 cm high; culms obtusely trigonous, glaucous green, leaved at base. Leaves 4 to a culm; blade compressed trigonous, glabrous, glaucous green; ligules ciliate; sheaths to 9 cm long, subloosely surrounding the culm, dorsally whitish green, the membranous, ventral side whitish cinnamon-coloured, ciliate with soft hairs on margins. Corymbs compound, to 7 cm long, subdensely bearing ca 30 spikelets; rays to 5, patent, compressed. Involucral bracts 1-3, the lowest leaflike, nearly 2 times as long as the corymb, the base dilated with broad brown ciliate hyaline margins, the second bract nearly equaling the corymb, the third setaceous. Spikelets ovoidal, to 6 mm long, terete, deeply rusty brown, wholly pubescent, densely many-flowered. Achenes obovate, biconvexed, 1 mm long, the surfaces yellow brown, shiny, finely cancellated.

In margins of marshes and open grass lands.

Fl. & Fr.: October – December.

Raghava Rao (1989) reported from Mahabubnagar district.

Fimbristylis quinquangularis (Vahl) Kunth, Enum. Pl. 2: 229. 1807; C.B.Clarke in Hook. f., Fl. Brit. India 6: 644. 1893; Fischer, Fl. Madras 3: 1659. 1931. *Scirpus quinquangularis* Vahl, Enum . Pl. 2: 279. 1806. *Fimbristylis miliacea* (L.) Vahl, Enum. Pl. 287. 1806; C.B.Clarke in Hook. f., Fl. Brit. India 6: 644. 1893; Fischer, Fl. Madras 3: 1660. 1931. *Scirpus miliaceus* L., Syst. Nat (ed. 10) 2: 868. 1759.

Annual, densely tufted, rhizomatous herb, to 70 cm high. Culms pentagonous, clothed at base with 4 sheaths; culm sheaths bladeless, subterete, pale green, the ventral side hyaline and lightly yellow brown, the orifice obliquely truncate, hyaline on margin, the uppermost sheath to 18 cm long, second to 10 cm. Leaf blade dorsiventrally flat, shorter than the culm, to 30 cm long, herbaceous, whitish green and 5-nerved on upper surface, smooth except on upper margins where spinulose scabrid; sheaths to 7 cm long, pale green, yellow-brownish, ventral side split as culm grows. Corymbs compound to 9 cm long; rays 7, patent, to 5 cm long, scabrous; secondary corymbs bearing to 30 spikelets. Involucral bracts 5, setaceous, much shorter than the rays. Spikelets solitary, ovoidal, to 5 cm long, rusty brown, ca. 40-flowered.

Common species usually found in marshy areas, along the sides of water courses and also a characteristic weed of the rice fields.

Fl. & Fr.: July – December.

Birsaipet RF (ADB), *GO* & *PVP* 4608; Ibrahimpet (NZB), *TP* & *BR* 6118; Mohammadabad RF (RR), *KH & MSM* 5583; Pakhal (WGL), *MHR* & *MCK* 17074, 17077; Sagar left canal (KMM), *MHR* & *MCK* 17028; Rayagiricheruvu (NLG), *KI* & *NVN* 8747; Aklaspur (KRN), *GVS* 27298 (MH & CAL); Narsapur (MDK), *KMS* 6720 (MH & CAL); Rollapadu (KMM), *RR* 108081 (BSID).

Fimbristylis schoenoides (Retz.) Vahl, Enum. Pl. 2: 286. 1806; C.B.Clarke in Hook. f., Fl. Brit. India 6: 634. 1893; Fischer, Fl. Madras 3: 1658. 1931. *Scirpus schoenoides* Retz., Observ. Bot. 5: 14. 1788.

Annual, caespitose, glaucous herb, to 70 cm high; culms filiform, obtusely angled, striate. Leaves few, blade filiform, shorter than the culm, to 16 cm long, ligule shortly hairy; sheaths pinkish-brown. Involucral bracts 3, stiff, filiform, not overtopping, to 1.5 cm long. Spikelets 3, ovoid-subglobose, 1 × 0.4 cm long, rhachilla strongly winged. Achenes compressed, obovoid, biconvex to 2 mm long, stipitate, striate.

This species is usually seen in sandy areas along river banks and also along the margins of rice fields.

Fl. & Fr.: September – December.

Sarvaipet (ADB), *GO* 4505; Bhadrakali lake (WGL), *MHR* & *MCK* 17091; Sagar left canal (KMM), *MHR* & *MCK* 17001; Gangapur (MDK), *TP* & *CP* 12080; Mohammadabad RF (RR), *MSM* 10483; Narsapur (MDK), *KMS* 6759 (MH); Bhadrachalam RF (KMM), *RCS* 104316 (BSID); Jakaram (WGL), *RKP* 105322 (BSID);

Fimbristylis sieberiana Kunth, Enum. Pl. 2: 237. 1837. *F. ferruginea* Decne., Nouv. Ann. Mus. Hist. Nat. Paris 3: 352. 1834 non (L.) Vahl, 1806.

Perennial, tufted, rhizomatous herb, rhizome small, woody; culms to 45 cm high, basally scabrid. Leaves flat, to 40 cm long, margins inrolled, scabrid; ligule short, hairy; sheaths purple brown, hairy. Inflorescence simple, to 3 cm long. Involucral bracts 3-5, lowest overtopping, to 5 cm long; rays 2, to 2 cm long. Spikelets solitary, terete, to 1 cm long, obtuse. Achenes obovoid, biconvex, stramineous, smooth, stipitate.

Occasional in open marshy places and in cultivated fields.

Fl. & Fr.: July – December.

Way to Potchera (ADB), *MHR* 14491; Gandipet (HYD), *KI* & *NVN* 18772; Katasintha (HYD), *KMS* 6014 (MH); Aklaspur (KRN), *GVS* 22532 (MH); Rusthempet (MDK), *KMS* 7996 (MH); Krishna river bank (NLG), *KMS* 9866 (MH).

Fimbristylis tenera Schult., Mant. 2: 57. 1825; C.B.Clarke in Hook. f., Fl. Brit. India 6: 642. 1893; Fischer, Fl. Madras 3: 1660. 193. *F. monticola* Hochst. ex Steud., Syn. Pl. Cyp. 3: 111. 1855.

Annual, slender, caespitose herb, to 30 cm high; culms filiform, obtusely trigonous, striate. Leaves erect, many at the base of the culm, shorter than the culm, filiform, glaucous-green, 15 cm long. Umbel subcompound, rays 4 with 2 spikelets on each. Involucral bracts very short, setiform. Spikelets to 1 cm long, oblong, acute, compressed, dark brown; rhachilla with deep pits. Achenes stipitate, globosely obovoid, obtusely trigonous, yellow or brown.

Usually seen in moist low lands, sandy soils, grass fields.

Fl. & Fr.: August – November.

Ankusapuram (ADB), *GO* 4337; Sardhana (MDK), *CP* 11744; Gandipet (HYD), *KI* & *NVN* 18770; Central University campus (RR), *MSM* 15177; Rudraram RF (RR), *TP* & *MSM* 12727.

Fimbristylis tetragona R. Br., Prodr. 226. 1810; C.B.Clarke in Hook. f., Fl. Brit. India 6: 631. 1893; Fischer, Fl. Madras 3: 1658. 1931.

Perennial, caespitose herb, to 35 high; culms slender, ribbed, quadrangular. Leaves reduced to membranous sheaths, grey. Spikelet solitary, hemispherical, or sub-globose, lowest glume hardly brachiform, coriaceous, deciduous; rhachilla thickly studded with deep tetragonal pits with raised edges. Achenes oblong, cylindrical, biconvex, tapering towards the base, stipitate, fulvous grey, to 2 mm long.

Not very common. It is seen growing in wet or moist open places, borders or rice fields, banks of ponds, edges of marshes and also in forest areas during monsoon.

Fl. & Fr.: July – December.

Jannaram (ADB), *TP & PVP* 4206; Potchera (ADB), *GO* & *DAM* 5059; Anantasagar (RR), *MSM* 10502; Shaikpalli (RR), *KH* & *MSM* 10931; Rachapalli (KRN), *GVS* 22261 (MH & CAL); Vazeedu (KMM), *VSR* 1000 (MH).

Fimbristylis woodrowii C.B.Clarke, Bull. Misc. Inform. 227. 1898; Kesava Murthy & Yoganarasimhan, Fl. Coorg, 509. 1990.

Annuals, stem 5-10 cm high, slender, tufted, flat. Leaves 3-10 cm long, flat, eligulate. Spikelets 1-3, clustered, each 0.4-1 cm long, lanceolate; bracts 2-3; glumes to 1 mm or slightly longer, ovate, acute with a sharp curved keel; stamens 1; styles 3-fid. Nut small, trigonous, longitudinally ribbed.

Rare in marshy places in Medak and Hyderabad districts (Murthy & *al.*, 2008; Ramana, 2010).

Fl. & Fr: August - October.

FUIRENA Rottb.

1. Perennials with decumbent rhizome; leaves glabrous; hypogynous scales and bristles absent:
    2. Stem trigonous .... **F. wallichiana**
    2. Stem pentagonous .... **F. umbellata**
1. Annuals with fibrous roots; stem sulcate; leaves pilose; hypogynous scales and bristles present:

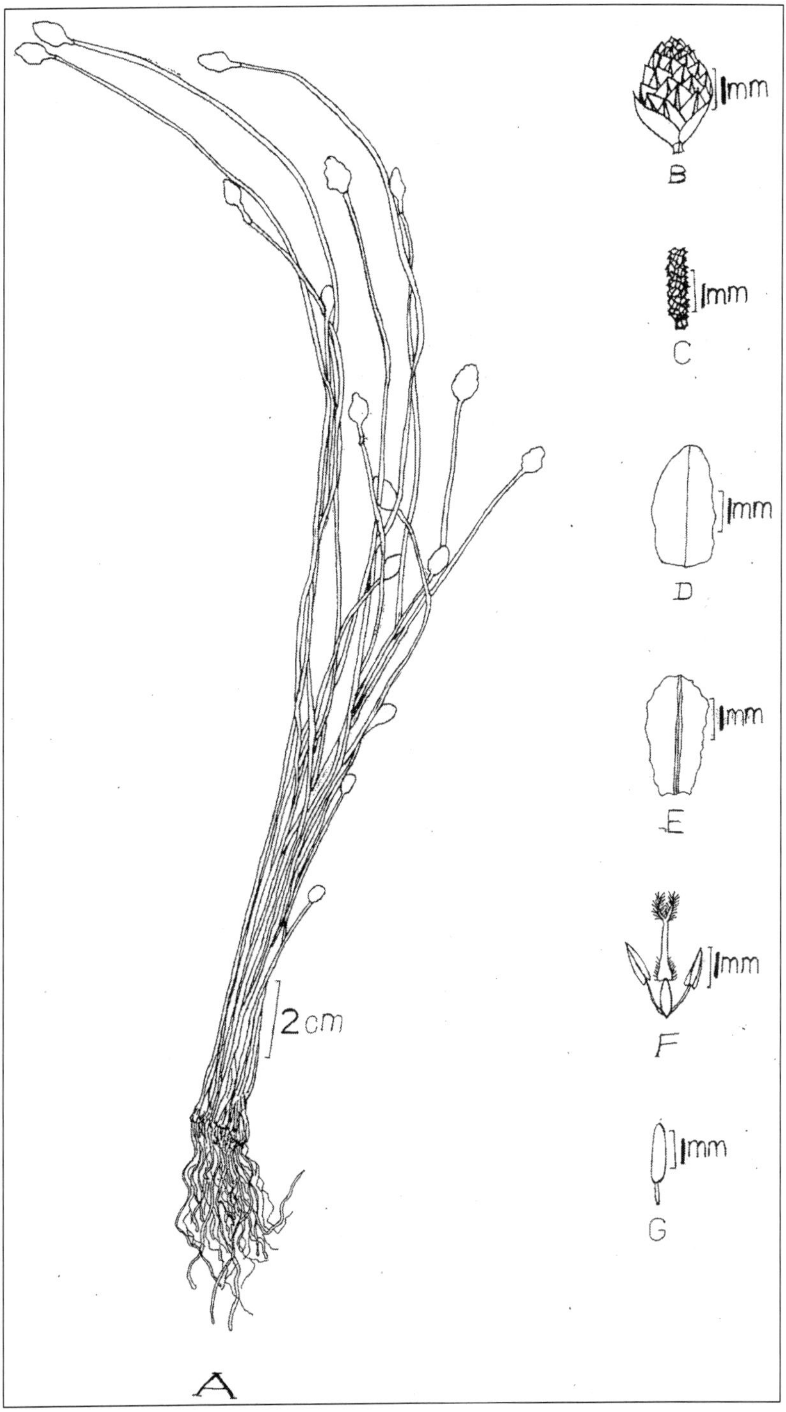

Figure 127: Fimbristylis tetragona R. Br.

A. Habit, B. Spikelet, C. Rhachilla, D. & E. Glumes, F. Pistil with stamens, G. Achene.

3. Scales quadrate or suborbicular:
    4. Inflorescence subcapitate; awn slender, pilose; bristles 4-6; claw of the petals (scales) as long as the blade .... **F. cilliaris**
    4. Inflorescence capitate; awn strong, thorn like, scabrid; bristles 3; claw of the petals shorter than the blade .... **F. capitata**
3. Scales lunate, trilobate .... **F. trilobites**

Fuirena capitata (Burm. f.) T. Koyama in Dssanayake & Fosberg, Rev. Handb. Fl. Ceylon 5: 151. 1985. *Scirpus capitatus* Burm. f., Fl. Ind. 21. 1768. non. L. 1753. *Fuirena uncinata* (Willd.) Kunth, Enum, Pl. 2: 184. 1837; C.B.Clarke in Hook. f., Fl. Brit. India 6: 666. 1893; Fischer, Fl. Madras 3: 1669. 1931. *Scirpus uncinatus* Willd., Sp. Pl. 1: 300. 1798.

Annual, tufted, pubescent herb with reddish brown fibrous roots. Culms subdensely tufted, to 20 cm high, obtusely subpentagonal, ribbed, 4 nodes including inflorescence, the internodes pilose, to 5 cm long. Leaves basal and cauline; blade linear-lanceolate, to 8 cm long, several-costate, flattish, herbaceous, subdensely pilose on both sides with white, spreading, tubercle based hairs; sheaths to 5 cm long, laxly clothing the culm internode, pilose. Basal leaves reduced to sheaths, those rusty brown. Inflorescence capitate, bearing 5 glomerules of spikelets. Spikelets sessile, squarrose, ellipsoidal, polygonal, bearing up to 20 glumes. Achenes rhombic-oval, triquetrous with concave sides, stipitate at base, maturing light brown.

Usually seen in wet or moist sandy fields, along marshy places in sticky muddy soils of rice fields.

Fl. & Fr.: December – March.

Vempalli (ADB), *GO* 4447; Shaikpalli (RR), *KH* & *MSM* 10934; Saleswaram (MBNR), *BSS* & *SKB* 33114.

Fuirena ciliaris (L.) Roxb., Fl. Ind. 1: 184. 1820. *Scirpus ciliaris* L., Mant. Pl. 182. 1771. *Fuirena glomerata* Lam., Tabl. Encycl. 1 : 150. 1791; C.B.Clarke in Hook. f., Fl. Brit. India 6 : 666. 1893; Fischer, Fl. Madras 3 : 1669. 1931.

Annual, slender erect, leafy herb, to 30 cms high; culms sub-cylindric below and trigonous above, 2 mm wide, sparsely hairy above, glabrous below. Leaves linear-lanceolate, acuminate, to 9 × 0.4 cm; sheaths long, closed striate, slightly pubescent. Inflorescence umbellate, 3-10 in a cluster in terminal and axillary spikelets in subcapitate heads, 1.8 × 1.8 cm. Involucral bracts 0. Spikelets oblong, obtuse, brown, squarrose, lowest glumes longest, empty ciliate, 0.8 × 0.3 cm, many-flowered.

Common species usually seen in wet open groups, river banks, margins of ponds and rice fields.

Fl .& Fr.: September – March

Jannaram (ADB), *TP & PVP* 4201, 4205; Sirnapalli river (NZB), *BR* & *GO* 9076; Pochammaralu (MDK), *BR* & *CP* 11474; Pocharam tank (MDK), *TP* & *CP* 12044; Mohammadabad (RR), *KH* & *MSM* 10924; Borapuram (MBNR), *BSS* & *SKB* 32293;

Figure 128: Fuirena ciliaris (L.) Roxb.

A. Habit, B. Spikelet, C. Rhachilla, D & E., Glumes, F. Pistil with stamens, G. Achene with hypogynous scales.

Kodimial (KRN), *GVS* 21836 (MH & CAL); Pakhal lake (WGL), *KMS* 11701 (MH & CAL); Khammam to Dhaniyayigudem (KMM), *RR* 108512 (BSID); Edupayalu RF (MDK), *RG* 113421 (BSID);Gandipet (HYD), *KI* & *NVN* 18774.

Fuirena trilobites C.B.Clarke in Hook. f. Fl. Brit. India 6: 666. 1893 & 111. Cyper. t. 5. 1909.

Annual, erect, caespitose herb, to 40 cm high; culms slender, striate, glaucous, leafy, sparsely hairy. Leaves to 11 cm long, linear-lanceolate, acuminate, 7-nerved, profusely hairy; sheaths 3 cm long, close, pubescent. Inflorescence cluster of 3-9 axillary and terminal spikelets in subcapitate heads; bracts absent but inflorescence invariably arising from leaf axil. Spikelets to 7 mm long, ovoid or oblong, obtuse, pale brown, squarrose, lowest glume empty, possess longest awn, densely ciliate. Achenes stipitate, triquetrous, acute, base trigonous flat, minutely tuberculate showing very faint horizontal striations.

Found growing in rice fields and water logged grass lands.

Fl. & Fr.: November – March.

Tada (NLR), *MCK* 23523; Darsi (PKM), *MCK* 22829; Buchireddypalem (NLR), *MCK* 22967.

Fuirena umbellata Rottb., Descr. Icon. Rar. 70. t. 19. f.3. 1773; C.B.Clarke in Hook. f., Fl. Brit. India 6: 666. 1893; Fischer, Fl. Madras 3: 1669. 1931.

Perennial herb with long, creeping rhizome; stems tufted, erect, 30-50 cm high, acutely 4-5-angular, glabrous. Leaves linear, 7-8 cm long, acute or acuminate, glabrous or hirsute on the margins in the basal part, with 5 proimnent nerves; cauline leaves distant. Inflorescence with a terminal, partial inflorescence and several axillary ones. Lower bracts similar to the leaves; spikelets in very dense clusters, ovoid to oblong-ovoid, 4-8 mm long, acute; glumes ovate or obovate, 2-2.5 mm long, membranous; stamens 3. Nut ellipsoid or obovate, 1-1.25 mm long, triquetrous, smooth, shining, stramineous to fuscous.

Rare in marshy localities.

Fl. & Fr.: September - January.

Karinmagar district, *NRR* & *TR* 86598 (MH).

Fuirena wallichiana Kunth, Enum, Pl. 2: 182. 1837; C.B.Clarke in Hook. f., Fl. Brit. India 6: 665. 1893; Fischer, Fl. Madras 3: 1669. 1931.

Perennial, robust, rhizomatous herb, to 60 cm high; culms many, 3 mm wide, rigid, trigonous, hollow, slightly pubescent towards apex. Leaves linear-lanceolate, acute with strong midnerve, 10 cm long, glaucous; sheaths 3 cm long, trigonous, striate, with a short ligule. Spikelets in corymbose terminal clusters, oblong, terete, many-flowered, 1.1 × 0.4 cm. Achenes turbinate, with conical pubescent apex, trigonous, pale yellow, 1 mm long.

Found along the banks of streams and tanks, common in most pasture lands.

Fl. & Fr.: Most of the year.

Adlur Yella Reddy (NZB), *TP* & *BR* 6415; Potchera (ADB), *GO* & *DAM* 5051; Shadnagar (MBNR), *SRR* 1711 (OU); Alampur (MBNR), *SRR* 3441 (OU).

KYLLINGA Rottb.

1. Rhizome short or absent, stems tufted:
   2. Plants annual .... **K. pumila**
   2. Plants perennial:
      3. Central spike ovoid, <6 mm long; siklelets oblong-elliptic; culms clearly thickened at base .... **K. bulbosa**
      3. Central spike oblong, 8-15 mm long; spikelets ovoid; culms hardly enlarged at base .... **K. odorata**
1. Rhizome long, stems solitary in linear rows along the length of the rhizome:
   4. Glumes not winged:
      5. Involucral bract short, spreading, < 6 cm long; rhizome slender; culms <30 cm long, distant .... **K. brevifolia**
      5. Involucral bract elongate, deflexed, >15 cm long; rhizome thick; culms 30-50 cm, juxtaposed .... **K. melanosperma**
   4. Glumes winged .... **K. nemoralis**

Kyllinga brevifolia Rottb., Descr. Icon. Rar. Pl. 13, t. 4, f. 3. 1773; C.B.Clarke in Hook. f., Fl. Brit. India 6: 588. 1893; Fischer, Fl. Madras 3: 1624, 1931.

Perennial, stoloniferous, rhizomatous herb; rhizome horizontal, creeping, clothed with brownish scales; culms many, spaced, arranged in a single row along the rhizome, to 25 cm high, slender, triquetrous, few-leaved at base. Leaf blade shorter than the culm, narrowly linear, flattish, plicate, herbaceous, scabrid on margin and on the abaxial midvein; sheaths brownish. Inflorescence a terminal, single globose head. Involucral bracts 3, foliaceous, very unequal in length, lowest to 8 cm long; head globose, pale green, densely bearing numerous spikelets. Spikelets lance-oblong, laterally compressed. Achenes obovate, laterally lenticular, brownish.

Occasional in grasslands in Nizamabad district.

Fl .& Fr.: July – December.

Nirmal – Adilabad road (ADB), *MHR* 14455; Kama Reddy (NZB), *BR* & *CPR* 9230; Patancheru (MDK), *CP* 11685; Pullajalamala (MBNR), *VBH* 84997 (BSID).

Kyllinga bulbosa P. Beauv., Fl. d' Oware & Benin 1: 11. t. 8. f. 1. 1804. *Kyllinga triceps* Rottb., Descr. Icon. Rar. 1 & t. 4. f. 6. 1773. *nom. illegit.*, C.B.Clarke in Hook. f., Fl. Brit. India 6: 587. 1893; Fischer, Fl. Madras 3: 1623. 1931.

Perennial, rhizomatous herb; culms to 15 cm high, densely tufted with short erect rhizome, obtusely trigonous, the base forming an ovoidal, corm-like enlargement clothed with brown fibres. Leaves few; blade flattish-plicate, sheaths stout, pale and basal ones tinged with light brown, sheaths disintegrate leaving a tuft of fibres. Inflorescence a head with 3 sessile spikes, pale greenish, central spike ovoid-globose, to 8 mm long, obtuse, the lateral ones globose. Involucral bracts 3, eventually reflexed, the longest to 10 cm long. Spikelets oblong, pale greenish, 1-flowered. Achenes oblong, laterally flattered, brownish.

Common species usually found in grass lands, lawns, and almost everwhere in the plains.

Fl. & Fr.: August – December.

Sirpur (ABD), *GO* & *PVP* 4969; Dichpalli (NZB), *BR* & *KH* 6440; Nagulabanda (MDK), *CP* 11323; Venkatapuram (MDK), *TP* & *CP* 14006; Aklaspur (KRN), *GVS* 25658 (CAL); Narsapur (MDK), *KS* 6639 (MH & CAL); Pakhal lake (WGL), *KMS* 13121 (MH & CAL); Billyguttalu (KMM), *RR* 105979 (BSID).

Note: *Kyllinga triceps* Rottb. is a superfluous illegitimate name.

Kyllinga melanosperma Nees in Wight, Contr. Bot. India 91. 1834; C.B.Clarke in Hook. f., Fl. Brit. India 6: 588. 1893; Fischer, Fl. Madras 3: 1624. 1931. *Cyperus melanospermus* (Nees) Suringar, Het. Gest. Cyperus Mal. Arching. 50, t. 2, f. 8. 1898.

Perennial, rhizomatous herb; rhizome decumbent, to 8 cm long, knotty, clothed with lance-ovate, dark brown scales, very aromatic; culms to 50 cm high, disposed in a row along the rhizome, close together, triquetrous with almost winglike angles. Leaves all reduced to bladeless sheaths; sheaths 4, membranous, to 20 cm long, greenish, tinged with purplish brown, the orifice with hyaline margin. Inflorescence a single globose head, to 12 mm long, greenish, densely many spiculose. Involucral bracts 3, to 15 cm long, spikelets elliptic-oblong. Achenes elliptic-oblong, apiculate, laterally flattened.

Though not common, this is occasionally found on forest margins, in swampy areas.

Fl. & Fr.: January – July.

Ramayampet (MDK), *BR* & *CP* 11504; Mallelatheertham (MBNR), *BSS* & *SKB* 33122; Sagar left canal (KMM), *MHR* & *MCK* 17006; Pathur (MDK), *TP* & *CP* 12119.

Kyllinga nemoralis (Forst. & Forst. f.) Dandy ex Hutchins. & Dalziel, Fl. W. Trop. Afr. 2: 487. 1936. *Thryocephalon nemoralis* Forst. & Forst. f. Char. Gen. Pl. 65. 1775. *Kyllinga monocephala* Rottb., Descr. Ic. Rar. Nov. Pl. 13, t. 4, f. 4, 1773; C.B.Clarke in Hook. f., Fl. Brit. India 6: 588. 1893; Fischer, Fl. Madras 3: 1624. 1931.

Perennial, laxly tufted, rhizomatous herb; rhizome long-creeping, culms to 20 cm high, along the rhizome, triquetrous, not thickened at base. Leaves many, longer than the stem; sheaths brown. Inflorescence a single head, densely bearing many spikelets, whitish. Involucral bracts 4, spreading, the lowest to 15 cm long. Spikelets ovate, elliptic, 2-flowered. Achenes oblong-obovate, biconvexed, maturing brownish.

Common species found on road sides, waste places, open grass lands, near canals, as herbaceous undergrowth on moist soil in Hyderabad and Ranga Reddi districts.

Fl. & Fr.: July – November.

Uppal (HYD), *KI* & *NVN* 18755, 18754; Centrl University campus (RR), *MSM* 15188.

Kyllinga odorata Vahl, Enum. Pl. 2: 382. 1906. subsp. **cylindrica** (Nees ex Wight) T. Koyama, Gard. Bull. Singapore 30. 161. 1977 & in Dassan. & Fosb., Rev. Handb. Fl. Ceylon 5: 244. 1985. *K. cylindrica* Nees ex Wight, Contr. Bot. Ind. 91. 1834; C.B.Clarke in Hook. f., Fl. Brit. India 6: 588. 1893; Fischer, Fl. Madras 3: 1624. 1931.

Perennial, tufted, rhizomatous herb; rhizome clothed with brown fibres; culms to 30 cm high, slender, triquetrous, hardly thickened at base. Leaves few, shorter than the culm; blades flattish, plicate; sheaths pale brownish, remain intact. Inflorescence a head with 3 spikes, whitish green; central spike cylindrical; lateral spikes elliptical, much shorter than the central spike, each densely bearing many spikelets. Involucral bracts 4, reflexed, the lowest to 10 cm long. Spikelets patent, elliptic. Achenes elliptic-obovate laterally flattened, maturing dark brown.

Usually grows in moist places and open grass lands.

Fl. & Fr.: May – October.

Ankusapuram (ADB), GO 4325.

Kyllinga pumila Michx., Fl. Fl. Bor. –Amer. 1: 28. 1803. *K. tenuifolia* Steud. Syn. Pl. Glumac. 2: 69. 1854.

Annual, caespitose herbs, not rhizomatous, culms 15-30 cm. Leaves flat or V-shaped, 4-20 × 1.5-3 mm. Spikes 1-3, pale greenish, bracts 4, horizontal to ascending; spikelets 50-150, lanceolate to ovate-lanceolate; floral scales pale brownish to transparent; stamens 2. Achenes light brown, stipitate, oblong

Occasional in Khammam and Warangal districts.

F. & Fr.: September – January.

Manuguru RF (KMM), *RCS* 104256 (BSID); Hanamkonda (WGL), *RKP* 105380.

LIPOCARPA R. Br. *nom. cons*

1. Spikelet prophyll and glume absent:
   2. Inflorescence terminal; stigmata 2; achenes elliptic to narrowly rhombic in cross section, stamen 1 .... **L. kernii**
   2. Inflorescence pseudolateral, stigmata 3, achenes subtriquetrous in cross section; stamens 1-2 .... **L. squarrosa**
1. Spikelet prophyll and glume present:
   3. Inflorescence pseudolateral, stigmata 2; spike 1, stamen 1 **L. hemisphaerica**
   3. Inflorescence terminal; stigmata 3; spike (1) 2 – 10; stamens 1-2:
      4. Plants robust, tufted perennials; spikelets 2-10; inflorescence white to creamy white with red dots, achenes trigonous in cross section, oblong to narrowly obovate .... **L. chinensis**
      4. Plants slender, tufted annuals; spikelets 2-4; inflorescence yellow to pale reddish brown; achenes subtrigonous to trigonous frontally obovate to subelliptical:
         5. Achenes conspicuously beaked, trigonous in cross section, frontally obovate to subelliptical; stamens 1-2; inflorescence yellowish to pale brownish (top reddish brown) .... **L. sphacelata**
         5. Achenes shortly beaked, subtriquetrous in cross section, frontally obovate, stamen usually 1; inflorescence pale to dark reddish brown with red stripes .... **L. reddyi**

Lipocarpha chinensis (Osbeck) Kern, Blumea, Suppl. 4: 167. 1958 & in Steenis, Fl. Males. 1, 7: 521. 1974. *Scirpus chinensis* Osbeck, Dagb. Osting. Rosa 220. 1757. *Lipocarpha argentea* (Vahl) R. Br. ex Nees, Linnaea 9: 287. 1835; C.B. Clarke in Hook. f., Fl. Brit. India 6: 667. 1893; Fischer, Fl. Madras 3: 1670. 1931.

Annual, caespitose herb with reddish brown fibrous roots, culms to 50 cm high, obtusely trigonous, whitish green. Leaves 4 to a culm; blade linear, rather soft and slightly spongy, whitish green; sheaths to 15 cm long, subloosely surrounding the culm, pale, the lower ones stained with reddish purplish colour. Inflorescence a head with 7 spikes. Involucral bracts 3, unequal, dilated at base, much surpassing the head, the longest up to 20 cm long. Spikes oblong-ellipsoidal, terete, to 8 mm long, rounded at apex, whitish, densely bearing numerous bracteoles. Achenes oblong, 1.5 mm long, trigonous, straight to weakly curved, straw-brown.

A common species usually seen in marshy areas, swamps and rice fields.

Fl. & Fr.: May – October.

Lipocarpha hemisphaerica (Roth) Goetghebeur, Wageningen Agric. Univ. Papers 89-1: 37. 1989; Ragan & *al.*, Rheedea 8: 99. 1998. *Scirpus hemisphaericus* Roth, Nov. Pl. Sp. 29. 1821. *S. isolepis* (Nees) Boeck, Linnaea 36: 498. 1870; C.B. Clarke in Hook. f., Fl. Brit. India 6: 663. 1893; Fischer, Fl. Madras 3: 1666. 1931. *Hemicarpha isolepis* Nees, Arnold Edinb. New Phil. J. 17: 263. 1834.

Small, annual herb, 2 - 15 cm high. Leaves up to 6 cm × 0.6 mm. Inflorescence pseudolateral, with 1 spike, 1-7 × 1-3 mm, spherical, ovoidal to conical; involucral bract 1, up to 10 cm long, erect; spikelet bract 0.5-0.95 × 0.4-0.7 mm, broadly obovate, shortly acminate to obtuse, smooth at the top, brown to red brown, often with pale to greenish midnerve; spikelet prophyll and glume 0.5-0.7 mm long; stamen 1; stigmas 2. Achene obovate with small style base remant, rounded to elliptical on cross section.

Rare in marshy places in Adilabad district.

Fl. & Fr.: September – November.

Nirmal (ADB), *VSR* 2093 (KU).

Lipocarpha kernii (Raymond) Goetghebeur, Wageningen Agric. Univ. Papers 1: 42. 1989; Ragan & *al.*, Rheedea 8: 99. 1998. *Rikliella kernii* (Raymond) J. Raynal, Adasonia ser. 2.13: 155. 1973. *Scirpus kernii* Raymond, Natur Canad 86: 230. 1959.

Small, tufted, annual herbs, 2-40 cm high. Leaves 1-3 to a culm, filiform, 16 cm × 2 mm, sub-acute at apex. Spikes 2-8, ovoid; involucral bracts 2-5, the largest up to 15 cm long; spikelet bract elliptical to obovate, long acuminate, scaberulous at the top, body yellowish green to pale brown; spikelet prophyll and glume absent; stamen 1; style 0.1 mm or shorter, with 2 branches. Achenes obovate with very small style base remnant, dorsiventrally flattened and elliptical to narrowly rhombic on cross section

Rare in moist places at low elevations in Adilabad district.

Fl. & Fr.: August – December.

Nirmal – Adilabad road (ADB), *MHR* 14485; Nirmal (ADB), *MHR* & *KI* 14577, *VSR* 2092 (KU), Mancherial (ADB), *VSR* 2150 (KU).

Lipocarpha reddyi Hooper, Kew Bull. 41: 427. 1986; Goetghebeur & Van den Borre, Wageningen Aric. Univ. Papers 89 – 1: 64. 1989; Ragan & *al.*, Rheedea 8: 100. 1998.

Slender, annual, tufted herbs; stem 5-15 × 0.4-0.6 mm. Leaves up to 4 cm × 1 mm. Inflorescence terminal, with 2-3 spikes, 3-6 × 2-3 mm, ovoidal to cylindrical; involucral bracts 2, the largest up to 3 cm, patent to reflexed. Spikelet bract 1.5-1.8 × 0.6-0.7 mm, spathulate, apical part 0.6-0.75 mm long, rather abruptly narrowed into a thin acumen, smooth to minutely scaberulous at the top, top of wings pale to dark red brown, often with red stripes, midnerve and acumen more pale, rarely as dark. Spikelets 2-4; stamens 1(2), anthers ca 0.5 mm long; style very short, with 3 short branches. Achenes subtriquetrous in cross section, frontally obovate, shortly beaked.

Rare in Nalgonda and Warangal districts.

Fl. & Fr.: September – December.

Annaram (NLG), *P.S. Reddy* 1232 (K); Hanmakonda area, Warangal (WGL), *P.S.Reddy* & *T.Rajagopal* 219 (K).

Lipocarpha sphacelata (Vahl) Kunth, Enum. Pl. 2: 283. 1806; C.B.Clarke in Hook. f., Fl. Brit. India 6: 667. 1893. *Hypaelyptum sphacelatum* Vahl, Enum. Pl. 2: 283. 1806. *Lipocarpha triceps* Nees in Wight, Contrib. Bot. Ind. 92. 1834; Fischer, Fl. Madras 3: 1670. 1931.

Annual, erect, caespitose herb, to 20 cm high; culms triquetrous. Leaves basal, canaliculated, shorter than stem, to 14 cm long; ligule 0. Involucral bracts 3, foliar, over-topping, to 7 cm long. Inflorescence terminal, capitate. Spikelets 3-5 in a cluster, globose or ovoid, many-flowered. Achenes ellipsoid, beaked, fuscous, stoutly stipitate and trigonous.

Occasional in damp grassy places and open wet fields and margins of swamps.

Fl. & Fr.: August – January.

Vempalli (ADB), *GO* 4450; Padivagu, East Narsapur RF (MDK), *BR* 11203; Gadirayal (RR), *MSM* 10531; Shaikpalli, Mohammadabad RF (RR), *KH* & *MSM* 10930; Aklaspur (KRN), *GVS* 25658 (MH).

Lipocarpha squarrosa (L.) Goetghebeur, Wageningen Agric. Univ. Papers 89-1; 71. 1989. *Scirpus squarrosus* L., Mant. Pl. 2: 181. 1771; C.B. Clarke in Hook. f., Fl. Brit. India 6: 663; 1893; Fischer, Fl. Madras 3: 1666. 1931. *Rikliella squarrosa* (L.) Raynal, Adansonia ser. 2. 13: 154. 1973.

Annual, slender, glabrous, tufted herb, to 25 cm high; culms filiform, 2 or 3 to a stem; sheaths pinkish-grey, short, tightly surrounding stem base. Inflorescence pseudolateral head with 1-3 spikes, deeply red brown. Involucral bracts 2, lower bract leaflike, erect, much surpassing the inflorescence, to 2.5 cm long, upper bract setaceous, slightly longer than the spikes. Spikes sessile, ovate-elliptic, squarrose, 6 mm long, deeply red brown with pale green awns, densely many-flowered. Achenes obovoid, trigonous, yellow, obtuse at apex.

Occasional in wet sandy soil at the margin of ponds, ditches and paddy fields.

Fl. & Fr.: August – December.

Vempalli (ADB), *GO* 4440; Gandhari RF (NZB), *BR* & *CP* 7224; Narsapur RF (MDK), *TP* & *CP* 14055; Gadirayal (RR), *MSM* 10527; Dharmapuram (RR), *KH* & *MSM* 10932; Aklapur (KRN), *GVS* 22493 (MH & CAL); Way to Pegarikutta, Narsapur (MDK), *KMS* 6616 (MH & CAL); Medak cheru (MDK), *RG* 113514;

PYCREUS P. Beauv.

1. Achenes transversely wrinkled .... **P. diaphanus**
1. Achenes smooth to puncticulate:
    2. Culms few to several nodose below the middle, the lower part decumbent or obliquely ascending, branching and rooting at lower nodes, culms clothed to considerably above the base .... **P. sanguinolentus**
    2. Culms not nodose above the base, erect from very base; culms with leaves only at the base:

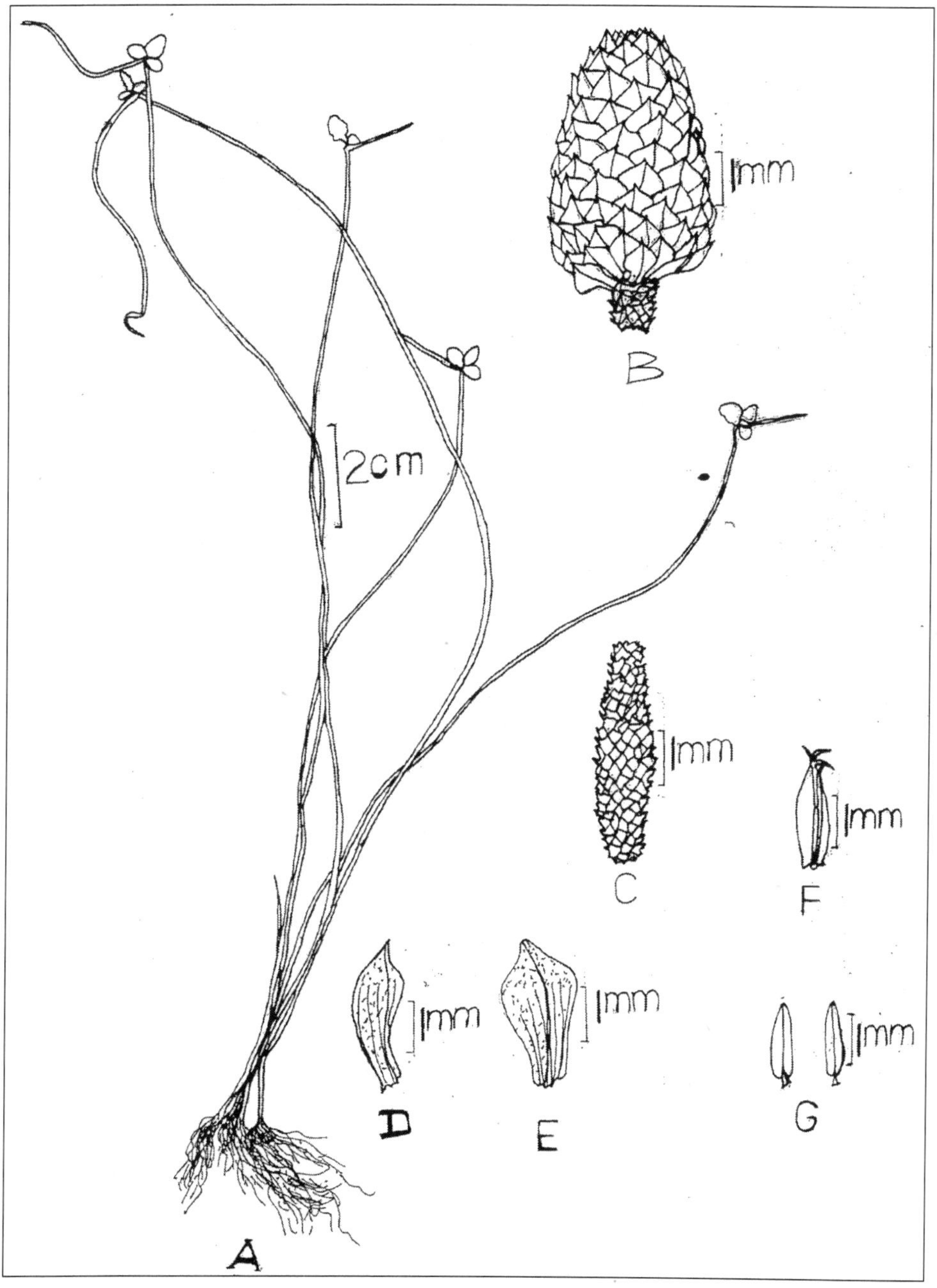

Figure 129: Lipocarpha sphacelata (Vahl) Kunth

A. Habit, B. Spikelet, C. Rhachilla, D & E. Glumes, F. Achene, G. Stamens.

3. Glumes distinctly cuspidate, apex retuse .... **P. pumilus**
3. Glumes not cuspidate nor retuse:
    4. Stems tufted:
        5. Achenes symmetric:
            6. Annuals; rhachilla straight, not at all winged; involucral bracts 2-4 .... **P. flavidus**
            6. Perennials; rhachilla winged; more or less zigzag; involucral bracts 8 .... **P. polystachyos**
        5. Achenes asymmetric, concave .... **P. sulcinux**
    4. Stems robust, solitary .... **P. puncticulatus**

Pycreus diaphanus (Schrad. ex Roem. & Schult.) Hooper & T. Koyama, J. Jap. Bot. 51: 316. 1976. *Cyperus diaphanus* Schrad. ex Roem. & Schult. Mant. 2: 477. 1824. *C. latespicatus* Boeck., Flora 42: 433. 1859. *Pycreus latespicatus* (Boeck) C.B.Clarke in Hook. f., Fl. Brit. India 6: 590. 1893.

Annual, rhizomatous herb, to 30 cm high, culms slender, stiff, obscurely angled, striate. Leaves shorter than the stem, rigid. Umbel simple, sometimes reduced to one head; involucral bracts 3, unequal, the longest sometimes reaching 10 cm long. Spikelets straw-coloured, shaded with chestnut brown, oblong-lanceolate, compressed, to 6-flowered. Achenes obovoid, black, shortly apiculate, slightly compressed.

Occasional in wet places, marshy grounds, along water streams and banks in Medak and Hyderabad districts.

Fl. & Fr.: August – January.

Narsapur town (MDK), *KMS* 6746 (MH & CAL); Uppal (HYD), *T.Rajagopal* 923 (OU).

Pycreus flavidus (Retz.) T. Koyama, J. Jap. Bot. 51: 313. 1976. var. **flavidus**. *Cyperus flavidus* Retz., Observ. Bot. 5: 13. 1788. *Pycreus globosus* (All.) Reichb. Fl. Germ. Excurs. 1: 140. 1830; Fischer, Fl. Madras 3: 1627. 1931. *Cyperus capillaris* Koenig ex Roxb., Fl. Ind. 1: 198. 1820. *P. capillaris* (Koenig ex Roxb.) Nees ex C.B.Clarke in Hook. f., Fl. Brit. India 6: 951. 1893. *Pycreus flavidus* (Retz.) T. Koyama var. *nilagiricus* (Hochst. ex Steud.) Karthik. in Kartik. & *al.* Florae Indicae enumeratio Monocotyledonae 65. 1989. *Cyperus nilagiricus* Hochst. ex Steud., Syn. Pl. Glum 2:2. 1855. *Pycreus capillaris* var. *nilagiricus* (Hochst. ex Steud.) C.B. Clarke, J. Linn. Soc. 21: 49. 1884; Fischer, Fl. Madras 3: 1627. 1931. *P. globosus* var. *nilagiricus* (Hochst. ex Steud.) C.B. Clarke, J. Linn. Soc. 36: 204. 1903.

Annual, densely tufted herb, to 40 cm high, obtusely trigonous. Leaves few, setaceous, shorter than the culm; blade slenderly linear, plicate; sheaths elongated, tinged with reddish-brown. Anthela simple with 4 unequal rays, at time congested in a head like cluster of spikelets; rays slender, the longer ones up to 5 cm long. Spikes bearing ca. 20 spikelets in short rhachis, ovoid, becoming subglobose cluster of spikelets in the upper part. Involucral bracts 2-4, erect-patent, the lowest one much surpassing the anthela, the longest up to 18 cm long. Spikelets spreading, linear, parallel-sided, strongly flattened, 20 mm long, densely 40-flowered, red brown; rhachilla straight, wingless, tetragonous. Achenes narrowly obovate, apiculate, lenticular and laterally flattened, 1 mm long, maturing dark brown.

Found in wet places in forests, near water falls and marshy localities.

Fl. & Fr.: July – November.

Potchera (ADB), *GO* & *DAM* 5054; Sirnapalli RF (NZB), *BR* & *PSPB* 9577; Mohammadabad RF (RR), *KH* & *MSM* 10923; 5573; Kodimial (KRN), *GVS* 21847 (MH & CAL); Rusthempet (MDK), *KMS* 7997 (MH & CAL); Warangal district, *P.S.Reddy* 751 (K); Kampally cheruvu (KMM), *MHR* & *MCK* 17048; Gandipet (HYD), *KI* & *NVN* 18775; Dornacheruvu, Mannanur beat (MBNR), *SRS* 10745 A (BSID).

Note: A detailed study of specimens in Kew from different regions of India made by Prasad and Simpson (2013) revealed continuous variation for a range of characters. Considering the high degree of variations within the collections of *Pycreus flavidus* in India Prasad and Simpson (2013) opined to treat the species *sensu lato* by merging all the infraspecific taxa described from India under the name *Pycreus flavidus* with relevant synonymy as given above.

Pycreus polystachyos (Rottb.) P.Beauv., Fl. De Oware & Benin 2: 48, t. 86, f. 2. 1807; C.B.Clarke in Hook. f., Fl. Brit. India 6: 592. 1893. var. **polystachyos.** *Cyperus polystachyos* Rottb., Descr. Ic. Rar. Nov. Pl. 39. t. 11, f. 1. 1773. *Pycreus odoratus* Urban, Symb. Antill 2: 164. 1900; Fischer, Fl. Madras 3: 1627. 1931.

Perennial erect, tufted rhizomatous herb, to 50 cm high; rhizome covered with purple scales; culms trigonous, smooth. Leaves shorter than the culms, flat, linear, gradually acuminate, scabrid on margins in the upper part; sheaths purple brown, 5 cm long. Inflorescence simple, reduced to a single head, 3 cm long and wide, rays sometimes present with spikelets spicately arranged. Involucral bracts 8, lower 5 much exceeding the inflorescence, others short; spikelets strongly compressed, yellow to straw-coloured, 20 mm long, linear-lanceolate, rhachilla flexuous, narrowly winged. Achenes oblong, biconvex, 1.2 mm long, dark brown, sharply apiculate.

This species is found in marshy areas, rice fields, along water streams, river banks etc.

Fl. & Fr.: September – December.

Nizamasagar project (NZB), *BR* & *CPR* 9245; Gandipet (HYD), *KI* & *NVN* 18773, 18764; Central University campus (RR), *MSM* 15178; Hamsanpally RF (MDK), *RG* 116444 (BSID).

Pycreus pumilus (L.) Nees ex C.B. Clarke in Hook. f., Fl. Brit. India 6: 591. 1893. Fischer, Fl. Madras 3: 1627. 1931. subsp. **pumilus.** *Cyperus pumilus* L., Cent. Pl. 2: 6. 1756. *Pycreus nitens* Nees, Nov. Act. Acad. Caes. Leopcarol. Nat. Cur. 19, Suppl. 1: 53. 1843; C.B. Clarke in Hook. f., Fl. Brit. India 6: 591. 1893.

Annual, densely tufted herb, to 15 cm high, culms slender, triquetrous. Leaves few, surpassing the culm, blade narrowly linear, soft, flat, light green, sheaths pale.

Anthela simple; involucral bracts 4, patent, lower 3 much surpassing the anthela, the longest 15 cm long; primary rays 4, patent, up to 3 cm long, slender; spikes subglobose, densely bearing many spikelets, the short rhachis up to 3 mm long. Spikelets spreading, oblong, 7 mm long, flattened, ca. 30-flowered, light greenish, rhachilla straight, wingless. Achenes elliptic, laterally flattened, lenticular, maturing brownish, contracted to truncate apex.

Occurs along the banks or in the drying beds of canals of soft sticky soil, marshy fields, sandy river sides.

Fl. & Fr.: August – November.

Potchera way (ADB), *MHR* 13292; Sagar left canal (KMM), *MHR* & *MCK* 17003; Narsapur tank (MDK), *BR* 11245 B; Parnasala RF (KMM), *RCS* 102474 (BSID); Aklaspar (KRN), *GVS* 25641 (MH & CAL).

Pycreus pumilus (L.) Nees ex C.B. Clarke subsp. **membranaceus** (Vahl) T.Koyama, Gard. Bull. Singapore 30: 151. 1977. *Cyperus membranaceus* Vahl. Enum. Pl. 2: 330. 1806.

Culms densely tufted, 5-16 cm high, smooth. Leaves shorter than culms or overtopping them, flat or channeled, acuminate, to 0.2 cm wide; sheaths pale brown. Inflorescence compound, to 5 cm, involucral bracts 3-5, unequal, overtopping, to 12 cm; spikes ovoid, 5-18-spikeletted, to 3 cm; spikelets oblong, compressed, subacute, 10-40-flowered, 0.4-1 cm; glumes ovate to elliptic, to 0.15 cm, membranaceous. Nut obovoid, biconvex, apiculate, stalked, brownish.

Rare in marshy lands of foot hills in Adilabad, Warangal and Ranga Reddi districts.

Fl. & Fr.: August – December.

Nirmal – Adilabad road (ADB), *MHR* 14473; Pakhal lake (WGL), *MHR* & *MCK* 17069, 17067; Rangammagudem (RR), *MSM* & *KH* 10475.

Pycreus puncticulatus (Vahl) Nees in Fl. Brasil. 2 (1): 10, in note 1842; C.B.Clarke in Hook. f., Fl. Brit. India 6: 593. 1893; Fischer, Fl. Madras 3: 1628. 1931. *Cyperus puncticulatus* Vahl, Enum, Pl. 2: 348. 1806.

Annual herb with brownish fibrous roots; culms robust, solitary, to 50 cm high; triquetrous, often some what enlarged at base. Leaves 3-5 to a culm, slightly overtopping the culm; leaf blade to 40 cm long, to 8 mm wide, subcoriaceous, glaucous green; sheaths to 20 cm long, broader than blade, pale brown. Anthela compound, lax, large, to 24 cm long and as broad. Involucral bracts 5, spreading, longer than the anthela, the lowest to 30 cm long; rays 10, unequal, patent slender, the longer ones to 14 cm long. Spikes lax, to 6 cm long, distantly bearing 20 mm long, compressed to 30-flowered, pale sanguineous. Achenes obovate, orbicular, laterally strongly flattened, rounded, emarginate at apex, stipitate at deltoid-contracted base, maturing orange-brown, densely puncticulate.

This species is usually found in the rice fields, swamps, marshy areas and tank margins.

Fl. & Fr.: August – December.

Hanamakonda (WGL), *P.S.Reddy* 494 (OU).

Pycreus sanguinolentus (Vahl) Nees ex C.B.Clarke in Hook. f., Fl. Brit. India 6: 590. 1893; Fischer, Fl. Madras 3: 1627. 1931. subsp. **sanguinolentus**. *Cyperus sanguinolentus* Vahl, Enum. Pl. 2: 351. 1806.

Perennial, rhizomatous herb, to 40 cm high; culms tufted, trigonous, clothed above the base, the lower part decumbent, 3 nodose, rooting and branching at nodes, the upper part erect. Leaves borne on the lower part of the culms, shorter than the culms; blade linear, plicate; sheaths loosely surrounding the culm, longer than the inter-node, pale green. Anthela simple with 3 short rays, 3 cm long; spikes ovoidal, densely bearing ca. 15 spikelets on short rhachis; ivolucral bracts 5, spreadidng, the lower 3 much surpassing the corymb, the longest to 12 cm long, spikelets narrowly ovate, 10 mm long, flattened, 16-flowered, sanguineous brown; rhachilla tetragonous, straight, not winged. Achenes broadly obovate, laterally lenticular, 1.2 mm long, maturing black.

Usually found in marshy localities, rice fields, swamps and along water courses.

Fl. & Fr.: July – December.

Gadirayal (RR), *MSM* 10528; Mohammadabad RF (RR), *KH* & *MSM* 10925, 10926; Dornacheruvu, Mannanur beat (MBNR), *SRS* 10745 B (BSID).

Pycreus sulcinux (Clarke) C.B.Clarke in Hook. f., Fl. Brit. India 6: 593. 1893. *Cyperus sulcinux* C.B.Clarke, J. Linn. Soc. 21: 56. 1884.

Perennial, slender, tufted herb. Spikelets longer, up to 40-fid; glumes more remote. Achenes, unsymmetric, its faces much compressed usually concave, narrow edge of achenes much flattened against the rhacheola.

Rare in marshy localities of Mahabubnagar district.

Fl. & Fr.: October – January.

Reported from Mahabubnagar district by S.R. Raghava Rao (1989).

QUEENLANDIELLA Domin.

Queenslandiella hyalina (Vahl) Ballard, Hooker's Icones Pl. 33: t. 3208. 1933. *Kyllinga hyalina* (Vahl) T.Koyama, J. Jap. Bot. 51: 313. 1976. *Cyperus hyalinus* Vahl, Enum. Pl. 2: 329. 1806. *Pycreus hyalinus* Domin., Biblioth. Bot. (20) 85: 417. 1915; Fischer, Fl. Madras 3: 1627. 1931.

An annual, densely tufted erect herb, up to 12 cm high. Leaves as long as or longer, flaccid, bracts 3-4, up to 15 cm. Inflorescence simple umbel, spikelets compressed, oblong, 6-10-flowered, glumes narrowly ovate-oblong. Nut oblique, ovate-oblong, much compressed, dark brown.

Rare in moist lands in Adilabad, Karimnagar and Hyderabad districts (Naqvi, 2001; Ramana, 2010).

Fl. & Fr.: August – October.

Ankusapuram (ADB), *GO* 4299.

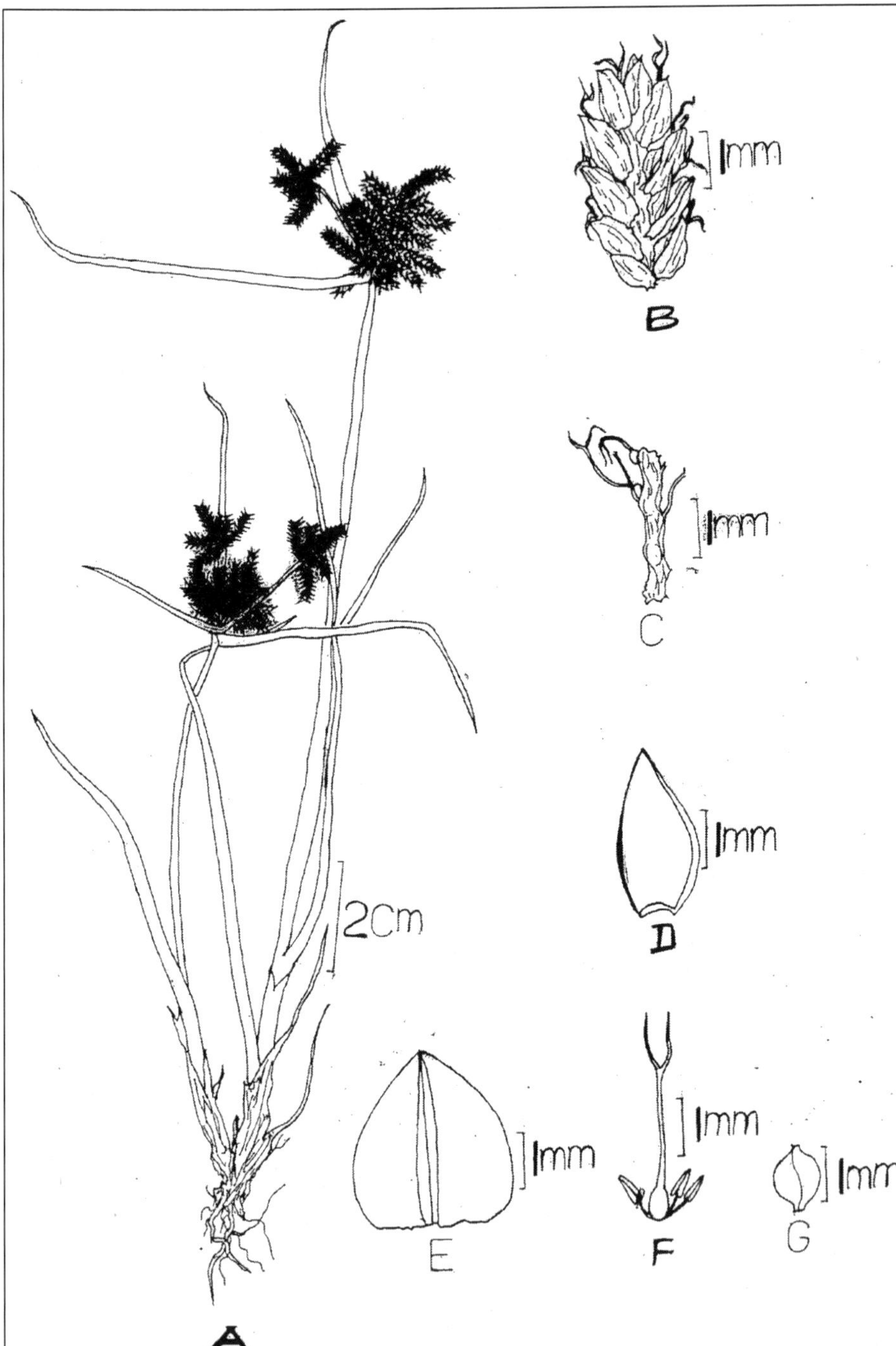

Figure 130: Pycreus sanguinolentus (Vahl) Nees ex C.B.Clarke

A. Habit, B. Spikelet, C., Rhachilla, D. Closed glume, E. Opened glume, F. Pistil with stamens, G. Achene.

RHYNCHOSPORA Vahl *nom. cons*

1. Stem leafy throughout; inflorescence paniculate consisting of a solitary or more corymbiform anthelas .... **R. gracillima**
1. Stem leafy only at the base; inflorescence capitate consisting of a solitary terminal subglobose head:
   2. Bristles all scabrid in the upper part only, the three inner ones plumose at the base .... **R. longisetis**
   2. Bristles all scabrid throughout, none plumose .... **R. wightiana**

Rhynchospora gracillima Thwaites, in Enum. Pl. Zeyl. 435. 1864; C.B.Clarke in Hook. f., Fl. Brit. India 6: 671. 1893; Fischer, Fl. Madras 3; 1672. 1931.

Annual, densely tufted herb, with yellow brown fibrous roots, to 50 cm high; culms trigonous, leaves at and above the base. Leaves 4 to a culm, setaceous; sheaths to 3 cm long, light green. Inflorescence to 20 cm long, bearing 3 very lax anthelas distantly; terminal anthela larger than the lateral ones, to 3 cm long, bearing 4 capillary rays, to 4 cm long, rays terminated by 2 spikelets; lateral anthelas simple. Involucral bracts setaceous, to 3 cm long, the sheathing base to 15 mm long. Spikelets solitary, peduncled, lanceolate, acuminate, to 7 mm long, 4-flowered, 2-fruited, brown. Achenes obovate-orbicular, rounded to short stipitate at base, truncate at apex, light grey, undulate, rugose with 8 ridges.

Occasional in the opening of secondary forest and wet grass lands.

Fl. & Fr.: August – December.

Rhynchospora longisetis R. Br., Prodr. 230. 1810; C.B.Clarke in Hook. f., Fl. Brit. India 6: 669. 1893.

Annual, slender, tufted herb, to 30 cm high; culms trigonous, smooth. Leaves flat or canaliculated, to 20 cm long, scaberulous at the tip. Inflorescence a terminal subglobose head. Involucral bracts 6, spreading, dilated and densely ciliate at the base, lowest up to 8 cm long; spikelets many, linear-lanceolate, acuminate, to 12 mm long, fulvous-ferrugineous, 2-flowered, lower female, upper male. Achenes sessile, oblong, compressed, bristly on the shoulders, to 3 mm long, brown; beak oblong-conical, to 2 mm long with a median groove on each side.

Though rare this species is seen in moist grounds in forests, in hilly areas, in wet grass lands.

Fl. & Fr.: September – December.

Lawal beat, Pasra (WGL), *RKP* 108153 (BSID); Nellippakka RF (KMM), *RCS* 99087 (BSID); Hyderabad (HYD), *MRS* 255 (DD).

Rhynchospora wightiana (Nees) Steud., Cyper. 148. 1855; C.B.Clarke in Hook. f., Fl. Brit. India 6: 669. 1893; Fischer, Fl. Madras 3: 1672. 1931. *Haplostylis wightiana* Nees, Nov. Act. Akad. Nat. Curr. 19 (Suppl. 1); 101. 1843.

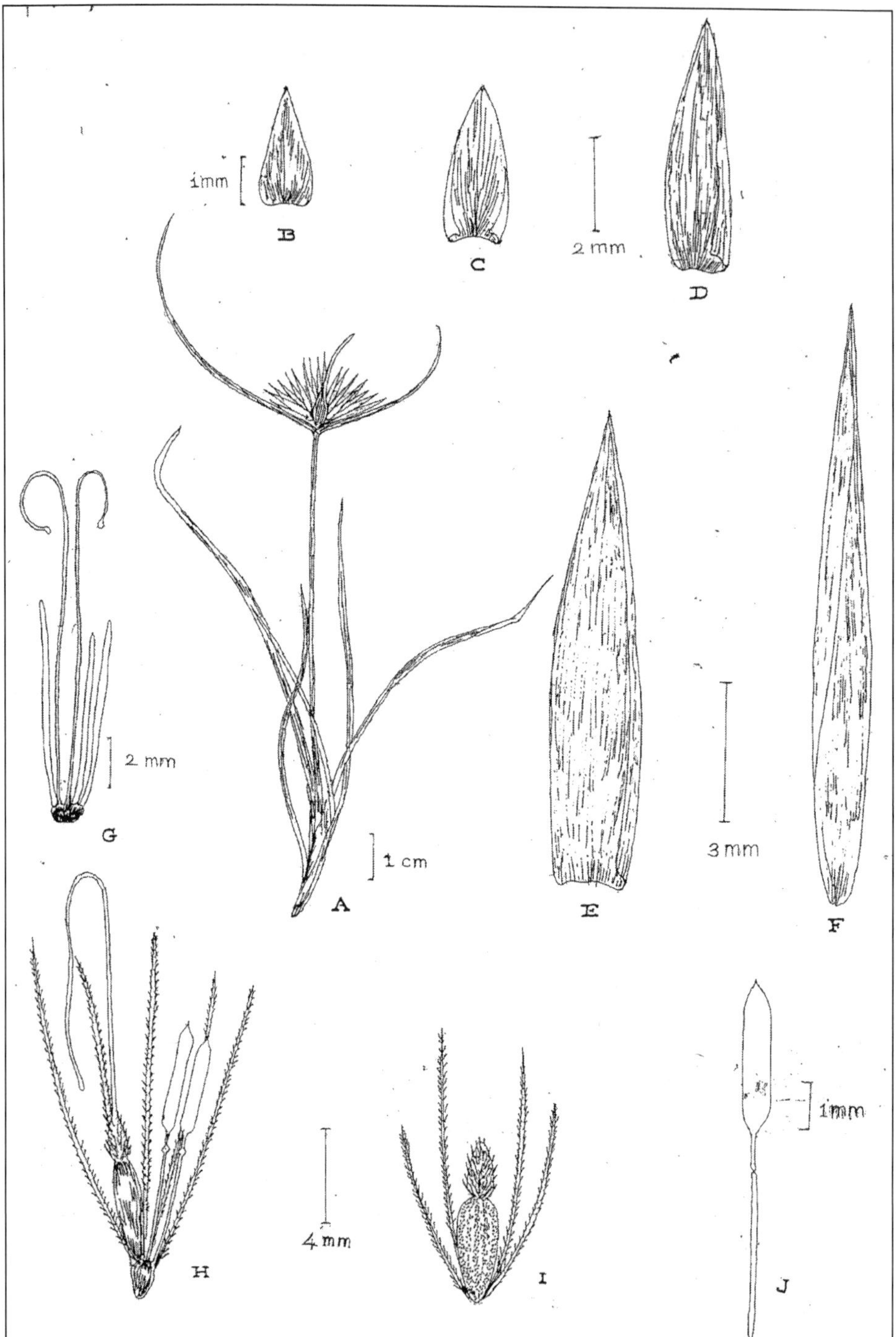

Figure 131: Rhynchospora wightiana (Nees) Steud.

A. Twig, B-F. Glumes, G. Pistil, H. Pistil with stamen, I. Nut with bristles, J. Stamen.

Perennial, erect, caespitose herb, to 35 cm high, culms slender, trigonous, glabrous. Leaves many, at the base of the stem only, to 14 cm long, linear, acuminate, grass like. Spikelets numerous in reddish-brown heads. Involucral bracts 6, unequal, the longest to 7 cm long, linear-lanceolate, acuminate, dilated and ciliate at the base. Spikelets to 1.5 cm long, lanceolate, acute, reddish-brown, smooth and shining. Achenes compressed, oblong, dark brown.

Occasionally found in rice fields, grass lands and in moist grounds in forests.

Fl. & Fr.: September – January.

Chennur (ADB), *GO* 4500; Annaram RF (NZB), *BR* & *CPR* 7177; Gandivet RF (NZB), *BR* 9545; Pocharam tank (MDK), *TP* & *CP* 12031; Mohammadabad RF (RR), *MSM* & *KH* 10480; Central University campus (RR), *MSM* 15168; Vazeedu (KMM), *VSR* 1005 (MH).

SCHOENOPLECTIELLA Lye

1. Culms flowering in the lower half, septate when dry; glumes not mucronate:
    2. Culms robust; sheaths lax; achenes with transverse wavy lines .... **S. articulatus**
    2. Culms slender; sheaths close; achenes with transverse undulate ridges:
        3. Perennial .... **S. senegalensis**
        3. Annual .... **S. praelongatus**
1. Culms flowering in the upper half, not septate (except *S. roylei*); glumes mucronate:
    4. Glumes multistriate, lax, inflated in fruit; stem septate when dry .... **S. roylei**
    4. Glumes not striate, imbricating; not inflated in fruit; stem not septate when dry .... **S. supinus**

Schoenoplectiella articulata (L.) Lye, Lidia 6: 20: 2003. *Scirpus articulatus* L., Sp. Pl. 47. 1753; C.B.Clarke in Hook. f., Fl. Brit. India 6: 656. 1893; Fischer, Fl. Madras 3: 1666. 1931. *Schoenoplectus articulatus* (L.) Palla, Bot. Jahrb. Syst. 10: 299. 1888.

Perennial, short living, robust herb, to 40 cm high; culms terete; 0.8 cm wide below, transversely septate at intervals of 1 cm, hollow, smooth, glabrous deeply green and slightly shiny when alive, clothed at base with 3 bladeless stramineous-brown lax sheaths; upper 2 basal sheaths cylindrical, to 12 cm long, herbaceous, pale green and later becoming yellow brownish, finely many-nerved, lower 1 sheath reduced and scale like. Inflorescence a pseudolateral head, globose or spherical, 4 cm in diameter, located at the midway portion of the (culm) stem or much lower down near the apex of upper-most sheath (depending upon the depth of the water at the locality as the inflorescence appear immediately above the water surface) densely bearing several to over 40 spikelets, lightly green. Involucral bract 1, stem like, to 40 cm long, terete, septate. Spikelets cylindrical or oblong, to 14 × 4 mm, pink-fulvous,

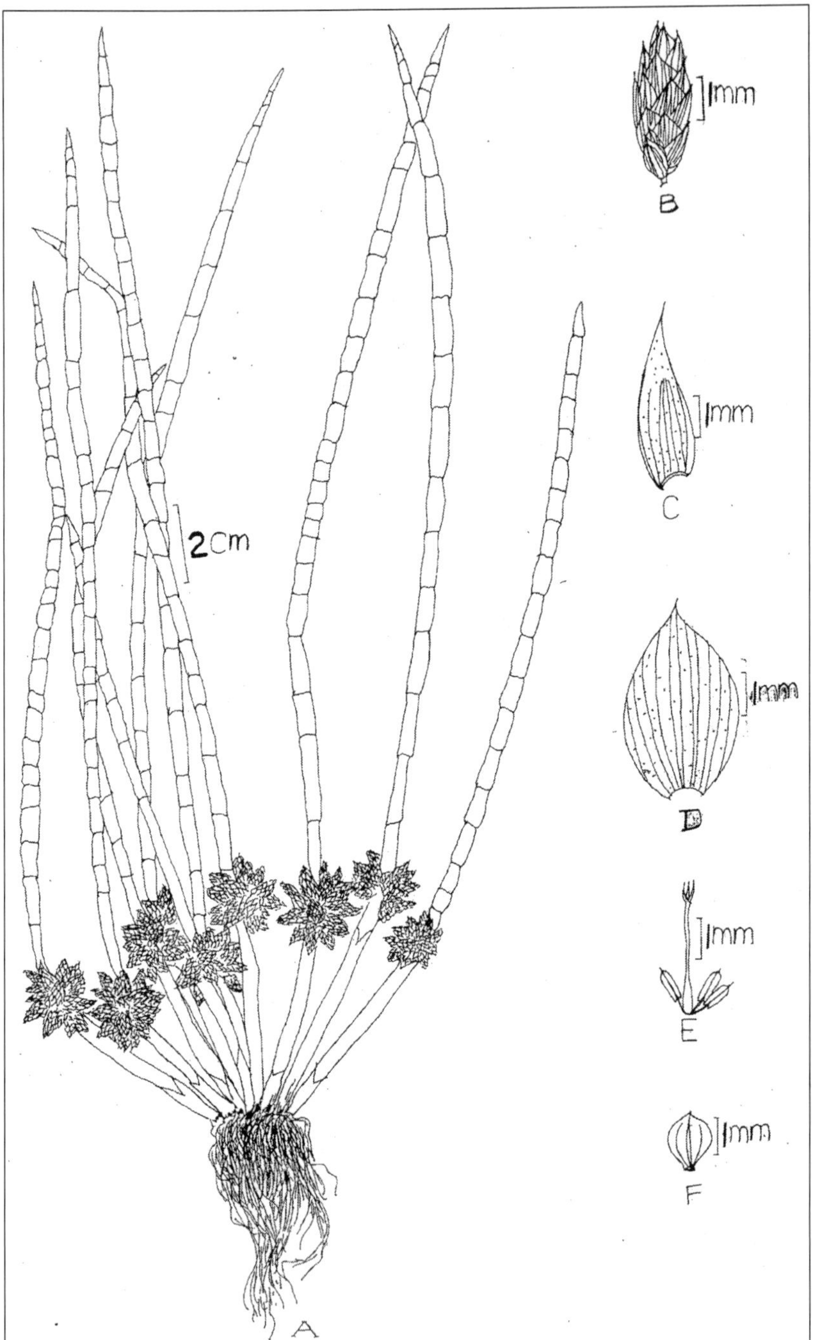

Figure 132: Schoenoplectiella articulata (L.) Lye [Syn: *Schoenoplectus articulatus* (L.) Palla]

A. Habit, B. Spikelet, C. Closed glume, D. Opened glume, E. Pistil with stamens, F. Achene.

acuminate. Achenes obovid, triquetrous, with transverse wavy lines and apiculate.

Common species usually found in marshy places, shallow waters and also along the margins of ponds.

Fl. & Fr.L August – December.

Kadam bank (ADB), *TP* & *PVP* 4145; Kunthala (ADB), *MHR* 13269; Indalwai (NZB), *TP* & *BR* 6405; Narsapur tank (MDK), *TP* & *CP* 11998; Kothapalli beat (RR), *KH* & *MSM* 10915; Mohammadabad RF (RR), *KH* & *MSM* 10916; Bhadrakali lake (WGL), *MHR* & *MCK* 17086, 17089; Lakkavarapu cheruvu (KMM), *MHR* & *MCK* 16689; Lingalacheruvu (KMM), *MHR* & *MCK* 17030; Vatavarlapalli (MBNR), *BSS* & *SKB* 33118; Rechapalli (KRN), *GVS* 22265 (MH & CAL).

Schoenoplectiella praelongata (Poir.) Lye, Lidia 6: 26. 2003. *Scirpus praelongatus* Poir. Encycl., 6: 764. 1805. *Schoenoplectus praelongatus* (Poir.) J. Raynal, Adansonia n.s., 16: 148. 1976.

Annual, tufted herb, 10-50 cm high; stem shorter than lowest bract, terete. Sheaths 2, upper as long as stem, lower shorter, mouth oblique, blades usually reduced to mucro, in upper sheath occasionally elongated. Inflorescence congested, spherical group of 7-30 sessile spikes, 10-25 mm diam; spikes ovoid, glumes 3.5-4.5 mm, margins scarious; perianth bristles 0; stamens 3, stigmas 3. Nut widely obovoid, trigonous, dark brown.

Occasional in Medak and Khammam districts.

Fl. & Fr.: September – January.

Narsapur (MDK), *RG* 116457 (BSID); Pocharam Wild Life Sancturay (MDK), *RG* 113600 (BSID); Pakhal (WGL), *RKP* 108290 (BSID); Bhadrachalam RF (KMM), *RCS* 98982 (BSID).

Schoenoplectiella roylei (Nees) Lye, Lidia 6: 26. 2003. *Isolepis roylei* Nees in Wight, Contrib. Bot. Ind. 107. 1834. *Schoenoplectus roylei* (Nees) Lye, Bot.Not. 124: 290. 1971. *Scirpus quinquefarius* Buch.-Ham. ex Boeck., Linnaea 36: 701. 1870; C.B. Clarke in Hook. f., Fl. Brit. India 6: 657. 1893.

Annual, erect, slender, tufted herb, to 30 cm high; culms terete, slender, slightly compressed, conspicuously transversely septate. Leaves absent. Spikelets in a single lateral cluster near the top of the stem, pale-brown, shining. Achenes triquetrous, obovoid, faintly transversely wavy, black.

Occasional in drying paddy fields, on muddy or wet soil and in marshes near canals and streams.

Fl. & Fr.: August – November.

Osmania University Campus (HYD), *T.Rajagopal* 932 (HH).

Schoenoplectiella senegalensis (Steud.) Lye, Lidia 6: 27. 2003. *Schoenoplectus senegalensis* (Hochst. ex Steud.) Palla Bot. Jahrb. 10: 299. 1888. *Isolepis senegalensis* Hochst. ex Steud., Syn. Pl. Glum. 2: 96. 1855. *Scirpus jacobii* Fischer, Kew Bull. 1931: 103. 1931; Fischer, Fl. Madras 3: 1666. 1931.

Perennial, erect, caespitose herb to 30 cm high; culms slender, subterete, transversely septate. Sheaths close, obliquely truncate. Inflorescence pseudolateral, capitate. Involucral bracts greater than 20 cm long, stem like. Spikelets ca. 20 in a cluster, ovoid-subglobose, 8 mm long, fulvous. Achenes obovoid, triquetrous, to 2 mm long, undulate, sharply angled, apiculate.

Occasional in marshy places.

Fl. & Fr.: August – December.

Navipet (NZB), *BR* & *PSPB* 9595; Narsapur tank (MDK), *BR* 11244; Mohammadabad RF (RR), *KH* & *MSM* 10918.

Schoenoplectiella supina (L.) Lye, Lidia 6: 27. 2003. *Schoenoplectus supinus* (L.) Palla, Bot. Jahrb. Syst. 10. 299. 1888. *Scirpus supinus* L., Sp. Pl. 49. 1753; C.B.Clarke in Hook. f., Fl. Brit. India 6: 655. 1893; Fischer, Fl. Madras 3: 1666. 1931.

Annual, tufted herb, to 30 cm high; culms trigonous, greenish, clothed at base with 2 sheaths only. Sheaths 5 cm long, membranous, pale green, the orifice obliquely truncate with a 2 mm long mucro, frequently the upper most sheaths born 3 cm above the base of the culm. Inflorescence pseudolateral, head like. Involucral bract 1, to 10 cm long, ventrally sulcate. Spikelets to ca. 15 to an inflorescence, wholly congested, ovate-oblong, to 6 mm long, pale green and becoming lightly straw-coloured, subdensely many-flowered. Achenes broadly elliptic 1.8 mm long, truly triangular, suddenly contracted at both ends, mucronate at apex, the sides transversely wrinkled, maturing black.

Common species usually found in wet grassy places, margins of ponds, rice fields.

Fl. & Fr.: August – December.

Kadam dam (ADB), *TP* & *PVP* 4143; Swarna project (ADB), *MHR* 14576; Potchera way (ADB), *MHR* 13297; Janakampalli (NZB), *TP* & *BR* 6324; Pathur (MDK), *TP* & *CP* 12107; Mahammadabad RF (RR), *KH* & *MSM* 10919; Sirnapalli RF (NZB), *BR* & *PSPB* 9576; Rusthempet (MDK), *KMS* 6786 (MH & CAL); Pakhal (WGL), *ANH* 15962 (MH).

SCHOENOPLECTUS (Reich.) Palla

1. Spikelets clustered on lateral or terminal heads:
    2. Hypogynous bristles present, retrorsely scabrid:
        3. Culms slender, terete; style 2-fid; achenes with transverse wavy lines .... **S. juncoides**
        3. Culms robust, sharply triquetrous; style 3-fid; achenes vertically lineolate with dark coloured patches .... **S. mucronatus**
    2. Hypogynous bristles absent .... **S. lateriflorus**
1. Spikelets many, stalked, umbellate .... **S. littoralis**

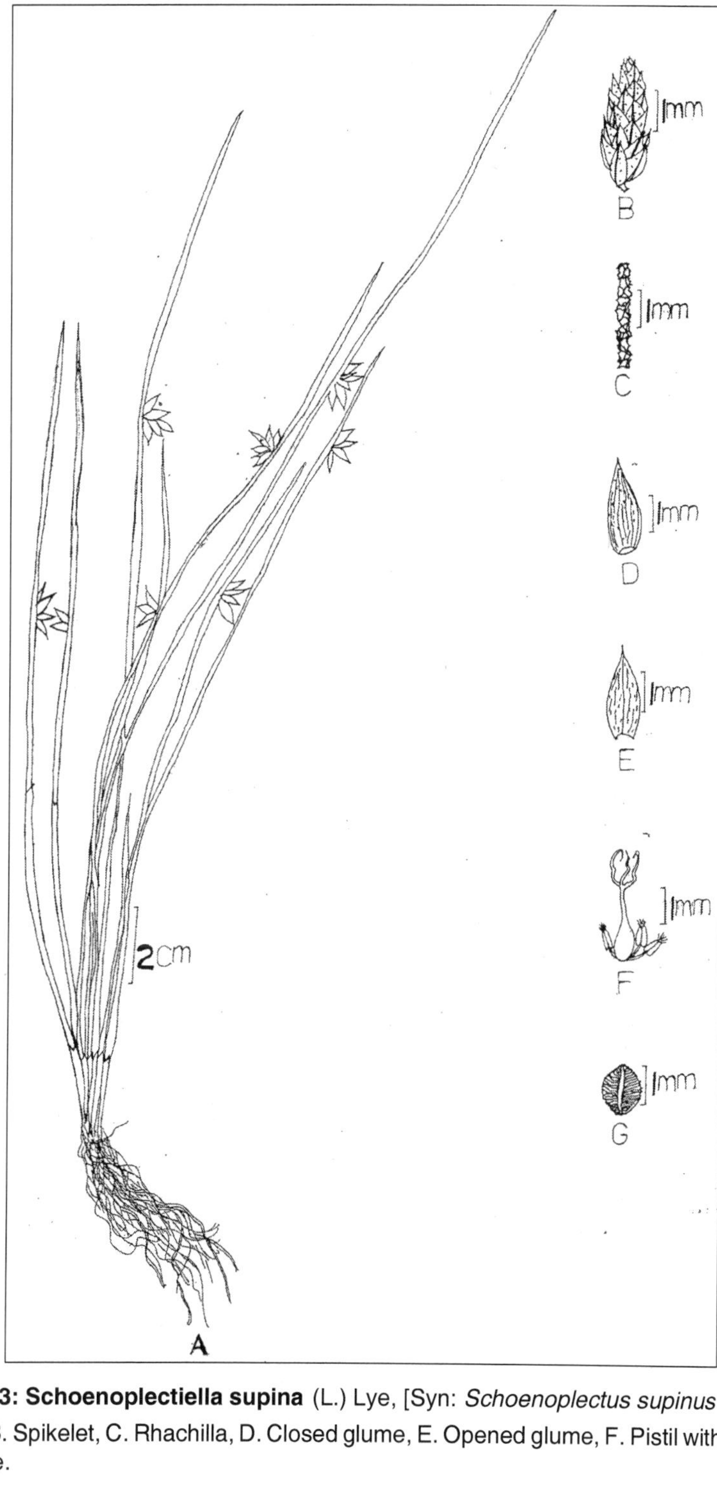

Figure 133: Schoenoplectiella supina (L.) Lye, [Syn: *Schoenoplectus supinus* (L.) Palla]
A. Habit, B. Spikelet, C. Rhachilla, D. Closed glume, E. Opened glume, F. Pistil with stamens, G. Achene.

Schoenoplectus juncoides (Roxb.) Palla, Bot. Jahrb. Syst. 10: 229. 1889. *Scirpus juncoides* Roxb., Fl. Ind. 1: 228. 1820. *S. erectus* C.B.Clarke in Hook. f., Fl. Brit. India. 6: 656. 1893; Fischer, Fl. Madras 3: 1666. 1931.

Annual, erect, caespitose herb without conspicuous rhizome, to 30 cm high; culms slender, terete, obtusely several-angled, light green, clothed at base with few sheaths only. Sheaths 3, the lower ones scale like, brownish, the upper one to 5 cm long, pale green, obliquely truncate. Inflorescence a pseudolateral head with 3-6 spikelets; involucral bracts to 13 cm long, rather suddenly subacute at apex, dilated at base. Spikelets oblong to ovoid-oblong, terete, 3 × 8 mm, straw-coloured, acute, densely many-flowered. Achenes broadly obovoid, unequally biconvex, 2 mm long, compressed, faintly transversely undulate, apiculalte.

Common in the margins of rice fields, river banks, swampy and marshy localities.

Fl. & Fr.: September – December.

Mohammadabad RF (RR), *KH* & *MSM* 10917; Kodimial (KRN), *GVS* 21859 (CAL), 20137 (MH); Vazeedu (KMM), *VSR* 1005 (MH); Pakhal (WGL), *RKP* 11600 (BSID).

Note: C.B.Clarke misapplied the same *Scirpus erectus* Poir. to a quite different species occurring in Madagascar, S.E. Asia and Tropical Australia and he was followed by several authors. The correct name of the species is *S. juncoides* Roxb. and the name *S. erectus* Poir. does not belong to its synonymy. *S. erectus* is related to *S. supinus* L. than to *S. juncoides* Roxb. It differs from *S. supinus* by the larger spikelets, large, more distinctly mucronate glumes, the bristly appendage of the connective, the bifid style, and the larger, convex, faintly wavy ridged elliptic or suborbicular achenes.

Schoenoplectus lateriflorus (J.F. Gmel.) Lye, Bot. Notis. 124: 290. 1971. *Scirpus lateriflorus* J.F. Gmel., Syst. Veg. ed. 13. 1: 127. 1791. *S. supinus* Roxb., Fl. Ind. 1: 219. 1820 non L. 1753; *S. supinus* var. *uninodis* sensu C.B.Clarke in Hook. f., Fl. Brit. India 6: 656. 1893 non (Del.) Asch. & Schweinf. 1887.

Annual, tufted herb, to 30 cm high; culms terete; sheaths often enclosing a long-stigmaed ovary or enlarged achenes at base. One leaf blade often developed. Involucral bract to 10 cm long, slender, canaliculated, obtuse. Spikelets 4-6, sessile, ovate-oblong, pale-brown, acute with glumes in 5 ranks; glumes elliptic, papery, strongly keeled above, apiculate, tinged dark red brown; hypogynous bristle lacking. Achenes broadly obovate, unequally triquetrous, to 2 mm long, sharply transversely ridged, black.

Common on plains, in rice fields, and other wet areas.

Fl. & Fr.: September – January

Reported from Karimnagar (Naqvi, 2001), Warangal (Satyanarayana Reddy, 1985) and Medak districts (Narasimha Rao, 1986).

Schoenoplectus litoralis (Schrad.) Palla, Sitzber. Zool. Bot. Ges. Wien 38: 49. 1888. *Scirpus litoralis* Scharad., Fl. Germ. 1: 42. t. 5, f. 7. 1806; C.B.Clarke in Hook. f., Fl. Brit. India 6: 659. 1893; Fishcer, Fl. Madras 3: 1667. 1931.

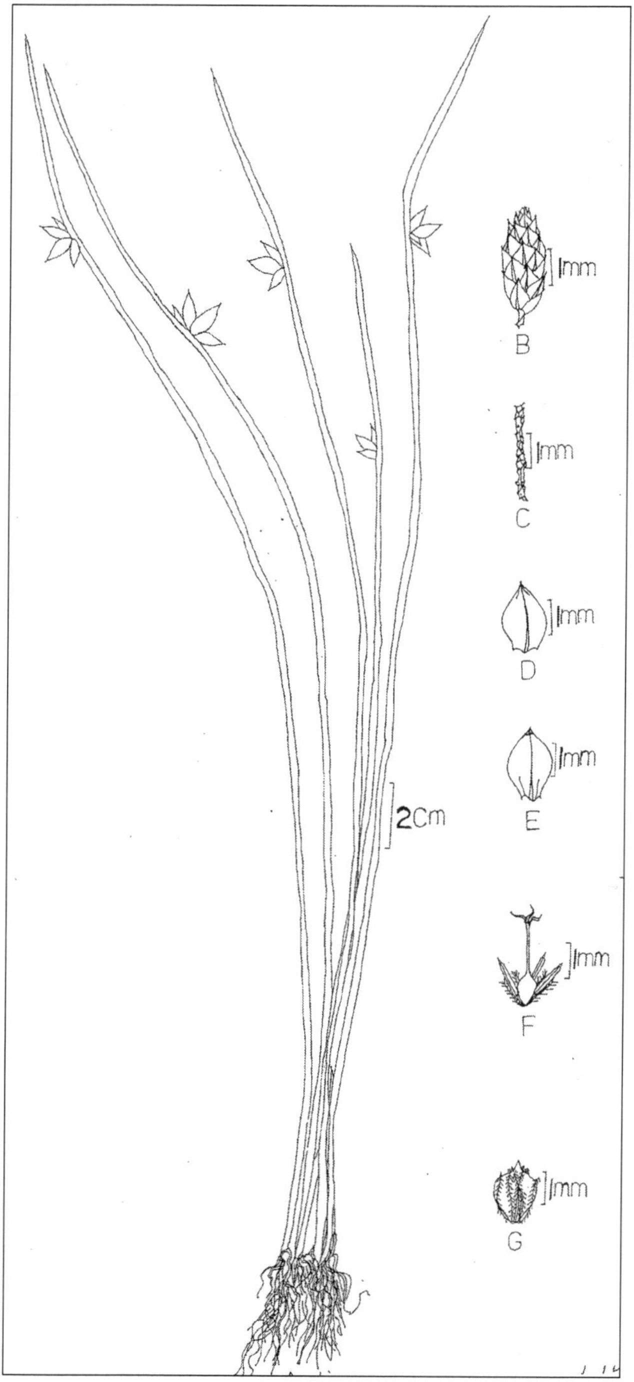

Figure 134: Schoenoplectus juncoides (Roxb.) Palla

A. Habit, B. Spikelet, C. Rhachilla, D & E. Glumes, F. Pistil with stamens and hypogynous bristles, G. Achene.

Perennial, rhizomatous, stoloniferous herb, the stolon terminated by a tuber; culms to 70 cm high, terete, but obtusely triangular below inflorescence, closed at base with blade-less sheaths. Basal sheaths membranous, cylindrical, to 15 cm long, greyish-brown, mouth oblique with hyaline margin. Inflorescence a psudolateral simple, umbel like corymb with 4 unequal ascending rays, to 3 cm long, terminated by a cluster of spikelets. Involucral bract 1, foliaceous, as long as corymb. Spikelets oblong-obovate, to 8 mm long, many-flowered, rust coloured. Achenes broadly obovate, biconvex, 2 mm long, faces a little shining, chestnut brown at maturity.

Rarely seen in swampy areas in Warangal, Adilabad and Nizamabad districts.

Fl. & Fr.: September – March.

Sone (ADB), *TP* & *PVP* 4048; Sreeramsagar (NZB), *BR* & *GO* 9226;

Schoenoplectus mucronatus (L.) Palla, Bot. Jahrb. Syst. 10: 229. 1889 subsp. **mucronatus.** *Scirpus mucronatus* L., Sp. Pl. 50. 1753; C.B.Clarke in Hook. f., Fl. Brit. India 6: 657. 1893; Fischer, Fl. Madras 3: 1666. 1931.

Perennial, rhizomatous herb, to 120 cm high; culms robust, sharply 3-angled, sides deeply green, striate when dry, surrounded at base with 2 to 3 sheaths tightly, the lowest sheath scale like, brown to chestnut-brown, the upper often septate-nodose below, mouth obliquely truncate, brown scarious. Inflorescence of hemispherical pseudolateral head with 10 spikelets. Involucral bract 1, culm like, sharply 3-angled, to 5 cm long. Spikelets sessile, stramineous-brown, ovate, to 13 mm long, densely many-flowered, terete. Achenes broadly obovate, obcompressed, triangular, 2 mm long, glossy, transversely wrinkled with dark coloured patches.

A rare species usually collected from forest areas, shallow waters, marshy places.

Fl. & Fr.: September – December.

Pathur (MDK), *TP* & *CP* 12113; Hussain sagar Dam (HYD), *KMS* 8036 (MH & CAL).

SCLERIA Bergius

1. Culms from woody rhizome, solitary or close together but not densely tufted; plants perennial:
    2. Achenes globose, as long as wide; hypogynium well developed .... **S. terrestris**
    2. Achenes elongated, conspicuously ovate, 1.5 times longer than the width: hypogynium much reduced, only strip like
        3. Robust plant, culms 90-150 cms high; inforescence corymbose-paniculate, dense .... **S. corymbosa**
        3. Slender plant, culms to 80 cm high; inforescence very close with spiciform branches .... **S. lithosperma**

1. Culms tufted with fibrous roots only, without conspicuous rhizome; plants annual:
    4. Hypogynium deeply 3-lobed:
        5. Lobes of hypogynium subcoriaceous, ovate-oblong:
            6. Achenes cancellated and often pubescent .... **S. parvula**
            6. Achenes irregularly lacunose, glabrous .... **S. multilacunosa**
        5. Lobe of hypogynium thin, narrowly lanceolate .... **S. tessellata**
    4. Hypogynium shallowly or hardly lobed .... **S. rugosa**

Scleria corymbosa Roxb., Fl. Ind. 2, 3: 574. 1832; Fischer, Fl. Madras 3: 1677. 1931. *S. ridleyi* C.B.Clarke in Hook. f. Fl. Brit. India 6: 686. 1893.

Perennial, rhizomatous robust herb, to 150 cm high; rhizome woody, horizontal, covered with reddish-brown scales; culms solitary, acutely triquetrous, many-noded. Leaves evenly distributed on the total length of culms; leaf blades broadly linear, to 60 × 20 mm, sub-coriaceous, light green, 3-costate, glabrous, abruptly narrowed to an obtuse apex, the margins smoothish for most length, scabrid towards leaf apices, sheaths elongated, to 20 cm long, somewhat loose, 3-sided, not wigned, the lower ones coloured with brown, the basal sheaths bladeless, reddish-brown; contraligule depressed-deltoid, to 6 mm long. Inflorescence to 80 cm long, cylindrical, leafy; lateral panicles at 3 to 5 nodes, decompound, corymbiform, dense, preduncles long exserted, unequal in length, acutely triquetrous. Bisexual and staminate spikelets intermingled; bisexual spikelets sessile, ovoid, 4-5 mm long, bearing a sessile staminate spikelet at axil of the uppermost glumes. Achenes ovoid, exceeding the subtending glumes.

Seshagiri Rao (2012) reported from Central University campus..

Fl. & Fr.: September – December.

Scleria lithosperma (L.) Sw., Prodr. 18: 1788; C.B.Clarke in Hook. f., Fl. Brit. India 6: 685. 1893; Fischer, Fl. Madras 3: 1677. 1931. subsp. **lithosperma.** *Scirpus lithospermus* L., Sp. Pl. 1: 51. 1753.

Perennial, erect, tufted herb, to 80 cm high; culms loosely tufted from short knotty rhizome, rigid, slender, 3-sided. Leaves dense or clustered at the middle of the stem; leaf blade narrowly linear, to 25 cm long, rigid, canaliculated, revolute on scabrous margins, deeply green and glaucescent when dried; sheaths 3-sided, to 7 cm long, brownish-reddish, not winged, pubescent, the basal ones bladeless, redish-purplish; contraligule convex, hairy. Panicles narrow, very loose, spicate, terminal ones to 15 cm long, axillary ones to 5 cm long, dark brown; involucral bracts leaf like, much longer than the inflorescence. Spikelets bisexual, solitary and clustered (in clustered condition 1 female, 1 or 2 male flowers together, others empty), ovate-turbinate. Achenes obovoid, to 3 mm long.

Common species usually found along hill slopes associated with other grasses, rocky surfaces and as the undergrowth of dry forests.

Fl. & Fr.: Throughout the year.

Ankusapuram (ADB), *GO* 4346.

Key to Subspecies

1. Nutlets smooth .... subsp. **lithosperma**
1. Nutlets transversely rugose or irregularly somewhat reticulate with wavy ridges .... subsp. **linearis**

Scleria lithosperma (L.) Sw. subsp. **linearis** (Benth.) T.Koyama in Dassanayake & Fosberg, Revis. Handb. Fl. Ceylon 5: 353. 1985. *Scleria lithosperma* var. *linearis* Benth., Fl. Austral. 7: 430. 1878.

Nutlets transversely rugose or irregularly somewhat reticulate with wavy ridges.

Rare weed in cultivated fields of Medak district.

Pathur fields (MDK), *TP* & *CP* 12118.

Scleria multilacunosa T. Koyama, Bull. Nat. Sci. Mus. 17: 71. f. 3. 1974.

Annual, caespitose herb with purplish fibrous roots, culms to 90 cm high, sharply triquetrous, 3 or 4 nodose, 1 or 2 leaved on the midway portion, clothed at base with 1 or 2 bladeless sheaths. Basal sheaths to 4 cm long, sharply triquetrous, without wings, pale-green and stained with purple, obliquely truncate at orifice. Leaves shorter than the culm; blade 15-30 cm × 3-8 mm, tricostate, fresh green, scabrous on margins and abaxial costa, gradually tapering to acute apex; sheaths to 6 cm long, greenish, sharply triquetrous, the angles scabrous; contraligule rounded-truncate, ciliate with short rusty brown hairs. Inflorescence with 2-4 partial panicles the lowest one distant from the remainder and on a long exserted peduncle, the remainder continguous at the culm apex; partial panicles subspiciform, to 2 cm long, branches bearing 3 to 5 pistillate spikelets and 1-3 staminate spikelets. Lowest leafy bract sheathing for 2 to 3 cm, the blade to 12 cm long, not surpassing the culm; blade of the second bract to 5 cm long, longer than its subtending partial panicle, ciliate at base, the third and fourth bracts much shorter. Staminate spikelets to 3.5 mm long, lanceolate, short peduncled. Pistiliate spikelets to 4 mm log, obdeltoid. Achenes globose, to 3 mm long, rounded at mucronate apex, the whitish ceramic surface irregularly lacunose with many shallow depressions of varying shape and size.

Occasional in wet grassy localities around bushes and herbs in the openings of the forests in Warangal district.

Fl .& Fr.: September – December.

Pakhal (WGL), *AR* 1195 (MH & KU).

Scleria parvula Steud., Syn. Pl. Glumac. 2: 174. 1855; C.B. Clarke in Hook. f., Fl. Brit. India 6: 686. 1893; Fischer, Fl. Madras 3: 1677. 1931.

Annual, erect, caespitose herb, to 25 cm high; culms slender, triquetrous, scabrid on angles, with dark or reddish brown fibrous roots. Basal sheaths bladeless, acutely triquetrous, reddish brown. Leaves with elongated blade spaced on the lower midway portion of the culm; blade linear, to 15 cm long, herbaceous, lightly green, scabrid on abaxial and vein and margins; sheaths to 8 cm long, loosely clothing the stem-internode, triquetrous; contraligule truncate or rounded, ciliate on margin. Inflorescence terminal and lateral, of 1-3 cm long, sparingly branched panicle.

Involucral bracts erect, leaf like, nearly-equalling to slightly surpassing the inflorescence. Staminate spikelets lanceolate; pistillate spikelets obdeltoid. Achenes globose to ellipsoidal, to 3 × 2 mm, cancellated with regularly arranged transversely rectangular lacunae, shining white, often pubescent, rounded at mucronate apex; hypogynium 3-lobed, pale green; lobes ovate.

Rarely seen in wet grass lands; margins of forests.

Fl. & Fr.: May – October.

Gandivedu fields (NZB), *BR* & *CPR* 7197; Vazeedu (KMM), *VSR* 1003 (MH).

Scleria rugosa R. Br., Prodr. 240. 1810. *S. zeylanica* sensu C.B.Clarke in Hook. f., Fl. Brit. India 6: 687. 1894. *S. flaccida* C.B.Clarke in Hook. f., *loc.cit.* 688 non Steud. 1855.

Annual, densely tufted herb, to 3.5 cm high; culms acutely 3-angled. Leaf blade narrowly linear, to 10 cm long, soft, 3-costate, abruptly narrowed to obtusish apex; sheaths to 3 cm long, triquetrous, the basal ones tinged with brown. Inflorescence consisting of 3 small contracted panicles, each to 2 cm long, with few clusters of spikelets; lateral panicles solitary on exserted peduncle; peduncles stout, compressed-triangular. Involucral bracts leaflike, elongated, the longer ones surpassing the total inflorescence. Pistillate spikelets and staminate spikelets intermingled. Pistillate spikelets broadly obovate, to 4 mm long. Staminate spikelets on short peduncle, to 2 mm long, lanceolate. Achenes shorter than the subtending glumes, globular, terete, white.

Occasionally seen in wet grass places, margins of marshy areas.

Fl. & Fr.: September – December.

Reported from Warangal district by Satyanarayana Reddy (1985) and Hyderabad district by Ramana (2010).

Scleria terrestris (L.) Fassett., Rhodora 26: 159. 1924. *Zizinia terrestris* L., Sp. Pl. 2: 991. 1753. *Scleria radula* Hance., Ann. Sci. Nat. Bot. 18. 232. 1862. *S. chinensis* Kunth. var. *biauriculata* C.B. Clarke in Hook. f., Fl. Brit. India 6: 690. 1893. *S. cochinchinensis* (Lour.) Druce, Bot. Exch. Club. Brit. Isles Rep. 4: 646. 1917; Fischer, Fl. Madras 3: 1678. 1931.

Perennial herb with short, creeping woody rhizome, to 150 cm high; culms elongated, triquetrous, densely leaved for the most length; leaf blades linear, to 40 cm long, to 20 mm wide, subcoriaceous, 3-costate; sheaths to 13 cm long, triquetrous, coloured with reddish brown, scabrid on angles. Inflorescence consisting of 4 compound panicles, generally spaced, to 40 cm long, 10 cm wide; partial panicles solitary, ovate pyramidal or broadly ovoid, dense, the branches ascending, scabrid, the peduncle long-exserted up to 20 cm long, compressed triangular, scabrous. Involucral bracts leaf like, slightly overtopping the inflorescence. Pistillate spikelets and staminate spikelets in groups of 2 or 3; pistillate spikelets sessile, broadly ovoid, to 6 mm long, staminate spikelets lanceolate, to 4 mm long, deeply red-brown. Achenes longer than the two subtending glumes, broadly obovoid-globose, to 3 mm long, shining white or finally dark purplish.

A rare species usually found in wet places in grasslands in Karimnagar district (Naqvi, 2001).

Fl. & Fr.: February – December.

Scleria tessellata Willd. Sp. Pl. 4: 315. 1805. *S. biflora* Roxb., Fl. Ind. 3:573. 1832; C.B.Clarke in Hook. f., Fl. Brit. India 6: 687. 1894. *S. tessellata* var. *biflora* (Roxb.) Blatt. & McCann, J. Bombay Nat. Hist. Soc. 37: 777. 1935.

Annual, loosly tufted herb, to 50 cm high; culms distinctly few nodose, triquetrous, glabrous. Leaves dense towards the base of culms, the upper ones much spaced; leaf blade linear, to 20 cm long, soft, 3-costate, sparsely pilose, abruptly narrowed to a short obtuse apex, scaberulous on upper margins; sheaths 3-sided, to 4 cm long, loosely surrounding the culm, scabrous on wing edges. Inflorescence with 4 panicles; terminal panicles larger than the lateral ones, to 4 cm long; lateral panicles much spaced, solitary, erect; peduncles slightly exerted from bract sheath, compressed-tringulalr, smooth. Involucral bracts leaf like, elongated, much longer than their subtending panicle but not overtopping the total inflorescence, long sheathing at base. Pistillate spikelets and staminate spikelets intermingled. Pistillate spikelets broadly obovoid, to 4 mm long. Achenes globose, short beaked at apex, ceramic white, finely trabeculate with regular vertical rows of square-pits, ferrugineous-pubescent on ridges between pit rows, disk deeply 3-lobed; lobes thin, narrowly lanceolate, brownish-yellow.

Usually found in wet grass lands and sedge swamps, wet forest scrub and as a weed in rice fields in Warangal district (Satyanarayana Reddy,1985; V.S.Raju & *al.*, 2008).

Fl. & Fr.: September – December.

POACEAE (*nom. alter.* GRAMINEAE)

KEY TO SUB-FAMILIES

1. Mature spikelets falling entire from their pedicels; all alike or differing in sex and structure, perfect spikelets with 2 heteromorphous florets, the upper bisexual, the lower male or barren .... **Panicoideae**

1. Mature spikelets breaking up, leaving more or less persistent glumes on the pedicel or if falling entire then not consisting of heteromorphous florets .... **Pooideae**

KEY TO THE TRIBES OF PANICOIDEAE

1. Male and female spikelets in separate inflorescence on the same plant or the female spikelets at the base of the inflorescence and male above them, lemma hyaline or membranous, thinner than the glumes .... **Maydeae**

1. Male and female spikelets in same inflorescence, all bisexual or with male or barren; male or barren spikelets near to the bisexual spikelet, lemma of the fertile floret indurated:

2. Spikelets usually in pairs or in threes, one sessile and other pedicelled, those of each pair homogamous or heterogamous, glumes more or less rigid and firmer than lemma, upper lemma usually awned or reduced to an awn .... **Andropogoneae**

2. Spikelets solitary, more or less similar, glumes membranous, lower lemma similar to upper glume in texture, upper lemma rigid, often chartaceous or crustaceous, awned or unawned .... **Paniceae**

KEY TO THE GENERA OF THE TRIBE MAYDEAE

1. Axes of female spikelets not fused; male and female spikelets in the same inflorescence:

   2. Female spikelets completely enclosed in a metamorphosed leaf- sheath through which the rhachis grows out and bears the male spikelets above .... **Coix**

   2. Female spikelets not enclosed in a false involucre as above .... **Chionachne**

1. Axes of female spikelets fused, male and female spikelets in different inflorescence .... **Zea**

KEY TO THE GENERA OF THE TRIBE ANDROPOGONEAE

1. Spikelets solitary .... **Dimeria**

1. Spikelets in pairs or in threes:

   2. Spikelets in each pair usually dissimilar, the sessile bisexual and pedicelled male or neuter:

      3. Inflorescence of solitary raceme:

         4. Spikelets awned:

            5. All spikelets heterogamous:

               6. Racemes supported by spathes:

                  7. Sessile spikelets pitted; perennial; nodes densely bearded .... **Dichanthium (foveolatum)**

                  7. Sessile spikelets not pitted; annual; nodes glabrous .... **Schizachyrium**

               6. Racemes not supported by spathes:

                  8. Ligule short membranous; inflorescence a terminal raceme .... **Sehima**

                  8. Ligule elongate, acute; inflorescence a panicle .... **Andropogon (fastigiatus)**

            5. Lowest pairs of spikelets homogamous .... **Heteropogon**

         4. Spikelets unawned:

            9. Sessile spikelet globose; lower glume pitted all over; pedicelled spikelets reduced .... **Hackelochloa**

9. Sessile spikelet not globose; lower glume not pitted; pedicelled; spikelets; well developed:

10. Ligule a short ciliate rim, sessile spikelet ovoid, keeled; pedicelled spikelet well developed, caryopsis ellipsoid .... **Rottboellia**

10. Ligule a frimbriate membrane; sessile spikelet oblong, not keeled; the pedicelled suppressed or rudimentary .... **Ophiuros**

3. Inflorescence of racemes more than one paired, digitate, fascicled, .corymbose or paniculate:

11. Spikelets in threes .... **Chrysopogon**

11. Spikelets in pairs:

12. Racemes supported by the spathes:

13. Involucral spikelets present:

14. Involucral spikelets with strong bristles, sessile, persistent .... **Themeda**

14. Involucral spikelets glabrous, pedicelled, deciduous .... **Iseilema**

13. Involucral spikelets absent:

15. Basal joints of the racemes bulbous .... **Apluda**

15. Basal joints of the racemes not bulbous:

16. Joints thickened upwards, toothed at apex .... **Andropogon (pumilus)**

16. Joints not thickened upwards, not toothed at apex .... **Cymbopogon**

12. Racemes not supported by spathes:

17. Inflorescence a panicle:

18. Spikelets with many pairs of spikelets arranged in whorled racemes; lodicules glabrous .... **Vetiveria**

18. Spikelets with a pair of spikelets arranged in the panicle; lodicules hairy .... **Sorghum**

17. Inflorescence not a panicle, digitate racemes:

19. Lower glume of sessile spikelet 2-toothed .... **Thelepogon**

19. Lower glume of sessile spikelet not toothed:

20. Upper lemma awned from the back, below the middle .... **Arthraxon**

20. Upper lemma awned from the apex or from the sinus:

21. Upper lemma well developed, awn arise from the sinus .... **Ischaemum**

21. Upper lemma reduced to a translucent base of an awn, awn arises from the apex of the lemma:

22. Pedicels and rhachis internodes with a translucent or balsamiferous median line:

23. Racemes 1-5-spikeleted ..... **Capillipedium**

23. Racemes more than 8-spikeleted .... **Bothriochloa**

22. Pedicels and rhachis internodes without a translucent median line .... **Dichanthium**

2. Spikelets in each pair usually similar:

24. Spikelets arranged in panicle, and enclosed in long, white, silky hairs:

25. Spikelets unawed

26. Panicle narrow, contracted, rhachis quite tough, all spikelets pedicelled .... **Imperata**

26. Panicle effuse, rhachis fragile, spikelets 2-nate, one sessile and other pedicelled .... **Saccharum**

24. Spikelets arranged in several pairs of racemes, and not enclosed in silky hairs:

25. Spikelets unawed .... **Spodioporyon**

27. Racemes supported by spathes:

28. Racemes few to many .... **Lophopogon**

28. Racemes solitary .... **Pogonatherum**

27. Racemes not supported by spathes:

29. Racemes 2-20, pedicelled spikelets well developed ...... **Eulalia**

29. Racemes 1-2, pedicelled spikelets reduced............**Apocopis**

KEY TO THE GENERA OF THE TRIBE PANICEAE

1. Spikelets unawned:

2. Inflorescence of spiciform racemes:

3. Ligule absent..**Echinochloa** *p.p.*

3. Ligule present:

4. Spikelets not in clusters, solitary, binate or ternate:

5. Lower glume well developed:
    6. Lower glume turned away from the axis:
        7. Spikelets flattened or depressed, upper lemma mucronate .... **Urochloa**
        7. Spikelets usually turgid, upper lemma acute .... **Paspalidium**
    6. Lower glume turned towards the axis .... **Brachiaria**
5. Lower glume usually absent, if present minute, scale like:
    8. Spikelets elliptic or elliptic-oblong .... **Digitaria**
    8. Spikelets orbicular, oblong or ovate .... **Paspalum**

4. Spikelets 1-6, clustered:
    9. Spikelets solitary or 2-3, enclosed in an involucre of bristles .... **Cenchrus**
    9. Spikelets 1-6, intermixed with small scale like empty glumes .... **Trachys**

2. Inflorescence of panicle:
    10. Spikelets covered by involucral bristles:
        11. Bristles solitary born by the prolongation of the branch beyond the terminal spikelet .... **Psuedoraphis**
        11. Bristles numerous born of highly modified bristle or spikelike branchlets:
            12. Spikelets ellipsoid, 2-3 mm long, bristles rigid, barbed, involucre persistent, upper lemma rugose .... **Setaria**
            12. Spikelets lanceolate, 6-7 mm long, bristles plumose, involucre deciduous, upper lemma smooth .... **Pennisetum**
    10. Spikelets not covered by involucral bristles:
        13. Spikelets without an annular callus at the base:
            14. Spikelets gibbose .... **Sacciolepis**
            14. Spikelets not gibbose:
        13. Spikelets with an annular callus at the base .... **Eriochloa**

1. Spikelets awned:
    15. Upper lemma glabrous along the margin, muticous, lower lemma awned or muticous; upper glume awned or aristate:
        16. Lower lemma awned:
            17. Upper glume entire, awned:

18. Culms spongy below, inflorescence a thyrsoid panicle, branches erect, appressed to the rhachis ... **Hymenachne**
18. Culms tough below, inflorescence a loose panicle of spiciform racemes .... **Echinochloa** *p.p.*
17. Upper glume emarginate or 2-lobed, aristate .... **Rhynchelytrum**
16. Lower and upper glume and lower lemma awned .... **Oplismenus**
15. Upper lemma densely ciliate along margins, shortly aristulate; lower lemma muticous; upper glume unawned, acuminate .... **Alloteropsis**

KEY TO THE TRIBES OF THE SUB-FAMILY POOIDEAE

1. Tall, woody, arborescent or shrubs, leaf blade usually with a petiole like base, articulate with the sheaths, lodicules 3 .... **Bambuseae**
1. Annual or perennial, herbs, leaf blade not articulate on the sheaths, lodicules 2:
    2. Inflorescence of solitary, digitate or scattered spikes borne on opposite sides of the rhachis, sessile:
        3. Spikelets on opposite sides of the rhachis of solitary spikes, ovary ciliate, lodicules hairy; caryopsis grooved longitudinally, cultivated .... **Triticeae**
        3. Spikelets secund on a tough rhachis, inflorescence of digitate or solitary spikes .... **Chlorideae**
    2. Inflorescence of panicles or spikelike, open or contracted, spikelets sessile or pedicelled:
        4. Spikelets with one fertile floret:
            5. Glumes minute or suppressed, fertile lemma and palea similar in texture .... **Oryzeae**
            5. Glumes well developed, fertile lemma and palea not similar:
                6. Spikelets with 2 or 3 florets, the lower male or barren, the upper bisexual:
                    7. Lower floret empty and without a palea, glumes up to half the length of the spikelet, spikelets falling along with pedicel .... **Thysanolaeneae**
                    7. Lower floret male or barren, palea present, glume as long as the spikelet, spikelets falling leaving the glumes on the rachis:
                        8. Glumes equal in size obtuse or mucronate, upper lemma unawned .... **Isachneae**
                        8. Glumes unequal, the lower acute, upper acuminate, upper lemma awned .... **Arundinelleae**

6. Spikelets with 1 fertile floret:
    9. Spikelets awned:
        10. Spikelets 1-flowered, awn single:
            11. Spikes solitary, cylindrical glumes persistent .... **Perotideae**
            11. Panicle many-branched, spikelets falling entire .... **Garnotieae**
        10. Spikelets many-flowered, awn trifid .... **Aristideae**
    9. Spikelets unawned:
        12. Inflorescence panicle; spikelets ellipsoid, terete .... **Sporoboleae**
        12. Inflorescence spike or spiciform racemes; spikelets laterally compressed .... **Zoysieae**
4. Spikelets with two or more fertile florets:
    13. Spikelets-awned, 2-3-flowered .... **Arundineae**
    13. Spikelets unawned, 4-many-flowered .... **Eragrosteae**

KEY TO THE GENERA OF TRIBE CHLORIDEAE

1. Spikelets usually awned:
    2. Inflorescence usually a solitary spike:
        3. Glumes awned .... **Melanocenchris**
        3. Glumes aristate or mucronate .... **Tetrapogon**
    2. Spikes with 2-20 racemes .... **Chloris**
1. Spikelets unawned:
    4. Spikes solitary, terminal :
        5. Spikelets more or less sunk in the rhachis, 1-2-flowered .... **Oropetium**
        5. Spikelets not sunk in the rhachis, only one-flowered .... **Microchloa**
    4. Spikes of 2-8 racemes, terminal, umbellate .... **Cynodon**

KEY TO THE GENERA OF TRIBE ERAGROSTEAE

1. Spikelets awned :
    2. Inflorescence of digitate racemes or spike:
        3. Spike solitary, cylindric, often interrupted at base .... **Elytrophorus**
        3. Digitate racemes 1-many:
            4. Racemes whorled or scattered, spikelets 8-20-flowered, caryopsis oblong.. **Acrachne**
            4. Racemes umbelled, spikelets 3-5-flowered, caryopsis subglobose .... **Dactyloctenium**

2. Inflorescence panicle or spiciform racemes:
    5. Lemmas entire .... **Dinebra**
    5. Lemmas emarginated, 2-4-fid at apex:
        6. Inflorescence of solitary spike, upper glume shortly 2-toothed at the apex, lemmas and paleas glabrous on the keels and nerves .... **Tripogon**
        6. Inflorescence a panicle, upper glume not toothed at the apex, lemmas and paleas pilose on keels and nerves . **Leptochloa** *p.p.*

1. Spikelets unawned:
    7. Inflorescence of panicle, lax or contracted:
        8. Upper glume 3-nerved .... **Eragrostis**
        8. Upper glume 1-nerved .... **Leptochloa** *p.p.*
    7. Inflorescence of digitate racemes or spikes:
        9. Spikes aggregate of dense globose heads .... **Coelachyrum**
        9. Spikes not in dense globose heads:
            10. Spikes solitary .... **Eragrostiella**
            10. Spikes 2-10 .... **Eleusine**

KEY TO THE GENERA OF TRIBE BAMBUSEAE

1. Spikelets 1-flowered; ovary appendage long, stiff, tapering, hollow .... **Cephalostachyum**
1. Spikelets (1- to) many flowered; ovary with short, solid apical appendage:
    2. Internodes solid, spikelets in globose congested heads, rhachilla internodes short, always unarmed, lodicules O, reed like shrub .... **Dendrocalamus**
    2. Internodes hollow, spikelets in heads on spicate branchlets, rhachilla internode elongate, sometimes armed, lodicules 3, tall trees, rarely climbing .... **Bambusa**

KEY TO THE GENERA OF TRIBE ORYZEAE

1. Spikelets awned .... **Oryza** *p.p.*
1. Spikelets unawned:
    2. Glumes absent or reduced, a hyaline rim .... **Leersia**
    2. Glumes present, minute, oblong-lanceolate .... **Oryza** *p.p.*

The following tribes are represented by only one genus in this Flora.

Aristideae .... **Aristida**
Arundinae .... **Phragmites**
Arundinelleae .... **Arundinella**
Garnotieae .... **Garnotia**
Isachneae .... **Isachne**
Perotideae .... **Perotis**
Sporoboleae .... **Sporobolus**
Thysanolaeneae .... **Thysanolaena**
Triticeae .... **Triticum**
Zoysieae .... **Tragus**

ACRACHNE Wight & Arn. ex Chiov.

Acrachne racemosa (B. Heyne ex Roem. & Schult.) Ohwi, Bull. Tokyo Sci. Mus. n. 18:1. 1947; Bor, Grass. India 487. 1960; Moulik, Grass. Bamb. India 2:584. 1997. *Eleusine racemosa* B. Heyne ex Roem. & Schult., Syst. Veg. 2:583. 1817. *E. verticillata* Roxb., Fl. Ind. 1:346. 1820; Hook. f., Fl. Brit. India 7:295. 1897; Fischer, Fl. Madras. 3:1840. 1934.

Annual, culms erect, tufted, to 70 cm high; nodes glabrous. Leaf sheaths glabrous; ligule hairy; leaf blade linear-lanceolate, ovate-acuminate, base rounded, acuminate, margins scaberulous. Racemes spiciform; spikelets ovate, 6-18-flowered; lower glume ovate, keeled, 3-nerved, acute and mucronate; upper glume ovate-lanceolate, keeled, acuminate, awned; lower lemma broadly ovate, 3-nerved, awned; upper lemma ovate, 2-nerved, awned; palea minute; lodicules 2; stamens 3. Caryopsis oblong, transversely rugose.

Common near streams in forests in Hyderabad, Nalgonda and Warangal districts.

Fl. & Fr.: July - December.

Nagarjunakonda valley (NLG), *KT* 9691 (CAL); Pakhal RF (WGL), *KMS* 13180.

ALLOTEROPSIS J. Presl

1. Perennial; leaf blade linear-lanceolate, base narrow, margins smooth; raceme rhachis pilose; spikelets 4 – 6 mm long .... **A. semialata**

1. Annual; leaf blade ovate-lanceolate, base cordate, margins pectinate; raceme rhachis glabrous; spikelets 2-4 mm long .... **A. cimicina**

Alloteropsis cimicina (L.) Stapf, in Prain, Fl. Trop. Afr. 9: 487.1919; Fischer. Fl. Madras 3: 1766. 1934; Bor, Grass. India 276. 1960; Moulik, Grass. Bamb. India 1: 68. 1997. *Mililum cimicinum* L., Mant. Alt. 184.1771. *Axonopus cimicinus* (L.) P. Beauv., Ess. Agrost. 12. 1812; Hook.f, Fl. Brit. India 7: 64. 1896.

Annual, culms erect, tufted, to 80 cm high. Leaf sheaths covered with tubercle based hairs; ligules fimbriate; leaf blade ovate-lanceolate, 2-10, 1-1.5cm, margins tubercle based hairy, acute. Racemes 4-10, digitate, terminal, 5-15 cm long; spikelets broadly elliptic, 2-4, 2mm, 2-flowered, lower staminate or barren, upper bisexual; lower glume ovate-lanceolate, 2, 1mm, acuminate, 3-nerved; upper glume broadly

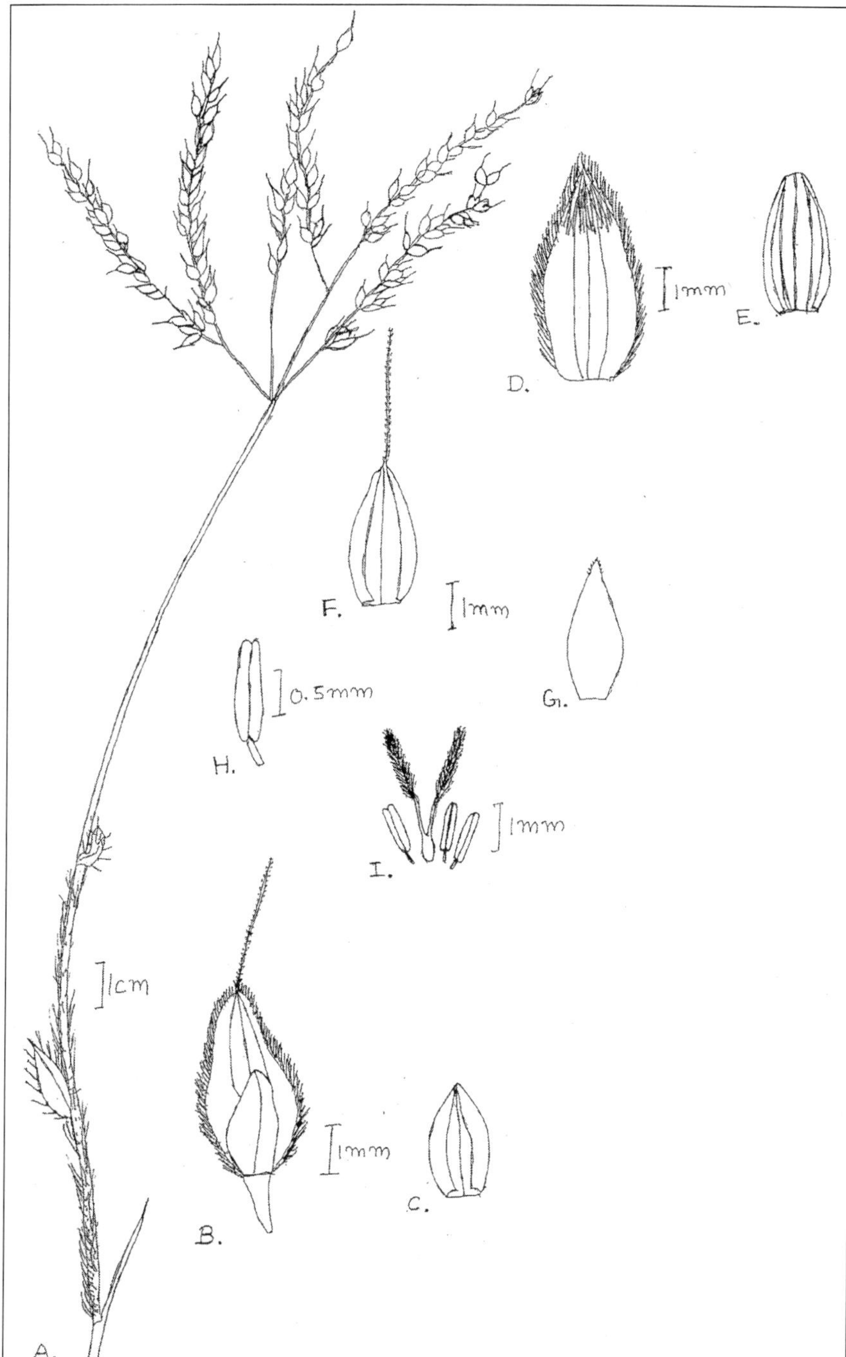

Figure 135: Alloteropsis cimicina (L.) Stapf

A. Twig, B. Spikelets, C. Lower glume, D. Upper glume, E. Lower lemma, F. Upper lemma, G. Palea, H. Stamen, I Pistil with stamens.

ovate, 2-4, 2mm, ciliate, acuminate, 5-nerved; lower lemma ovate, 1.5-3 mm, acuminate, 5-nerved; lower palea elliptic, 1.5-2, 0.5 mm, hyaline, bilobed at apex; upper lemma elliptic, 2-3, 1mm, aristate, margin ciliate; upper palea ovate, 2-3, 1mm, 2-keeled, chartaceous, papillose; lodicules 2; stamens 3; stigmas plumose. Caryopsis ovate.

Common as undergrowth in forests and in open dry habitats in all districts.

Fl. & Fr.: June – March.

Rebbana (ADB), *GO* & *PVP* 4841; Mamidipalli (NZB), *BR* 6429; Manchippa RF (NZB), *BR* 6031, 6015; Pocharam tank (MDK), *TP* & *MSG* 12049; Rudraram RF (RR), *MSM* 12715; Western side of Pakhal (WGL), *KMS* 13120 (CAL); Aklaspur (KRN), *GVS* 20235 (CAL); Hyderabad (HYD), *MRS* 93 (DD); Boddupalle (NLG), *BVRS* 647 (NU); 10 km towards north of Pasra (WGL), *RKP* 108219 (BSID); Manuguru RF (KMM), *RCS* 104257 (BSID).

Alloteropsis semialata (R.Br.) Hitch. in Contrib. US Nat. Herb. 12, 210. 1909; Bor, Grass. India 276. 1960; Moulik, Grass. Bamb. India 1: 70 1997. *Panicum semialatum* R.Br., Prodr. Fl. Nov. Holl. 192. 1810. *Axonopus semialatus* (R.Br.) Hook. f., Fl. Brit. India 7: 64. 1896.

Perennials; culms erect, tufted, to 130 cm high; basal sheaths covered with silky bases; nodes bearded. Leaf sheaths 4-20 cm, mouth hairy, basal ones densely silky hairy; ligules truncate, 0.3 mm, membranous or fimbriate; leaf blade 10-20 × 0.4-1 cm, linear-lanceolate, margin scabrid, green or glaucous. Racemes 3-5, sub-digitate, rhachis and pedicels hairy; spikelets ovate, 4-6 × 3 mm, awned; lower glume ovate, 3.5 × 2 mm, aristate, chartaceous, 3-nerved; upper glume broadly ovate, 5-5.5 × 2 mm, chartaceous, 5-nerved, winged, ciliate; lower lemma ovate-lanceolate, 4.5 × 2.5 mm, coriaceous, 3-nerved; palea minute; upper lemma oblong-lanceolate, 4.5 × 2 mm, coriaceous, keeled; lodicules 2; stamens 3, anther 1 mm; ovary oblong. Caryopsis ovoid.

Rare in forests, usually in dried pond beds and sandy soils in Mahabubnagar district.

Fl. & Fr.: July – November.

Appapur (MBNR), *BR* & *AMR* 26802.

ANDROPOGON L.

1. Inflorescence of solitary raceme .... **A. fastigiatus**
1. Inflorescence of two racemes .... **A. pumilus**

Andropogon fastigiatus Sw., Prodr. Veg. Ind. Occ. 26. 1788; Hook. f., Fl. Brit. India 7: 167. 1896. *Diectomis fastigiata* (Sw.) Kunth in Humb. & Bonpl., Nov. Gen. & Sp. 1: 193; t. 64. 1816; Bor, Grass. India 135. 1960; Moulik, Grass. Bamb. India 1: 275. 1997.

A perennial, tufted grass; culms up to 50 cm high, decumbent below, fastigiately branched upwards, leafy. Leaf sheaths terete; ligule elongate, acute; blade 7-30 × 0.1-

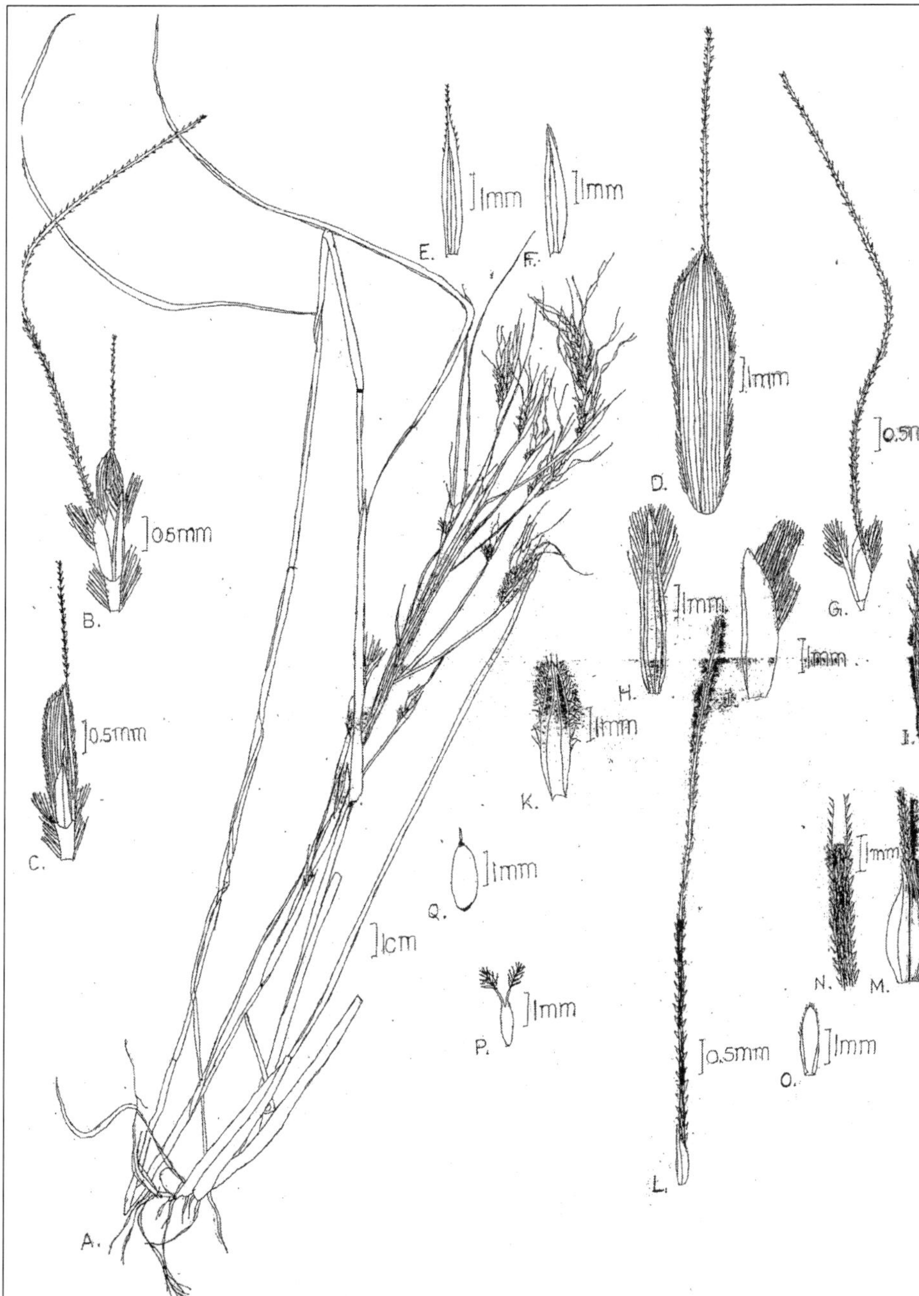

Figure 136: Andropogon fastigiatus Sw. (Syn.: *Diectomis fastigiata* Kunth)

A. Twig, B. Spikelets, C. Pedicelled, spikelet, D. Lower glume, E. Upper glume, F. Lemma, G. Sessile spikelet, H & I. Lower glume front and back views, J. Upper glume, K. Lower lemma, L & M. Upper lemma side and front views, N. Part of awn, O. Palea, P. Pistil, Q. Caryopsis.

0.2 cm, narrowly linear, acuminate, scabrid. Raceme single, enclosed by spathe, 2-3.5 cm long, fragile; joints much shorter than the spikelets, villous with white hairs, apex strongly toothed; spathes 7.5 cm, linear-lanceolate. *Sessile spikelet:* 5 mm, oblong; callus with long hairs; lower glume linear-lanceolate, 6-nerved, bicuspidate, keeled, margins thick, apex ciliate; upper glume shorter, cymbiform, 3-nerved, bidentate, lobes obtuse, awned, apex ciliate; lower lemma linear-oblong, 2-keeled, 2-nerved, margins ciliate; upper lemma cymbiform, 2-toothed, awned; palea minute, hyaline; lodicules 2; stamens 3. *Pedicelled spikelets:* 7 mm, neuter; lower glume ovate-lanceolate, 2-cuspidate, awned, ciliate, many-nerved; upper glume much smaller, membranous, awned; lower lemma very small, oblong, hyaline.

Occasional in Hyderabad and Medak districts.

Fl. & Fr.: August– December.

Pocharam tank (MDK), *TP* & *MSG* 12052; Hyderabad, *MRS* 246 (DD).

Andropogon pumilus Roxb., Fl. Ind. 1: 277. 1820; Hook.f., Fl. Brit. India 7: 170, 1806; Fischer, Fl. Madras 3: 1753. 1934; Bor, Grass. India 93. 1960; Moulik, Grass. Bamb. India 1: 209. 1997.

An annual, stoloniferous grass; culms up to 30 cm high, tufted, slender, nodes glabrous. Leaf sheaths 1-3 cm, loose, glabrous; ligule membranous; blade 3-7 cm, linear, base cordate, apex acuminate, glabrous. Racemes 2, subtended by a narrow glabrous spathe; racemes 2 cm, digitate pedunculate; spikelets in pairs, sessile bisexual and pedicelled staminate or neuter. *Sessile spikelet*: 4 mm. Lower glume linear, 5-nerved, bidentate; upper glume boat-shaped, 3-nerved; lower lemma linear-oblong, keeled, upper lemma awned; palea minute or absent; lodicules 2; stamens 3; stigma plumose. Caryopsis narrowly ovate. *Pedicelled spikelet:* 3 mm, muticous, similar to sessile spikelet.

Common grass on drier areas of Adilabad, Hyderabad, Medak and Ranga Reddi, districts.

Fl. & Fr.: Sept. – December

Bendala (ADB), *GO* 4480; Choutkoor fields (MDK), *TP & MSG* 11928; Hyderabad (HYD), *MRS* 197 (DD).

APLUDA L.

Apluda mutica L., Sp. Pl. 92. 1753; Fischer, Fl. Madras 3: 1750. 1934; Bor, Grass. India 93. 1960; Moulik, Grass. Bamb. India 1: 305. 1997. *Apluda varia* Hack. subsp. *aristata* (L.) Hack., Monogr. Phan. 6: 196. 1889; Hook.f., Fl. Brit. India 7: 150. 1896. *A. aristata* L., Amoen. Acad. 4. 303. 1759; Fischer, Fl. Madras 3: 1750. 1934.

A leafy, perennial, tufted grass; culms erect, up to 3 m high, glabrous, nodes glabrous. Leaf sheaths 3.5 cm long, glabrous; ligule membranous; blade 6-20 × 0.2-0.5 cm, linear-lanceolate, base narrowed, apex cuspidate, glabrous. Panicle 3-90 cm long, terminal; spikelets in threes, one sessile, bisexual, the rest pedicelled, one staminate and the other rudimentary. *Sessile spikelet*: 2-flowered, lower male, upper bisexual, 6 mm; lower glume oblong, coriaceous, many-nerved, margins scabrid; upper glume

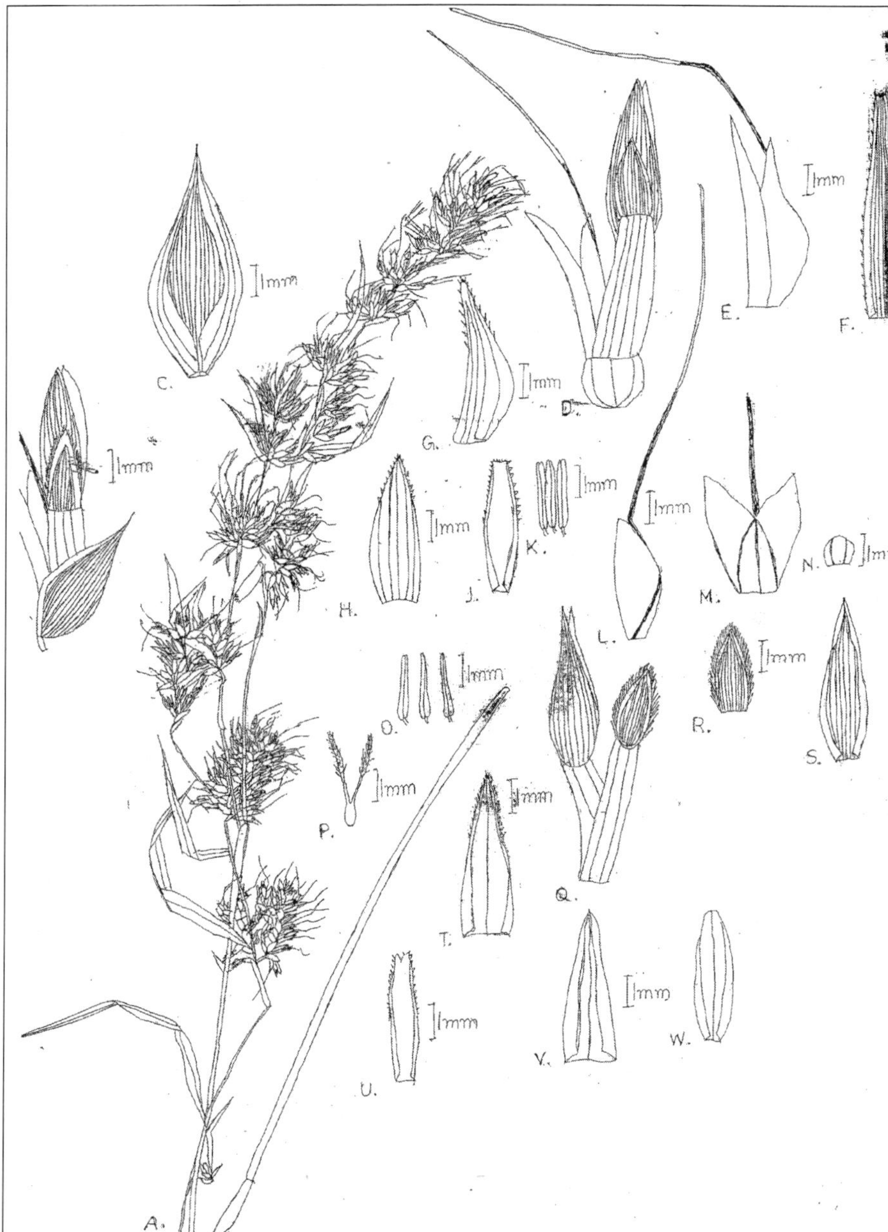

Figure 137: Apluda mutica L.

A. Twig, B. Spikelets with bract, C. Bract, D. Spikelets without bract, E. Sessile spikelet, F. Lower glume, G. Upper glume, H. Lower lemma, J. Palea, K. Stamens, L & M. Upper lemma side and front views, N. Palea, O. Stamens, P. Pistil, O. Pedicelled spikelet, R. Lower glume, S. Upper glume, T. Lower lemma, U. Palea, V. Upper lemma, W. Palea.

oblong, acuminate, coriaceous, 5-nerved, apex hairy; lower lemma broadly ovate, chartaceous, 5-nerved, apex scabrid; palea oblong, 2-nerved, apex scabrid, male; upper lemma hyaline, deeply bidentate, awned; palea small, ovate, bisexual; stamens 3, stigmas plumose. Caryopsis ellipsoid. *Pedicelled spikelet*: 5 mm, lanceolate, bisexual, awnless; glumes unequal, many-nerved, lower glume ovate, coriaceous, hairy; upper glume lanceolate, coriaceous, apex scabrid; lower lemma lanceolate, many-nerved, chartaceous, apex scabrid; palea oblong, shortly bidentate, apex, scabrid; upper lemma oblong, many-nerved, chartaceous; palea oblong; stamens 3, stigmas plumose.

Common along the hill slopes of all districts.

Fl. & Fr.: September – December.

Ankusapuram (ADB), *GO* 4348; Laxmipur RF (ADB), *TP* & *PVP* 3431; Rajampet (NZB), *BR* & *KH* 9673; Choutkoor (MDK), *TP* & *MSG* 11910; Lingampalli (RR), *MSM* 12750; Thumukunta (MBNR), *BR* & *MV* 40726; Hyderabad (HYD), *MSR* 219 (DD); Umamaheswaram (MBNR), *SRS* 111373 (BSID); Nilagiri view point (MBNR), *SRS* 109716(BSID); Nagarjunakonda Valley (NLG), *KT* 9686 (CAL).

Note A very much variable plant, occurring in awned and unawned forms. Dr. N. L. Bor upholds Hackel's opinion that the varieties pass into one another by numerous intermediates and for that matter several other equally distinct varieties could be set up.

APOCOPIS Nees

Apocopis courtallumensis (Steud.) Henrard, Blumea 4: 524. 1942; Bor, Grass. India 95. 1960; Moulik, Grass. Bamb. India 1: 329. 1997. *Andropogon courtallumensis* Steud., Syn. Pl. Glumac. 1: 377. 1854. *Apocopis wightii* Nees ex Thwaites, Enum. Pl. Zeyl. 365. 1864; Hook. f., Fl. Brit. India 7: 142. 1896; Fischer, Fl. Madras 3: 1825. 1934.

A perennial, densely tufted grass; culms up to 60 cm high. Leaf sheaths glabrous, softly hairy; blade 1.5-13 × 0.1-0.4 cm, narrowly lanceolate, flat, more or less hairy from tuberculate based hairs. Racemes usually 2, 2-4 cm long; rhachis and pedicels rufous hairy; spikelets paired, closely imbricate on racemes, sessile well developed, pedicelled reduced to pedicel; lower glume obcordate-obovate, truncate, yellow with a red-brown apex or dark brown with a yellow tip, glabrous-densely rufous hairy; upper glume narrowly lanceolate, 2-keeled, inflected, scabrid; lower lemma oblong, apex ciliate; upper lemma linear, 2-dentate, awned, geniculate; palea very short, convolute; stamens 2; ovary narrowly oblong-elliptic, stigma plumose, brownish.

In drier localities of Hyderabad district.

Fl. & Fr.: September – December.

ARISTIDA L.

1. Glumes muticous .... **A. adscensionis**
1. Glumes aristate:
    2. Panicle effuse, pyramidal with spreading branch .... **A. hystrix**

Figure 138: Aristida adscensionis L.

A. Twig, B. Spikelet, C. Lower glume, D. Upper glume, E. Lemma, F. Palea, G. Stamen, H. Pistil with stamens.

2. Panicle contracted:
   3. Awn with a column:
      4. Articulation in between the top of the column *i.e.*, just below the awns .... **A. mutabilis**
      4. Articulation in between the top of the lemma and the base of the column .... **A. funiculata** var. **funiculata**
   3. Awn sessile:
      5. Lateral awns very much reduced .... **A. redacta**
      5. Lateral awns well developed .... **A. setacea**

Aristida adscensionis L., Sp. Pl. 82. 1753; Hook. f., Fl. Brit. India 7: 221. 1896; var. **adscensionis**: Bor, Grass. India 407. 1960; Moulik, Grass. Bamb. India 2: 584. 1997. *A. depressa* Retz., Observ. Bot. 4. 22. 1786; Fischer, Fl. Madras 3: 1809. 1934.

A perennial, tufted grass; culms up to 40 cm high, erect or geniculately ascending, glabrous; nodes glabrous. Leaf sheaths 25 – 40 cm long, glabrous; ligule a ridge of fine soft hairs; leaf blade 2-4 × 0.1-0.2 cm, narrowly linear, convolute, base cordate, apex sub-acute. Panicle contracted, 9-16 cm long; spikelet 1-flowered, linear, 0.6-1 cm; lower glume membranous, narrowly ovate, 1-nerved, 1-keeled on back; upper glume membranous, linear-lanceolate, acute, minutely aristate; lemma elliptic, awned trifid, middle awn longer than lateral awns; palea narrowly linear, hyaline; lodicules 2; stamens 3, stigmas plumose. Caryopsis as long as the lemma, linear.

Common in plains and slopes of Nizamabad, Medak, Warangal, Hyderabad and Nalgonda districts.

Fl. & Fr.: August-September.

Manchippa RF (NZB), *BR* & *PSPB* 9618; Pocharam (MDK), *TP* & *MSG* 12023; Anantasagar (RR), *MSM* & *KH* 11009; Hyderabad (HYD), *MRS* 267 (DD); Kambalpally (NLG), *BVRS* 295 (NU); Kudurpalli (KRN), *SLK* 70740 (NBG).

Aristida funiculata Trin. & Rupr., Sp. Gram. Stip. 159. 1842; var. **funiculata**; Bor, Grass. India 410. 1960; Fischer, Fl. Madras 3: 1809. 1934; Hook. f., Fl. Brit. India 7: 226. 1896; Moulik, Grass, Bamb. India 2: 424. 1997.

Vern.: *Cheepuru gaddi.*

An annual, tufted grass; culms up to 50 cm high, geniculately ascending; nodes glabrous. Leaf sheaths 4-8 cm long, glabrous; ligule a small ciliate membrane; blade 6-16 × 0.1-0.2 cm, linear, convolute, base cordate, apex acuminate. Panicle lax, narrow, 22 cm long; spikelets linear, 0.8-1 cm, pedicelled, 1-flowered; upper glume membranous, linear to narrowly ovate, 1-nerved, 1-keeled, hyaline, arista elongate; lemma truncate, 3-nerved, awn trifid, subequal; palea hyaline, narrowly obovate; lodicules 2; stamens 3, stigmas plumose. Caryopsis linear, spiny.

Common grass along the hill slopes and waste places of Nizamabad, Nalgonda, Hyderabad and Warangal districts.

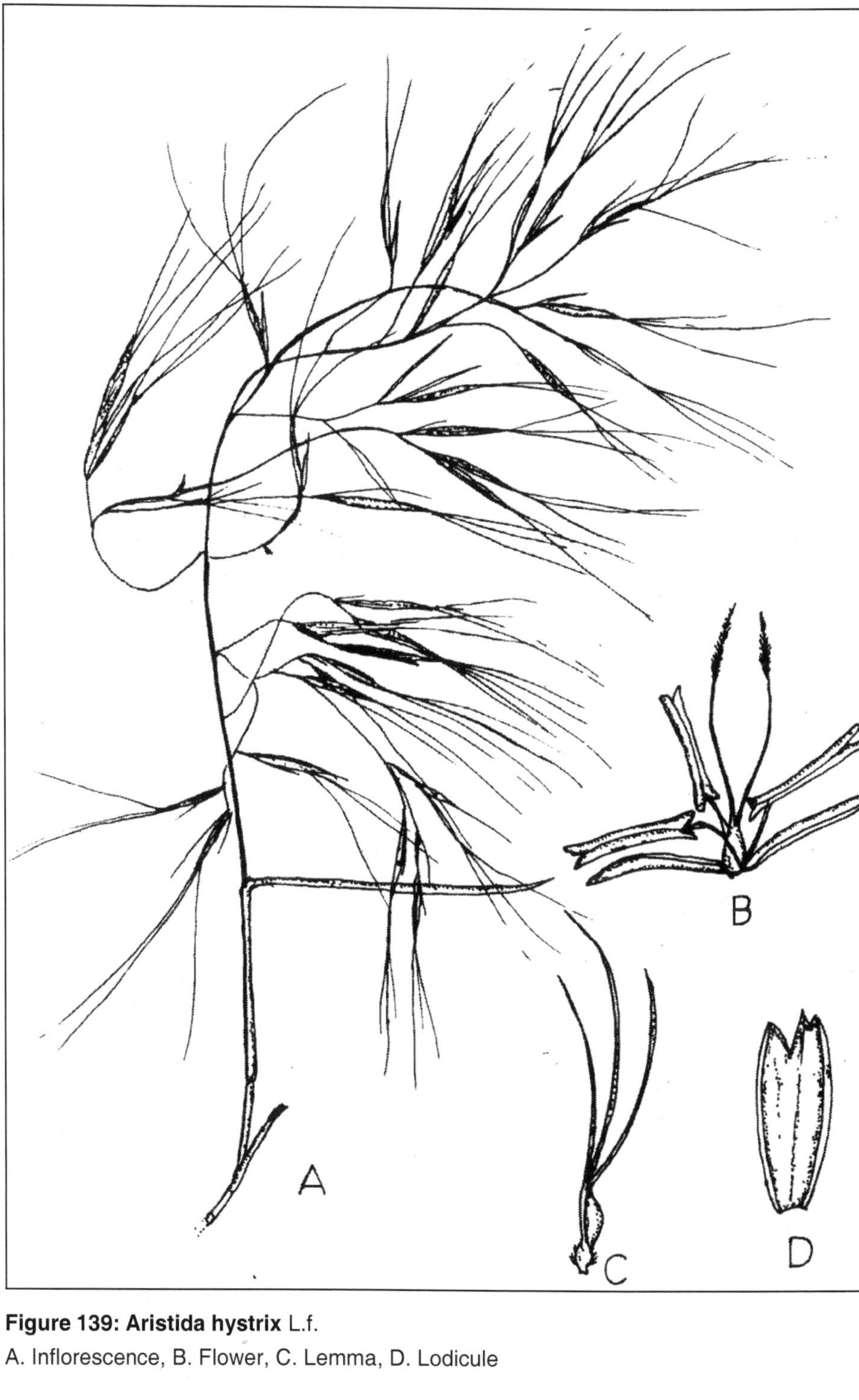

Figure 139: Aristida hystrix L.f.

A. Inflorescence, B. Flower, C. Lemma, D. Lodicule

Fl. & Fr.: August– September.

Nasrullabad beat (NZB), *TP* & *BR* 6155; Hyderabad (HYD), *MRS* 268 (DD); Nagarjunasagar (NLG), *KMS* 9739 (CAL).

Aristida hystrix L.f., Suppl. Pl. 113. 1781; Hook. f., Fl. Brit. India 7: 225. 1896; Fischer, Fl. Madras 3: 1809. 1934; Bor, Grass. India 410. 1960; Moulik, Grass, Bamb. India 2: 424. 1997.

Vern.: *Dollubagaddi.*

A perennial, tufted grass; culms up to 60 cm high, diffuse with creeping rootstock, ascending; nodes glabrous. Leaf sheaths 2-4 cm long, densely pilose at the mouth; ligule of soft hairs; blade 3-9 × 0.1-0.2 cm, linear, base cordate, sparsely pilose on the margins, apex acuminate. Panicle lax, pyramidal, 10-20 cm, terminal; spikelet 1.5-2.5 cm, linear; glumes linear, aristate; lower lemma coriaceous, narrowly ovate, 3-nerved, 3-awned, central longer than lateral, awns 3-4 cm long, scabrid on margins; palea hyaline, narrowly ovate, nerveless; lodicules 2; stamens 3, stigmas plumose. Caryopsis linear, prickly.

Common grass in plains and lower slopes in all districts.

Fl. & Fr.: September – December.

Gowraram RF (NZB), *TP* & *BR* 6380; Choutkoor (MDK), *TP* & *MSG* 11937; Anantasagar RF (RR), *MSM* & *KH* 11017; Upparapalli (HYD), *MRS* 266 (DD); Vempalli (KRN), *GVS* 20180 (MH); Nagarjuna Konda valley (NLG), *KT* 9689 (CAL); Cherukupally (NLG), *BVRS* 659 (NU); Ramanagutta (KMM), *RR* 107993 (BSID).

Aristida mutabilis Trin. & Rupr., Sp. Gram. Stip. 150. 1842; Hook. f., Fl. Brit. India 7: 226. 1896; Fischer, Fl. Madras 3: 1809. 1934; Bor, Grass. India 411. 1960; Moulik, Grass. Bamb. India 2: 426. 1997.

An annual, tufted grass; culms up to 40 cm high; nodes glabrous. Leafsheths 3-7 cm long, glabrous; ligule membranous; blade 5-12 × 0.2-0.5 cm, linear, convolute, base cordate, apex acuminate, glabrous. Panicle lax, sub cylindric, 15 cm long, spikelets 1-1.5 cm, linear, 1-flowered, lower glume 1-keeled, scabrous, upper glume 6 mm long, shortly awned; lemma cylindric, scaberulous, callus shortly bearded, awn trifid, subequal; stamens 3, stigmas plumose. Caryopsis narrowly cylindric.

Occasional grass in plains of Adilabad district.

Fl. & Fr.: July – September

Jaipur RF (ADB), *PVP* 9433.

Aristida redacta Stapf, Kew Bull. 1892: 85. 1892; Hook.f., Fl. Brit. India 7: 227. 1896; Fischer, Fl. Madras 3: 1809. 1934; Bor, Grass. India 412. 1960; Moulik, Grass. Bamb. India 2: 426. 1997.

A perennial, tufted grass; culms up to 50 cm high, simple or proliferously branched; nodes glabrous. Leaf sheaths 3-8 cm long, glabrous; ligule shortly and densely ciliate; blade 8-15 × 0.2 cm, linear, convolute, base cordate, rough margins, apex acuminate, hairy above, smooth below. Panicle effuse, 25 cm; spikelets 10 mm, linear, pedicelled; glumes linear, subequal, shortly aristate, 1-nerved, 1-keeled, keel scabrous; lemma chartaceous, callus hairy, awn trifid, subequal; lodicules 2; stamens 3, stigmas plumose.

Rare grass in Hyderabad district.

Fl. & Fr.: September – April.

Hyderabad, *MRS* 216 (DD).

Aristida setacea Retz., Observ. Bot. 4: 22. 1786; Hook. f., Fl. Brit. India 7: 225; 1896; Fischer, Fl. Madras 3: 1809. 1934; Bor, Grass. India 412. 1960; Moulik, Grass, Bamb. India 2: 426. 1997.

Vern.: *Gaddiparakalu, Cheepurugaddi.*

A perennial, tufted, grass; culms up to 2 m high, erect or geniculately ascending from a woody base; nodes glabrous. Leaf sheaths 2-10 cm long, glabrous; ligule of short hairs; blade 10-25 × 0.1-0.3, linear, convolute, base cordate, apex acuminate, glabrous. Panicle contracted, 30 cm, branches erect; spikelets 1-1.5 cm, linear, 1-flowered, pedicelled; glumes linear, convolute, awned; lemma linear, awn trifid, subequal; palea hyaline, linear to narrowly ovate, 1-2-nerved; lodicules 2; stamens 3; stigma plumose. Caryopsis linear, prickly.

Common in plains of all districts.

Fl. & Fr.: September – April.

Neelwai RF (ADB), *TP* & *GO* 5436; Vempalli (ADB), *GO* & *PVP* 4775; Lingampet (NZB), *BR* & *CPR* 9238; Gowraram RF (NZB), *TP* & *BR* 6381; Pocharam (MDK), *TP* & *MSG* 12022; Choutkoor fields (MDK), *TP* & *MSG* 11941; Aklaspur (KRN), *GVS* 20223 (CAL); Hyderabad (HYD), *MRS* 220 (DD); on way to Umamaheswaram (MBNR), *SRS* 111309 (BSID); Vatavarlapalli (MBNR), *VBH* 84968 (BSID); way to Gangaram (WGL), *PVS* 76990 (BSID); Jakaram (WGL), *RKP* 105277 (BSID).

ARTHRAXON P. Beauv.

1. Stamens 2:
    2. Pedicelled spikelets, at least in the upper part of the raceme developed, spikelets up to 3 mm .... **A. lancifolius**
    2. Pedicelled spikelets not developed, pedicels more or less reduced, spikelets up to 5 mm .... **A. hispidus**
1. Stamens 3:
    3. Lower glume of sessile spikelet smooth dorsally .... **A. depressus**
    3. Lower glume of sessile spikelet with broad based spinules or spicules or tuberculate hairs:
        4. Lower glume of sessile spikelet pectinately spinulose conspicuous on keels from base to apex:

5. Pedicelled spikelet present .... **A. lanceolatus**
5. Pedicelled spikelet absent .... **A. castratus**

4. Lower glume of sessile spikelet faintly pectinately spinulose on keels from base to apex .... **A. prionodes**

Arthraxon castratus (Griffith) Naray. ex Bor in Kanj. & *al*., Fl. Assam 5: 376. 1940 & 99. 1960; Welzen, Blumea27: 263. 1981. *Andropogon castratus* Griffith, Not. Pl. Asiat. 3: 89. 1851. *Arthraxon rudis* (Steudel) Hochst., Flora 39: 188. 1856; Hook.f., Fl. Brit. India 7: 144. 1896; Fischer, Fl. Madras 3: 1728. 1934.

Culms 30-60 cm, glabrous, nodes bearded. Leaves ovate, lanceolate, to 6 cm long, Racemes 2-5, to 7 cm; sessile spikelets 6 mm long. Lower glume lanceolate, 5-7-nerved, apex spicules without; upper glume elliptic, 3-nerved, 1-keeled, spiculed.

In Karimnagar district, frequent in hill-forests, in crevices of stones (Naqvi and Raju 1998a).

Fl. & Fr.: July – March.

Arthraxon depressus Stapf ex Fischer, Bull. Misc. Inform. Kew 1933: 349. 1933 & in Fl. Madras 3: 1728. 1934; Bor, Grass. Ind. 425. 1960; Moulik, Grass. Bamb. India 1: 259. 1997.

Annual or perennial grass; culms glabrous, geniculate, nodes glabrous. Leaf sheaths up to 2.5 cm, ligule up to 2 mm, blade 2.5-3.5× 0.8-1.5 cm, tuberculate-pilose. Racemes 2-4, up to 5 cm, joints 3-4 mm, pubescent. *Sessile spikelet:* up to 6.5 mm, base cuneate; glumes unequal, lower glume 6 mm, 7-nerved, rounded without, margin not reflexed, smooth; upper glume up to 6.3 mm, obscurely 3-nerved, apex acuminate; lemmas pilose on margin; lower lemma up to 5.5 mm; upper lemma up to 4.8 mm; awn up to 8 mm; palea up to 2.5 mm; anthers 3. *Pedicelled spikelet:* reduced into a 2.2-3.5 mm long pilose stipe.

Common in hills in Karimnagar district.

Fl. & Fr.: September – March.

Shanigram (KRN), *A.H.Naqvi* 960955 (KU).

Arthraxon hispidus (Thunb.) Makino, Bot. Mag. Tokyo 26: 214. 1912; Fischer, Fl. Madras 3: 1728. 1934; Bor, Grass. India 99. 1960; Welken, Blumea 27: 266. 1981; Moulik, Grass. Bamb. India 1: 259. 1997. *Phalaris hispida* Thunb., Fl. Jap. 44. 1784. *Arthraxon ciliaris* P. Beauv., Ess. Agrost. 111, 152. 11. f. 6. 812; Hook. f., Fl. Brit. India 7: 147. 1896. *A micans* (Nees) Hochst., Flora 39: 188. 1856; Bor, Grass. India 101. 1960. *A. nudus* (Steud.) Hochst. *loc.cit.* Bor, *loc.cit. A. submutius* (Nees ex Steud.) Hochst. *loc.cit.* Hook.f., *loc.cit.* 144; Bor, *loc.cit.* 102. *A. breviaristatus* Hack. in DC. Monogr. Phan. 6: 350. 1889; Hook.f., *loc.cit.*; Bor, *loc.cit.* 99. *A. intermis* Hook. f. *loc.cit.* 145. 1896; Bor, *loc.cit.* 100. *A. quartinianus* (A. Rich.) Nash in N. Amer. Flora 17: 99. 1812; Bor, *loc.cit.* 102; Fischer, *loc.cit.* 3: 1728. 1934.

An annual, tufted grass; culms up to 35 cm high, prostrate and rooting base; nodes shortly bearded. Leaf sheaths 1.5-2.5 cm, hirsute in the upper part with tuberculate based hairs; ligule membranous with ciliate apex; blade 2.5-5 × 0.5-2 cm, ovate-lanceolate, broad, base amplexicaul, apex acuminate, tuberculate based hairs on both surfaces, ciliate towards the base. Racemes 1-9, 25 cm long, terminal; rhachis fragile; spikelets 3-4 mm, solitary, oblong-lanceolate, laterally compressed; glumes subequal; lower lanceolate, sub chartaceous, scaberulous towards apex, 7-9-nerved; upper obliquely lanceolate, mucronate, 3- or 5-nerved; lower lemma oblong, hyaline, 2-nerved; upper lemma narrowly linear-lanceolate, awned; palea absent; stamens 2, stigmas plumose. Caryopsis linear. Pedicelled spikelet not developed or reduced.

Occasional in Hyderabad district.

Fl. & Fr.: September – February.

Arthraxon lanceolatus (Roxb.) Hochst. var. **echinatus** (Nees) Hack. Mon. Androp. 348. 1889. *Bathratherum echinatum* Nees, Edinb. New Phil. J. 18: 181. 1835. *Arthraxon echinatus* (Nees) Hochst. Flora 39: 188. 1856; Fischer, Fl. Madras 3: 1728. 1934; Bor, Grass. India 99. 1960. *A. spathaceus* Hook. f., Fl. Brit. India 7: 145. 1896; Moulik, Grass. Bamb. India 1: 259. 1997.

An annual or perennial, tufuted grass; culms up to 40 cm high, geniculately ascending from the base; nodes glabrous. Leaf sheaths 2-3 cm long, tuberculate ciliate-glabrous; ligule fimbriate; blade 3-4.5 × 0.5-1 cm; ovate-lanceolate, base amplexicaul, margin tuberculate ciliate, apex acuminate, glabrous. Racemes 3.5-5 cm, terminal. *Sessile spikelet:* 0.5-0.6 cm, ovate; glumes unequal speculate; lower glume linear, coriaceous, 6-nerved, echinulate on the back; upper glume membranaceous, glabrous, margin ciliate hairy, 3-4-nerved; lemmas linear, ciliate, upper awned, awn geniculate; palea with notched apex; lodicules 2; stamens 3, stigmas plumose. Caryopsis narrowly linear. *Pedicelled spikelet:* narrowly ovate; lower glume coriaceous, few scabrous hairs on the dorsal side, narrowly ovate, margin inflexed, 9-nerved; upper glume membranous, narrowly ovate, margin inflexed, ciliate hairy, 3-nerved; lower lemma hyaline, glabrous, narrowly elliptic; upper lemma membranous, narrowly ovate, 1-nerved; palea absent; stamens 3; lodicules 2.

Occasional along the streams and moist places of Medak, Hyderabad and Mahabubnagar districts.

Fl. & Fr.: September – March.

Pullavagu (MDK), *TP* & *MSG* 12088; Lingampalle (RR), *MSM* 12764; Borapuram (MBNR), *BSS* & *SKB* 32287 (BSID); Yenkur (KMM), *RR* 108572, 108571 (BSID).

Arthraxon lancifolius (Trin.) Hochst., Flora. 39. 188. 1856; Bor, Grass. India 100. 1960; Fischer, Fl. Madras 3: 1729. 1934. *Andropogon lancifolius* Trin., Mem. Acad. Sci. Petersb. Ser. 6, 2: 271. 1832. *Arthraxon microphyllus* (Trin.) Hochst., Flora 39: 188. 1856; Hook. f., Fl. Brit. India 7: 147. 1896. *Andropogon microphyllus* Trin., Mem. Acad. Sci. Petersb. Ser. 6, 2: 275. 1832.

An annual, tufted grass; culms up to 60 cm high, decumbent at base; nodes slightly bearded. Leaf sheaths 0.1-0.15 cm long, glabrous; ligule membranous; blade 1.5-3 × 0.4-0.8 cm, ovate-lanceolate, base amplexicul, apex acute. Panicle densely flowered, spicate; rhachis capillary. Sessile spikelets 3 mm, laterally compressed, ciliate, narrowly lanceolate; lower glume lanceolate, minutely puberulous, 2-cuspidate, 4-nerved; upper glume narrowly lanceolate, tapering into a very fine awn-like point; lower lemma shorter than upper glume, lanceolate, epaleate; upper lemma ovate-lanceolate, hyaline with an awn, awn geniculate; palea absent; lodicules 2; stamens 2. Caryopsis linear. Pedicelled spikelet ovate to narrowly ovate; lower glume membranous, glabrous, ovate, 5-nerved, apex two toothed; upper glume membranous, glabrous,1-nerved.

Common grass of damp habitats growing in colonies in rock crevices of Adilabad, Nizamabad, Medak and Mahabubnagar districts.

Fl. & Fr.: August– December.

Sathenapalle beat (ADB), *TP* & *PVP* 4079; Ankusapuram (ADB), *GO* 4339; Thallapalli vagu (NZB), *TP* & *BR* 6223; Ramayampet (MDK), *BR* & *CP* 11506.

Arthraxon prionodes (Steud.) Dandy in Andr., Fl. Pl. Sudan 3: 399. 1956; Bor, Grass. India 101. 1960. *Andropogon prionodes* Steud., Syn. Pl. Glumac. 1: 383 1854. *Arthraxon lanceolatus* sensu Hook. f., Fl. Brit. India 7: 143. 1896 *p.p.* non (Roxb.) Hochst. 1856.

A perennial, tufted grass; culms up to 60 cm high, hairy, trailing; nodes glabrous. Leaf sheaths slightly pubescent; blade 4-20 × 2-7 cm, lanceolate-ovate, base cordate, apex acuminate, tuberculate ciliate on margin. Racemes 2-8, 7 cm long; rhachis slightly hairy; spikelets 5 mm long, linear-lanceolate. Lower glume hyaline, 5-7-nerved, acute, pectinately spinulose on keels, faint; upper glume lanceolate, 3-nerved; lower lemma shorter, linear-oblong, acute; upper lemma hyaline, awned; stamens 3, stigmas plumose.

In Hyderabad and Mahabubnagar districts.

Fl. & Fr.: September – December.

Farahabad (MBNR), *MVS* & *MB* 38004.

ARUNDINELLA Raddi

1. Upper lemma with two setae, one on each side of the awn at the tip
    2. Perennials .... **A. setosa**
    2. Annuals .... **A. tuberculata**
1. Upper lemma devoid of setae, with a single awn:
    3. Panicles effuse .... **A. pumila**
    3. Panicles contracted .... **A. ciliata**

Arundinella ciliata (Roxb.) Nees ex Miq., Nieuwe Verh. Eerste K.l. Kon. Ned. Inst. Wetensch. Amsterdam (ser. 3) 4 : 30.1851; Bor, Grass. India 421. 1960. *A. agrostoides* Hook.f. Fl. Brit. India 7 : 71. 1896, non Trin. 1836. *A. holcoides* acut. Non (Kunth) Trin. : Fischer, Fl. Madras 3 : 1801. 1934.

Annual grass; culms 25-40 cm high, sparsely tuberculate-pilose; nodes bearded. Leaf-blades oblong, 5-10 x 0.5-0.6 cm; sheaths tubeculate pilose; ligule to 2 mm. Panicles oblong, contracted, 5-15 cm; spikelets oblong, to 3 mm (excluding awn); glumes 5-nerved, tuberculate pilose without, to one third at base; lower glume to 3 mm, awned; upper glume 3-5 mm; lower lemma 2-2.5 mm, acute; palea 1-1.5 mm. Caryopsis ellipsoid 0.5-1mm.

In rocky situations in Hyderabad district (Ramana, 2010).

Fl. & Fr. : October-January.

Arundinella pumila (Hochst. ex A. Rich.) Steud., Syn. Pl. Glum. 1, 114. 1854; Bor, Grass. India 423. 1960; Fischer, Fl. Madras 3: 1801. 1934. *Acratherum pumilum* Hochst. ex A. Rich., Tent. Fl. Abyss. 2. 414. t.100. 1851. *Arundinella tenella* Nees ex Steud. *loc.cit.* 115; Hook. f., Fl. Brit. India 7 :71. 1896.

An annual, slender grass, culms erect, up to 40 cm high, sparsely tuberculate-pilose; nodes sparsely bearded. Leaf sheaths 2-8 cm long, glabrous with ciliate margins; ligule a fringe of hairs; blade 4-12 x 0.5-1.3 cm, ovate-lanceolate, membranous, base narrowed, apex acuminate, veins prominent, with long soft hairs. Panicle effuse, capillary, branched; spikelets 2 mm, pedicelled; glumes ovate-lanceolate, thinly membranous, lower 3-nerved, acute; upper 5-nerved, acuminate; lower lemma ovate, obtuse, paleate; upper lemma lanceolate, awned; stamens 3. Caryopsis narrow, ellipsoid.

Occasional grass in Medak district.

Fl. & Fr.: July-November.

Arundinella setosa Trin., Gram. Panic. 63. 1826; Hook. f., Fl. Brit. India 7: 70 1896; Bor, Grass. India 424. 1960; Moulik, Grass. Bamb. India 1: 61. 1997; var. **setosa**

Vern.: *Pathi oopagaddi.*

A perennial, stoloniferous grass; culms erect, up to 70 cm high, glabrous at the base, slender; nodes glabrous, Leaf sheaths 4-9 cm long, loose, tuberculate-pilose; ligule membranous, ciliate at apex; blade 5-12 x 0.4-0.6 cm, base narrow, apex acuminate. Panicle 20 cm, effuse, terminal; spikelets 3.5 mm, elliptic-ovate, pedicelled, 1- 2-flowered; glumes oblong, 5-nerved, acuminate, tuberculate-pilose, upper awned; lemma oblong-lanceolate, 3-nerved, both paleated, upper sparsely scaberulous, awned; paleas small, narrowly lanceolate, keeled, obtuse; lodicules 2; stamens 3, stigmas plumose. Caryopsis oblong.

Occasional in Nizamabad, Medak and Hyderabad districts.

Fl. & Fr.: Jul-December.

Sirnapalli RF (NZB), *BR* & *PSPB* 9579; Narsapur RF (MDK), *TP* & *MSG* 11970.

Note: This grass can be distinguished from other species of *Arundinella* by its spikelets being glabrous and length 4 mm.

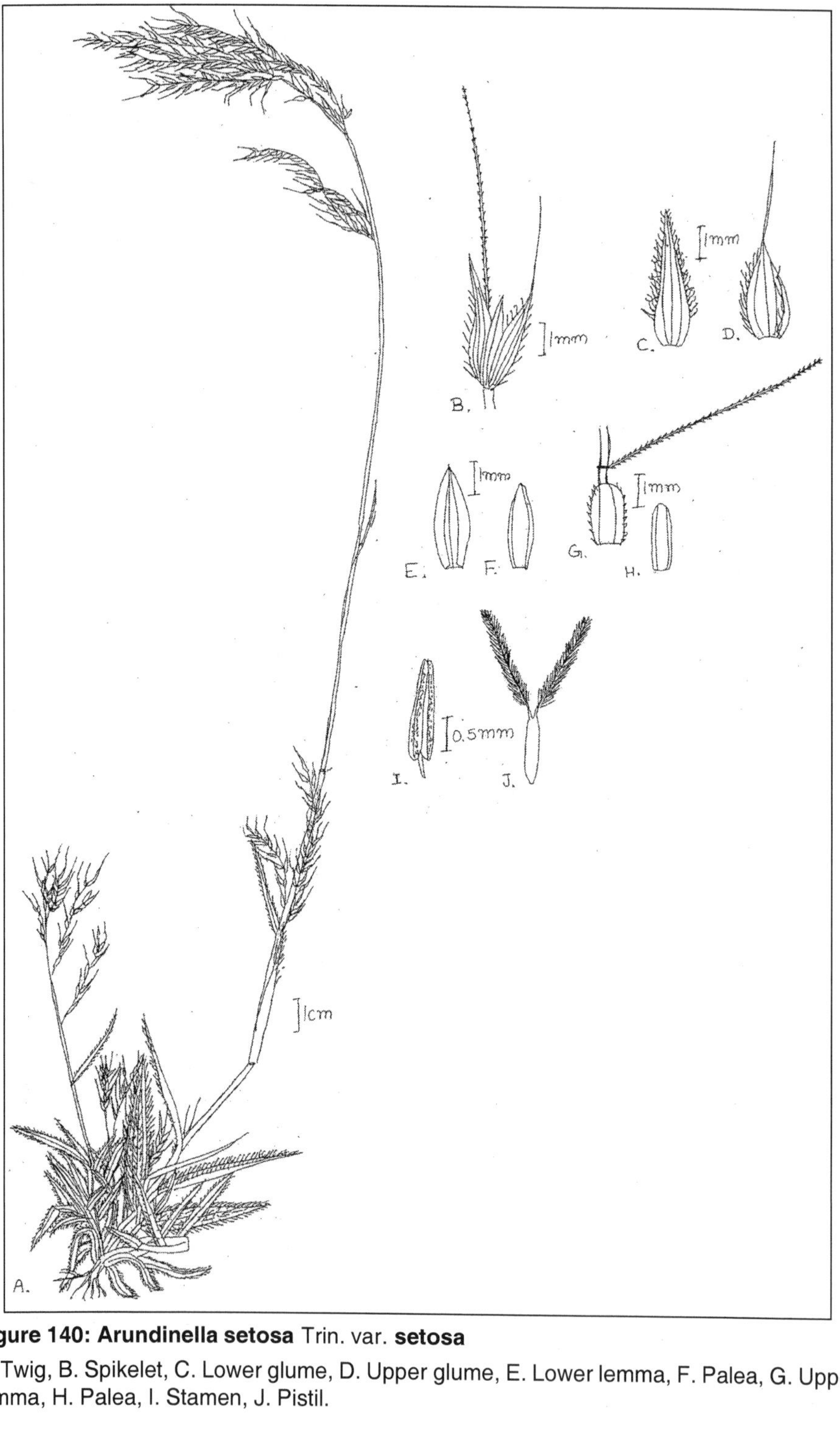

Figure 140: Arundinella setosa Trin. var. **setosa**

A. Twig, B. Spikelet, C. Lower glume, D. Upper glume, E. Lower lemma, F. Palea, G. Upper lemma, H. Palea, I. Stamen, J. Pistil.

Arundinella tuberculata Munro ex Lisboa, J. Bombay Nat. hist. Soc. 5: 344. 1891; Hook. f., Fl. Brit. India 7: 69. 1896; Bor, Grass India. 425. 1960; Moulik, Grass. Bamb. India 1: 61. 1997.

Annuals, culms erect or decumbent at the base, to 120 cm high; nodes glabrous. Leaf sheaths 4-12, covered with tubercle based hairs, margins villous. Ligules a ridge of hairs. Panicles 15-35 cm; spikelets ovoid, acute, 10-12 x 1-1.5 mm; lower glume oblong-lanceolate, 4-4.5 x 1-1.5 mm, 3-nerved, chartaceous, acuminate, covered with tubercle based hairs; upper glume ovate, acute, 3-3.5 x 1-1.5 mm, awn 3 mm, 5-nerved, covered with tubercle based hairs. Florets 2; lower floret male or neuter, upper bisexual; lower lemma, 3-3.5 x 1-1.5 mm, elliptic-oblong, 5-nerved, 2-keeled, glabrous, subcoriaceous; lower palea oblong-lanceolate, 2.5 x 0.7 mm, hyaline, 2-keeled; upper lemma lanceolate, acute, 2.2 x 0.7 mm, notched, awned with 2 setae, 1.7 mm on either side, awn 10.5 mm, subcrustaceous when ripe, brown, 2-keeled, keels scabrid.; upper palea lanceolate, 2.5 x 0.7 mm, 2-nerved, 2-keeled, hyaline; stamens 3, filament 0.3 mm, anthers 0.9 x 0.2mm; ovary 0.2 x 0.1 mm, styles 0.5 mm, stigmas 0.5 mm; lodicules 2, 0.1 x 0.1 mm, ovate, hyaline.

Dry hills of Mahabubnagar district; occasional in drier localities (Sadasivaiah & *al.*, 2008).

Fl. & Fr.: October – February.

Vatavarlapalli (MBNR), *BSS & SKB* 30573.

Note: Endemic to India.

BAMBUSA Schreber *nom. cons.*

Bambusa bambos (L.) Voss, Vilm. Blumengartn. ed. 3 1: 1189. 1895. *Arundo bambos* L., Sp. Pl. 1: 81. 1753. *Bambusa arundinacea* (Retz.) Willd., Sp. Pl. ed. 4. 2: 245. 1799; Fischer, Fl. Madras 3: 1859. 1934; Hook. f., Fl. Brit. India 7; 395. 1896; Moulik, Grass. Bamb. India 1: 22. 1997. *Bambosa arundinacea* Retz., Observ. Bot. 5: 24. 1788.

Vern.: *Mulla veduru.*

A large thorny, tufted shrub; culms up to 30 m high, stout, erect, internodes hollow; nodes prominent, the lower producing horizontal, armed with 2-3 stout, recurved spines; leaf sheaths 1- 4 cm, glabrous; ligule narrow, entire or fringed with pale hairs; blade 10-15 x 1-15 cm, lanceolate, appressed hairy, base rounded, margin ciliate-scabrous, apex acuminate, midrib narrow, nerves 4-6. Panicles compound, sub-secund recemosely arranged spikes; spikes 1-3 cm long, in loose clusters, lanceolate, acute, glabrous; glumes 2 or 0, ovate-lanceolate, acute, many-nerved; lemmas 3-7, the upper most 1-3 male or neuter, 3-nerved, obtuse; palea truncate, with 2 ciliate keels; lodicules 3; stamens 6; style short. Caryopsis elliptic, dark.

Rare in forests in Adilabad and Karimnagar districts, perhaps introduced by the forest department, and got naturalized.

Fl. & Fr.: Very infrequent.

BOTHRIOCHLOA Kuntze

1. Lower racemes shorter than the rachis .... **B. bladhii**
1. Lower racemes longer than the rachis:
   2. Lower glume of the sessile spikelets pitted:
      3. Stoloniferous grass, nodes sparsely bearded, racemes digitate .... **B. pertusa**
      3. Tufted grass, nodes densely bearded, spikes umbellate .... **B. kuntzeana**
   2. Lower glume of the sessile spikelets not pitted .... **B. pseudoischaemum**

Bothriochloa bladhii (Retz.) S.T.Blake, Proc. Roy. Soc. Queensland 80: 62. 1969. *Andropogon bladhii* Retz., Observ. Bot. 2: 27. 1781. *Dichanthium glabrum* (Roxb.) Jain & Deshp., Bull. Bot. Surv. India 20: 134. 1978 (1979) var. *glabrum;* Deshp. in Fasc. Fl. India 15: 13. 1984. *Andropogon glaber* Roxb., Fl. Ind. 1: 267, 1820. *A. intermedius* auct. non R. Br. 1810; Hook.f., Fl. Brit. India 7: 175. 1896. *Amphilophis glabra* (Roxb.) Stapf in Fl. Trop. Afr. 9: 172 1917; Fischer, Fl. Madras 3: 1732. 1934. *Bothriochloa glabra* (Roxb.) A. Camus, Ann. Soc. Lyon, 1930; Bor, Grass. India 107. 1960; Moulik, Grass.Bamb. India 1: 455. 1997.

A perennial, tufted, coarse grass; culms up to 1 m high; nodes bearded. Leaf sheaths 10-20 cm, glabrous; ligule truncate; blade 15-30 × 0.4-0.6 cm, linear, base truncate, apex acuminate, glabrous, midrib prominent. Panicle 10-15 cm long, narrowly oblong; lower branches shorter than upper. Spikelets 6 mm, 2-nate. *Sessile spikelet:* 3 mm, oblong, bisexual; lower glume coriaceous, narrowly ovate, glabrous, margins inflexed, 7-9-nerved, apex obtuse; upper glume membranaceous, narrowly ovate, glabrous, margins inflexed, 3-nerved, apex acuminate; lower lemma oblong, nerveless, hyaline; upper narrow, awned; lodicules 2; stamens 3, stigmas plumose. Caryopsis oblong, slightly compressed. *Pedicelled spikelet:* 3 mm, oblong, male or barren, similar to sessile but awnless.

Occasional in the forests of Adilabad district.

Fl. & Fr.: July – November.

Alampalli (ADB), *TP* & *PVP* 4169.

Bothriochloa kuntzeana (Hack.) Henr., Blumea 3: 456. 1940; Bor, Grass. India 108. 1960.; Moulik, Grass. Bamb. India 1: 266. 1997. *Dichanthium kuntzeana* (Hack.) Jain & Deshp., Bull. Bot. Surv. India 20: 134. 1978 (1979). *Andropogon kuntzeana* Hack. in DC., Monogr. Phan. 6: 478. 1889; Hook. f., Fl. Brit. India 7: 175. 1896. *A. pertusus* (L.) Willd. var. *longifolia* Hack. in DC. Monogr. Phan. 6: 482. 1889; Hook. f., Fl. Brit. India 7: 174. 1896. *Amphilophis kuntzeana* (Hack.) Haines, Bot. Bih. Orissa 1031. 1924; Fischer, Fl. Madras 3: 1732. 1934.

A perennial, tufted grass; culms up to 30 cm high, terete, branches from base; nodes densely bearded. Leaf sheaths terete, glabrous; ligule membranous; blade 10-40 × 0.3-1 cm, linear, cauline, margins scaberulous, apex acuminate, glabrous. Spikes corymbose, 30 cm long, densely fascicled; joints and pedicels plumosely ciliate with long hairs. *Sessile spikelet:* linear-lanceolate, 5 mm, bisexual; callus with long hairs; lower glume membranous, 7-9-nerved, pitted, glabrous; upper glume lanceolate, apex acumimate, margins ciliate; lower lemma lanceolate, glabrous; upper lemma with a geniculate awn; lodicules 2; stamens 3. *Pedicelled spikelet:* linear-lanceolate, 5 mm, male; lower glume linear, 9-11-nerved, keels ciliate, scabrid at apex; upper glume lanceolate, 5-nerved; lower lemma ovate-oblong, ciliate, nerveless; upper lemma linear, small or some times absent.

Occasional in Hyderabad district.

Fl. & Fr.: August– December.

Hyderabad, *MRS* 111 (DD).

Bothriochloa pertusa (L.) A. Camus, Ann. Soc. Lyon. 1930, n.s. 76: 164. 1931; Bor, Grass. India 109. 1960; Moulik, Grass. Bamb. India 1: 267. 1997. *Dichanthium pertusum* (L.) Clayton, Kew Bull. 32: 4. 1977; Deshp. in Fasc. Fl. India 15: 22. 1984. *Holcus pertusus* L., Mant. Alt. 301. 1771. *Andropogon pertusus* (L.) Willd., Sp. Pl. 4: 922. 1806; Hook.f., Fl. Brit. India 7: 173. 1896. *Amphilophis pertusus* (L.) Nash ex Stapf, Agric. New W. Ind. 15: 179. 1916; Fischer, Fl. Madras 3: 1731. 1934.

A perennial, stoloniferous grass; culms up to 50 cm high, ascending; nodes sparsely hairy. Leaf sheaths 2-6 cm long, glabrous; ligule membranous; blade 3-10 × 0.2-0.4 cm, linear, base truncate, apex acuminate, glabrous. Racemes 2-6.5 cm long, slender, digitate, silky hairy; spikelets 6 mm, 2-nate, pitted. *Sessile spikelet*: 4 mm, oblong-lanceolate, pitted; callus bearded; lower glume elliptic-oblong, bearded below, dorsally pitted, subchartaceous, 7-9-nerved; upper glume ovate-lanceolate, 3-5-nerved, finely pointed at the tip; lower lemma lanceolate, nerveless, hyaline; upper lemma reduced to a slender awn; lodicules 2; stamens 3; stigmas plumose. Caryopsis oblong. *Pedicelled spikelet*: 4 mm, male or barren, similar to sessile spikelet, but awnless.

Common in plains and hills of Adilabad, Nizamabad, Medak, Warangal, Nalgonda and Mahabubnagar districts.

Fl. & Fr.: July – January.

Mancheryal (ADB), *GO & PVP* 4939; Kamareddy (NZB), *BR* 9276; Chutkoor (MDK), *TP* & *MSG* 11918; Narsapur RF (MDK), *TP* & *MSG* 1966; Pakhal RF (WGL), *KMS* 11668 (CAL); Madikonakutta-Narsapur (MDK), *KMS* 6730 (CAL); Nagarjunakonda Valley (NLG), *KT* 9767 (CAL); Yelandu – Karepally (KMM), *RR* 105952 (BSID); Mysamma temple (MBNR), *SRS* 108622 (BSID); Guddaegunda, Koilkonda (MBNR), *SRS* 110769 (BSID).

Bothriochloa pseudoischaemum (Nees ex Stued.) Henr., Blumea 3: 457. 1940; Bor, Grass. India 109. 1960; Moulik, Grass. Bamb. India 1: 267. 1997. *Dichanthium pseudoischaemum* (Nees ex Steud.) Join & Deshp., Bull. Bot. Surv. India. 20-134. 1978 (1979); Deshp. in Fasc. Fl. India 15: 23. 1984. *Andropogon pseudoischaemum* Nees ex Steud. Syn. Pl. Glum. 1: 380. 1854; Hook. f., Fl. Brit. India 7: 172. 1896. *A. ischaemum* sensu Roxb., Fl. Ind. 1: 264. 1820. (non L. 1753). *Amphilophis pseudoischaemum* (Nees ex Steud.) Fischer, Fl. Madras 3: 1731. 1934.

A perennial, tufted grass; culms up to 2 m high, branching from the base; nodes bearded. Leaf sheaths 5-15 cm long, hispid at mouth; ligule membranous, ciliate; blade 15-30 × 0.5-1 cm, linear, base rounded, apex acuminate, scaberulous. Panicle 5-10, 8 cm long, joints and pedicels pilose; spikelets 1 cm, 2-nate. *Sessile spikelet:* 5 mm, elliptic-lanceoalte, bisexual; lower glume elliptic-lanceolate, silky, 2-keeled, subchartaceous, bidentate, 5-7-nerved; upper glume lanceolate, apex aristate, 3-nerved; lower lemma lanceolate, nerveless, hyaline; upper lemma reduced to an awn; stamens 3; stigmas plumose. Caryopsis oblong, compressed. *Pedicelled spikelet:* 5 mm, elliptic-lanceolate, male or barren, similar to sessile spikelet, but awnless.

Common on hills of Hyderabad district.

Fl. & Fr.: July – September.

Hyderabad, *MRS* 247 (DD).

BRACHIARIA (Trin.) Griseb.

Note: Some authors treat most of the species of *Brachiaria* under *Urochloa*. This view has not been followed here as this concept is yet to receive wider acceptance.

1. Spikelets imbricate or approximate:
    2. Rhachis flat, margins scaberulous .... **B. distachya**
    2. Rhachis trigonous or terete:
        4. Panicle linear, racemes closely appressed to the rhachis ... **B. eruciformis**
        4. Panicle oblong-lanceolate, branches erect:
            5. Leaves ovate; lemma stiff-setose at tip .... **B. semiundulata**
            5. Leaves lanceolate or linear; lemma other than above:
                6. Spikelets up to 2.5 cm long .... **B. reptans**
                6. Spikelets 3-5 mm long .... **B. ramosa**
1. Spikelets distant:
    7. Leaves flat, lanceolate, rounded at base with bulbous based hairs .... **B. kurzii**
    7. Leaves narrowly linear, base narrow without bulbous based hairs .... **B. remota**

Brachiaria distachya (L.) Stapf in Prain, Fl. Trop. Afr. 9: 565. 1919; Fischer, Fl. Madras 3: 1769. 1934; Bor, Grass. India 285. 1960; Moulik, Grass. Bamb. India 1: 76. 1997. *Panicum distachyon* L., Mant. Pl. 183. 1771; Hook.f., Fl. Brit. India 7: 37. 1896. *Urochloa distachya* (L.) Nguyen, Novosti Sist. Vyssh. Rast. 1966: 13. 1966.

Vern.: *Koranna gaddi.*

An annual, stoloniferous grass; culms up to 40 cm high, rooting at nodes; nodes glabrous. Leaf sheaths 1.5-3 cm, margins ciliate; ligule a fimbriate membrane; blade 2-4 × 0.3-0.5 cm, linear-narrowly lanceoalate, base cordate, margins cartilaginous, apex acute, pubescent-glabrous. Racemes with 2 or 3 branches, 4 cm long; rahis flat, ribbon like; peduncle hirsute. Spikelets 3 mm. ellipsoid, second, glabrous; glumes unequal, 5-nerved, lower glume small, ovate; upper oblong, acute, lemma 3-nerved, paleated; lower lemma oblong-ovate, margin incurved; palea oblong, obtuse; upper lemma ovate, coriaceous; palea similar to upper lemma; lodicules 2; stamens 3; stigmas plumose. Caryopsis ovoid, compressed.

Common in waste lands and cultivated fields of Adilabad, Ranga Reddi, Hyderabad and Karimnagar districts.

Fl. & Fr.: August- October.

Ankusapuram (ADB), *GO* 4340; Malkalcheruvu (RR), *MSM* & *KH* 10568.

Brachiaria eruciformis (Sm.) Griseb. in Ledeb. Fl. Ross. 4. 469. 1856; Bor, Grass. India 283. 1960; Fischer, Fl. Madras 3: 1769.1934; Moulik, Grass. Bamb. India 1: 76. 1997. *Panicum eruciformis* Sm. in Sibth & J.E. Smith, F1. Graeca 1.44 1806. *P. isachne* Roth ex Roem. & Schult., Syst. Veg. 2.458. 1817; Hook. f., Fl. Brit. India 7: 28. 1896.

Vern.: *Domakalu gaddi.*

An annual grass, culms up to 40 cm high, geniculately ascending; nodes densely pubescent. Leaf sheaths 2-4.5 cm long, slightly pubescent; ligule hairy; blade 1.5-4.5 x 0.2-.0.4 cm, lanceolate, base rounded, apex acuminate, glabrous. Panicle 5.5 cm long, subsecund, racemosely arranged spikes; rhachis triquetrous, winged; spikelets ellipsoid, 2 mm, densely pubescent, subacute-obtuse; glumes dissimilar, lower a minute scale: upper broadly ovate, densely pubescent, 5-nerved; lower lemma lanceolate, slightly pubescent, 3-nerved; palea truncate; upper lemma crustaceous, smooth, 3-nerved; palea similar to upper lemma, nerveless; stamens 3; stigma plumose. Caryopsis ovoid, 1.5 mm dark.

Common in fields and moist localities of all districts.

Fl. & Fr: August- December.

Kadthala (ADB), *GO & DAM* 5025; Poosampalli (MDK), *TP* & *MSG* 11956; NSF fields (MDK) *BR* & *CP* 11669; Mekavaripalli (RR), *TP* & *MSM* 11046.

Brachiaria kurzii (Hook.f.) Camus, in Lecomte, Fl. Ge. del' Indo-Chine 7 : 438. 1922; C. Fischer, Fl. Madras 3 : 1770. 1934; Bor, Grass. Burma, Ceylon, India & Pakistan 283. 1960. *Panicum kurzii* Hook.f. Fl. Brit. India 7 : 38. 1896.

Annual grasses, 40-60 cm high; culms decumbent, hispid. Leaf-blades flat, lanceolate, rounded at base, bristly with bulbous based hairs, 4-6 x 0.8-1.00 cm.; sheaths ciliate. Racemes few, 3-4 cm long; rachis triquetrous. Spikelets elliptic or elliptic-oblong, acute, 2.5-3 mm long; nerved on the upper glume well marked. Lower glume elliptic; upper glume elliptic to oblong; upper lemma crustaceous, margins incurved. Caryopsis compressed.

Rare along the hedges of open lands and cultivated fields in Hyderabad district (Ramana, 2010).

Fl. & Fr. : August-December.

Brachiaria ramosa (L.) Stapf in Prain, Fl. Trop. Afr. 9: 542. 1919; Fischer, Fl. Madras 3: 1770. 1934; Bor, Grass. India 284. 1960; Moulik, Grass. Bamb. India 1: 78. 1997. *Panicum ramosum* L., Mant. Pl. 29. 1767; Hook.f., Fl. Brit. India 7: 36. 1896. *Urochloa ramosa* (L.) Nguyen, Novosti Sist. Vyssh. Rast. 1966: 13. 1966.

An annual, tufted grass; culms up to 50 cm high; nodes pubescent. Leaf sheaths 0.4-0.8 cm long, glabrous, margin hairy; ligule membranous with a row of hairs; blade 6-12 × 0.4-1 cm, linear-lanceolate; base rounded, margin scaberulous, apex acuminate, slightly pubescent, densely hairy near the base. Panicle 9-12 cm long; rhachis triquetrous, scaberulous; spikelets 0.3 cm, broadly elliptic-oblong, acuminate, pubescent or glabrescent, mostly paired at raceme base and solitary towards apex; glumes membranous, acuminate, puberulous; lower glume broadly ovate, 5-nerved, upper ovate-lanceolate, 7-nerved; lower lemma similar to upper glume; upper lemma ovate, sub crustaceous, smooth, 3-nerved; palea similar to upper lemma, nerveless; lodicules 2; stamens 3; stigma plumose. Caryopsis ovoid.

Common in cultivated fields and waste lands of all districts.

Fl. & Fr.: August–September

Vempalli (ADB), *GO* & *PVP* 4759; Gangapur RF (MDK), *TP* & *MSG* 12072; East of Rechepalle (KRN), *GVS* 20163 (CAL); Pakhal RF (WGL), *KMS* 1371 (CAL); Hyderabad, *MRS* 272 (DD); Ramanagutta (KMM), *RR* 107979 (BSID); Boddupalle (NLG), *BVRS* 649 (NU).

Brachiaria remota (Retz.) Haines, Bot. Bih. Orissa 1005. 1924; Fischer, Fl. Madras 3: 1770. 1934; Bor, Grass. India 285. 1960; Moulik, Grass. Bamb. India 1: 78. 1997. *Panicum remotum* Retz. Observ. Bot. 4: 17. 1786; Hook. f., Fl. Brit. India 7: 38. 1896.

An annual, tufted grass; culms up to 50 cm high, weak, decumbent at base; nodes glabrous Leaf sheaths 9 cm long, glabrous, ciliate margins towards upwards; ligule membranous; blade 5-10 x 0.3-0.7 cm, linear, base rounded, apex acuminate, glabrous. Panicle open, 10 cm long; branches alternate, spreading, filiform; spikelets 5 mm, ellipsoid, distant, sparsely hairy; glumes membranous, acuminate, lower broadly ovate; upper ovate-lanceolate; lower lemma similar to upper glume; upper lemma subcrustaceous, ovate; palea similar to its lemma; lodicules 2; stamens 3; stigmas plumose. Caryopsis ellipsoid.

Occasional along wastelands usually under shade in Adilabad, Karimnagar and Medak districts.

Fl. & Fr: August-October.

Sirpur (ADB), *GO & PVP* 4912; Vempalli (ADB), *GO & PVP* 4774; Rudraram RF (RR), *TP* & *MSM* 12720; Rechapalle (KRN), *GVS* 20153 (MH);

Brachiaria reptans (L) C. A. Gardner & C.E Hubb. in Hook. Icon. Pl. Sub. t 3363. 1938; Bor, Grass. India 285. 1960; Moulik, Grass. Bamb. India 1: 79. 1997. *Panicum reptans* L., Syst. Nat. (ed. 10) 870. 1759. *Panicum prostratum* Lam., Tabl. Encycl. 1: 171. 1791; Hook. f., Fl. Brit. India 7: 33. 1896. *Urochloa reptans* (L.) Stapf in Prain, Fl. Trop. Afr. 9: 601. 1920; Fischer, Fl. Madras 3: 1775. 1934.

An annual grass; culms up to 40 cm, prostrate, rooting at the nodes; nodes bearded. Leaf sheaths 2-3 cm long, glabrous; ligule membranous; blade 4-6 × 0.7-1 cm, ovate-lanceolate, base amplexicaul, basally margins ciliate, apically scaberulous, apex acuminate, glabrous, midrib prominent. Panicle of 5-9 racemes, up to 7 cm long, densely crowded spikelets on trigonous rhachis; spikelets 0.2 cm, ovoid, paired, alternate, shortly pedicelled, rarely one reduced; lower glume broadly ovate-sub orbicular, 1-nerved; upper glume apiculate, 7-nerved; lower lemma 5-nerved, paleated, male; upper lemma indurated, apiculate, 3-nerved; palea similar to lemma, bisexual; lodicules 2; stamens 3; stigmas plumose. Caryopsis broadly oblong, flattened

Common in cultivated and fallow fields in Adilabad and Karimnagar districts.

Fl. & Fr.: June-September.

Jaipur RF (ADB), *GO & PVP* 4959; Kudurupalli (KRN), *SLK* 70785 (NBG); Cherukupally (NLG), *BVRS* 661 (NU).

Brachiaria semiundulata (Hochst.) Stapf in Oliver, Fl. Trop. Africa 9: 556. 1919; Fischer, Fl. Madras 3: 1770. 1934; Bor, Grass. India 285. 1960. *Panicum semiundulatum* Hochst. Flora 24: 715. 1841.

Culms to 30 cm. Leaves ovate, leaf blade to 3.5 × 1 cm. Racemes to 4 cm, branches of racemes ascending, rachis pilose. Spikelets ovoid, to 2 mm, paired, second, densely villous, acute, mucronate.

Occasional in wet fields in Karimnagar district (Naqvi, 2001).

Fl. & Fr.: July – January.

CAPILLIPEDIUM Stapf

1. Culms weak, decumbent and trailing, much geniculately branched .... **C. filiculme**
1. Culms erect, robust, branched .... **C. huegelii**

Capillipedium filiculme (Hook. f.) Stapf in Hook. f., Ic. Pl. Sub. tab. 3085. 1922; Fischer, Fl. Madras 3: 1730. 1934; Bor, Grass. India 111. 1960; Moulik, Grass. Bamb. India 1: 269. 1997. *Andropogon filiculme* Hook. f., Fl. Brit. India 7: 18 1896. *Dichanthium filiculme* (Hook. f.) Jain & Deshp. & Bull. Bot. Surv. India 20: 134. 1978 (1979).

A perennial tufted grass; culms up to 1 m high, decumbent or geniculately ascending below, filiform, glabrous; nodes bearded with white hairs. Leaf sheaths 3-10 cm long; ligule membranous; blade 2-10 × 0.4-0.6 cm, lanceolate, base narrow, apex acuminate. Panicle 10 cm long; branches few, filiform, scented; spikelets 3 mm, 2-nate; pedicels 2 mm, filiform. *Sessile spikelet:* 3 mm, oblong, lanceolate, acute, sparsely hairy; lower glume coriaceous, elliptic, 2-keeled, 5-7-nerved; upper glume coriaceous, narrowly elliptic, 1-keeled, 3-nerved, apex acute; lower lemma orbicular-ovate, obtuse, hyaline, small; upper lemma reduced to a geniculate awn, hairy; lodicules 2; stamens 3, stigmas plumose. Caryopsis elliptic. Pedicels capillary, 1.8-2 mm long. *Pedicelled spikelet:* 4 mm, oblong-lanceolate, unawned, male; lower glume oblong-lanceolate, sparsely hairy; upper glume broadly oblong; hairy in the upper portions; lower lemma small, ovate, hyaline; palea absent; stamens 3.

Occasional in the plains and forests of Adilabad district.

Fl. & Fr.: August– January.

Sathenapalli beat (ADB), *TP* & *PVP* 4282.

Capillipedium huegelii (Hack.) Stapf in Hook. Ic. Pl. Sub. tab. 3685 1922; Fischer, Fl. Madras 3: 1730. 1934; Bor, Grass. India 111. 1960; Moulik, Grass. Bamb. India 1: 270. 1997. *Andropogon huegelii* Hack. in DC. Monogr. Phan. 6: 492. 1889; Hook. f., Fl. Brit. India 7: 180. 1896. *Dichanthium huegelii* (Hack.) Jain & Deshp., Bull. Bot. Surv. India 20: 135. 1978 (1979); Deshp. in Fasc. Fl. India 15: 15. 1984.

A perennial, tufted grass; culms up to 1 m high, branching from the base, red; nodes bearded. Leaf sheaths 5-15 cm long, ciliate at mouth, glabrous; ligule membranous; blade 10-25 × 0.6-1 cm, linear, flat, base narrow, apex acuminate, glabrous. Panicle 9 cm long; branches capillary with bearded axis; spikelets 6 mm, 2-nate. *Sessile spikelet:* 3 mm, oblong-ovate, bisexual; lower glume ovate, villous below the middle, a dorsal pit, 11-nerved, chartaceous; upper glume boat-shaped, obscurely 3-nerved, strongly keeled on back, margins inflexed, glabrous; lower lemma nerveless, hyaline; upper lemma stipiform, awned; lodicules 2; stamens 3; stigmas plumose. Caryopsis ovate, slightly compressed. *Pedicelled spikelet:* 3 mm, male or barren, similar to sessile spikelet, but awnless.

Occasional in the forests of Mahabubnagar district.

Fl. & Fr.: August- December.

CENCHRUS L.

1. Involucral bristles retrorsely barbellate, rigid, spinose; lower glume absent .... **C. biflorus**
1. Involucral bristles antrorsely barebellate, flexuous, not spinose; lower glume present .... **C. ciliaris**

Cenchrus biflorus Roxb., Fl. Ind. 1: 238. 1820; Hook. f., Fl. Brit. India 7:89. 1896; Bor, Grass. India 287. 1960; Moulik, Grass. Bamb. India 1: 81. 1997. *C. barbatus* Schumach., Beskr. Guin. Pl. 43. 1827; Fischer, Fl. Madras 3: 1749. 1934. *C. catharticus* Delile, Index Seminu. Hort. Reg. Bot,. Monspel. 1838: 4. 1839; Hook.f., Fl. Brit. India 7: 90. 1896.

An annual, erect or ascending grass; culms up to 30 cm high; nodes glabrous. Leaf sheaths 1.5-5.5 cm long, glabrous, keeled, ciliate near the mouth; ligule pubescent ring; blade 4-9 x 0.2- 0.5 cm, linear-lanceolate, base rounded, apex acuminate, glabrous. Racemes 5-10 cm long, solitary, cylindric; upper branches reduced to subesessible custers, involucral bristles rigid, apex retrorsely barbellate; outer bristles acicular, erect, spreading; inner bristles ciliate at base, grooved at base; spikelets 5 mm, subglobose; lower glume 0; upper glume lanceolate, acute, 3-nerved; lower lemma lanceolate, 5-nerved; palea containing a male floret or empty; upper lemma scaberulous on nerves, palea similar to its lemma, bisexual; lodicules absent; stamens 3. Caryopsis oblong, hard.

Common in sandy or dry soils of Nalgonda district.

Fl. & Fr.: July – Nov.

Vijayapuri (NLG), *KMS* 9851 (CAL); Nagarjunasagar (NLG), *BVRS* 525 (NU).

Cenchrus ciliaris L., Mant. Pl. 302. 1771; Fischer, Fl. Madras 3: 1749. 1934; Bor, Grass. India 287. 1960; Moulik, Grass. Bamb. India 1: 82. 1997. *Pennisetum cenchroides* Rich. in Pers. Syn. Pl. 1: 72. 1805, *nom. superfl;* Hook. f., Fl. Brit. India 7: 88. 1896.

A perennial, tufted grass; culms up to 50 cm high, erect or decumbent, much branched. Leaf sheaths 4-6 cm long, keeled, glabrous, ciliate near the mouth; ligule a pubescent ring; blade 7-15 × 02-0.5 cm, linear-lanceolate, base rounded, apex acuminate, glabrous-sparsely tuberculate pilose. Racemes 6-10 cm long, solitary, cylindric; involucral bristles antrorsely barbellate; outer filiform, not ciliate, inner flattened, thickened base, densely ciliate, connate below into a ring; spikelets 6 mm, oblong-lanceolate, 1-3 in each involucral; lower glume ovate, hyaline, 1-nerved, upper glume ovate-lanceolate, 3-5-nerved; lemmas ovate-lanceolate, acuminate, 5-nerved; paleas lanceolate, lower usually male, upper bisexual; lodicules absent; stamens 3. Caryopsis oblong, apex beaked.

Cultivated as fodder crop, sometimes found as an escape.

Fl. & Fr.: July – January.

CEPHALOSTACHYUM Munro

Cephalostachyum pergracile Munro, Trans. Linn. Soc. London. 26: 141. 1868; Hook.f., Fl. Brit. India 7: 413. 1897; Naithani & *al.*, Indian Forester 137: 528. 2011. *Schizostachyum pergracile* (Munro) R. B. Majumdar in Karthikeyan & *al.*, Fl. Ind. Enum. Monocot. 282. 1989.

A deciduous, arboreous, caespitose bamboo; culms erect, glaucous-green, somewhat white-puberulous below the nodes. 10-13 m high, 5-7.5 cm thick; internodes 30-45 cm long; walls very thin; culm-sheaths much shorter than internodes, 10-15 cm long, 15-20 cm broad, densely covered with black, stiff, deciduous hairs, afterward polished chestnut-brown; blade 5 cm long, ovate, cordate, cuspidate, auricles rounded, edges with long stiff curved white bristles; ligule very narrow, entire. Leaves linear-lanceolate, 15-35 × 2.5-3.25 cm, base rounded or cuneate, petiole 5 mm long; sheaths glabrous. Inflorescence a large panicle with verticals of long drooping, filiform spikes, bearing distant broad heads of spikelets supported by small chaffy sheathing bracts; rhachis very slender, wiry, 3.25-5 cm between the clusters. Spikelets in bracteate clusters, 1.25-1.75 cm long, without glumes, but 1-2 sterile florets, followed by a fertile floret, ending in a terminal sterile floret or filiform, lemma 1.25-1.75 cm long, ovate-lanceolate, many-nerved, densely pale-hairy, mucronate; palea as long as the lemma, 2-keeled, the keels close together, ciliate, apex deeply bifid, mucronate; lodicules 2, narrow, 5 mm long, lanceolate, somewhat obtuse and ciliate at tip, 3-5-nerved; stamens with narrow filaments; anthers purple; ovary smooth, sub-globular at base; style 3-cornered, stigmas 2-3, stout, recurved. Caryopsis obovate-oblong, shining, 1.25 cm long, grooved on one side, ending in a 1.25 cm long beak.

Common in moist and dry mixed deciduous forests of Sukumamidi North and South beats in Khammam district (Naithani & *al.*, 2011). Frequently found near streams associated with *Dendrocalamus strictus.*

CHIONACHNE R. Br.

Chionachne gigantea (J.Koenig) Veldkamp, Blumea 47: 559. 2002. *C. koenigii* (Spreng) Thwaites, Enum. Pl. Zeyl. 357. 1864; Fischer, Fl. Madras 3:1706. 1934; Bor, Grass. India 263. 1960; Moulik, Grass. Bamb. India 1: 172. 1997. *Coix koenigii* Spreng in L., Syst. Veg. 1: 239. 1824. *Polytoca barbata* (Roxb.) Stapf ex Hook. f., Fl. Brit. India 7: 102. 1896.

A perennial, tufted grass; culms up to 2 m high; nodes bearded. Leaf sheaths 2-6 cm long, glabrous, margin ciliate; ligule membranous; blade oblong-lanceolate, base rounded, margin scaberulous, apex acuminate, glabrous, midrib prominent. Racemes terminal, 5 cm long, umbellate. Female spikelets at base. Male spikelets at the top of the racemes; spathes 3 cm, oblong, apex setose, spatheoles 1.5 cm, open, with small setaceous apex. Male spikelet: 2-flowered, one pedicelled other sessile, rhachis not articulate. Spikelets 8 mm, oblong; lower glume shallowly concave, narrowly winged, 7-8-nerved; upper glume chartaceous, narrowed, 6-nerved; lemmas similar, ovate-lanceolate, membranous to chartaceous, hyaline, 0-4-nerved; paleas shortly aristate; stamens 3. Female spikelet: solitary, oblong, 4 mm, dorsiventrally compressed; glumes dissimilar, lower glume obovate, coriaceous, margins incurved; upper ovate, membranous; lemmas hyaline; lodicules 2; stigmas plumose. Caryopsis orbicular, compressed.

Common on hill slopes in the forests of Adilabad, Hyderabad and Mahabubnagar districts.

Fl. & Fr.: July December.

Sone (ADB). *TP & PVP* 4013; way to Tekupalli (KMM), *PVS* 84052 (BSID).

CHLORIS Sw.

1. Spikes recemosely arranged .... **C. roxburghiana**
1. Spikes umbelled:
    2. Perfect florets 2; culms puberulous below the inflorescence .... **C. bournei**
    2. Perfect florets 1; culms glabrous below the inflorescence:
        3. Empty lemma above, the fertile floret rudimentary to well developed:
            4. Spikes more than 8 cm long, spikelets dorsally compressed .... **C. dolichostachya**
                5. Lowest lemma with a crown of spreading hairs at the apex, margins shortly ciliate .... **C. virgata**
                5. Lowest lemma without a crown of spreading hairs at the apex, margin with long hairs .... **C. gayana**

3. Empty lemma above, the fertile lemma well developed:
    6. Sterile lemmas 2, 3-awned, keels of the fertile lemma densely ciliate on the back .... **C. barbata**
    6. Sterile lemmas 3, 4-awned, keels of the fertile lemma glabrous on the back .... **C. montana**

Chloris barbata Sw., Fl. Ind. Occid. 1: 200. 1797; Hook. f., Fl. Brit. India 7: 292. 1896; Fischer, Fl. Madras 3: 1838. 1934; Bor, Grass. India 465. 1960; Moulik, Grass. Bamb. India 1: 563. 1997.

An annual, tufted grass; culms up to 60 cm high, creeping at base; nodes glabrous. Leaf sheaths 4-9 cm long, glabrous; ligule membranous; blade linear-lanceolate, base rounded, apex acuminate, glabrous. Spikes terminal, digitate, 2 - 7 cm long; spikelets 3 mm, 3-flowered, solitary perfect and 2 imperfect or sterile florets above it; lower glume membranous, narrowly ovate, 1-nerved, upper glume membranous, ovate-elliptic, 3-nerved, densely ciliate on the margin; fertile lemma obovate, densely stiff ciliate on margins, awned; both sterile lemmas boat-shaped, awned; lodicules 2; stamens 3, stigmas plumose. Caryopsis oblong, compressed.

Common in plains and open forests of all districts.

Fl. & Fr.: July – December.

Mosra forest Nursery (NZB), *TP* & *BR* 6080; Patancheruvu (MDK), *BR* 11318; Mohammadabad (RR), *MSM* & *KH* 10536; Krishna river Bank (NLG), *KMS* 9859 (CAL); ICRISAT, Patancheruvu (MDK), *LJGV* 2834 (CAL); Nagarjuna Konda valley (NLG), *KT* 9765 (CAL); Hyderabad (HYD), *MRS* 242 (DD).

Chloris bournei Rang. & Tadul., J. Indian Bot. 2: 189. 1921; Fischer, Fl. Madras 3: 1838. 1934; Bor, Grass. India 466. 1960.

An annual, tufted grass; culms up to 90 cm high; nodes glabrous. Leaf sheaths 5-10 cm long, glabrous; ligule membranous; blade 3-15 × 0.2-0.4 cm, linear-lanceolate, glabrous. Spikes 3-8, 9 cm long, digitate: peduncle puberulous below the spikes; spikelets 7-8 mm, elliptic, perfect florets 2, 4-awned; empty lemmas 2; glumes unequal, lower glume elliptic, membranous; upper glume elliptic; fertile lemma obovate, back scaberulous, both margins and keels ciliate, awned; palea narrow; fertile lemma II broader than the first, awned: empty lemmas 2, sub-globose, awned; lodicules 2; stamens 3; stigmas plumose. Caryopsis narrowly elliptic.

Occasional in Warangal district.

Fr. & Fl.: June-November.

Chloris dolichostachya Lagasca, Gen. Sp. Pl. 5. 1816; Fischer, Fl. Madras 3: 1838. 1934; Bor, Grass. India 466. 1960. *C. incompleta* Roth, Nov. Pl. Sp. 60, 1821; Hook. f., Fl. Brit. India 7: 290. 1896.

A perennial, tufted grass; culms up to 1 m high, procumbent, rooting at base; nodes glabrous. Leafsheaths 3-7 cm long, sparsely hairy; ligule obscure; blade 0.7-1.5 x 0.4-0.7 cm, base narrow, margin scaberulous, apex acuminate, scabrid, midrib prominent. Spikes arranged digitately, drooping, terminal; spikelets narrowly oblong. 2-flowered; glumes lanceolate, with prominent midrib, aristate; fertile lemma oblong-lanceolate, ciliate on the back, bidentate apex. awned, nerved; palea oblong, keels scaberulous, callus sparsely ciliate; sterile lemma reduced to an awn; stamens 3. Caryopsis oblong.

Very rare in open forests in Warangal district.

Fl. & Fr.: November–February.

Pakhal (WGL), *KMS* 1164 (MH).

Chloris gayana Kunth ex Stapf in Dyer, Fl. Cap. 7: 642. 1900; Bor, Grass. India 466. 1960.

A perennial, stoloniferous, introduced grass; culms up to 1 m high, simple or branched below: nodes glabrous. Blades 15-40 x 0.2-0.8 cm, linear, flat, apex acuminate. Spikes 6-15, 15 cm long, umbellate, sessile, greenish or brownish; spikelets 3-4-flowered, perfect floret one, 3 mm, 2-awned; glumes very unequal; fertile 1st lemma lanceolate, pallid, bidentate ciliate along the marginal nerves and keel, shortly bearded below the tip, awned; upper florets sterile, 2nd lemma projecting from the side of the lowest lemma, lanceolate, scaly-ciliate on the margins, awned; 3rd lemma reduced to a scabrid oblong, or clavate, awnless scale; 4th lemma reduced to a clavate awnless scale or sometimes absent.

Common in field bunds and along roadsids in Hyderabad district (Ramana, 2010).

Fl. & Fr.: August-December.

Chloris montana Roxb., Fl. Ind. 1: 331.1820: Hook. f., Fl. Brit. India 7: 292. 1896; Fischer, Fl. Madras 3: 1838. 1934; Bor, Grass. India 466. 1960.

A perennial, tufted grass; culms up to 80 cm high; nodes glabrous. Leaf sheaths 2-5 cm; ligule membranous, short: blade 3-12 x 0.2-0.4 cm linear, involute, glabrous or pilose, apex acuminate. Spikes 4-6, 5-6 cm long, erect, digitate, terminal; spikelets 4 mm, unilateral, 4-awned; lower glume membranous, oblong, keeled, mucronulate; upper glume keeled, aristate, scabrid on keels; fertile lemma I ciliate on margin, glabrous at back, awned; palea narrow, oblong, keeled, hyaline; sterile lemmas 3, awned; lodicules 2; stamens 3; stigma plumose, purplish brown. Caryopsis fusiform.

Occasional in the plains and waste places of Hyderabad, Mahabubnagar and Nalgonda districts.

Fl. & Fr.: August- January.

Nagarjuna Sagar (NLG), *KMS* 7756 (CAL); Boddupalle (NLG), *BVRS* 648 (NU); Hyderabad (HYD), *MRS* 195 (DD).

Chloris roxburghiana Schult., Syst. Veg. 2. Mant. 329. 1827; Bor, Grass. India 468. 1960. *C. polystachya* Roxb., Fl. Ind. 1, 332, 1820; Hook. f., Fl. Brit. India 7: 292. 1896, non Lagasca (1816); Fischer, Fl. Madras 3: 1838. 1934.

A perennial, tufted grass; culms up to 1 m high, erect or occasionally geniculately ascending; nodes glabrous. Leaf sheaths 3-8 cm long, glabrous, ciliate at mouth; ligule membranous; blade 9-22 x 0.3-0.6 cm, linear, folded, base rounded, apex acuminate. Spikes 9 cm long, racemosely arranged, terminal, compact, feathery heads of numerous (12-30) spikes; spikelets 0.25 cm long, 3-flowered, 3-awned; glumes elliptic-obovate, strongly 1-nerved, scaberulous on nerves (back side), upper with mucronate apex. Florets 3, lower bisexual, others sterile with lemmas only; perfect lemma oblong-obovate, 1-nerved, ciliate on keels and margins, awned; palea as long as the lemma, elliptic; sterile lemmas 2, broadly ovate, apex truncate, awned; stamens 3; stigmas plumose. Caryopsis linear-oblong.

Occasional in plains and foot hills of Adilabad, Karimnagar, Medak and Nizamabad districts.

Fl. & Fr.: July – December.

Sarvaipet (ADB), *GO* 4515 Patharajampet (NZB), *BR & CPR* 7162; Choutkoor fields (MDK), *TP & MSG* 11912.

Chloris virgata Sw., Fl. Ind. Occ. 1: 200, 1397; Hook. f., Fl. Brit. India 7: 291. 1896; Fischer, Fl. Madras 3: 1838. 1934; Bor, Grass.India 468. 1960.

An annual, leafy, stoloniferous grass; culms up to 60 cm high, geniculately ascending or creeping, rooting at the nodes; nodes glabrous. Leaf sheaths 2-10 cm long, glabrous; ligule membranous; blade 3-10 x 0.1-0.4 cm, narrowly linear, base rounded, margin sparsely tuberculate base hairs, scaberulous, apex acuminate, glabrous. Spikes 4-8, 6 cm long, digitate, crowded at the apex of peduncle; spikelets 3 cm long, ovate-lanceolate, perfect only one; rhachis hairy; glumes unequal, membranous, awned. Florets 2, lower fertile, upper sterile, epaleate; fertile lemma ciliate along the margins, keels with crown of hairs, awned; callus pilose; sterile lemma I emerging from the side of fertile lemma, oblong, awned; sterile lemma II oblanceolate, clavate, awnless; stamens 3, stigmas plumose, purplish. Caryopsis ellipsoid, compressed.

Along field bunds and waste lands in Hyderabad and Medak districts.

Fl. & Fr.: January-November.

Borapuram (MBNR), *BR* & *BSS* 32286 (BSID); Narsapur (MDK), *KMS* 6764 (CAL); Hyderabad (HYD), *MRS* 108 (DD).

CHRYSOPOGON Trin. *nom. cons.*

1. Pedicels half the length of the sessile spikelets or longer:
    2. Pedicels glabrous or nearly so .... **C. aciculatus**
    2. Pedicels villous with rusty or pale hairs .... **C. orientalis**
1. Pedicels shorter than half the length of the sessile spikelet .... **C. fulvus**

Chrysopogon aciculatus (Retz.) Trin., Fund. Agrost. 188. 1820; Fischer, Fl. Madras 3: 1738. 1934; Bor, Grass. India 115. 1960; Moulik, Grass. Bamb. India 1: 240. 1997. *Andropogon aciculatus* Retz., Observ. Bot. 5: 22. 1789; Hook. f., Fl. Brit. India 7: 188. 1896.

A perennial, tufted grass; culms up to 60 cm high, geniculately ascending below, nodes glabrous. Leaf sheaths 3-5 cm, glabrous; ligule a minute fimbriate rim; blade 2-7 x 0.4-0.6 cm, oblong, flat, base rounded, margins scabrid, apex acute. Panicle narrow, up to 10 cm long; spikelets 2-nate. *Sessile spikelet*: narrowly ovate; lower glume flat, 3-nerved, stiff hispid on the back, bidentate; upper glume lanceolate, folded, scabrid on the back, 3-nerved, apex aristate; lower lemma linear-lanceolate, sparsely pilose; upper lemma narrow, hyaline awned, paleated; lodicules 2; stamens 3. Caryopsis laterally compressed. *Pedicelled spikelet*: lanceolate, 5 mm; pedicel glabrous; lower glume thin, lanceolate, 3-nerved, apex acuminate; upper glume short, acuminate, ciliate; lower lemma oblong, hyaline; upper lemma smaller, broader, ciliate; palea hyaline, elliptic, nerveless, apex truncate; lodicules 2; stamens 3.

Occasional in Adilabad district.

Fl. & Fr.: July – October.

Sirpur (ADB), *GO* 4417.

Chrysopogon fulvus (Spreng.) Chiov., F1. Somala 327. 1929; Bor, Grass. India 166. 1960; Moulik, Grass. Bamb. India 1: 240. 1997. *Pollinia fulva* Spreng., Pl. Minus Cognit. Pug. 2: 10. 1815. *Chrysopogon montanus* Trin. ex Spreng., Entadeck. 2: 93. 1821; Fischer, Fl. Madras 3: 1739. 1934. *Andropogon monticola* Roem. & Schult., Syst. Veg. 2. 665. 1827; Hook. f., Fl. Brit. India 7: 192. 1896.

A perennial, tufted grass; culms up to 40 cm high, erect or geniculately ascending; nodes glabrous. Leaf sheaths 1-6, glabrous, ciliate near mouth; ligule membranous; blade 6-15 x 0.2-0.5, linear, base truncate, margin sparsely scaberulous, apex acuminate. Panicle 7 cm long, terminal; spikelets 7.5 cm, in whorls, 3-flowered, 2-pedicelled, 1 sessile. *Sessile spikelet*: 6 mm, narrowly elliptic, bisexual; callus villose, lower glume narrow-oblong, keeled, 4-nerved, margins scaberulous, apically hispid; upper glume boat-shaped, 3-nerved, densely hispid on back, awned; lower lemma elliptic, ciliate, 2-nerved, apex obtuse; upper lemma minute, awned, awn geniculate; palea narrowly linear; lodicules 2; stamens 3, stigmas plumose. Caryopsis linear. *Pedicelled spikelet*: 6 mm, narrowly elliptic, male or barren; pedicel villose; lower glume oblong, margins scaberulous, 7-nerved, chartaceous below; upper glume elliptic, 3-nerved; lemma ovate, 2-nerved, ciliate; palea minute, narrowly lanceolate; lodicules 2; stamens 3.

Common in plains and hilly tracts of most districts.

Fl. & Fr: July-November.

Ankusapuram (ADB), *GO* 4345; Navipet (NZB), *BR* & *PSPB* 9594; Sri Ramasagar Project (NZB), *BR* 7297; Choutkoor fields (MDK), *TP & MSG* 11934; Nagarjuna Konda Valley (NLG), *KT* 9666 & 9757 (MH); Hyderabad (HYD), *MRS* 109 (DD);

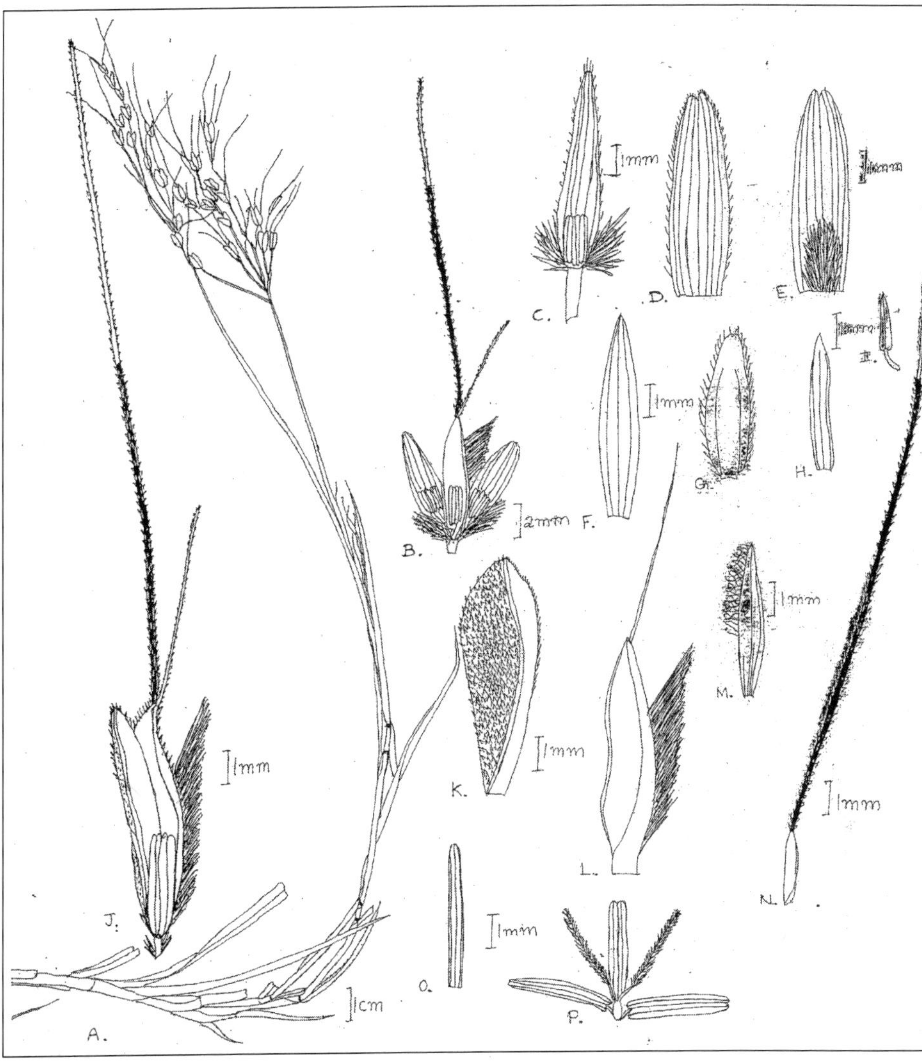

Figure 141: Chrysopogon fulvus (Spreng.) Chiov.

A. Twig, B. Spikelets, C. Pedicelled spikelet, D & E. Lower glume front and back views, F. Upper glume, G. Lemma, H. Palea, I. Stamen, J. Sessile spikelet, K. Lower glume, L. Upper glume, M. Lower lemma, N. Upper lemma, O. Palea, P. Pistil with stamens.

Umamaheswaram (MBNR), *SRS* 111339 (BSID); Farahabad chowrasta (MBNR), *SRS* 111423 (BSID); Koriguttalu (KMM), *RR* 106040 (BSID); Yenkur (KMM), *RR* 108573 (BSID).

Chrysopogon orientalis (Desv.) A. Camus in Lecomte, Fl. Indo-Chine 7: 332. 1922; Fischer, Fl. Madras 3: 1739. 1934; Bor, Grass. India 118. 1960. *Rhaphis orientalis* Desv., Opusc. Sci. Phys. Nat. 69. 1831.

Densely tufted grass, matty at base; culms to 90 cm. Leaves basal and cauline; leaf blade oblong, to 10 cm long, sheaths shortly ciliate at collar. Panicles terminal, to 15 cm long, branches with reddish-brown; hairs at tip; glumes of sessile spikelets 5 mm long, apically hirsute and muticous; of pedicelled spikelets glumes 8 mm long, awn to 1.5 cm.

Occasional in grasslands in Karimnagar district (Naqvi, 2001).

Fl. & Fr.: September – December.

COELACHYRUM Hochst. & Nees

Coelachyrum lagopoides (Burm.f.) Senaratna, Grass. Sri Lanka 79.1956; Bor, Grass. India 488. 1960. *Cynosurus lagopoides* Burm.f., Fl. Ind. 29. 1768. *Coelochyropsis lagopoides* (Burm.f.) Bor, Ann. Natur. Hist. Mus. Wien.75:25. 1972; Gayathri & Pullaiah in Pullaiah, Fl. Andhra Pradesh 3: 1157. 1997; Moulik, Grass. Bamb. India 2:585.1997. *Eleusine lagopoides* (Burm. f.) Merr., Philipp. J. Sci. 19:339.1921; Fischer, Fl. Madras 3:1840.1934. *E. brevifolia* (Willd.) R. Br. ex Hook. f., Fl. Brit. India 7:294.1896.

Perennials, culms creeping or erect, to 50 cm, rooting at nodes; nodes glabrous. Leaf sheaths glabrous. Ligules membranous. Leaf blade oblong-lanceolate, margins ciliate. Spikes numerous, crowded into a globose head; spikelets elliptic-ovate, many–flowered; lower glume lanceolate, 1-nerved, keeled, chartaceous; upper glume ovate-lanceolate, chartaceous, acute, 1-nerved. Florets 5-8, bisexual; upper lemma elliptic-lanceolate, 3-nerved, ciliate, hyaline, keels winged; palea oblong, coriaceous, base bearded; lodicules 2; stamens 3. Caryopsis trigonous or ovate, transversely rugose.

Occasional in plains and forest peripheries in Nalgonda district.

Fl. & Fr.: July-December.

Vijayapuri (NLG), *KMS* 9850 (CAL).

Note: The genus *Coelochyropsis* Bor (1960) was treated as synonym to *Coelachyrum* by Clayton & Renovoize (1986), same concept is followed in the present study.

COIX L.

1. Leaf blade sparsely to densely tuberculate hairy; spikelets 5-6 mm broad; lower glume of the male spikelets narrowly or hardly winged .... **C. lacryma-jobi**
1. Leaf blade glabrous; spikelets 2-3 mm broad; lower glume of the male spikelets broadly winged:

2. Culms floating and creeping, tuberculate-based hairs on the upper surface of blade; involucre abruptly constricted at the neck into a beak .... **C. aquatica**

2. Culms erect, tuberculate-based hairs absent on the upper surface of the blade; involucre not abruptly constricted at the neck into a beak .... **C. gigantea**

Coix aquatica Roxb., Fl. Ind. 3: 571. 1832; Hook. f., Fl. Brit. India 7: 100. 186; Bor, Grass. India 264. 1960.

A floating and creeping, perennial grass; culms up to 3 m high, succulent, floating roots at the nodes. Leaf sheaths compressed, 4-7 cm long; ligule membranous; leaf blade linear, 30 – 90 × 2.5-3 cm, margins hispid, apex acute, glandular hairs on the upper surface. Male spikes drooping, many-lobed; spikelets 3-seriate, centre one pedicelled, female, lower 2 sessile; lower glume elliptic-lanceolate, many-nerved, keels broadly winged; upper glume similar, without wings, 7-9-nerved. Florets 2; lower lemma oblong-lanceolate, as long as the upper glume, hyaline, 3-5-nerved; lower palea similar, 2-nerved, 2-keeled; lodicules 2; stamens 3. Female spikelet usually solitary near base, ovate, yellowish white, completely enclosed by bract forming a bony involucre; lower glume oblong-lanceolate; upper glume narrowly oblong, chartaceous, hyaline, keeled. Florets 2, lower empty, upper female; lower lemma oblong-lanceolate, 5-nerved, epaleate; involucre turbinate, perforated, finely ciliate at apex; ripe involucre abruptly constricted at the neck into a beak.

Occasional along stream banks, margins of ponds in Mahabubnagar district.

Fl. & Fr: Oct-January

Coix gigantea Koen. ex Roxb., Fl. Ind. 3: 570. 1832; Fischer, Fl. Madras 3: 1705. 1934; Bor, Grass. India 264. 1960; Moulik, Grass. Bamb. India 1: 174. 1997. *C. lacryma-jobi* L. var. *giganatea* (Koen. ex Roxb.) Stapf ex Hook. f., Fl. Brit. India 7: 100. 1896.

A perennial, erect, tufted grass; culms up to 2 m high. Leaf sheath compressed, 10-15 cm long, glabrous; ligule membranous; leaf blade 12-50 x 0.25-2 cm, linear-flat, base rounded, apex acuminate, glandular based hairs absent above. Panicle dioecious; racemes 3 cm long, spiciform; spikelets 3-nate, female spikelets completely enclosed in a globose basal bract through which the rhachis grows out and bears the male spikelet. *Male spikelet:* narrowly elliptic to narrowly obovate; lower glume membranous, narrowly elliptic, 15-17-nerved, 2-keeled, keels broadly winged, apex 2-toothed; upper glume membranous, narrowly elliptic, margins inflexed, 10-12-nerved, 1-keeled, apex acute; lower lemma hyaline, narrowly elliptic, margins inflexed, 3-nerved; palea hyaline, narrowly elliptic, margins inflexed, 2-nerved, 2-keeled; lodicules 2; stamens 3. *Female spikelet:* globose or ovoid, bony. Caryopsis broadly ovate, flat.

In marshy localities of Karimnagar and Hyderabad districts.

Fl. & Fr.: October – January.

Rechapalli (KRN), *GVS* 25696 (MH); Adikmet (HYD), *MRS* 88012 (DD).

Coix lacryma-jobi L., Sp. Pl. 972. 1753; Hook. f., Fl. Brit. India 7: 100. 1896, Fischer, Fl. Madras 3: 1705. 1934; Bor, Grass. India 264. 1960; Moulik, Grass. Bamb. India 1: 175. 1997.

A tall, perennial leafy, monoecious grass; culms up to 1 m high, rooting at lower nodes; nodes thick, glabrous. Leaf sheaths 4-6 cm long, glabrous; ligule a very narrow membrane; blade 15-30 x 1-2 cm, convolute, base cordate, margins spinulosely serrate, apex acuminate, midrib thick, glabrous. Racemes 5 mm long, axillary and terminal, lower spikelets solitary, female, male spikelets protrude through the apex of female spikelets; peduncle very long, up to 9 mm. *Female spikelets:* enclosed in a hardened, polished involucre, 9 mm; glumes broadly ovate, acuminate, margin incurved, coriaceous, lemmas ovate, acuminate, sub-coriaceous; palea ovate-oblong, sub-coriaceous; styles basally united, stigmas long, plumose. Caryopsis broadly ovate, obtuse. *Male spikelet*: Racemes up to 40 cm long, spikelet 9 mm, oblong, 2-flowered; lower glume oblong, obtuse, coriaceous, winged, wings scabrid; upper glume oblong, sub-coriaceous; lower lemma oblong, chartaceous; palea oblong; upper lemma lanceolate; palea hyaline, lanceolate; stamens 3.

In moist localities of Adilabad, Medak and Hyderabad districts.

Fl. & Fr.: October – February.

Sone (ADB), *TP* & *PVP* 4012; Pitlabongaram (ADB), *GO* & *PVP* 4565; Gangapur (MDK), *TP* & *MSG* 12087.

CYMBOPOGON Spreng.

1. Sessile spikelet without an awn:
    2. Sessile spikelets linear – lanceolate; lower glume usually shallowly concave in the lower half .... **C. citratus**
    2. Sessile spikelets oblong – oblanceolate; lower glume flat on the back in the lower half .... **C. nardus** var. **nardus**
1. Sessile spikelets awned:
    3. Lower glume of the sessile spikelet with a deep slot-like depression occupying most of its length .... **C. gidarba**
    3. Lower glume of sessile spikelet without a slot, either flat, convex, shallowly concave or with a deep narrow slit in the lower half:
        4. Lower glume of the sessile spikelet with a narrow slit-like groove in the lower half which appears as a rib on the inner surface:
            5. Blade cordate and amplexicaul at the base, linear lanceolate, acuminate .... **C. martini**
            5. Blade narrow and rounded at the base .... **C. caesius**
        4. Lower glume of the sessile spikelet flat in the lower half:
            6. Pedicel of the lowest pedicelled spikelet in the sessile raceme not swollen:

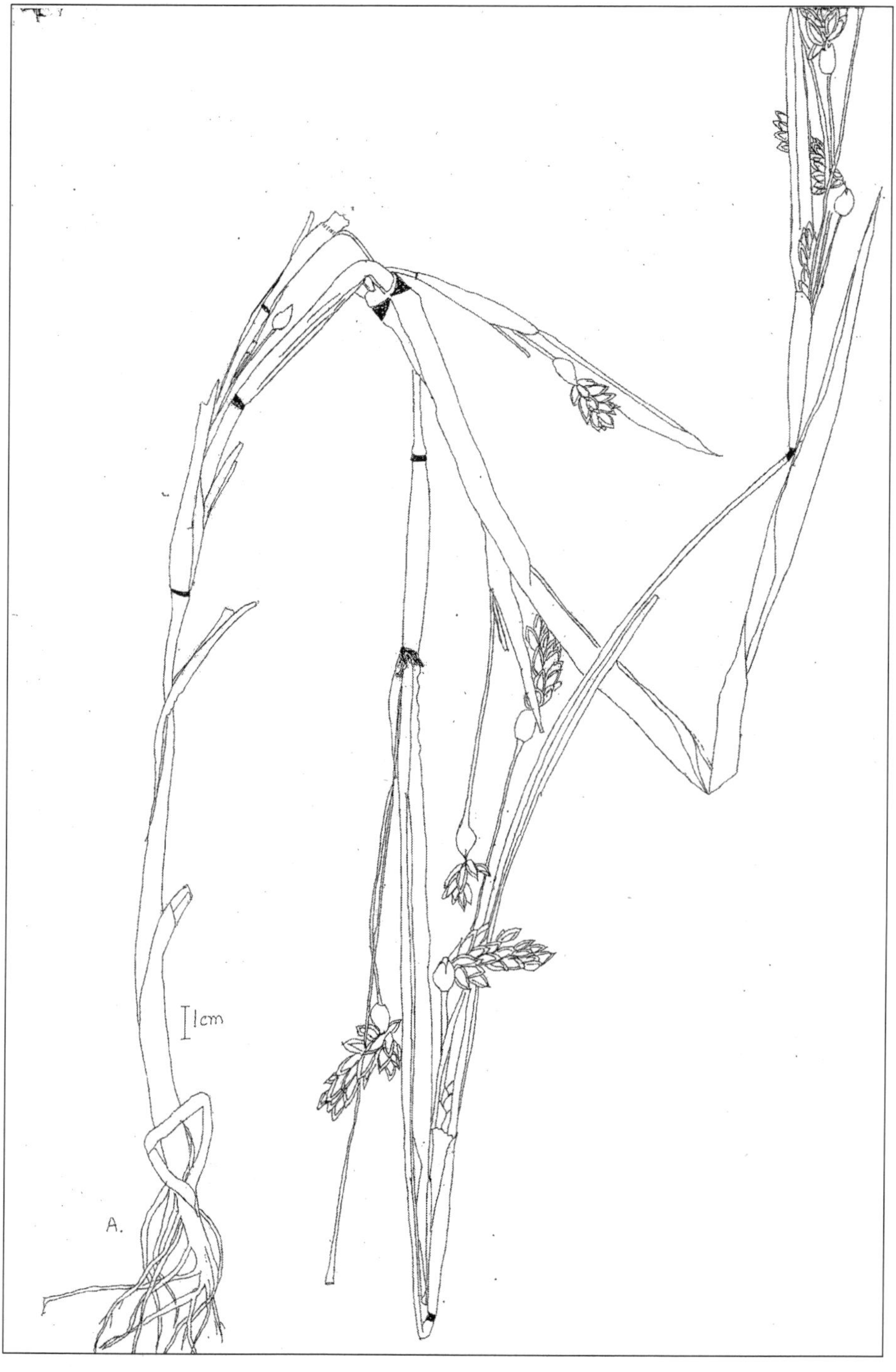

Figure 142: Coix lacryma-jobi L.

A. Twig.

7. Inflorescence very congested .... **C. nardus** var. **confertiflorus**
7. Inflorescence lax and loose .... **C. flexuosus** var. **flexuosus**

6. Pedicel of the lowest pedicelled spikelet in the sessile raceme swollen .... **C. coloratus**

Cymbopogon caesius (Hook. & Arn.) Stapf, Bull. Misc. Inform. Kew 1906: 341 & 366. 1906; Fischer, Fl. Madras 3: 1757. 1934; Bor, Grass. India 125. 1960; Moulik, Grass. Bamb. India 1: 212. 1997. *Andropogon caesius* Nees ex Hook. & Arn. in Hook. Bot. Beechey. Voy. 244. 1838. *A. schoenanthus* L. subsp. *gencinus* var. *caesius* Hack. in DC., Monogr. Phan. 6. 610. 1889; Hook.f., Fl. Brit. India 7 : 205. 1896.

A perennial, tufted grass; culms up to 2 m high, erect or ascending, more or less wiry; nodes hairy. Leaf sheaths shorter than the internodes, 5-10 cm long, glabrous; ligule rounded-truncate, scarious; leaf blade 5-25 x 0.5-0.9 long, linear-lanceolate, base rounded, apex tapering with a long setaceous point, glabrous. Spatheate panicle up to 30 cm long; spathes 5 cm long, narrow-lanceolate; spatheoles 2 cm long; narrow-lanceolate; racemes up to 6 mm long, 2-nate, rhachis white-villous along the margins. *Sessile spikelet*: oblong, 4 mm, bisexual; lower glume flat in the upper half and with a deep v-shaped groove in the lower half which appears as a rib on the inner surface, 2-6-nerved, 2-keeled, keels broadly winged, membranous; upper glume oblong-lanceolate, sparsely ciliate, 3-nerved, 1-keeled above middle, broadly winged. Florets 2, lower barren, upper bisexual, epaleate; lower lemma oblong-lanceolate, truncate, faintly 2-nerved; upper lemma stiptiform, ciliate, deeply bifid, awned, awn geniculate, epaleated; lodicules 2; stamens 3; stigmas plumose. Caryopsis oblong. *Pedicelled spikelet*: linear-lanceolate, 4 mm, male or barren; lower glume oblong-lanceolate, 7-10-nerved; upper glume 3-nerved, hyaline. Florets 2, lower barren, upper male; lemmas oblong, 2-nerved, truncate; epaleated; stamens 3.

Common in plains and forests in Adilabad, Nizamabad, Hyderabad, Ranga Reddi, Mahabubnagar and Warangal districts.

Fl. & Fr.: August-March.

Boath (ADB), *GO* & *PVP* 4634; Gandhari RF (NZB), *TP* & *BR* 6334; Lingampalli (RR), *MSM* 12739; Hyderabad (HYD), *MRS* 236 (DD); Erloo to Mahabubnagar (MBNR), *SRS* 104515 (BSID); Guddaegudem RF (MBNR), *SRS* 110760 (BSID).

Cymbopogon citratus (DC.) Stapf, Bull. Misc. Inform. Kew 1906: 357. 1906; Fischer, Fl. Madras 3: 1756. 1934; Bor, Grass. India 126. 1960; Moulik, Grass. Bamb. India 1: 213. 1997. *Andropogon citratus* DC., Cat. Hort. Monsp. 78. 1813; Hook.f., Fl. Brit. India 7: 210. 1896.

Vern. : *Vasana gaddi*. English: Lemon grass.

A perennial, densely tufted grass; culms up to 70 cm high; internodes usually coated with wax. Leaf sheaths loose at base, cinnamon-coloured inside; ligule membranous; blade in dense tuft from short annular rhizome, 45-70 cm long, rough on both sides. Panicle pyramidal; spikelets paired, sessile and pedicelled, awnless or reduced to a bristle not visible outside the spikelet, 2-nate. *Sessile spikelet:* linear-lanceolate, 6 mm; lower glume linear-lanceolate, acuminate, concave in lower half on back side, 2-keeled; upper glume lanceolate, acute at apex, 1-keeled. Florets 2, lower barren, epaleate, upper bisexual; lower lemma hyaline; upper lemma lanceolate, hyaline, entire or 2-fid at apex; upper palea absent or minute; stamens 3. *Pedicelled spikelet:* male, elliptic, deciduous along with sessile spikelets; glumes acute, muticous at apex; lemmas 2, enclosed by glumes.

Cultivated, grown for its aromatic oil.

Fl. & Fr.: July – December.

Cymbopogon coloratus (Hook. f.) Stapf, Bull. Misc. Inform. Kew 1906: 321. 1906; Fischer, Fl. Madras 3: 1756. 1934; Bor, Grass. India 126. 1960; Moulik, Grass. Bamb. India 1: 213. 1997. *Andropogon nardus* L. var. *coloratus* Hook. f., Fl. Brit. India 7: 206. 1896.

Vern.: *Boda gaddi.*

A tufted perennial, grass; culms up to 2 m high; nodes bearded. Leaf sheaths 5-15 cm, basal tomentose, upper glabrous; ligule fimbriate membrane; blade 10-40 × 0.4-0.8 cm, flat, linear – lanceolate, base rounded apex acuminate, scabrid. Panicle dense, up to 8 cm, spathes 2 cm long; spatheoles up to 1.5 cm; racemes 4 cm, 2-nate, lowest pedicel of sessile racemes swollen, densely hairy along the margins. *Sessile spikelet:* oblong, 5 mm; lower glume coriaceous, oblong to elliptic-lanceolate, 2-nerved, 2-keeled, keels winged above middle, wings serrulate; upper glume oblong-lanceolate, 3-nerved, midnerve prominent, hyaline, sparsely hairy, minutely aristate. Florets 2, both epaleate, lower barren, upper bisexual; lower lemma narrowly oblong-lanceolate, ciliate, keeled, hyaline; upper lemma narrowly oblong, 2-lobed, lobes ciliate, awned from sinus, awns straight to geniculate; lodicules 2; stamens 3; stigmas plumose. *Pedicelled spikelet:* oblong, male or barren, 5 mm; glumes keeled: lemmas hyaline, epaleated; stamens 3.

Common in the forests of Medak, Ranga Reddi, Warangal and Hyderabad districts.

Fl. & Fr.: August– December.

Narsapur tank (MDK), *TP* & *MSG* 12017; Lingampalle (RR), *MSM* 12735; Pakhal RF (WGL), *KMS* 11642 (CAL); Hyderabad, *MRS* 119 (DD); Umamaheswaram (MBNR), *SRS* 111333 (BSID); Sivampet RF (MDK), *RG* 116473 (BSID).

Cymbopogon flexuosus (Nees ex Steud.) Wats. in Atkins., Gag. N. W. Prov. Ind. 392. 1882; var **flexuosus**; Fischer, Fl. Madras 3: 1756. 1934; Bor, Grass. India 127. 1960. Moulik, Grass. Bamb. India 1: 213. 1997. *Andropogon flexuosus* Nees ex Steud., Syn. Pl. Glum 1. 388. 1854. *A. nardus* L. var. *flexuosus* Hack. in DC. Monogr. Phan. 6, 603. 1889; Hook. f., Fl. Brit. India 7: 207. 1896.

A perennial, tufted grass; culms erect, up to 2.5 m high; nodes glabrous. Leaf sheaths 6-15 cm long, glabrous; ligule membranous; blade 15-50 × 1-1.5 cm, lanceolate, base attenuate, margins scaberulous, apex acuminate, glabrous. Panicle decompound, very lax, up to 30 cm long; spathes 3 cm; spatheole to 1.5 cm; racemes 2 cm; spikelets

2-nate. *Sessile spikelets*: 3 mm, oblong, bisexual; lower glume oblong, chartaceous, back with 1-2 slot like depressions, nerveless; upper glume boat-shaped, chartaceous, 3-nerved, 1-keeled, keel narrowly winged. Florets 2, lower barren, upper bisexual, both epaleate; lower lemma ovate-lanceolate, membranous, hyaline, nerveless; upper lemma oblong-lanceolate, acuminate. 2-lobed, awned, awn geniculate; stamens 3, stigmas plumose. *Pedicelled spikelet:* male or barren, oblong, acumiate; pedicel hairy; lower glume ovate-lanceolate, 10-nerved, 2-keeled, keels winged; upper glume 3-nerved; lemma hyaline, ciliate, epaleated; stamens 3.

Common in waste lands, hill tops and in open forests of Ranga Reddi, Karimnagar, Nalgonda and Hyderabad districts.

Fl. & Fr.: July – December.

Mekavaripalli (RR), *TP* & *MSM* 11052; Kodimial (KRN), *GVS* 25681 (CAL); Nagarjunakonda valley (NLG), *KT* 9800 (CAL); Kambalpally (NLG), *BVRS* 200 (NU); Osmania University campus (HYD), *MRS* 88001 (DD).

Cymbopogon gidarba (Ham. ex Steud.) Haines, Bot. Bih. Orissa 1048. 1924; Fischer, Fl. Madras 3: 1757. 1934; Bor, Grass. India 128. 1960; Moulik, Grass. Bamb. India 1: 214. 1997. *Andropogon gidarba* Ham. ex Steud., Syn. Pl. Glumac. 1: 387. 1854; Hook.f., Fl. Brit. India 7: 208. 1896.

An aromatic, tufted, perennial; culms up to 75 cm high. Leaf sheaths terete, glabrous; ligule short, truncate; blade 4-40 × 0.4-0.6 cm, linear, margin ciliate at base, apex acuminate, glabrous. Panicles narrow, slender; branches simple; spikelets 6 mm long, 2-nate, joints and pedicels shortly villous. *Sessile spikelet*: narrowly ovate, pale or dark green, bisexual; lower glume chartaceous, narrowly ovate, deeply concave on dorsal surface below the middle, margins inflexed, 2-keeled, puberulous or quite glabrous in the slot; upper glume cymbiform, 3-nerved, acute, broadly winged, wings hyaline; lower lemma lanceolate, hyaline, 2-nerved, apex obtuse, palea absent; upper lemma linear, bidentate, ciliate, awned, palea hyaline, broadly ovate; lodicules 2; stamens 3, stigmas plumose. *Pedicelled spikelet:* longer than sessile spikelet, lanceolate, acute; lower glume chartaceous, narrowly ovate, margins inflexed, 7-nerved, apex acute; upper glume membranous, narrowly ovate, margins inflexed, ciliate, 3-nerved; lower lemma hyaline, narrowly ovate, margins ciliate, 2-nerved, palea hyaline, upper lemma absent; lodicules 2; stamens 3.

Occasional in open grasslands in Mahabubnagar and Warangal districts.

Fl. & Fr.: July – December.

Cymbopogon martinii (Roxb.) Watson in Kinson E.T. Gaz., N.W. Prov. India 392. 1882; Fischer, Fl. Madras 3: 1756. 1934; Bor, Grass. India 129. 1960. Moulik, Grass. Bamb. India 1: 215. 1997. *Andropogon martinii* Roxb., Fl. Ind. 1: 280. 1820. *A. schoenanthus* L. var. *martinii* (Roxb.) Hook. f., Fl .Brit. India 7: 204. 1896.

A perennial, tufted, sweet scented grass; culms up to 2 m high, straw-coloured, glabrous. Leaf sheaths 7-15 cm, glabrous; ligule 3-4 mm, membranous; blade 10-25 × 1-1.5 cm, flat, linear-lanceolate, base cordate and amplexicaul, apex acuminate, midrib prominent. Panicle up to 25 cm long; spathes 3 cm; spatheoles 1.5 cm; lower most pedicel swollen; rhachis pilose along the margins; spikelets 2-nate. *Sessile spikelet:* bisexual, oblong, 4 mm; lower glume chartaceous, oblong to elliptic-lanceolate, 2-keeled, keels winged, slightly bidentate; upper glume boat-shaped, 1-keeled, ciliate, 3-nerved. Florets 2, both epaleate, lower barren, upper bisexual; lower lemma narrow-lanceolate, ciliate; upper lemma narrow, deeply bidentate, awned, epaleated; lodicules 2; stamens 3, stigmas plumose. *Pedicelled spikelet:* male or barren, oblong; pedicel hairy; lower glume oblong to elliptic-lanceolate; 9-nerved, slightly keeled; upper glume ovate-lanceolate, 3-nerved, hyaline, ciliate; lemmas hyaline, nerveless, ciliate, epaleate; stamens 3.

Occasional in forests of Medak, Ranga Reddi, Warangal, Karimnagar, Hyderabad and Mahabubnagar districts.

Fl. & Fr.: July – December.

Choutkoor (MDK), *TP* & *MSG* 11924; Rangammagudem (RR), *MSM* 10580; Hyderabad, *MRS* 212 (DD).

Cymbopogon nardus (L.) Rendle in Cat. Welw. Afr. Pl. 2: 155. 1899 var. **nardus;** Fischer, Fl. Madras 3: 1756. 1934; Bor, Grass. India 130. 1960; Moulik, Grass. Bamb. India 1: 216. 1997. *Andropogon nardus* L., Sp. Pl. 1046. 1753; Hook. f., Fl. Brit. India 7: 205. 1896.

A perennial, tufted grass; culms up to 1.5m high, root stock stout, leafy. Leaf sheaths ca 13 cm, lower ones fibrous at maturity, more or less ribbed; ligule membranous, lacerate with fimbriate apex, brownish yellow; leaf blade linear-lanceolate, 17-46 × 1.4-1.6 cm, base rounded to truncate, apex acuminate. Panicle large, subpyramidal, interrupted; branches loosely or closely packed, drooping; spathes imbricate, lanceolate, 2 cm; spikes with 4-5 pairs of spikelets; joints and pedicels ciliate. *Sessile spikelet:* 5 mm, lanceolate, awnless; lower glume lanceolate, acute, back flat, 2-nerved, 2-keeled, keels narrowly winged, wings with ciliolate or serrulate margins, margins incurved; upper glume oblong-lanceolate, boat-shaped, chartaceous, 3-nerved, 1-keeled, keel narrowly winged from middle above, margins incurved. Florets 2, lower barren, upper bisexual; lower lemma narrowly oblong-lanceolate, hyaline to brownish, margins infolded, ciliate towards apex; lower palea absent; upper lemma narrowly oblong, 2-lobed at apex, awn from sinus, geniculate; upper palea oblong, membranous, hyaline, nerveless; lodicules 2; stamens 3. Caryopsis oblong-lanceolate. *Pedicelled spikelet:* narrowly oblong-elliptic, acute at apex, unawned; lower glume ovate-lanceolate to oblong-lanceolate, brownish, prominently 11-nerved, 2-keeled, keels narrowly winged, wings serrulate, margins incurved; upper glume oblong-lanceolate, 3-nerved, margins incurved, ciliate towards apex. Florets 2, both epaleate, lower barren, upper bisexual; lower lemma narrowly oblong-lanceolate, nerveless, margins ciliate towards apex; upper lemma narrowly oblong, hyaline, some times reduced; stamens 2.

Only known in cultivation.

Fl. & Fr.: July – August.

Cymbopogon nardus (L.) Rendle var. **confertiflorus** (Steud.) Stapf ex Bor, J. Bombay Nat. Hist. Soc. 51: 905. 1953. *Andropogon confertiflorus* Steud., Syn. Pl. Glumac. 1: 385. 1854. *Cymbopogon confertiflorus* (Steud.) Stapf, Kew Bull. 1906: 318. 1906.

Aromatic tall grass, about 1 m high.

Occasional in hill forests, on clayey soil in Karimnagar district (Naqvi, 2001).

Fl. & Fr.: September – March.

Key to Varieties

1. Spikelets awned .... var. **confertiflorus**
1. Spikelets unawned .... var. **nardus**

CYNODON Rich. *nom. cons.*

1. Lemmas longer than the upper glume:
    2. Stoloniferous; ligule a ring of hairs; spikes 2-6, up to 6 cm long; hairs on the lemma not clavellate .... **C. dactylon**
    2. Not stoloniferous; ligule membranous; spikes numerous, up to 10 cm long; hairs on the lemma some times clavellate .... **C. radiatus**
1. Lemmas equal or slightly longer than the upper glume .... **C. barberi**

Cynodon barberi Rang. & Tadul., J. Bombay Nat. Hist. Soc. 24: 846. 1916; Fischer, Fl. Madras 3: 1836. 1934; Bor, Grass. India 469. 1960; Moulik, Grass. Bamb. India 2: 565. 1997.

A perennial, tufted grass; culms up to 30 cm high; nodes glabrous. Leaf sheaths 2-4 cm long, glabrous; ligule ciliate; leaf blade 2-6 × 0.5-1.5 cm, flat, lanceolate, base truncate, apex acute. Spikes 3-5, up to 6 cm long, digitate; spikelets 3 mm long, closely appressed to the rhachis; spikelets sessile or subsessile, elliptic; lower glume membranous, boat-shaped, 1-nerved, 1-keeled, keel scabrous, apex acute; upper glume membranous, narrowly ovate, 1-nerved, 1-keeled, keel scabrous, apex acute; lemma narrowly boat-shaped, keels with clavellate hairs; palea linear, 2-nerved, 2-keeled; lodicules 2; stamens 3; stigmas plumose. Caryopsis oblong.

Occasional in Warangal, Hyderabad, Mahabubnagar and Ranga Reddi districts.

Fil. & Fr.: August– October.

Mekavaripalli (RR), *TP* & *MSM* 11054; Vicarabad (RR), *MRS* 116 (DD).

Note: This grass may be confused with *Cynodon dactylon,* but can be easily distinguished by the absence of stoloniferous stems, acute leaves, upper glume always equal to or longer than the lemma, presence of clavellate hairs on the keels and margins of the lemma and on the keels of the palea.

Cynodon dactylon (L.) Pers., Syn. Pl. 1: 85. 1805; Hook.f., Fl. Brit. India 7: 288. 1960; Fischer, Fl. Madras 3: 1835. 1934; Bor, Grass. India 469. 1960; Moulik, Grass. Bamb. India 2: 566. 1997. *Panicum dactylon* L., Sp. Pl. 58. 1753.

Vern.: *Garika gaddi.*

A perennial, stoloniferous grass; culms creeping by scaly rhizomes or by flat stolons forming matted tufts, decumbent, later erect, 10-30 cm high; nodes glabrous. Leaf sheaths 0.3-3 cm long, glabrous; ligule ciliate; leaf blade 1-10 × 0.2-0.3 cm, distichous, linear, base rounded, apex acuminate, softly puberulous. Spikes 4 cm long, 3-4, digitate; spikelets subsessile, 3 mm long, laterally compressed; lower glume lanceolate, 1-nerved, 1-keeled, keel scabrous, apex acute; upper glume membranous, narrowly ovate, 1-nerved, 1-keeled, keel scabrous, apex acute; lemma oblong, aristate, membranous, keels ciliate; palea linear, as long as the lemma; stamens 3, stigmas plumose. Caryopsis oblong.

Very common in plains and lower hills in all districts.

Fl. & Fr.: August–October.

Sattenapalle (ADB), *TP* & *PVP* 4151; Pittabonguram (ADB), *GO* & *PVP* 4566; Mingaram (NZB), *TP* & *BR* 6312; Malkalcheruvu (RR), *MSM* 10570; Nagarjuna konda valley (NLG), *KT* 9637 (CAL); Cherukupally (NLG), *BVRS* 669 (NU); Pakhal (WRL), *ANG* 15976 (CAL); Adikmet (HYD), *MRS* 275 (DD).

Cynodon radiatus Roth ex Roem. & Schult., Syst. Veg. 2: 411. 1817. *C. arcuatus* J.S. Presl in Presl, Reliq. Haenk. 1: 290. 1830; Bor, Grass. India 489. 1960; Moulik, Grass. Bamb. India 2: 565. 1997. *C. dactylon* (L.) Pers. var. *intermedius* (Rang. & Tadul.) Fischer, Fl. Madras 3: 1836. 1934. *C. intermedius* Rang. & Tadul., J. Bombay Nat. Hist. Soc. 26: 309. 1918.

A perennial, stoloniferous grass; culms straggling, up to 30 cm high, roots near the nodes; nodes glabrous. Leaf sheaths 2-5 cm, glabrous; ligule membranous; blade 6-12 × 0.3-0.6 cm, lanceolate, base cordate, margin finely serrulate, apex acute, glabrous. Digitate racemes 4, terminal; racemes 4-10 cm long; spikelets subsessile, 25 mm, elliptic, acuminate; lower glume boat-shaped, narrowly ovate, 1-nerved, 1-keeled, keel scabrous, apex acute; upper glume membranous, boat-shaped, narrowly ovate, 1-nerved, 1-keeled, keels scabrid, apex acute; lemma hyaline, curved, 3-nerved, 1-keeled, keels covered with clavate hairs, apex obtuse; palea involute, 2-nerved, scaberulous back on nerve; lodicules 2; stamens 3, stigmas plumose. Caryopsis elliptic.

Occasional in the forests and plains of Adilabad, Medak, Hyderabad, Mahabubnagar and Warangal districts.

Fl. & Fr.: August–October.

Sattenapalle (ADB), *TP* & *PVP* 4098; Bheemaram (ADB), *TP* & *GO* 5464; Narsapur tank (MDK), *TP* & *MSG* 12002; Pakhal (WGL), *ANH* 15960 (MH).

DACTYLOCTENIUM Willd.

1. Stoloniferous; spikes 2-6 .... **D. aegyptium**
1. Non-stoloniferous; spikes solitary .... **D. aristatum**

Dactyloctenium aegyptium (L.) P. Beauv., Ess. Agrost. Expl. Pl. 15.1812; Fischer, Fl. Madras 3:1840.1934; Bor, Grass. India 489. 1960; Moulik, Grass. Bamb. India 2:585.1997. *Eleusine aegyptia* (L.) Desf., Fl. Atlant. 1:85. 1798; Hook. f., Fl. Brit. India 7:295. 1896. *Cynosurus aegyptius* L., Sp. Pl. 72.1753.

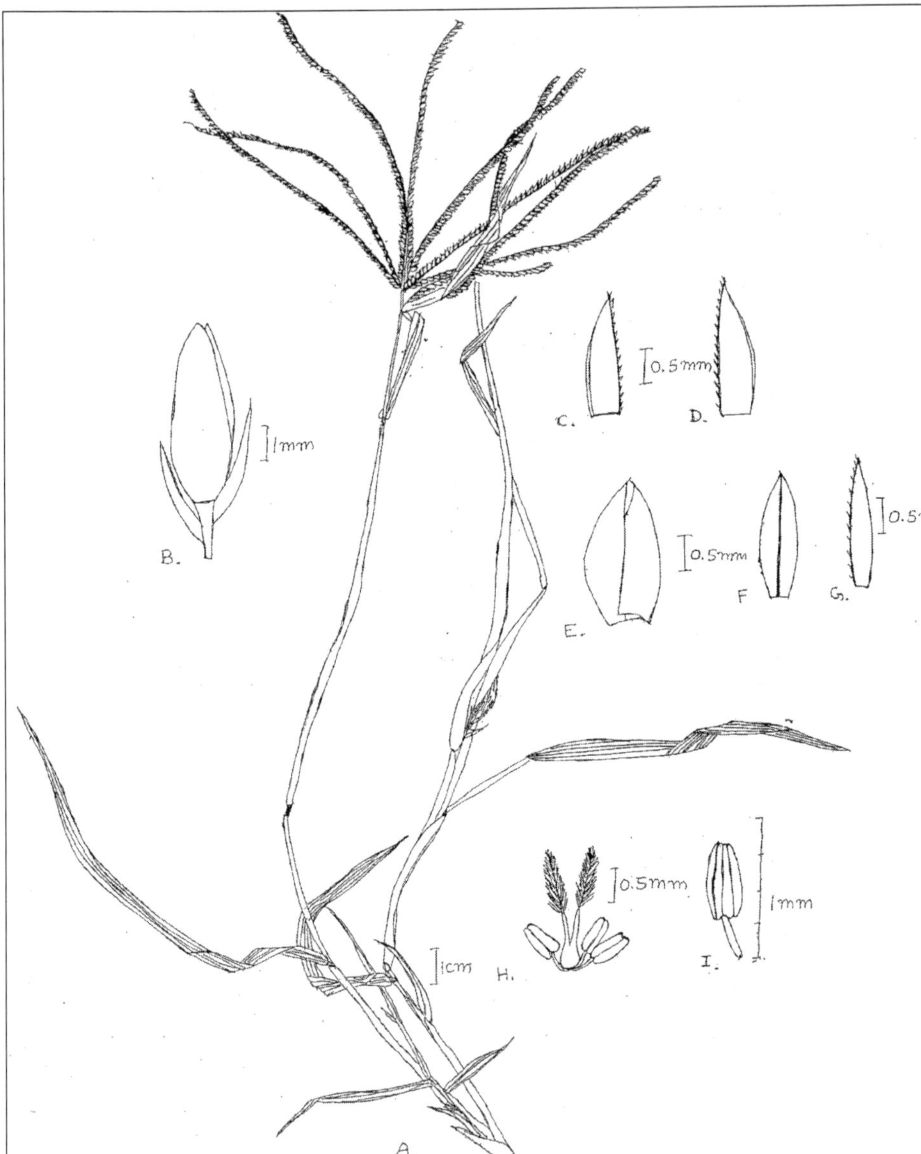

Figure 143: Cynodon radiatus Roth ex Roem. & Schult. (Syn.: *Cynodon arcuatus* C.B. Presl.)

A. Twig, B. Spikelet C. Lower glume, D. Upper glume, E. Lower lemma, F. Upper lemma, G. Palea, H. Pistil with stamens, I. Stamen.

Annuals, culms stoloniferous, to 70 cm, erect or ascending, rooting at nodes; nodes glabrous. Leaf sheaths glabrous; ligules membranous; leaf blade linear-lanceolate, 5-20 × 0.25-0.5 cm, acuminate, glabrous, sparsely ciliate on the margins. Spikes 2-6, digitate; spikelets oblong, 3-5-flowered; lower glume ovate, boat-shaped, 1-nerved, keeled, scabrid; upper glume obovate, 1-nerved, mucronate, scabrid; lemma ovate-lanceolate, mucronate, keeled; palea lanceolate, acuminate, 2-nerved, 2-keeled; lodicules 2; stamens 3. Caryopsis obovate, transversely rugose.

Common in forest plains, margins and found as a weed in crop fields and waste lands throughout.

Fl. & Fr.: June – September.

Alampalle (ADB), *TP* & *PVP* 4171; Mosra Forest Nursery (NZB), *TP* & *BR* 6069; Choutkoor (MDK), *TP* & *MSG* 11904; Mohammadabad RF (RR), *MSM* & *KH* 10551; Borapuram (MBNR), *BSS* & *SKB* 29567 & 32291; Narsapur (MDK), *KMS* 6765 (CAL); Upperpalli (HYD), *MRS* 2774 (DD); Subedari (WGL), *RKP* 105239 (BSID); Thupakulagudem (WGL), *PVS* 84010 (BSID); Mamillagudem (KMM), *RR* 105934 (BSID); Ramanagutta (KMM), *RR* 107993.

Dactyloctenium aristatum Link, Hort. Berol. 1:59. 1827; Bor, Grass. India 489. 1960; Moulik, Grass. Bamb. India 2:587. 1997. *Eleusine aristata* Hook. f., Fl. Brit. India 7:296.1896.

Annuals, culms erect or geniculately ascending, tufted, to 10 cm, rooting at nodes; nodes glabrous. Leaf sheaths hairy; ligules fimbriate, membranous; leaf blade linear-lanceolate, 2.5-8 × 0.3-0.6 cm, base cordate or rounded, margins covered with bristles, apex acuminate, glabrous. Spike solitary; spikelets ovate, 2-3-flowered; lower glume ovate, 1-nerved, 1-keeled, chartaceous; upper glume ovate-elliptic, aristate, scabrid, chartaceous, awned, awn geniculate; lower lemma ovate-lanceolate, membranous, keeled, acute; upper lemma akin to lower lemma; paleas ovate-lanceolate, chartaceous, keeled; lodicules 2; stamens 3. Caryopsis globose, rugose.

Rare in open forests and peripheries in Adilabad, Karimnagar and Ranga Reddi districts.

Fl. & Fr.: June – November.

Wankidi Roadside (ADB), *TP* & *GO* 5495; Central University campus (RR), *MSM* 15176.

DENDROCALAMUS Nees

Dendrocalamus strictus (Roxb.) Nees, Linnaea 9: 476. 1834; Hook. f., Fl. Brit. India 7: 404. 1896; Fischer, Fl. Madras 3: 1858. 1934. *Bambusa stricta* Roxb., Pl. Coromandel t. 80. 1798.

A perennial, unarmed, tufted grass; culms up to 15 m high, erect from branched root stock, solid or only with a small cavity; culm sheaths brown, glabrous-hairy, triangular, 5-10 × 6-14 cm; nodes swollen, the lower node often rooting. Leaf sheaths 3-15 cm, striate, glabrous; auricle short, ciliate with few waxy deciduous hairs; ligule narrow, serrate; blade 5-15 × 1.5-3 cm, linear, lanceolate, erect, stiff; shortly petiolate.

Inflorescence a compound panicle; spikelets in congested globose heads in long spikes, heads 1-5-3 cm wide; spikelets 1 cm long, spiny, 1-3-flowered, all bisexual; glumes ovate, spinescent, many-nerved; lemmas ovate, ending in a sharp spine, surrounded by ciliate tufts of hairs; palea ovate-obovate, emarginate, the lower ones 2-keeled, the upper most keelless; lodicules absent; stamens 6; stigmas plumose. Caryopsis ovoid-subglobose.

Very common in the forest of all districts.

Fl. & Fr.: Very rare. July – December.

Sadarmatt (ADB), *TP* & *PVP* 4104; Sirpur RF (ADB), *GO* 4424; Gannaram (NZB), *TP* & *BR* 6397; NSF (MDK), *BR* & *CP* 11650; Central University campus (RR), *MSM* 15172.

DICHANTHIUM Willement

1. Racemes solitary .... **D. foveolatum**
1. Racemes 2-many:
   2. Lower glume of pedicelled spikelets not armed with marginal bulbous based hairs:
      3. Nodes with short hairs, ligule a short ciliate membrane; lower glume of sessile spikelets elliptic; sessile spikelet 4-5 mm long .... **D. caricosum**
      3. Nodes with long hairs, ligule with long hairs; lower glume of sessile spikelets obovate; sessile spikelets 3-4 mm long . **D. annulatum**
   2. Lower glume of pedicelled spikelets armed with, marginal, bulbous based bristles .... **D.maccannii**

Dichanthium annulatum (Forssk.) Stapf in Prain Fl. Trop. Afr. 9: 178. 1917; Fischer, Fl. Madras 3: 1741. 1934; Bor, Grass. India 133. 1960; Deshp. in Fasc. Fl. India 15: 5. 1984; Moulik, Grass. Bamb. India 1: 271. 1997. *Andropogon annulatus* Forssk., Fl. Aegypt-Arab. 173. 1775; Hook. f., Fl. Brit. India 7: 196. 1896.

A perennial, tufted grass; culms up to 70 cm high, erect or ascending, nodes bearded. Leaf sheaths 5-10 cm long, glabrous, ciliate near the mouth; ligule membranous; blade 6-20 × 0.3-0.5 cm, linear, base rounded, apex acuminate. Racemes 6 cm long, subdigitately fascicled; peduncles glabrous; spikelets 4 mm, 2-nate. *Sessile spikelet:* 4 mm, bisexual, elliptic-oblong, subacute, ciliate; lower glume membranous, narrowly obovate, margins scaberulous, 7-9-nerved, 2-keeled; upper glume membranous, narrowly elliptic, margins inflexed, ciliate, 3-nerved; lower lemma hyaline, narrowly ovate, margins ciliate, nerveless; upper lemma narrow, with a scabrid, slender awn; palea absent; lodicules 2; stamens 3; stigmas plumose. Caryopsis oblong- slightly compressed. *Pedicelled spikelet:* 4 mm, male or barren; lower glume elliptic-oblong, 7-11-nerved keeled; upper glume narrow, 3-nerved; lower lemma ciliate; upper lemma small or obsolete.

Common in plains and hills of most districts.

Fl. & Fr.: All the year round.

Manchippa RF (NZB), *BR* & *PSPB* 9619; Narsapur RF (MDK), *TP* & *MSG*, 11993; Pakhal (WGL), *ANH* 15978 (CAL); Amberpet (HYD), *MRS* 264 (DD).

Dichanthium caricosum (L.) A. Camus, Bull. Mus. Hist. Nat. Paris 27: 549. 182; Fischer, Fl. Madras 3: 1741. 1934; Bor, Grass. India 134. 1960; Deshp. in Fasc. Fl. India 15: 7. 1986. *Andropogon caricosus* L., Sp. Pl. 2: 1480. 1763; Hook. f., Fl. Brit. India 7: 196. 1896. *Dichanthium aristatum* (Poir.) C.E. Hubb., Kew Bull. 1939. 654. 1939; Bor, Grass. India 134. 1960; Gayathri & Pullaiah in Pullaiah, Fl. Andhra Pradesh, 3: 170. 1997. *Andropogon aristatus* Poir. in Lam., Encycl. 1: 585. 1810. *Dichanthium pallidum* (Hook.f.) Stapf ex Fischer, Fl. Madras 3: 1741. 1934; Bor, Grass. India 135. 1960; Moulik, Grass. Bamb. India 1: 273. 1997. *Apocopis pallida* Hook.f., Fl. Brit. India 7: 143. 1896.

A perennial, tufted grass; culms up to 1 m high, ascending; nodes bearded or with short hairs. Leaf sheaths 3-10 cm long, glabrous; ligule shortly ciliate membrane; blade 4-20 × 0.3-0.5 cm, linear, base rounded, apex acuminate, glabrous-sparsely pubescent. Racemes 1-6, up to 8 cm long, both terminal and axillary; joints and pedicels of racemes with short hairs; spikelets 4 mm. 2-nate. *Sessile spikelet:* 2 mm, oblong, bisexual; lower glume oblong-obovate, 9-11-nerved, truncate, pilose; upper glume elliptic, 3-nerved, hairy; lower lemma lanceolate, nerveless, hyaline; upper narrow, awned; upper lemma hyaline to the base of awn, linear, 1-nerved, awned, geniculate; palea hyaline, narrowly ovate, nerveless, apex acuminate; lodicules 2; stamens 3, stigma plumose. Caryopsis obtuse, slightly compressed. *Pedicelled spikeletes:* 2 mm, obovate, male or barren; lower glume obovate, 17-19-nerved, 2-keeled, keels slightly winged, scabrous; upper glume and lemmas similar to sessile spikelet; lodicules 2; stamens 3.

Common in damp places and cultivated fields of most districts.

Fl. & Fr.: July – December.

Birsaipet (ADB), *GO* & *PVP* 4618; Manchippa RF (NZB), *BR* & *PSPB* 9620; Choutkoor fields (MDK), *TP* & *MSG* 11909; Gangapur RF (MDK), *TP* & *MSG* 12086; Patancheruvu (MDK), *BR* & *CP* 11317; ICRISAT, Patancheruvu (MDK), *LJGV* 2592 (CAL); Mohammadabad RF (RR), *MSM* & *KH* 10554; Hyderabad (HYD), *MRS* 243 & 217 (DD).

Note: Deshpande (1984) reduced *Dichanthium aristatum* (Poir.) C.E.Hubb. to the synonymy of *D. caricosum* (L.) A. Camus.

Dichanthium foveolatum (Del.) Roberty, Monogr. Sysy. Andropog. Du Globe 163. 1960. *Eremopogon foveolatus* (Del.) Stapf in Prain Fl. Trop. Afr. 9. 183. 1917; Fischer, Fl. Madras 3: 1751. 1934; Bor, Grass. India 148. 1960. *Andropogon foveolatus* Del. Fl. d' Egypte. 16. t. 2. 1812; Hook.f., Fl. Brit. India 7: 168. 1896.

Vern.: *Namalachetta*.

A perennial, tufted grass; culms up to 80 cm high, shortly rhizhomatous below, slender; nodes bearded. Leaf sheaths 4-6 cm, scabrid-glabrous; ligule ciliate, hyaline;

blade mostly basal, 5-15 × 0.1-0.3 cm, flat, base rounded, apex acuminate, glabrous. Raceme solitary, up to 5 cm long, subtended by a spatheole; spikelets 6 mm, 2-nate, one sessile, other pedicelled. *Sessile spikelet*: 4 mm, bisexual, ovate-oblong; lower glume oblong-ovate, 2-keeled, 5-nerved, pitted, scabrid on margin; upper glume ovate-lanceolate, 5-nerved, hairy near apex; lower lemma linear, scabrid on margin; upper lemma narrow, awned; stamens 3, stigma plumose. Caryopsis obovoid. *Pedicelled spikelet:* 3 mm, oblong-ovate, male or barren; glumes oblong-lanceolate, hairy, lower pitted; lemma oblong, obtuse, scabrid; stamens 3.

Common in waste lands and cultivated fields of most districts.

Fl. & Fr.: September- January.

Sattenapalle (ADB), *TP* & *PVP* 4088; Manchippa RF (NZB), *TP* & *BR* 6040; Choutkoor (MDK), *TP* & *MSG* 11930; Aklaspur (KRN), *GVS* 20236 (CAL); Nagarjunakonda valley (NLG), *KT* 9688 (CAL); Cherukupally (NLG), *BVRS* 662 (NU).

Dichanthium maccannii Blatt., J. Bombay Nat. Hist. Soc. 32: 357. 1927; Bor, Grass. India 135. 1960; Moulik, Grass. Bamb. India 1: 273. 1997.

Annual, tufted grass, up to 45 cm high, nodes densely bearded. Leaf sheath terete, upper side covered with bulbous based hairs; ligule hairy; leaf blade linear-ovate, 2-10 × 0,2-0.6 mm, covered with bulbous based bristles. Racemes 1-2, lower 3-4 pairs of spikelets all male. *Sessile spikelet:* narrowly ovate; lower glume narrowly ovate-oblong, 7-nerved, apex obtuse; upper glume narrowly ovate, apex acuminate; lower lemma narrowly oblong, nerveless, palea absent. Upper lemma awned, awn geniculate. *Pedicelled spikelet:* narrowly obovate, lower glume narrowly obovate, covered with submarginal bulbous based bristles on dorsal side, 10-nerved; upper glume narrowly ovate, margins inflexed, ciliate, 3-nerved; lower lemma narrowly oblong, margins ciliate, palea narrowly ovate.

Warangal. It is known from single collection only (Reddy, 1983; Sudhakar Reddy & *al.*, 2001).

Fl. & Fr.: August – November.

DIGITARIA Haller

1. Inflorescence a true panicle, as long as broad with spikelets dispersed throughout the panicle on solitary, long pedicels .... **D. tomentosa**
1. Inflorescence of spikelike racemes, these solitary, conjugate, digitate or verticillate in whorls, never truly effuse; spikelets binate, ternate or in small groups:
    2. Spikelets binate:
        3. Margins of the rhachis smooth .... **D. radicosa**
        3. Margins of the rhachis serrate:
            4. Spikelets homomorphous:
                5. Spikelets glabrous to sparsely pubescent, lower glume absent, upper glume less than 1/3 as long as the spikelets .... **D. setigera**

5. Spikelets various pubescent, lower glume present, upper glume more thin 2/3 or little longer than the spikelet .... **D. ciliaris**

4. Spikelets heteromorphous .... **D. bicornis**

2. Spikelets ternate:

6. Annuals; upper glume distinct:

7. Racemes 2, rhachis straight, flat; hairs on the spikelets verrucose; caryopsis ellipsoid .... **D. longiflora**

7. Racemes 3-10, rhachis capillary, trigonous; hairs on the spikelets clavate; caryopsis oblong:

8. Tips of pedicels with a circle of stiff, overtopping hairs or spicules up yo 1 mm .... **D. ternata**

8. Tips of pedicels without overtopping hairs or spicules .... **D. abludens**

6. Perennials; upper glume represented by a tuft of hairs .... **D. stricta**

Digitaria abludens (Roem. & Schult.) Veldkamp, Blumea 21 : 53. 1773; Moulik, Grass. Bamb. India 1: 90. 1997. *Panicum abludens* Roem. & Schult., Syst. Veg. 2:457. 1817. *Paspalum granulare* Trin. ex Spreng., Neue Entd. 2 : 47. 1821. *Digitaria granularis* (Spreng.) Henr., Mongr. Digitaria 302. 1950; Bor, Grass. India 301. 1960. *Paspalum pedicellare* Trin. ex Hook. f., Fl . Brit. India 7 : 19. 1896. *Digitaria pedicellaris* (Hook, f.) Prain, Bengal Pl. 1182. 1903; Fischer, Fl. Madras 3: 1765. 1934.

An annual, tufted grass; culms up to 50 cm high, slender, filiform; nodes glabrous. Leaf sheaths 1-3 cm, keeled, glabrous; ligule membranous; blade 3-9 × 0.2 - 0.5 cm, base rounded, apex acuminate, glabrous. Spikes 3-10, 9 cm long, digitate; rhachis capillaris, trigonous; pedicels filiform, with clavate hairs; spikelets ternate, 1.5 mm; lower glume absent; upper glume elliptic, 5-nerved, hairy along the margins; lower lemma elliptic-lanceolate, 3-nerved, hairy; upper lemma ovate, subchartaceous, 3-nerved, apiculate; palea similar to upper lemma; lodicules 2; stamens 3; stigmas plumose. Caryopsis oblong, tightly enclosed in lemma and palea

Abundant in plains, near water sources of Adilabad, Nizamabad, Hyderabad, Warangal and Mahabubnagar districts.

Fl. & Fr.: July – October.

Ankusapuram (ADB), *GO* 4318 & 4347; Rebbana (ADB), *GO* & *PVP* 4892; Mancheryal (ADB), *GO* & *PVP* 4979; Potchera (ADB), *GO* & *DAM* 5058; Bellampalli (ADB), *PVP* 9459; Arsapally (NZB), *BR* & *KH* 6486; Anantagiri (RR), *TP* & *MSM* 11065; Vikarabad (RR), *MRS* 84 (DD); Pakhal (WGL), *KMS* 13128 (MH).

Digitaria bicornis (Lam.) Roem. & Schult. in L . Syst. Veg. 2 : 470. 1817; Bor, Grass. India 299. 1960; Moulik, Grass. Bamb. India 1: 90. 1997. *Paspalum bicorne* Lam., Tabl. Encycl. 1: 176. 1791. *P. heteranthum* Hook. f., Fl . Brit. India 7 : 16. 1896. *Digitaria biformis* Willd., Enum. Hort. Berol. 92. 1809; Bor, *loc. cit.*

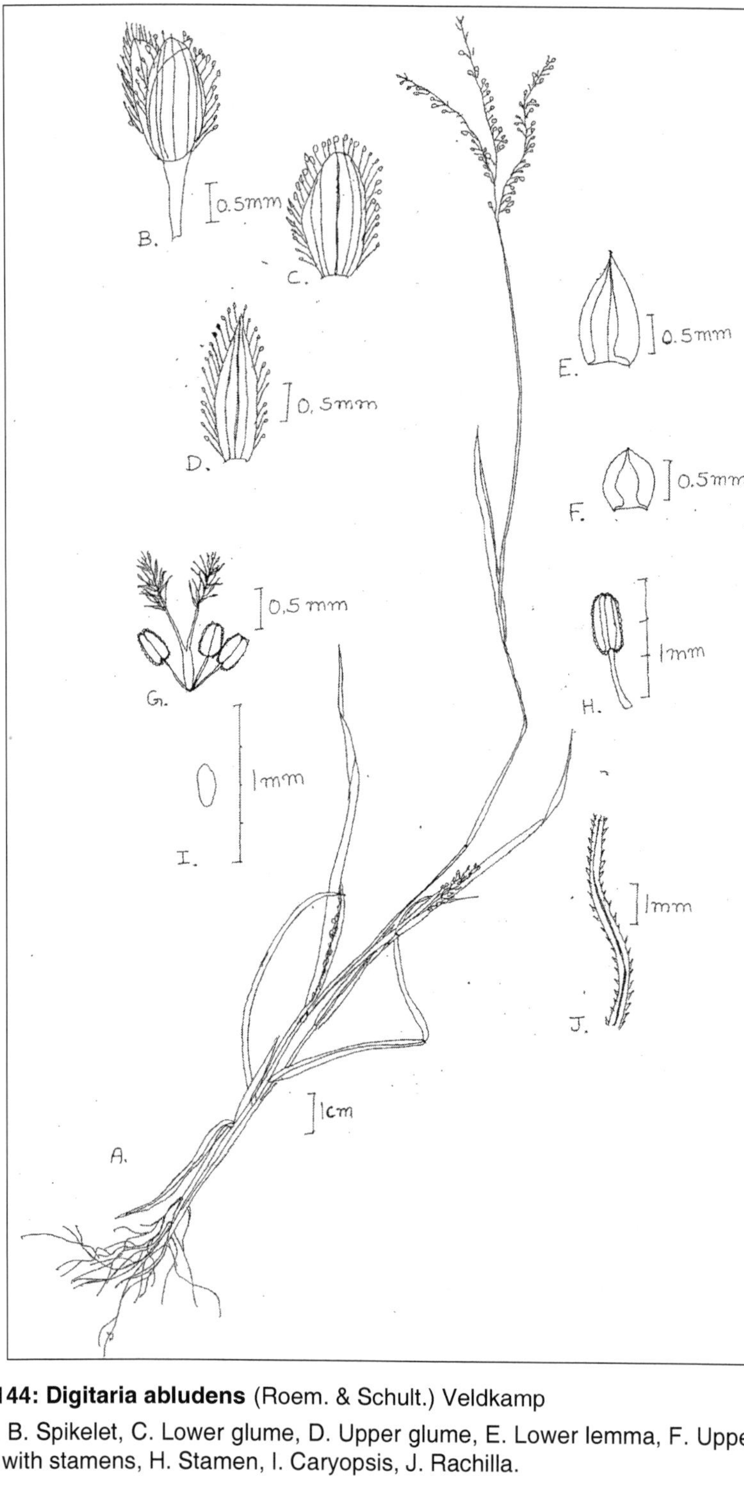

Figure 144: Digitaria abludens (Roem. & Schult.) Veldkamp

A. Twig, B. Spikelet, C. Lower glume, D. Upper glume, E. Lower lemma, F. Upper lemma, G. Pistil with stamens, H. Stamen, I. Caryopsis, J. Rachilla.

An annual, tufted grass; culms up to 50 cm high, base geniculately ascending, rooting at the nodes; nodes sparsely bearded, glabrescent. Leaf sheaths 2-8 cm, ligule membranous; blade 2-6 x 0.4-0.6 cm, oblong, base cordate, apex acuminate, glabrous. Racemes 2-9, 10 cm long, digitate; rhachis serrulate; spikelets 3 mm, lanceolate, 2-nate, heteromorphous. *Sessile spikelet:* 3 mm, glabrous-pilose; lower glume scaly; upper glume elliptic, 3-5-nerved, pubesent. Florets 2, lower barren, epaleate, upper bisexual, paleate; lower lemma elliptic, 7-nerved, mid and lateral nerves prominent, membranous; upper lemma elliptic-lanceolate, 3-nerved, chartaceous; palea similar to upper lemma; stamens 3; stigmas plumose. Caryopsis ellipsoid, lightly enclosed by hardened lemma and palea.

Very variable grass commonly occurring in plains of Adilabad, Nizamabad Medak, Karimnagar, Hyderabad and Nalgonda districts.

Fl. & Fr : July-November.

Mallampet RF (ADB), *TP & GO* 5451; Ghanapur (NZB), *BR & KH* 7113; Kamareddy (NZB), *BR* 9277; Nagulabanda (MDK), *CP* 11361; Sivanoor (MDK), *BR & CP* 11550; Nagarjuna Konda valley (NLG), *KT* 9684 (CAL).

Digitaria ciliaris (Retz.) Koel., Descr. Gram 27. 1802; Moulik, Grass. Bamb. India 1: 90. 1997. *Panicum ciliare* Retz., Observ. Bot. 4: 16. 1786. *Digitaria marginata* Link *var. fimbriata* (Link) Stapf in Prain, Fl ., Trop. Afr. 9: 440. 1919; Fischer, Fl. Madras 3: 1764. 1934. *D. adscendens* (Kunth) Henr., Blumea 1: 92. 1934; Bor, Grass. India 298. 1960. *Panicum adscendens* Kunth, Nov. Gen. Pl. 1 : 102. 1821.

An annual, tufted grass; culms up to 90 cm high, usually ascending from geniculate or prostrate base; branching from lower nodes; nodes glabrous. Leaf sheaths 4-12 cm long, glabrous; ligule membranous, truncate; leaf blade 4-10 × 0.4-0.6 cm, linear- lanceolate, base cordate, apex acute, glabrous. Racemes 4-9, 10 cm long, digitate; rhachis serrate, trigonous; spikelets 3 mm, elliptic-lanceolate, 2-nate, appressed to the rhachis, glabrous- sparsely ciliate, homomorphous; lower glume reduced to a triangular scale; upper glume linear-lanceolate, 3-5-nerved, pubescent; lower lemma oblong-lanceolate, 5-nerved, marginal nerves pubescent, palea absent or very minute; upper lemma oblong-lanceolate, obscurely 3-nerved, margins inflexed; lodicules 2; stamens 3, stigmas plumose. Caryopsis oblong.

Very common in moist plains and hills of all districts.

Fl. & Fr.: July-October.

Sattenapalle beat (ADB), *TP & PVP* 4084; Madikonakutta (MDK), *KMS* 6729 (CAL); Way to Rusthampet (MDK), *KMS* 6807 (CAL); Bhupatipur (KRN), *GVS* 20204; Near Hussainsagar dam (HYD), *KMS* 8022 (CAL); Pillalamarri (NLG), *BVRS* 44 (NU).

Digitaria ciliaris (Retz.) Koeler subsp. **ciliaris** var. **fimbrita** (Link) Jain & Doli Das, Indian Forester 99(9): 572. 1973. *Panicum ciliare* Retz. Obs. Bot. 4: 16. 1786. *Paspalum sanguinata* Lam. var. *ciliare* (Retz.) Hook.f., Fl. Brit. India 7: 15. 1896. *Digitaria adscendens* (Kunth) Henr. subsp. *adscendens* var. *criniformis* Henr., Mongr. Gen. Digitaria 999. 1950; Bor, Grass. India 299. 1960.

Annuals, culms erect or decumbent, tufted, to 80 cm, rooting at lower nodes, glabrous. Leaf sheaths 2-10 cm, glabrous, keeled; ligule 2 mm, truncate, membranous; leaf blade 5-10 × 0.2-1 cm, lanceolate or linear-lanceolate, base cordate, acuminate, scabrid or villous. Racemes 2, rarely 3, digitate, rhachis serrate; pedicels triquetrous, scabrid; spikelets 2, homomorphous, sessile and pedicelled, ovate-lanceolate, 3.5 × 2 mm, hairy; lower glume absent; upper glume lanceolate, 3 × 1 mm, chartaceous, 3-nerved, hairy; lower lemma ovate-lanceolate, 3.5 × 1.7 mm, chartaceous, 9-nerved, hairy; upper lemma elliptic-lanceolate, 3.3 × 1 mm, coriaceous, 3-nerved; palea elliptic-lanceolate, 3.2 × 1 mm, crustaceous, 2-keeled, 2-nerved; lodicules 2, ovate; stamens 3, anthers 1 mm; ovary 0.5 mm. Caryopsis ovate-oblong, 1.5 mm.

Common in forest plains and margins in Mahabubnagar (A.M.Reddy & *al.*, 2009) and many other districs of Telangana

Fl. & Fr.: August – December.

Sarvaipet (ADB), *GO* & *PVP* 4610; Jaipur RF (ADB), *PVP* 9434; Mosra forest nursery (NZB), *TP* & *BR* 6078; Kamareddy (NZB), *BR* 9278; NSF Fields (MDK), *BR* & *CP* 11675; Mohammadabad (RR), *MSM* 10593; Appapur (MBNR), *BR* & *AMR* 25679.

Digitaria longiflora (Retz.) Pers., Syn. Pl. 1: 85. 1805; Fischer, Fl. Madras 3: 1765. 1934; Bor, Grass. India 302. 1960; Moulik, Grass. Bamb. India 1: 93. 1997. *Paspalum longiflorum* Retz., Observ. Bot. 4: 5. 1786/87; Hook. f., Fl. Brit. India 7:17. 1896.

An annual, tufted grass; culms up to 60 cm high, ascending, rooting at the nodes; nodes glabrous. Leaf sheaths 2-4 cm long, glabrous; ligule membranous; blade 2-7 x 0.2-0.5 cm, linear- lanceolate, base rounded, apex acuminate, glabrous. Racemes 2-3, digitate, 2-5.7 cm long; rhachis serrate; spikelets 2 mm long, elliptic, homomorphous, sparsely pubescent; lower glume absent; upper glume oblong-ovate, 3-5-nerved, hairy; lower lemma similar to upper glume; upper lemma ovate-oblong, subchartaceous, nerveless; palea similar to upper lemma; lodicules 2; stamens 3, stigmas plumose. Caryopsis ellipsoid.

Occasional in the plains of Adilabad, Hyderabad, Mahabubnagar and Warangal districts.

Fl. & Fr.: July-October.

Ankusapuram (ADB), *GO* 4338 & 4354; Jaipur RF (ADB), *PVP* 9435; Vikarabad (RR), *MRS* 94 (DD); Hyderabad (HYD), *MRS* 236 (DD).

Digitaria radicosa (C.B. Presl) Miq., Fl. Ned. Ind. 3: 437 1857; Moulik, Grass. Bamb. India1: 94. 1997. *Panicum radicosum* C.B. Presl, Rel. Haenk. 1: 297. 1830. *Digitaria timorensis* (Kunth) Hal., J. Bot. (Morot) 4 : 138. 1890; Bor, Grass. India 306. 1960. *Panicum timorense* Kunth, Enum. Pl. 1: 83. 1833. *Paspalum sanguinale* Lam. var. *debile* auct. non Hook. f., Fl. Brit. India 7:16. 1896.

An annual, tufted grass; culms up to 30 cm high, decumbent or ascending at base; nodes glabrous. Leaf sheaths 2-5 cm long, loose, glabrous; ligule membranous; blade 3-5 x 0.2-0.4 cm, linear-oblong, base narrow, apex acute, sparsely pilose. Racemes 2 or 3, 7 cm long; rhachis smooth; spikelets narrowly lanceolate, 3 mm, homomorphous, 2-nate, pubescent; lower glume reduced to a scale; upper glume lanceolate, hairy, 3-nerved, apex acute; lower lemma lanceolate-oblong, 5-nerved, marginal nerves pubescent; upper lemma elliptic-lanceolate, 3-nerved; palea similar to upper lemma; stamens 3; stigmas plumose. Caryopsis oblong.

Occasional grass in plains of Adilabad district.

Fl. & Fr.: July-December.

Sattenapalle RF (ADB), *TP & PVP* 4098.

Note: This species in closely related to *Digitaria ciliaris* but it can be distinguished by the slender, weak, trailing culms, few and slender racemes, smooth edged rhachis and narrow lanceolate spikelets.

Digitaria setigera Roth, Syst. Veg. 2 : 474. 1817; Bor, Grass. India 305. 1960; Moulik, Grass. Bamb. India 1: 96. 1997. *Paspalum sanguinale* Lam var. *extensum* Hook. f., Fl. Brit. India 7 : 15. 1896; Fischer, Fl. Madras 3: 1772. 1934.

An annual, stoloniferous grass; culms up to 50 cm, creeping, rooting at the lower nodes; nodes glabrous. Leaf sheaths 2-6 cm long, loose, glabrous; ligule membranous; blade 3-10 x 0.2-0.6 cm, linear-oblong, base rounded, apex acuminate, glabrous. Racemes 3-12, 10 cm long, digitate; rhachis serrate, basally with stiff hairs; hairy near joints and pedicels; lower glume obsolete; upper glume triangular in sessile; linear in pedicelled spikelets, 3-nerved, pubescent; lower lemma lanceolate, pubescent on marginal nerves; upper lemma elliptic-lanceolate; lodicules 2; stamens 3; stigmas plumose. Caryopsis ellipsoid.

Occasional weed of cultivation and road sides in Medak district.

Fl. & Fr.: July-October.

Choutkoor fields (MDK), *TP* & *MS* 11917.

Digitaria stricta Roth, Syst. Veg. 2 : 474. 1817 var. **denudata** (Link) Henr., Monogr. Digitaria 175. 1950; Bor, Grass. India 306. 1960; Moulik, Grass. Bamb. India 1: 97. 1997. *D. denudata* Link, Hort. Berol. 1 : 222. 1827. *Paspalum royleanum* Nees ex Thwaites, Enum. Pl. Zeyl. 358. 1864; Hook. f., Fl. Brit. India 7: 18 . 1896.

A perennial, tufted grass, culms up to 50 cm high; nodes glabrous. Leaf sheath 2-5 cm, appressed hairy; ligule membranous; blade 6-16 x 0.2-0.4 cm, linear, flat, sparsely hairy along the margins. Racemes 2-10, erect, 10 cm long; rhachis filiform, trigonous; spikelets 1.6 mm, broadly ellipsoid, secund, tomentose with minute clavellete hairs; lower glume absent; upper glume reduced to the tuft of hairs. Florets 2, lower barren, upper bisexual; lower lemma broadly elliptic, thinly membranous, 5-nerved; upper lemma ovate, coriaceous, 3-nerved, acute with inflexed margins; upper palea cartilaginous; stamens 3. Caryopsis oblong.

Rare grass in Ranga Reddi, Medak and Mahabubnagar districts.

Fl. & Fr.: August-January.

Patancheruvu (MDK), *LJGVM 2S42* (CAL); Vikarabad (RR), *MRS* 324 (DD).

Digitaria ternata (A.Rich.) Stapf, Fl. Cap. 7: 376. 1898; Bor, Grass. India 305. 1960; Moulik, Grass. Bamb. India 1: 97. 1997. *Cynodon ternatus* A. Rich., Tent. Fl. Abyss. 2: 405. 1851. *Paspalum ternatum* (A.Rich.) Hook.f., Fl. Brit. India 7: 17. 1896.

Annuals; culms erect or ascending, to 120 cm high, nodes glabrous. Leaf sheaths 5-14 cm, glabrous, ciliate at mouth. Ligules hairy, membranous, 0.5 mm; leaf blade 5-15 × 0.5-1 cm, linear-lanceolate, base rounded. Rhachis villous; pedicels 2.4 mm, scabrid. Spikelets ovate, 2.5-3 ×1.3 mm, clavate, hairy; lower glume absent; upper glume linear-lanceolate, 1.7 × 0.6 mm, hyaline; lower lemma elliptic or oblong-lanceolate, 2.8 × 1.2 mm, chartaceous, scabrid, 7-nerved; upper lemma oblong-lanceolate, 2.5 × 1 mm, crustaceous, keeled; lodicules 2, ovate; stamens 3, anthers 0.6 mm; ovary elliptic, styles 0.5 mm, stigmas 0.8 mm. Caryopsis oblong.

Common in forest plains and margins in Mahabubnagar district (A.M.Reddy & *al.*, 2009).

Fl. & Fr.: August – December.

Appapur (MBNR), *BR* & *AMR* 25677.

Digitaria tomentosa (Willd.) Henr., Blumea 1: 100. 1934; Bor, Grass. India 306. 1960. *Milium tomentosum* Koen. ex Willd., Ges. Naturf. Freunde Berlin Neue Schriften 4: 220. 1803. *Panicum subeglume* Trin., Mem. Acad. Imp. Sci. St. Petersb. Ser. 6, Sci. Math. 3 : 292. 1834; Hook.f., Fl. Brit. India 7 : 51. 1896; Fischer, Fl. Madras 3: 1783. 1934.

A perennial, tufted grass, culms up to 80 cm high, geniculately ascending at base; nodes sparsely hairy. Leaf sheaths 4-9 cm, loose, sparsely pubescent-pilose; ligule membranous; blades 7-15 x 0.3-0.7 cm, oblong, flat, base rounded, margins ciliate, apex acuminate, sparsely pubescent to tuberculate based hairs. Panicle lax, large, up to 30 cm long, spreading; rachis capillary, pilose at branch axis; spikelets 2 mm, narrowly lanceolate, solitary, pubescent. Lower glume obsolete; upper glume elliptic-lanceolate, pubescent; upper lemma elliptic-lanceolate, pilose; upper lemma elliptic-oblong, coriaceous; palea similar to upper lemma; stamens 3, stigmas plumose. Caryopsis oblong.

Occasional grass in forests in Warangal district.

Fl. & Fr.: July – October.

Thupakulagudem (WGL), *PVS* & *NRR* 76999 (BSID).

DIMERIA R.Br.

1. Upper glume winged:
    2. Upper glume winged all along:
    2. Upper glume winged below the tip only .... **D. orissae**
        3. Racemes solitary .... **D. avenacea**
        3. Spikes 2 .... **D. kanjirapallilana**
1. Upper glume not winged .... **D. ornithopoda**

Dimeria avenacea (Retz.) Fischer, Kew Bull. 1932 : 72. 1932 & Fl. Madras 3 : 1713. 1934; Bor, Grass. Burma, Ceylon, India & Pakistan 139. 1960. *Anthoxanthum avenacum* Petz., Observ. Bot. 3: 8. 1783. *Dimeria pusilla* Thw. Enum. Pl. Zeyl. 369. 1864; Hook.f., Fl. Brit. India 7 : 103. 1896.

Annual herbs; culms tufted, 20- 35cm high. Leaf blades narrow, apex abruptly acuminate; sheaths tuberculate-pilose; ligule membranous. Racemes solitary, spike like; rachis not jointed, trigonous, winged; callus bearded. Spikelets oblong, laterally compressed. Glumes coriaceous, sparsely ciliate, upper winged, keels ciliate awned; lemmas hyaline; upper awned. Caryopsis linear to elliopsid.

Rare on rocky soil in open grasslands in Hyderabad district (Ramana, 2010).

Fl. & Fr. : November-February.

Dimeria kanjirapallilana K.C. Jacob, J. Bombay Nat. Hist. Soc. 47: 48. 1947; Manilal & Sivaraj., Fl. Calicut 231. 1982; Sreekumar & V.J. Nair, Fl. Kerala Grasses 91. 1991.

An annual tufted grass, culms up to 25 cm high, erect, occasionally branched near the base; nodes sparsely bearded or glabrous. Leaf sheaths 3-4 cm long, glabrous, keeled; ligule membranous, ovate, acute; blade linear, 2-6 × 0.2-0.4 cm, rounded at base, bulbous based hairs along the margins, acuminate at apex. Spikes 2, 3-8 cm long, divaricate; rhachis flat, zig-zag, margins sparsely ciliate, narrowly winged; spikelets sessile, 4.5 mm, oblong-lanceolate; callus villous; lower glume linear-lanceolate, 2.5 mm, chartaceous, sparsely hairy at the apex, acuminate; upper glume lanceolate, 3 mm, chartaceous, hairy at the apex, margins hyaline, lower lemma hyaline, 2 mm, blong, 1-nerved; upper lemma hyaline, 3 mm (excluding the awn), bifid, awned at the sinus, awn 1 cm long; stamens 2, anthers oblong, 0.9 mm, filament 0. 5 mm; pistil 1.75 mm. Caryopsis linear.

Occasional in the humus soil and rocky areas of Medak district. Associated with *Drosera* species.

Fl. & Fr.: October-January.

Pocharam (MDK), *TP & MSG* 12024.

Dimeria orissae Bor, Kew Bull. 1952: 579. 1953 & Grass. India 142. 1960; S. Moulik, Grass. & Bamb. India 1: 283. 1997.

Tufted annuals; culms erect, 25 – 60 cm high, nodes hairy. Leaf blades linear-lanceolate, 3 - 10 × 0.3 – 0.5 cm, sparsely tubercule-based hairy; ligule truncate, membranous, 0.5 mm long; leaf sheath 1-3 cm long. Racemes 2 – 3, 3 – 6 cm long; peduncle up to 20 cm long; rachis triquetrous, flat on back, 0.5 – 0.7 mm wide, winged, ciliate along margins. Spikelets elliptic-oblong, 4.2 – 4.7 × 0.5 – 0.7 mm, hairy; callus bearded; pedicel truncate, 0.25-0.3 mm; lower glumes linear-lanceolate, 3.5 – 3.7 × 0.3 – 0.4 mm, acute at apex, ciliate, keeled throughout, keel narrowly winged just below the tip; upper glumes linear-lanceolate, 4.2 – 4.5 × 0.5 - 0.6 mm, acute at apex, ciliate throughout, keeled, keel narrowly winged just below the tip; florets 2, epaleate, lower one barren, upper bisexual; lower lemma oblong, hyaline, 1.7 – 2 × 0.1 – 0.2 mm;

upper lemma linear-lanceolate, 3.2 – 3.5 × 0.8 – 1 mm, hyaline, cleft at apex into 2 lobes, awn 11 – 15 mm long; anthers 2, 0.75 mm. Caryopsis linear, 2 mm long.

On grassy patches in Adilabad district.

Fl. & Fr.: September – October.

Kawal Tiger Reserve, Indanpalli Range (ADB), *Alok Chorghe, Chandra Mohan & P. S. Annamma* 3933 (BSID); Thadlapet Range (ADB), *Alok Chorghe, Chandra Mohan & P. S. Annamma* 3962 (BSID).

Dimeria ornithopoda Trin., Fund. Agrost. 167. t. 14. 1820.

Key to Varieties

1. Spikelets 2 mm long; anthers 0.5 mm .... var. **ornithopoda**
1. Spikelets 3 mm long; anthers 0.7 mm .... var. **khasiana**

Dimeria ornithopoda Trin. var. **ornithopoda;** Hook. f., Fl. Brit. India 7: 104. 1896; Fischer, Fl. Madras 3: 1713. 1934; Bor, Grass. India 142. 1960; Moulik, Grass. Bamb. India 1: 283. 1997.

An annual, tufted grass; culms up to 40 cm long, erect or ascending. Leaf sheaths 2.5 cm, keeled, tubercle-based hairy; ligule membranous; blade 4-6 x 0.3-0.8 cm, lanceolate, apex acuminate, densely pubescent. Racemes 2 or 3, 6 cm long, terminal digitate; rhachis trigonous; spikelets 2mm, oblong; lower glume subcoriaceous, 1-nerved, 1-keeled, scabrellate on keel and dorsal surface, upper glume subcoriaceous, oblong, sparsely hairy, keel hairy at upper side, mucronate. Florets 2, lower barren, upper bisexual; lower lemma oblanceolate, hyaline, delicate, nerveless; upper lemma narrowly elliptic, awned; lodicules 2; stamens 2. Caryopsis ellipsoid.

Rare along road sides, on bunds of fields and hill slopes in Mahabubnagar district.

Fl. & Fr.: August-January.

Irlapenta (MBNR), *BR* & *BSS* 32290 (BSID).

Dimeria ornithopoda Trin. var. **khasiana** Bor, Kew Bull. 7 : 576. 1952 & Grass. India 144. 1960.

An annual, tufted grass; culms up to 20 cm high. Leaf sheaths keeled, tuberculate pubescent; ligule membranous; blade 2.5-7.5 x 0.2-0.4 cm, linear, pubescent. Racemes 2 or 3, 5 cm; rhachis trigonous; spikelets 4 mm, oblong; lower glume mucronate; upper glume apically elongate-bristly; upper lemma awned. Caryopsis ellipsoid.

Occasional in Mahabubnagar district.

Fl. & Fr.: August-January.

DINEBRA Jacq.

Dinebra retroflexa (Vahl) Panz., Denkschr., Acad. Wiss. Muenchen. 270. t. 12. 1814; Fischer, Fl. Madras 3:1841. 1934; Bor, Grass. India 491. 1960; Moulik, Grass. Bamb. India 2:589.1997. *Cynosurus retroflexus* Vahl, Symb. Bot. 2:20.1791. *Dinebra arabica* Jacq., Fragm. 77.t.121. f. 1.1809; Hook. f., Fl. Brit. India 7:297. 1896.

Annuals, culms erect or ascending, tufted, leafy, to 60 cm high; nodes slender, glabrous. Leaf sheaths glabrous; ligules membranous; leaf blade linear-lanceolate, base cordate, apex acuminate, glabrous. Inflorescence elliptic-oblong to pyramidal; spikelets 1-3-flowered; lower glume elliptic-lanceolate, 1-nerved, 1-keeled, coriaceous, scaberulous, awned; upper glume elliptic-lanceolate, 1-nerved, 1-keeled, awned or cuspidate, persistent. Florets 2 or 3, upper most reduced into lemma, lemmas narrow, ovate, hyaline, sub acute or obtuse, 1-nerved; paleas ovate, hyaline, 2-keeled, sparsely ciliate on the keels; lodicules 2; stamens 3. Caryopsis ovoid, trigonous.

Occasional in forests and common in crop fields in all districts.

Fl. & Fr.: August-September.

Surzapur (ADB), *TP* & *PVP* 4110; Moosra (NZB), *TP* & *BR* 6062; Sirikonda RF (NZB), *BR* 9699; Poosapalle (MDK), *TP* & *MSG* 11954; NSF (MDK), *BR* 11673; Narsapur (MDK), *KMS* 7989 (CAL); Anantasagar (RR), *MSM* & *KH* 11015; Nagarjunakonda valley (NLG), *KMS* 9852 (CAL); Amberpet (HYD), *MRS* 271 (DD); Achaaryapur – Koilkonda (MBNR), *SRS* 109766 (BSID); Garlapadu RF (MBNR), *SRS* 109085 (BSID).

Note: This grass looks like some species of *Eragrostis,* but it can be distinguished by the lax, loose, grey panicles, larger spikelets and lobed as well as awned lemmas with villous nerves.

ECHINOCHLOA P. Beauv.

1. Ligule present .... **E. picta**
1. Ligule absent:
    2. Spikelets awned:
        3. Spikelets up to 4 mm, upper glume up to 4 mm .... **E. crusgalli**
        3. Spikelets up to 6.5 mm, upper glume up to 5 mm .... **E. oryzoides**
    2. Spikelets unawned:
        4. Wild grass; racemes usually distant .... **E. colona**
        4. Cultivated grass; racemes densely crowded .... **E. frumentacea**

Echinochloa colona (L.) Link, Hort. Berol. 2: 209. 1833; Fischer, Fl. Madras 3: 1776. 1934; Bor, Grass. India 308. 1960; Moulik, Grass. Bamb. India 1: 98. 1997. *Panicum colonum* L., Syst. Nat. (ed. 10) 810. 1759; Hook. f., Fl. Brit. India 7:32. 1896.

An annual, tufted grass; culms up to 70 cm high, branching from lower nodes; nodes glabrous. Leaf sheaths 5-15 cm long, glabrous; ligule obsolete; blade 5-25 x 0.3 - 0.6 cm, linear-lanceolate, base cordate, apex acuminate, glabrous. Racemes pyramidal, spikes suberect, usually distant, 5-10, rhachis angular, scaberulous; spikelets 2.5 cm, broadly ovate, acute or sub- cuspidate, crowded in rows, secund; glumes unequal; lower glume half the length of the upper glume, broadly ovate-orbicular, 3-nerved, clasping at base, scaberulous on the nerves; upper glume broadly ovate, mucronate, 5-7-nerved, puberulous. Florets 2, lower barren to male, upper bisexual; lower lemma broadly ovate, cuspidate, 5-7-nerved, nerves usually prominent, lateral nerves hispid, margins infolded; lower palea oblong to elliptic;

upper lemma broadly ovate, boat-shaped, obtuse; palea oblong; lodicules 2; stamens 3; stigmas plumose. Caryopsis broadly ellipsoid.

Very frequent in cultivated fields of all districts.

Fl. & Fr.: Almost round the year.

Kadam dam site (ADB), *TP & PVP* 4148; Rabbana (ADB), *GO & PVP* 4850; Malkapur thanda (NZB), *TP* & *BR* 6098; Dichpally (NZB), *BR & KH* 6445; Gangapur fields (MDK), *CP* 11793; Naraspur (MDK), *KMS* 6721 (CAL); Anantasagar (RR), *MSM* & *KH* 11016; Pakhal (WGL), *KMS* 13187 (CAL); Kodimial (KRN), *GVS* 20118 (MH); Nagarjuna Konda valley (NLG), *KT* 9834 (CAL); Moosi river (HYD), *MRS* 317(DD); Garlapadu RF (MBNR), *SRS* 109090 (BSID); Billyguttalu (KMM), *RR* 10669, 105996 (BSID); Subedari (WGL), *RK* 105238 (BSID).

Echinochloa crusgalli (L.) P. Beauv., Ess. Agrost. 53. 161. 1812; Fischer, Fl. Madras 3 : 1777. 1934; Bor, Grass. India 310. 1960; Moulik, Grass. Bamb. India 1: 99. 1997. *Panicum crusgalli* L., Sp. Pl. 56. 1753; Hook. f., Fl. Brit. India 7 : 30. 1896.

An annual, tufted grass; culms 70 cm high, branching from lower nodes; nodes glabrous. Leaf sheaths 5-25 cm, glabrous; ligule obsolete; blade 5-20 x 0.4-0.8 cm, linear-lanceolate, base cordate, apex acuminate, glabrous. Spike like panicle 10 cm long, racemes spreading; rhachis triquetrous, scabrid. Spikelets paired, similar, pedicelled to subsessile, broadly ovoid, 5 mm, crowded, acute-cuspidate, awned; lower glume broadly ovoid or orbicular, margin ciliate, half as long as the upper glume, 3-5-nerved, scabrid on nerves; upper glume broadly ovate-oblong, concave, cuspidate, 5-nerved, hispid on back. Florets 2, lower male, upper bisexual; lower lemma ovate-oblong, flat or depressed on the back, awned, hispid at apex; palea ovate-oblong; lodicules 2; stamens 3; stigmas plumose. Caryopsis broadly ellipsoid.

Common in rice fields and wet places of most districts.

Fl. & Fr.: Almost through the year especially in rainy season.

Mandora village (NZB), *TP & BR* 6226; Narsingi (MDK), *BR & CP* 11530; Mohammadabad (RR), *MSM* 10537; Kodimial (KRM), *GVS* 20126 (MH); Along Moosi river bank (HYD), *KMS* 5930 (CAL).

Echinochloa frumentacea (Roxb.) Link, Hort.Berol. 1: 204. 1827; Bor, Grass. India 311. 1960. *Panicum frumentaceum* Roxb. Fl. Ind. 1: 307. 1820, non Salisb. 1796. *P. crusgalli* L. var. *frumentaceum* Hook.f., Fl. Brit. India 7: 31. 1986. *Echinochloa colonum* var. *frumentacea* Ridley, Fl. Malay Penin. 5: 223. 1925; Fischer, Fl. Madras 3: 1777. 1934.

Annual, culms to 1 m high, slender, erect or decumbent. Leaf blade to 20 × 2 cm; ligule absent. Panicles much robust, nodding; spikes thickened, spikelets to 4 mm long; lower glume ovate, cuspidate, unawned.

Cultivated and wild, often near and in paddy fields in Karimnagar (Naqvi, 2001) and Warangal districts.

Fl. & Fr.: July – October.

Thupakulagudem (WGL), *PVS* & *NRR* 76971 (BSID).

Echinochloa oryzoides (Ard.) Fritsch., Verh. K.K.Zool. Bot. Ges. Wien 41: 742. 1891; Naqvi & Raju, J. Econ. Taxon. Bot. 22: 451-458. 1998.

Culms up to 1 m high. Leaf sheaths up to 20 cm long, ligule 0; blade 20-30 × 0.8-1 cm. Inflorescence up to 15 cm long, racemes irregularly 2-rowed, 1.5-4.5 cm; spikelets 3.5-4 mm long, hispid; lower glume 1.5-2 mm, with a short mucro; outer glume up to 4.5 mm, awn up to 1.5 mm; lower lemma up to 4 mm, awn up to 1.5 cm; upper lemma up to 3.8 mm, aristate.

Frequent weed of paddy fields in winter in Karimnagar and Hyderabad districts.

Fl. & Fr.: October – January.

Laxmipur (KRN), *A. Naqvi* 930104 (KU).

Echinochloa picta (J.Koen.) Michael, Philipp. J. Weed Sci. 5: 18. 1978; Naqvi & Raju, J. Econ. Taxon. Bot. 22: 451-458. 1998. *Echinochloa stagnina* (Retz.) P. Beauv., Ess. Agrost. 53. 161 & 171. 1812; Fischer, Fl. Madras 3: 1777. 1934; Bor, Grass. India 311. 1960. *Panicum stagninum* Retz., Observ. Bot. 5: 17. 1789. *P. crusgalli* sensu Hook. f., Fl . Brit. India 7: 30. 1896 *p.p.* non L. 1753.

An annual stoloniferous grass; culms up to 50 cm high, erect to decumbent, rooting at the nodes; nodes brownish, glabrous. Leaf sheaths 5-10 cm long, loose, glabrous; ligule a fringe of stiff hairs; blade 4-15 × 0.4-0.8 cm, linear-lanceolate, base cordate, apex acuminate, midrib prominent. Spikelike panicle 5-10 cm long, racemes interrupted; rhachis angular, scabrid; spikelets paired, 2-4-rowed, ovate, crowded, acutely acuminate, awned, secund, pubescent; lower glume broadly ovate, bidentate, 5-nerved, scabrid, margins ciliate at apex; upper glume broadly ovate, acuminate, 5-7-nerved, spinulose on nerves, hairy at apex and margins; lower lemma ovate, hairy and scabrid at apex, awned; palea oblong-ovate, scabrid margin. Florets 2, lower male, upper bisexual; lower lemma broadly ovate, boat-shaped with 1.5-4 mm long awn at apex, acuminate, prominently 5-7-nerved, awn twisted, lower palea ovate-elliptic, hyaline, 2-nerved, 2-keeled, keels ciliate; upper lemma ovate, scabrid along the margins, acute, chartaceous; palea similar to upper lemma; lodicules 2; stamens 3, stigmas plumose. Caryopsis broadly ellipsoid.

Occasional in rice fields and marshy areas of Mahbubnagar, Karimnagar (Naqvi and Raju, 1998) and Nalgonda districts.

Fl. & Fr.: September - January.

Rampur (KRN), *A.H.Naqvi* 931011 (KU); Krishna river bank (NLG), *KMS* 9823 (CAL); Erloo to Mahabubnagar (MBNR), *SRS* 104509 (BSID); Dornacheru, Mannanur (MBNR), *SRS* 107458 (BSID).

Note: Michael (1978) and Lazarides (1980) consider that the plants of India and SE Asia mentioned under *Echinochloa stagnina* in various Floras is actually *E. picta* (J.Koen.) Michael, a species closely allied to *E. stagnina* (Retz.) P. Beauv. Noltie (2000) treats both these species as conspecific. This view is followed here.

ELEUSINE Gaertn.

1. Spikes slender, narrow, straight; wild species .... **E. indica**
1. Spikes stout, broad, curved at apex; cultivated species .... **E. coracana**

Eleusine coracana (L.) Gaertn., Fruct. Pl. 1:8. 1789; Hook. f., Fl. Brit. India 7:294. 1896; Fischer, Fl. Madras 3:1839; Bor, Grass. India 492. 1960; Moulik, Grass. Bamb. India 2: 590. *Cynosurus coracanus* L., Syst. Nat. (ed. 10) 875. 1759.

Annual, erect, tufted grass, to 1m high; nodes glabrous. Leaf sheaths glabrous; ligule hairy; leaf blade linear, convolute, overlapping the culm, acuminate, base truncate, glabrous. Spikes 4-7, suberect with their ends or whole spike sometimes incurved; spikelets ovoid, glabrous, 3-5-flowered; glumes oblong, margins scaberulous, acute; lemma broadly ovate; palea ovate; stamens 3. Caryopsis globose, dark red brown, often black.

Commonly cultivated; often run wild.

Fl. & Fr.: July–September.

Eleusine indica (L.) Gaertn., Fruct. 1:8.1789; Hook. f., Fl. Brit. India 7:293.1896; Fischer, Fl. Madras 3:1839.1934; Bor, Grass. India 493. 1960; Moulik, Grass. Bamb. India 2:590.1997. *Cynosurus indicus* L., Sp. ed. 1.72.1753.

Annuals or perennials, culms erect, tufted, to 90 cm high; nodes glabrous. Leaf sheaths glabrous; ligules membranous; leaf blade linear or linear-lanceolate, 9 - 32 × 0.4-0.6 cm, base cordate, tip blunt. Spikes 2-7, digitate or sub-digitate; spikelets 2-5-flowered; lower glume oblong-lanceolate, chartaceous, keeled, 1-3-nerved, acute, scabrid on the keels; upper glume ovate-oblong, chartaceous, keeled, 5-7-nerved; lemmas ovate oblong, chartaceous, 3-nerved; paleas elliptic-oblong, hyaline, 2-keeled, 2-nerved; lodicules 2; stamens 3. Caryopsis ovate.

Common in open places usually near water courses in all districts.

Fl. & Fr.: July–December.

Alampalle (ADB), *TP* & *PVP* 4172; Potchera water falls (ADB), *GO* & *DAM* 5053; Mosra Forest Nursery (NZB), *TP* & *BR* 6081; Ponnala fields (MDK), *CP* 11367; Nagarjunakonda valley (NLG), *KT* 9691 CAL); Near Hussainsagar dam (HYD), *KMS* 8041 (CAL); Narsapur (MDK), *KS* 6785 (CAL); Lingampalle (RR), *MSM* 10732; Amangal to Shadnagar (MBNR), *SRS* 104574 (BSID).

ELYTROPHORUS P. Beauv.

Elytrophorus spicatus (Willd.) A. Camus in Lecomte, Fl. Gen. de 1' Indo-Chine 7: 547. 1923; Fischer, Fl. Madras 3:1843. 1934; Bor, Grass. India 493. 1960; Moulik, Grass. Bamb. India 2: 592. 1997. *Dactylis spicata* Willd. in Ges. Naturf. Freunde Berlin Neue Schrift. 3: 416. 1801.

An annual, decumbent grass; culms tufted, up to 30 cm high; nodes glabrous. Leaf sheaths short, overlapping, glabrous, loose; ligule fringed hairs; blade narrow, 5-15 x 0.2-0.7 cm, glabrous, base rounded, apex acute-acuminate. Catkin like spikes with globose cluster arranged on the rhachis of the inflorescence; spikelets short, 5 mm, 4-6-flowered, laterally compressed; glumes membranaceous, linear to narrowly ovate, 1-keeled, 1-nerved, keels scaberulous running into short awn; lemmas ovate-lanceolate, 3-keeled, 3-nerved, awned; palea broad, truncately 3-lobed, lobes ciliate, keels winged; lodicules 2; stamens 1-3; stigmas plumose. Caryopsis narrowly oblong.

Occasional in ditches and wet places of Hyderabad, Mahabubnagar, Medak and Nizamabad districts.

Fl . & Fr : August- January

Nagaram (NZB), *BR* 9710; Pocharam RF (MDK), *TP & MSG* 12055; Hyderabad (HYD), *MRS* 218 (DD); Ramancheru, Perugunta (MDK), *RG* 96288 (BSID).

ERAGROSTIELLA Bor

1. Leaves from the basal sheaths reaching to the base of the inflorescence .... **E. walkeri**
1. Leaves from the basal sheaths very much shorter, never reaching the base of the inflorescence:
   2. Keels of the palea broadly-winged; leaves filiform, often inrolled, up to 20 cm long .... **E. bifaria**
   2. Keels of the palea narrowly-winged; leaves linear, up to 6 cm long .... **E. brachyphylla**

Eragrostiella bifaria (Vahl) Bor, Indian Forester 66: 270.1940 & Grass. India 494. 1960; Moulik, Grass. Bamb. India 2:594.1997. *Poa bifaria* Vahl, Symb. Bot. 2: 19. 1791. *Eragrostis bifaria* (Vahl) Wight ex Steud., Nom. Bot. ed. 2.1: 562. 1840; Stapf in Hook. f., Fl. Brit. India 7: 325.1896; Fischer, Fl. Madras 3:1828. 1934.

Annuals or perennials, culms erect, tufted, to 90 cm high; nodes glabrous. Leaf sheaths glabrous; ligules membranous or fimbriate; leaf blade linear or acicular, often rolled, base narrow, apex acute, glabrous. Inflorescence spikes; spikelets slightly compressed, ovate-lanceolate or oblong-lanceolate, 5-30-flowered; lower glume oblong–lanceolate, chartaceous, 1-nerved, 1-keeled, keel scabrid; upper glume ovate–lanceolate, chartaceous, 1-nerved; lemmas broadly ovate, chartaceous, 3-nerved; paleas elliptic–ovate, hyaline, 2-keeled, winged, ciliate; lodicules 2; stamens 3. Caryopsis globose.

Common in grass lands, plains and forest peripheries in Adilabad, Nizamabad, Medak, Krimnagar, Ranga Reddi, Nalgonda and Hyderabad districts.

Fl. & Fr.: October – December.

Ankusapuram (ADB), *GO* 4335; Arsapally (NZB), *BR* & *KH* 6487; Nagulabanda (MDK), *CP* 11360; Anantagiri RF (RR), *MSM* 11076; Aklaspur (KRN), *GVS* 25637 (CAL); Nagarjunakonda vlley (NLG), *KT* 9663 (CAL); Hyderabad, *MRS* 305 (DD); Koriguttalu (KMM), *RR* 106045 (BSID).

Eragrostiella brachyphylla (Stapf) Bor, Indian Forester 66: 270. 1940 & Grass. India 494. 1960; Moulik, Grass. Bamb. India 2: 594. 1997. *Eragrostis brachyphylla* Stapf in Hook. f., Fl. Brit. India 7: 327. 1896; Fischer, Fl. Madras 3: 1828. 1934.

Perennials, culms erect, tufted, to 50 cm high, slender; nodes glabrous. Leaf sheaths compressed, glabrous; ligule a ciliate rim; leaf blade linear, 2-6 × 0.1-0.2 cm, base rounded, apex acute, glabrous. Inflorescence solitary raceme; spikelets ovoid-oblong, secund, 5-18-flowered; lower glume linear-lanceolate, acute, 1-nerved, keel scaberulous; upper glume lanceolate, acute, 1-nerved, keel more or less smooth; lemma broadly ovate, 3-nerved, keeled, margins hyaline; palea obovate, keels ciliate, margins inflexed, hyaline; lodicules 2; stamens 3. Caryopsis ellipsoid.

Common in waste lands, forest plains and on hill slopes in Adilabad, Nizamabad, Ranga Reddi, Hyderabad, Karimnagar and Warangal districts.

Fl. & Fr.: July-November.

Vempalli (ADB), *GO* & *PVP* 4771; Kamareddy (NZB), *BR* 9279; Anantagiri RF (RR), *TP* & *MSM* 11076; Pakhal (WGL), *KMS* 13160 (CAL); Manuguru RF (KMM), *RCS* 104255 (BSID); Bandirevu RF (KMM), *RCS* 104300 (BSID).

Eragrostiella walkeri (Stapf) Bor, Indian Forester 66: 270. 1940 & Grass. India 495. 1960; Moulik, Grass. Bamb. India 2:594. 1997. *Eragrostis walkeri* Stapf in Trin., Fl. Ceyl. 5: 298. 1900; Fischer, Fl. Madras 3:1828. 1934.

Perennial, culms erect, tufted, to 80 cm high, tuft of leaf sheaths at the base; nodes glabrous. Leaf sheaths glabrous; ligules ciliate, membranous; leaf blade filiform, convolute, 2.5-6.8 × ca. 0.1 cm, base rounded, glabrous, margin scabrous, apex acuminate. Inflorescence terminal, solitary racemes; rhachis flattened; spikelets secund, ovate, subsessile, margin scaberulous, 2-8-flowered; lower glume linear-lanceolate, acute, 1-nerved; upper glume boat-shaped, keeled, acute, 1-nerved; lemmas broadly ovate, margin hyaline, 3-nerved; palea spatheolate, keels ciliate, margin inflexed, hyaline; stamens 3. Caryopsis ellipsoid.

Common in forest plains and on open slopes in Mahabubnagar district.

Fl. & Fr.: July-December.

Amrabad (MBNR), *BSS* & *SKB* 32210.

ERAGROSTIS Wolf

1. Spikelets breaking up from above downwards; rhachis fragile:
    2. Lemma ciliate on the margins .... **E. ciliata**
    2. Lemma not ciliate on the margins:
        3. Keels of the palea ciliate:
            4. Panicle spiciform or compact and dense:
                5. Lemma ciliate on the keels; annuals .... **E. ciliaris**
                5. Lemma not ciliate on the keels; perennials:

6. Panicle very dense, cylindric; lemma rounded at apex .... **E. riparia**
6. Panicle more or less dense, interrupted; lemma acute at apex .... **E. coarctata**

4. Panicle effuse:
7. Plants viscous .... **E. viscosa**
7. Plants not viscous .... **E. tenella**

3. Keels of the palea not ciliate:
8. Panicle lax, open thyrsiform .... **E. aspera**
8. Panicles not thyrsiform .... **E. japonica**

1. Spikelets breaking up from below upwards; rhachis tough:
9. Annuals:
10. Lemmas up to 1.5 mm long:
11. Plants glandular:
12. Lowest branches of the panicle fascicled or sub whorled .... **E. pilosa**
12. Lowest branches not whorled .... **E. minor**
11. Plants eglandular .... **E. gangetica**
10. Lemmas over 1.5 mm long:
13. Spikelets 2 mm or more wide, ovate or ovate-oblong:
14. Plants with or without glands; upper glume 3-nerved, paleas persistent .... **E. cilianensis**
14. Plants eglandular; upper glume 1-nerved, paleas deciduous .... **E. unioloides**
13. Spikelets 0.75-2 mm wide, lanceolate .... **E. tremula**
9. Perennials:
15. Lemmas up to 1.8 mm long:
16. Spikelets 1.7-2.5 mm wide, paleas deciduous .... **E. atrovirens**
16. Spikelts 1-1.2 mm wide; paleas persistent .... **E. nutans**
15. Lemmas 2 mm or longer .... **E. tenuifolia**

Eragrostis aspera (Jacq.) Nees, Fl. Afr. Austr. 911. 408. 1841; Stapf in Hook. f., Fl. Brit. India 7:314.1896; Fischer, Fl. Madras 3:1825.1934; Bor, Grass. India 501. 1960; Moulik, Grass. Bamb. India 2:597.1997. *Poa aspera* Jacq., Hort. Bot. Vindob 332.t.56.1776.

Annuals, culms erect, tufted, to 70 cm high. Leaf sheaths glabrous; ligules ring of hairs; leaf blade 7-20 × 0.2-0.8 cm, linear-lanceolate, base rounded, margins serrulate, apex acuminate. Panicles lax, branches numerous, filiform; spikelets linear-oblong, 4-15-flowered, breaking up from above downwards; lower glume ovate-lanceolate, chartaceous, acute, 1-keeled, 1-nerved; upper glume ovate-lanceolate, acute, chartaceous, bifid, purple at apex, 1-keeled, 1-nerved; lemmas broadly ovate, markedly truncate, not ciliate on margins; paleas obovate, keeled, scabrid, hyaline, 2-nerved; lodicules 2; stamens 3; stigma feathery. Caryopsis sub-globose.

Occasional in forest plains, waste lands and ground nut fields in Nizamabad and Hyderabad districts.

Fl. & Fr.: July – November.

Eragrostis atrovirens (Desf.) Trin. ex Steud., Nom. Bot. ed. 2.1:562.1840; Bor, Grass. India 503. 1960; Moulik, Grass. Bamb. India 2: 597. 1997. *Poa atrovirens* Desf., Fl. Atlant. 1:73. t.14. 1798.

Perennials, culms erect, tufted, to 1.3 m high; nodes glabrous. Leaf sheaths glabrous; ligules membranous; leaf blade 10-30 × 0.3-0.4 cm, linear–lanceolate, base rounded or cordate, apex acuminate, scabrellate. Panicles contracted, often lax, 5-20 cm long; spikelets short-pedicelled, ovate, 3-30-flowered; lower glume lanceolate, chartaceous, 1-keeled, keel scabrid, 1-nerved; upper glume ovate–lanceolate, chartaceous, 1-keeled, 1-nerved; lemmas ovate, chartaceous, acute, 3-nerved; paleas lanceoloate, keeled, hyaline, keels ciliate, 2-nerved; lodicules 2; stamens 3; stigma plumose. Caryopsis ellipsoid.

Common along stream banks and in marshy areas throughout the State.

Fl. & Fr.: July - February.

Kadam river bank (ADB), *TP* & *PVP* 4142; Sarvaipet (ADB), *GO* 4503; Sirnapalli river (NZB), *BR* & *GO* 9077; Pochammralu (MDK), *BR* & *CP* 11497; Narsapur tank (MDK), *TP* & *MSG* 12011; Pocharam tank (MDK), *TP* & *MSG* 12039; Mohammadabad (RR), *MSM* 10538; Appapur (MBNR), *AMR* 25700; Near Neelampalli, Mahadevpur (KRN), *NRR* & *TRS* 83712 (BSID); Mallelatheertham (MBNR), *SRS* 111414 (BSID).

Eragrostis cilianensis (Alt.) Vignolo-Lutati ex Janchen, Mitt. Naturwiss. Vereins Univ. Wien n.s. 5: 110. 1907; Fischer, Fl. Madras. 3:1827.1934; Bor, Grass. India 503.1960; Moulik, Grass. Bamb. India 2: 598. 1997. *Poa cilianensis* Alt., Fl. Pedem. 2:246. t. 91. f. 2. 1785. *Eragrostis major* (L.) Host., Gram. Austr. 4:14 & 24.1809; Stapf in Hook f., Fl. Brit. India 7: 320. 1896.

Annuals, culms erect or ascending, tufted, to 70 cm high, leafy; nodes glabrous. Leaf sheath hairy at mouth; ligule ciliate; leaf blade 5-12 × 0.4-0.6 cm, linear-lanceolate, base cordate, margins glandular, apex acuminate. Panicles contracted, often lax, 10-20 cm long, green; branches filiform, scaberulous; pedicels filiform, scaberulous; spikelets oblong, 10-25-flowered; lower glume oblong–lanceolate, chartaceous, 1-keeled, keels scaberulous, 1-nerved; upper glume ovate-lanceolate, acute, chartaceous, 3-nerved; lemmas ovate, chartaceous, 3-nerved, glands on the keels; paleas oblong, 2-keeled, 2-nerved, keels ciliate; lodicules 2; stamens 3; stigma plumose, purple. Caryopsis sub-globose.

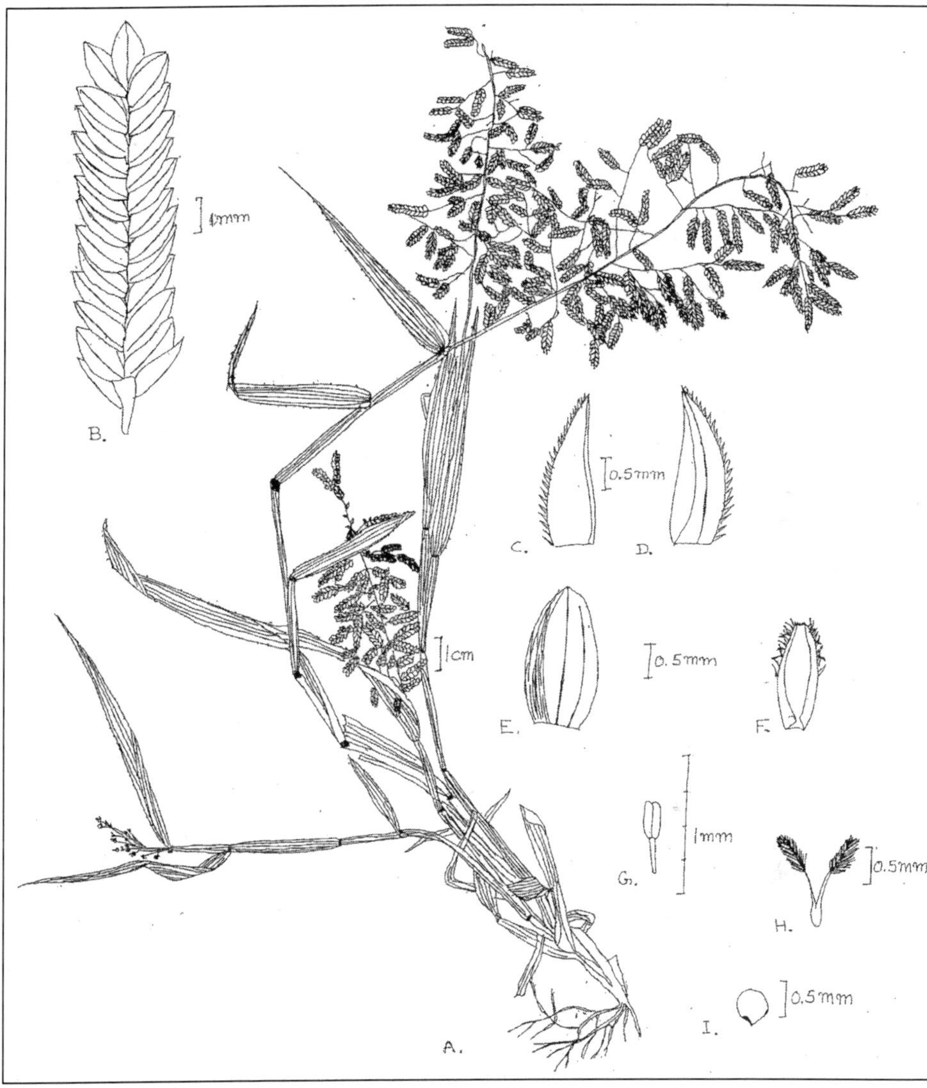

Figure 145: Eragrostis cilianensis (Alt.) Vignolo-Lutati ex Janchen

A. Twig, B. Spikelet, C. Lower glume, D. Upper glume, E. Lemma, F. Palea, G. Stamen, H. Pistil, I. Caryopsis.

Occasional along the forest margins in shady localities in Adilabad, Ranga Reddi, Hyderabad and Mahabubnagar districts.

Fl. & Fr.: September – December.

Sattenapalle (ADB), *TP* & *PVP* 4152; Vikarabad (RR), *MRS* 105 (DD).

Eragrostis ciliaris (L.) R. Br. in Tuckey Narr. Exp. Congo, App. 478.1818, var. **ciliaris**; Stapf in Hook. f., Fl. Brit. India 7:314.1896; Fischer, Fl. Madras 3: 1825. 1934; Bor, Grass. India 506. 1960; Moulik, Grass. Bamb. India 2: 598. 1997. *Poa ciliaris* L., Syst. Nat. ed. 10. 2: 875. 1759.

Annuals, clums erect, tufted, to 60 cm high, slender; nodes glabrous. Leaf sheaths ciliate at mouth; ligules short silky hairy; leaf blade 2-5 × 0.2-0.4 cm, linear, base rounded, apex acuminate, glabrous. Panicles spiciform or compact, cylindric, 2-6 cm long; spikelets compressed, crowded, ovate, 3-9-flowered; lower glume ovate-acute, chartaceous, 1-keeled, keel scabrid, 1-nerved; upper glume ovate-lanceolate, chartaceous, 1-keeled, keel scabrid, 1-nerved; lemmas broadly ovate, chartaceous, 3-nerved, ciliate; palea obovate, truncate, soft ciliate, 2-keeled, 2-nerved; lodicules 2; stamens 3. Caryopsis ovoid.

Common in plains and forest margins in Ranga Reddi district.

Fl. & Fr.: August - November.

Eragrostis ciliata (Roxb.) Nees, Agrost. Bras. 512.1829; Stapf in Hook.f., Fl. Brit. India 7:373.1896; Fischer, Fl. Madras. 3:1825.1934; Bor, Grass. India 506. 1960. *Poa ciliata* Roxb., Fl. Ind. 2: 256. 1820.

Annuals or perennials, tufted, erect, to 80 cm high, geniculate; nodes glabrous. Leaf sheaths glabrous, bearded at mouth; ligule a few hairs; leaf blade 5-15 × 0.2-0.4 cm, linear-lanceolate, flat or convolute, long ciliate towards base, margins inrolled, apex acuminate, glabrous. Panicles compact, cylindrical, spiciform, 4-6 cm long; spikelets compressed, ovate-lanceolate, 6-12-flowered, bisexual; lower glume ovate, chartaceous, 1-nerved, acute, keeled; upper glume ovate, chartaceous, 1-nerved, acute; lemma obovate, chartaceous, 3-nerved, ciliate; palea ovate-lanceolate, hyaline, 2-keeled, 2-nerved, long ciliate; stamens 2; stigma plumose, yellow to green. Caryopsis ovate-elliptic.

Occasional in open places in Karimnagar, Medak, Hyderabad, Nalgonda and Mahabubnagar districts.

Fl. & Fr.: May-September.

Aklaspur (KRN), *GVS* 20237 (MH); Krishna river bank (NLG), *KMS* 9819 (MH).

Eragrostis coarctata Stapf in Hook. f., Fl. Brit. India. 7:313.1896; Fischer, Fl. Madras 3:1825.1934; Bor, Grass. India 507.1960; Moulik, Grass. Bamb. India 2:601.1997.

Perennials, culms erect, tufted, to 1.1 m high, geniculate; nodes glabrous. Leaf sheaths terete, ciliate at mouth; ligules hairy; leaf blade 4-12 × 0.2-0.5 cm, linear–lanceolate, base rounded, apex acuminate, glabrous to sparsely tuberculate hairy.

Panicles compact, dense, cylindrical, 5-12 cm long; spikelets short pediceled, oblong, grey coloured, 5-12-flowered, breaking up from above downwards; lower glume lanceolate, chartaceous, 1-keeled, keels scaberulous, 1-nerved; upper glume ovate-lanceolate, chartaceous, 1-keeled, keel scaberulous, 1-nerved; lemmas broadly ovate, chartaceous, 3-nerved, ciliate on the margins; paleas oblong, 2-keeled, 2-nerved, ciliate on the keels; lodicules 2; stamens 3; stigma plumose. Caryopsis ovoid.

Occasional in stream beds and marshy areas in forests in Ranga Reddi, Hyderabad, Karimnagar and Mahabubnagar districts.

Fl. & Fr.: December - July.

Mohammadabad (RR), *MSM* 10550; Aklaspur (KRN), *GVS* 20236 (CAL); Upparapally (HYD), *MRS* 260 (DD); Farahabad (MBNR), *MVS* & *MB* 38026; Lakshmipuram RF (KMM), *RCS* 98910 (BSID); Rathamhutta hills (KMM), *RCS* 99059 (BSID).

Eragrostis gangetica (Roxb.) Steud., Syn. Pl. Glum. 1: 266. 1854; Fischer, Fl. Madras 3:1826.1934; Bor, Grass. India 508. 1960; Moulik, Grass. Bamb. India 2:602.1997. *Poa gangetica* Roxb., Fl. Ind. 1:341.1820. *Eragrostis stenophylla* Hochst. ex Miq. in Ver. Konink. Nederl. Inst. 3, 4, 39. 1851; Stapf in Hook. f., Fl. Brit. India 7: 318.1896.

Annuals, culms erect, tufted, to 80 cm high, slender; nodes glabrous. Leaf sheaths terete, more or less ribbed, hairy at mouth; ligules membranous; leaf blade 5-15 × 0.2-0.4 cm, linear-lanceolate or linear, base cordate, margins infolded and villous hairy at base, apex acute or acuminate. Panicles lax, 5-15 cm long, branches slender; spikelets ovate-oblong, 6-15-flowered, grey; lower glume ovate, chartaceous, 1-keeled, keel scaberulous, 1-nerved; upper glume oblong-lanceolate, chartaceous, 1-keeled, keel scaberulous, 1-nerved; lemmas ovate-lanceolate, chartaceous, 3-nerved; paleas elliptic-lanceolate, 2-keeled, 2-nerved, keels ciliate; lodicules 2; stamens 3; stigma plumose. Caryopsis orbicular.

Common in forest plains throughout the State.

Fl. & Fr.: July – January.

Satnella (ADB), *GO* & *PVP* 4222; Velutla RF (NZB), *BR* 9525; Mudheli RF (NZB), *BR* & *CP* 7206; Pocharam tank (MDK), *TP* & *MSG* 12042; Choutkoor (MDK), *TP* & *MSG* 11948; Saleswaram (MBNR), *BSS* & *SKB* 30584; Appapur (MBNR), *BSS* & *SKB* 32297; Kodimial (KRN), *GVS* 20132; Pakhal (WGL), *KMS* 11762 (CAL) & *ANH* 15911 (CAL).

Eragrostis japonica (Thunb.) Trin., Mem. Acad. Sci. Peterst., Ser. 6. 1: 405. 1831; Fischer, Fl. Madras 3: 1826. 1934; Bor, Grass. India 509. 1960; Moulik, Grass. Bamb. India 2:603. 1997. *Poa japonica* Thunb., Fl. Jap. 51. 1784. *Eragrostis interrupta* sensu Stapf in Hook. f., Fl. Brit. India. 7: 316. 1896. incl. vars., non (R.Br.) P.Beauv. 1812. *E. diarrhena* (Schult.) Steud., Syn. Pl. Glumac. 1: 266. 1854; Fischer, Fl. Madras 3: 1826. 1934; Bor, *loc. cit. Poa diarrhena* Schult., Mant. 2: 616. 1824. *Eragrostis diplochoides* Steud., Syn. Pl. Glumac. 1: 268. 1854; Bor, *loc. cit.* 508.

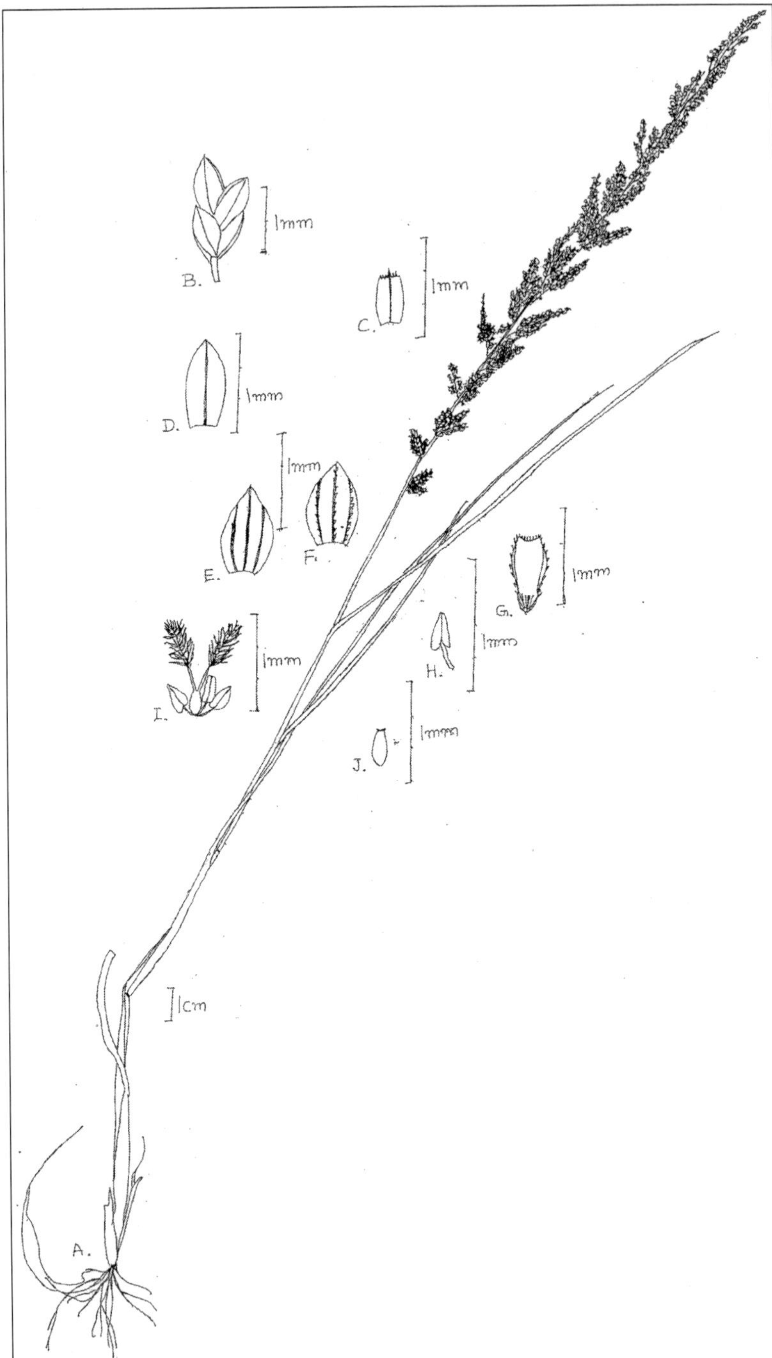

Figure 146: Eragrostis japonica (Thunb.) Trin

A. Twig, B. Spikelet, C. Lower glume, D. Upper glume, E & F. Lemma front & Back views, G. Palea, H. Stamen, I. Pistil with stamens, J. Caryopsis.

Annual or perennials, culms erect, tufted, to 1.3 m high, ascending; nodes glabrous. Leaf sheaths 5-12 cm, glabrous; ligule membranous; leaf blade 10-20 × 0.2-0.4 cm, linear-lanceolate, involute, base rounded, apex acuminate, glaucous. Panicles contracted, dense, spiciform, 15-35 cm long; pedicels capillary, scaberulous; spikelets oblong-lanceolate, 5-15-flowered; lower glume ovate-lanceolate, chartaceous, 1-keeled, keels scaberulous, 1-nerved; upper glume obovate, chartaceous, 1-keeled, keel scaberulous, 1-nerved; lemmas obovate, chartaceous, 3-nerved; paleas oblong-lanceolate, 2-keeled, 2-nerved, keels ciliate; stamens 2. Caryopsis ellipsoid.

Occasional in stream beds and moist places of forests in Adilabad, Nizamabad, Medak, Karimnagar and Mahabubnagar districts.

Fl. & Fr.: October-June.

Sone (ADB), *TP* & *PVP* 4057; Gandhari (NZB), *TP* & *BR* 6347; Choutkoor fields (MDK), *TP* & *MSG* 11905; Pathur fields (MDK), *TP* & *MSG* 12117; Pocharam RF (MDK), *TP* & *MSG* 12056 & 12065; Uttanoor (MBNR), *MVS* & *DV* 40722.

Eragrostis minor Host., Icon. Descr. Austr. 4. 15. 1809; Stapf in Hook. f., Fl. Brit. India 7:321. 1896. *E. poaeoides* P. Beauv., Ess. Agrost. 162. 1812; Fischer, Fl. Madras. 3: 1827. 1934; Bor, Grass. India 512. 1960.

Annuals, culms erect, tufted, to 45 cm high, leafy; nodes glabrous. Leaf sheaths 2-5 cm long, ciliate at mouth, glabrous; ligule a pubescent ridge; leaf blade 7-15 × 0.1-0.2 cm, linear, base narrow, usually glandular along the margin, apex acuminate, glabrous. Panicle lax, open, 5-25 cm long, branches filiform, false whorled at base; spikelets ovate-oblong, 6-16-flowered; glumes subequal, ovate, acute, 1-nerved; lemma broadly ovate, acute, 3-nerved; palea obovate-oblong, keels scabrid or slightly ciliate; lodicules 2; stamens 3. Caryopsis broadly oblong, dorsally rounded or slightly flattened.

Rare in crop fields and waste lands in Adilabad, Hyderabad, Medak and Nalgonda districts.

Fl. & Fr.: July – December.

Bellampalli (ADB), *PVP* 9445; Choutkoor fields (MDK), *TP* & *MSG* 11946; Pocharam RF (MDK), *TP* & *MSG* 12062; Golkonda (HYD), *MRS* 303 (DD); Nagarjunakonda valley (NLG), *KT* 9687 (CAL).

Eragrostis nutans (Retz.) Nees ex Steud., Nom. Bot. ed. 2.563. 1840; Fischer, Fl. Madras 3:1826.1934; Bor, Grass. India 511. 1960; Moulik, Grass. Bamb. India 2: 605. 1997. *Poa nutans* Retz., Observ. Bot. 4:19.1786. *Eragrostis elegantula* Nees ex Steud., Syn. Pl. Glum. 1: 266. 1854; Stapf in Hook. f., Fl. Brit. India 7: 318. 1896.

Perennials, culms erect, tufted, to 90 cm high, slender; nodes glabrous. Leaf sheaths 1-5 cm long, glabrous; ligules hairy; leaf blade 6-15 × 0.3-0.5 cm, linear-lanceolate, glabrous, base rounded, apex acuminate. Panicles lax, up to 12 cm long, black or grey coloured; spikelets oblong-lanceolate, 6-15-flowered; lower glume ovate, chartaceous, 1-keeled, keels minutely ciliate, 1-nerved; upper glume oblong, chartaceous, glabrous, 1-nerved; lemmas ovate, chartaceous, 3-nerved; palea obovate, 2-keeled, 2-nerved, ciliate on the keels; stamens 3; stigma plumose. Caryopsis ellipsoid.

Common in moist habitats in Karimnagar, Medak and Ranga Reddi districts.

Fl. & Fr.: June – April.

Kondapur (RR), *MSM* 10589; Pegarikutta, Narsapur (MDK), *KMS* 6655 (CAL); Aklaspur (KRN), *GVS* 22511 (MH).

Eragrostis pilosa (L.) P. Beauv., Ess. Agrost. 71: 162.175.1812; Stapf in Hook. f., Fl. Brit. India 7:323.1896; Fischer, Fl. Madras 3: 1827. 1934; Bor, Grass. India 512. 1960; Moulik, Grass. Bamb. India 2:606. 1997. *Poa pilosa* L., Sp. Pl. 68.1753.

Annuals, culms erect, tufted, to 1.2 m high, slender; nodes glabrous. Leaf sheaths 3-12 cm long, glabrous; ligules fimbriate, membranous; leaf blade linear-lanceolate, 5-25 × 0.2-0.4 cm, base rounded, margins smooth or scaberulous, apex finely acuminate. Panicles lax, 5-25 cm long, pedicels capillary, filiform; spikelets ovate-lanceolate or linear-oblong, 4-10-flowered; lower glume ovate-lanceolate, chartaceous, 1-keeled, 1-nerved; upper glume lanceolate, chartaceous, 1-nerved; lemmas oblong-lanceolate, chartaceous, 3-nerved; paleas ovate, keeled, 2-nerved, keels scabrid; lodicules 2; stamens 3; stigma plumose, purplish yellow. Caryopsis ovate.

Occasional in forest margins and plains in Adilabad, Hyderabad, Warangal, Ranga Reddi and Mahabubnagar districts.

Fl. & Fr.: August – December.

Bheemaram RF (ADB), *TP* & *GO* 5465; Narsapur (MDK), *TP* & *MSG* 12018; Choutkoor (MDK), *TP* & *MSG* 11949; Kusumasamudram (RR), *MSM* 11029; Appapur (MBNR), *AMR* 25699; Upparpalli (HYD), *MRS* 304 (DD).

Eragrostis riparia (Willd.) Nees, Agrost. Bras. 512.1829; Fischer, Fl. Madras 3: 1825.1934; Bor, Grass. India 513. 1960; Moulik, Grass. Bamb. India 2: 606. 1997. *Poa riparia* Willd., Ges. Natur. Freunde Berlin, Neue, Schrift. 4: 185. 1803. *Eragrostis tenella* P. Beauv. var. *riparia* Stapf in Hook. f., Fl. Brit. India 7: 315. 1896.

Perennials, culms erect, tufted, to 90 cm high, ascending; nodes glabrous. Leaf sheaths 1.5-5 cm long, glabrous, hairy at mouth; ligules a row of hairs; leaf blade 2-4 × 0.2-0.4 cm, linear-lanceolate, sparsely hairy, base rounded, apex acute, hairy. Panicles spiciform, dense, 2.5-7.5 cm long; spikelets ovate, oblong, 5-10-flowered; lower glume ovate-linear, chartaceous, 1-keeled, keel scaberulous, 1-nerved; upper glume oblong, chartaceous, 1-keeled, keel scaberulous, 1-nerved; lemmas oblong, chartaceous, 3-nerved; paleas oblong, 2-keeled, 2-nerved, keels ciliate; lodicules 2; stamens 3; stigma plumose, greenish brown. Caryopsis elliptic.

Occasional in marshy areas especially near sandy moist soils in Nalgonda district.

Fl. & Fr.: May-November.

Krishna river bank (NLG), *KMS* 9819 (CAL) .

Eragrostis tenella (L.) P. Beauv. ex Roem. & Schult., Syst. Veg. 2:576.1871. var. **tenella**; Bor, Grass. India 513. 1960; Moulik, Grass. Bamb. India 2: 608. 1997. *Poa tenella* L., Sp. Pl. 69. 1753. *Eragrostis plumosa* (Retz.) Link, Hort. Berol. 1: 192. 1827; Fischer, Fl. Madras 3: 1826. 1934. *Eragrostis tenella* (L.) P. Beauv. ex Roem. & Schult. var. *plumosa* (Retz.) Stapf in Hook. f., Fl. Brit. India. 7: 315. 1896.

Annuals, culms erect, tufted, to 80 cm high, slender; nodes glabrous. Leaf sheath 1.5-6 cm long, hairs at mouth; ligules membranous; leaf blade linear-acuminate, base rounded, acuminate. Panicles 20 cm long, effuse; spikelets oblong-ovate, 5-9-flowered; lower glume lanceolate, chartaceous, 1-nerved; upper glume ovate-lanceolate, chartaceous, 1-nerved; lemmas oblong-ovate, chartaceous, 3-nerved; paleas lanceolate or elliptic-lanceolate, 2-keeled, 2-nerved, keels ciliate; stamens 3. Caryopsis ovate-elliptic.

Common in hill slopes, waste places and plains throughout the State.

Fl. & Fr.: April – November.

Ankusapuram (ADB), *GO* 4349; Choutkoor fields (NZB), *TP* & *MSG* 11906; Manchippa RF (NZB), *TP* & *BR* 6025; Saleswaram (MBNR), *BSS* & *SKB* 34801; Aklaspur (KRN), *GVS* 22491 (MH); Pakhal (WGL), *KMS* 11667 (MH); Katasintha (HYD), *KMS* 6008 (MH); Nagarjunakonda (NLG), *BVRS* 102 (NU).

Key to Varieties

1. Panicle loose with spreading branches; axis more or less hairy at nodes .... var. **tenella**
1. Panicle contracted with ascending branches; axis glabrous .... var. **insularis**

Eragrostis tenella (L.) P. Beauv. ex Roem. et Schult. var. **insularis** C.E.Hubbard, Kew Bull. 654. 1939; Bor, Grass. India 513.1960; Moulik, Grass. Bamb. India 2: 608.1997.

Annuals, culms erect, tufted, to 30 cm high, slender; nodes glabrous. Leaf sheaths 2-4 cm, hairy at mouth; ligule membranous, fimbriate; leaf blade 4-10 × 0.3 cm, linear-lanceolate, base rounded, acuminate, glabrous. Panicles 5-15 cm, contracted; spikelets oblong-ovate, 3 × 1 mm, 5-9-flowered; lower glume lanceolate, 0.5 × 0.3 mm, chartaceous, 1-nerved; upper glume ovate-lanceolate, 0.8 × 0.3 mm, chartaceous, 1-nerved; lemmas broadly ovate, 1 × 0.8 mm, chartaceous, 3-nerved; paleas lanceolate or elliptic-lanceolate, 1 × 0.5 mm, hyaline, 2-keeled, 2-nerved, keels ciliate. Caryopsis ovate-elliptic.

Rare in plains in shady localities in Medak, Mahabubnagar (Sadasivaiah & *al.*, 2013) and Hyderabad districts (Ramana, 2010).

Fl. & Fr.: April – November.

Kollapur (MBNR), *BR* & *BSS* 32918; Subedari (WGL), *RKP* 105235 (BSID); Medak (MDK),

Eragrostis tenuifolia Hochst. ex Steud., Syn. Pl. Glum. 1:268. 1854; Stapf in Hook. f., Fl. Brit. India 7:322.1896; Fischer, Fl. Madras 3: 1827. 1934; Bor, Grass. India. 514.1960; Moulik, Grass. Bamb. India 2: 609. 1997.

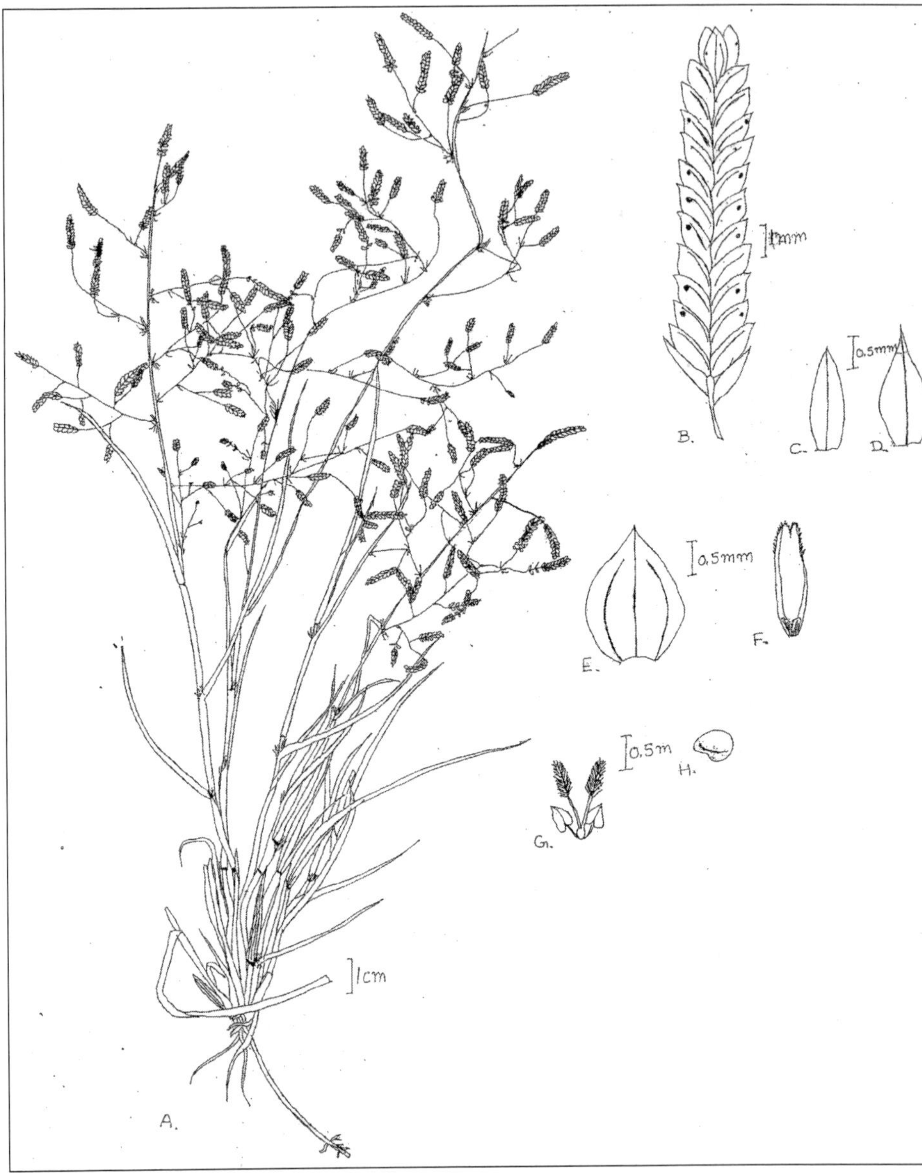

Figure 147: Eragrostis tremula Hochst. ex Steud.

A. Twig, B. Spikelet, C. Lower glume D. Upper glume, E. Lemma, F. Palea, G. Pistil with stamens, H. Caryopsis.

Perennials, culms erect, slender, tufted, to 60 cm high; nodes glabrous. Leaf sheaths 2-14 cm long, glandular, sparsely hairy; ligule ciliate; leaf blade 4-7 × 0.1-0.3cm, linear glaucous, glabrous, base rounded, apex acuminate, glabrous. Panicles lax, 5-15 cm long; spikelets linear-oblong, 6-13-flowered, breaking up from below upwards, pedicels capillary, glandular; lower glume minute, ovate, linear, 1-nerved, lobed; upper glume ovate-oblong, 1-nerved; lemmas ovate-elliptic, 3-nerved, mucronate; palea oblanceolate, 2-nerved, 2-keeled, keels scabrid, hyaline, persistent; lodicules 2; stamens 3; stigma plumose, greenish to brown. Caryopsis oblong.

Rare in open dry places and along road sides in Adilabad, Medak and Hyderabad districts.

Fl. & Fr.: July – October.

Ankusapurm (ADB), *GO* 4349.

Eragrostis tremula Hochst. ex Steud., Syn. Pl. Glum. 1:269.1854; Stapf in Hook. f., Fl. Brit. India 7:320.1896; Fischer, Fl. Madras 3: 1827. 1934; Bor, Grass. India 514. 1960; Moulik, Grass. Bamb. India 2: 609. 1997.

Annuals or perennials, culms erect, tufted, to 90 cm high, ascending; nodes glabrous. Leaf sheaths 1.5-5 cm long, bearded at mouth; ligule fimbriate; leaf blade 5-15 × 0.2-0.5 cm, linear, flat, base cordate, apex acuminate, glabrous. Panicles up to 20 cm long, effuse, filiform; pedicels capillary, scaberulous; spikelets oblong, 15-50-flowered, breaking up from below upwards; lower glume oblong, chartaceous, 1-nerved, keeled; upper glume oblong, chartaceous, 1-keeled, 1-nerved; lemmas ovate-oblong, chartaceous, 3-nerved; paleas elliptic-oblong, 2-keeled, 2-nerved, keels scabrid; lodicules 2; stamens 3. Caryopsis globose.

Occasional in forest plains in Medak, Hyderabad, Karimnagar and Mahabubnagar districts.

Fl. & Fr.: June - November.

Pocharam (MDK), *TP* & *MSG* 12034; Ramayampeta (MDK), *BR* & *CP* 11520; Kudurpalli (KRN), *SLK* 70768 (NBG); Guddaeguda – Koilkonda (MBNR), *SRS* 110797 (BSID).

Eragrostis unioloides (Retz.) Nees ex Steud., Syn. Pl. Glum. 1: 264. 1854; Fischer, Fl. Madras. 3: 1826.1934; Bor, Grass. India 515. 1960; Moulik, Grass. Bamb. India 2: 609. 1997. *Poa unioloides* Retz., Observ. Bot. 5:19.1789. *Eragrostis amabilis* auct. non Wight & Arn. ex Nees 1838: Stapf in Hook. f., Fl. Brit. India. 7:317.1896.

Annuals or perennials, culms erect or ascending, tufted, to 90 cm high; nodes glabrous. Leaf sheaths 3-5 cm long, glabrous; ligule membranous, leaf blade 5-10 × 0.2-0.3 cm, linear-oblong, glabrous, base cordate, apex acuminate. Panicles lax, 5-20 cm long, purplish, branches filiform; spikelets ovate, 10-30-flowered, breaking up from below upwards; lower glume ovate-oblong, chartaceous, 1-keeled, keel scaberulous, 1-nerved; upper glume ovate-oblong, chartaceous, 1-keeled, keel scaberulous, 1-nerved; lemmas broadly ovate, chartaceous, 3-nerved; paleas oblong-elliptic, 2-keeled, 2-nerved, keels scabrid; lodicules 2; stamens 3; stigma plumose, purplish brown. Caryopsis obovoid.

Common in wet lands, streams and in moist habitats throughout the State.

Fl. & Fr.: July – April.

Alampalli (ADB), *TP* & *PVP* 4170; Mosra RF (NZB), *TP* & *BR* 6111; Annaram (NZB), *BR* & *CPR* 7190; Narsapur tank (MDK), *TP* & *MSG* 12001 & *KMS* 6707 (CAL); Mohammadabad RF (RR), *MSM* & *KH* 10438, 10533; Pakhal RF (WGL), *KMS* 11662 (CAL); Bandirevu RF (KMM), *RCS* 99100 (BSID); Kudurupalli (KRN), *SLK* 70764 (NBG); Guddaeguda – Koilkonda (MBNR), *SRS* 116774 (BSID).

Eragrostis viscosa (Retz.) Trin., Mem. Acad. Sci. Petersb. Ser. 6.1:397. 1830; Fischer, Fl. Madras 3:1826. 1934; Bor, Grass. India 515. 1960; Moulik, Grass. Bamb. India 2: 611.1997. *Poa viscosa* Retz., Observ. Bot. 4:20.1786. *Eragrostis tenella* var. *viscosa* Stapf in Hook.f., Fl. Brit. India 7: 315. 1896.

Annuals, culms erect or ascending, tufted, viscous below, to 70 cm high; nodes glabrous. Leaf sheaths 1.5-6 cm long, glabrous; ligule membranous; leaf blade 3-10 × 0.2-0.4 cm, linear, convolute, glabrous, base rounded, apex acuminate. Panicles up to 20 cm long, narrowly cylindric, glandular; spikelets oblong-lanceolate, 4-20-flowered, often purplish; lower glume linear, chartaceous, 1-keeled, keel scaberulous, 1-nerved; upper glume oblong, chartaceous, 1-keeled, keel scaberulous, 1-nerved; lemmas ovate, chartaceous, 3-nerved; paleas linear, ciliate on the keels, 2-keeled, 2-nerved; lodicules 2; stamens 3; stigma plumose, brownish. Caryopsis ovoid.

Common in wastelands and forest plains throughout the State.

Fl. & Fr.: April – November.

Sone (ADB), *TP* & *PVP* 4020; Choutkoor field (MDK), *TP* & *MSG* 11915; Rangammagudem RF (RR), *MSM* & *KH* 10584; Hyderabad (HYD), *MRS* 301 (DD); Kambalpally (NLG), *BVRS* 296 (NU); Ramavaram (KMM), *RR* 112584 (BSID).

ERIOCHLOA Kunth

Eriochloa procera (Retz.) C.E. Hubb., Kew Bull. 1930: 256. 1930; Fischer, Fl. Madras 3: 1767. 1934; Bor, Grass. India 312. 1960; Moulik, Grass. Bamb. India 1: 102. 1997. *Agrostis procera* Retz., Observ. Bot. 4:19. 1786. *Eriochloa polystachya* Hook. f., Fl. Brit. India 7: 20. 1896.

A perennial, tufted grass; culms up to 50 cm high, creeping at base; nodes glabrous. Leaf sheaths 4-15 cm long, loose, glabrous; ligule fimbriate; blade 5-20 x 0.4-0.7 cm, linear, flat, base rounded, apex acuminate, glabrous. Panicle up to 10 cm long, lax; spikelets 3 mm, ellipsoid; lower glume obscure; upper glume elliptic, equal to spikelet, pilose, 3-5-nerved; lower lemma lanceolate-elliptic, equal to spikelet, pilose, male or empty; upper lemma crustaceous, ovate, apical barbellate mucro at apex; palea crustaceous, ovate; lodicules 2; stamens 3, stigmas plumose. Caryopsis oblong-ellipsoid, dorsally compressed.

Common grass near stream and marshy places of all districts.

Fl. & Fr.: August- October.

Sarvaipet (ADB), *GO & PVP* 4820; Nizamsagar project (NZB), *BR & CPR* 9243; Yedapally-Bodhan Road (NZB), *BR & KH* 6478; Mosra (NZB), *TP & BR* 6067; Narsapur

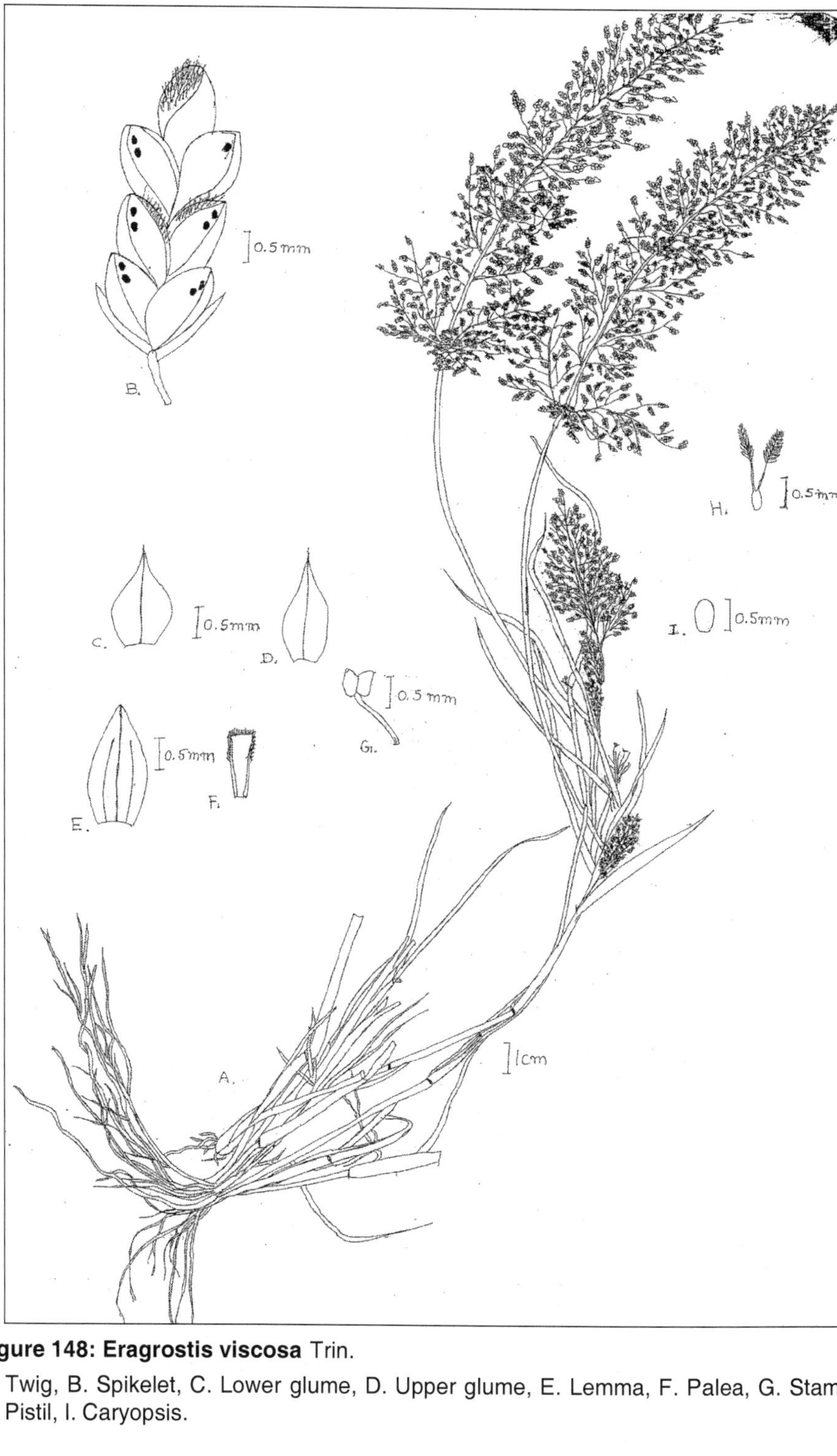

Figure 148: Eragrostis viscosa Trin.

A. Twig, B. Spikelet, C. Lower glume, D. Upper glume, E. Lemma, F. Palea, G. Stamen, H. Pistil, I. Caryopsis.

tank (MDK), *TP & MSG* 12020; Pupple vagu (MDK), *TP & MSG* 12089; Mohammadabad (RR), *MSM* 10594; Hyderabad, *MW* 5315 (DD); Mudwin, Amangal RF (MBNR), *SRS* 108695 (BSID); Rachakonda (NLG), *BVRS* 546 (NU).

EULALIA Kunth

Eulalia trispicata (Schult.) Henr., Blumea 3 : 453. 1940; Bor, Grass. India 157. 1960. *Andropogon trispicatus* Schult. Syst. Veg. 2, Mant. 452. 1824. *A. tristachyus* Roxb., Fl . Ind, 1: 261. 1820 non Kunth 1816. *Eulalia tristachya* (Roxb.) Kuntze, Revis. Gen. Pl. 2: 775. 1891; Fischer, Fl. Madras 3: 1715. 1934. *Pollinia argentea* Trin., Bull. Sci. Acad. Petersb. 1: 71. 1836; Hook. f., Fl. Brit. India 7: 111. 1896.

A perennial, tufted grass; culms up to 1.5 m high; nodes glabrous. Leaf sheaths 15 cm long, glabrous; ligule membranous, hairy; blades 10-25 x 0.3-0.5 cm, linear-lanceolate, glabrous-sparsely hairy. Racemes 2-12, 15 cm long, sub-digitate, capillary; rachis and pedicels densely villous with whitish silvery hairs; spikelets 4 mm, 2-nate, oblong, compressed. *Sessile spikelets:* Glumes subequal, subchartaceous; lower glume narrowly elliptic, 2-nerved, 2-keeled, hairy below the middle; upper glume boat-shaped, 1-nerved, 1-keeled, ciliate along the margin; lemmas hyaline, unequal; lower oblong epaleated, male; upper stipiform, bi-dentate, awned, bisexual; stamens 3; stigmas plumose. Caryopsis oblong. *Pedicelled spikelet:* 4 mm, similar to sessile spikelet.

Occasional in grasslands, near river-banks in Karimnagar district (Naqvi, 2001).

Fl. & Fr.: September –December.

GARNOTIA Brongn.

Garnotia tenella (Arn. ex Miq.) Jan., Fedder Repert. 17: 86. 1921. *Berghausia tenella* Arn. ex Miq . in Verh. Konink. Nederl. Inst. 3, 4: 34. 1851. *Garnotia stricta* sensu Hook. f., Fl. Brit. India 7: 243. 1896; Fischer, Fl. Madras 3: 1812. 1934; Bor, Grass. India 569. 1960.

An annual, tufted grass; culms up to 60 cm high, geniculately ascending and branched at base; nodes bearded. Leaf-sheaths 4-9 cm long, glabrous; ligule a narrow ciliate membrane, blades 4-15 x 0.3-0.5, linear-lanceolate, flat, base rounded, apex acuminate, hirsute above. Panicle up to 15 cm long; branches distant at base, close at top, in dense clusters, appressed to the rachis; spikelets 4 mm, oblong-linear, paired, sparsely hairy at the base; glumes ovate-lanceolate, 3-nerved, apex acuminate, hairy; lemma lanceolate, 3-nerved, apex bidentate, awned; palea oblong; stamens 3, stigmas plumose. Caryopsis linear, dorsally compressed.

Occassional grass in Nallamalais in Mahabubnagar district.

Fl.& Fr.: September-January.

Umamaheswaram (MBNR), *SRS* 109137, 111376 (BSID).

HACKELOCHLOA Kuntze

Hackelochloa granularis (L.) Kuntze, Revis. Gen. Pl. 2 . 776. 1891; Fischer, Fl. Madras 3: 1758. 1934; Bor, Grass. India 159. 1960; Moulik, Grass. Bamb. India 1: 193. 1997. *Mnesithea granularis* (L.) Koning & Sosef, Blumea 31: 295. 1986. *Cenchrus* *granulans* L., Mant. 2, App. 575. 1771. *Manisuris granularis* (L.) L.f., Nov. Gram. Gen. 40. 1779; Hook. f., Fl. Brit. India 7: 159. 1896.

An annual, tufted grass; culms up to 40 cm high, tuberculate pilose; nodes bearded. Leaf sheaths 2-6 cm long, hispid; ligule membranous; leaf blade 4-10 x 0.4-1 cm, lanceolate, flat, base rounded, apex acute. Racemes up to 4 cm long, spiciform, both terminal and axillary; rhachis fragile and hollow; spikelets 3 mm, globose, 2-nate. *Sessile spikelet:* globose, 2 mm; lower glume globose, crustaceous, pitted and tuberculate outside; upper glume broadly ovate, chartaceous, concave, 1-nerved; lower lemma ovate-obtuse, nerveless; upper lemma broadly ovate; palea ovate; lodicules 2; stamens 3, stigmas plumose. Caryopsis ovoid. *Pedicelled spikelet:* 3 mm, narrowly globose; lower glume obovate, incurved, winged; upper glume navicular, keeled on back; lemma often wanting, when present upper lemma either male or empty; palea ovate; stamens 3.

Common in the hills and weed in cultivated fields of all districts.

Fl. & Fr.: July-December.

Wankidi RF (ADB), *PVP* 9479; Bhimnagar RF (NZB), *TP & BR* 6215; Kamareddy (NZB), *BR & KH* 7141; Madikonakutta (MDK), *KMS* 6736 (CAL); Pakhal RF (WGL), *KMS* 13172 (CAL); Tandur (RR), *TP* & *MSM* 12706; Vikarabad (RR), *MRS* 323 (DD).

HETEROPOGON Pers.

Heteropogon contortus (L.) P. Beauv. ex Roem. & Schult., Syst. Veg. 2: 836. 1817; Fischer, Fl. Madras 3: 1743. 1934; Bor, Grass. India 163. 1960; Moulik, Grass. Bamb. India 1: 226. 1997. var. **contortus**. *Andropogon contortus* L., Sp. Pl. 1045. 1753; Hook. f., Fl. Brit. India 7: 199. 1896.

Vern.: *Kaserigaddi, Eddigaddi.*

An annual, tufted grass; culms up to 60 cm high, erect or decumbent below; nodes glabrous. Leaf sheaths 5-10 cm, glabrous; ligule membranous, fimbriate; leaf blade 4-10 × 0.2-0.4 cm, lanceolate, flat, base rounded, apex acute, scaberulous. Racemes solitary, 6 cm long, awns forming a twisted spire; lower spikelets homogamous, unawned, either neuter or male; upper spikelets heterogamous, awned, 2-nate, sessile bisexual, pedicelled male or barren; pedicel glabrous; callus pungent, bearded with brown hairs. *Sessile spikelets:* 7 mm, linear, hispidulous, female; callus bearded with brown hairs; lower glume linear-oblong, hispidulous, many-nerved; upper glume linear, obtuse, margin hyaline, dark brown, hispidulous, 3-nerved; lower lemma oblong, truncate, nerveless, epaleate; upper lemma linear, stipitiform, awned, epaleate; lodicules 2; stamens 3; stigma plumose. Caryopsis linear. *Pedicelled spikelet:* oblong, hispidulous; lower glume lanceolate, dorsally hispidulous with long bulbous based hairs, winged, serrulate; upper glume oblong-lanceolate, margin hyaline; lemmas hyaline, nerveless, hairy; lodicules 2; stamens 3.

Very common in plains and waste lands of all districts.

Fl. & Fr.: July - November.

Sone (ADB), *TP & PVP* 4056; Nasrullabad beat (NZB), *TP & BR* 6146; Ramayampet (MDK), *BR* & CP 11541; Peddakunta (RR), *MSM* & *KH* 10571; Pakhal RF (WGL), *KMS*

11641 (CAL); Bottchalami (NLG), *KMS* 9789 (CAL); Bhupatipur (KRN), *GVS* 25708 (CAL); Way to Pagarikutta east (MDK), *KMS* 67101 (CAL); Hyderabad (HYD), *MRS* (DD).

HYMENACHNE P. Beauv.

Hymenachne amplexicaulis (Rudge) Nees, Fl. Bras. Enum. Pl. 2: 276. 1829. *Panicum amplexicaule* Rudge, Pl. Guian 21. 1805. *Hymenachne acutigluma* (Steud.) Gilliland, Gard. Bull. Singapore 20: 314. 1963. *Panicum acutiglumum* Steud., Syn. Pl. Gram. 66. 1854. *Hymenachne pseudointerrupta* Müll. in Bot. Zietung (Berlin) 19: 333. 1861; Fischer, Fl. Madras 3: 1784. 1934; Bor, Grass. India 313. 1960; Moulik, Grass. Bamb. India 1: 104. 1997. *Panicum myuros* sensu Hook. f., Fl . Brit. India 7: 39. 1896, non Lam. 1791.

An annual, tufted grass; culms up to 1 m high, root stock spongy, prostrate, leafy. Leaf sheaths 15 cm long, glabrous, ligule membranous; blade 16-25 × 1-2 cm, linear, flat, base cordate, margins serrulate, apex acuminate. Panicle 35 cm long, dense, contracted; branches appressed to the rhachis, winged, scabrid; spikelets 6 mm, oblong, secund, grouped; glumes aristate, lower glume broadly ovate, keels ciliate, 3-nerved; upper glume narrowly lanceolate, 5-nerved, aristate, ciliate; lower lemma longer than upper glume, lanceolate, 5-nerved, awned, ciliate on nerves; upper lemma small, linear, 2-nerved; palea having bisexual floret, 2-keeled; lodicules 2; stamens 3; styles distinct; stigma plumose, brown. Caryopsis ellipsoid-obovoid.

On sands and near water courses of Karimnagar district.

Fl. & Fr.: September-January.

Mukunuru (KRN), *NRR & TRS* 86599 (MH).

Note: It looks like *Sacciolepis interruptea* in the field, because of its interrupted, spiciform panicles and creeping or trailing, spongy culms. But it can be easily distinguished from *Sacciolepis* by the rigid culms, lanceolate, acuminate spikelets, glumes and lemmas acuminate or aristate.

IMPERATA Cirillo

Imperata cylindrica (L.) Raeusch., Nomencl. Bot. ed. 3 : 10. 1797; Bor, Grass. India 169. 1960; Moulik, Grass. Bamb. India 2: 311. 1997. var. **cylindrica.** *Lagurus cylindricus* L., Syst. Nat. ed. 10. 878. 1759. *Imperata arundinacea* Cirillo, Pl. Rar. Neap. 2 : 26, t. 11. 1792; Hook. f., Fl. Brit. India 7: 160. 1896. *I. cylindrica* var. *major* (Nees) C.E . Hubb. ex C . E . Hubb. & Vaughan Grass Maur. 96. 1940. *I. cylindrica* var. *koenigii* (Retz.) T. A. Durand & Schinz, Consp. Fl. Afric. 5: 694; Fischer, Fl. Madras 3: 1707. 1934.

Vem.: *Dharbagaddi.*

A perennial, tufted grass; culms up to 70 cm high; nodes bearded. Leaf sheaths overlapping, 1.5-2 cm long, glabrous, ciliate near mouth; ligule membranous; leaf blade mostly basal, 6-20 × 0.2-0.5 cm, lanceolate, base rounded, apex acuminate, glabrous. Spike like panicle, up to 15 cm long, very dense; callus bearded; spikelets 3 mm, 2-nate, all alike, unawned, covered with long silky hairs; lower glume lanceolate, hyaline, long pilose outside, 5-7-nerved, apex bidentate; upper glume lanceolate, hyaline, 3-5-nerved, apex fringed; lower lemma oblong, margin scaberulous, hyaline; upper lemma lanceolate, margin scaberulous, hyaline; palea truncate, inrolled; lodicules absent; stamens 2, stigmas plumose. Caryopsis oblong.

Common in hedges and waste lands of Nizamabad, Ranga Reddi, Hyderabad and Nalgonda districts.

Fl. & Fr.: July-November.

Sarvapur (NZB), *TP & BR* 6377; Moosi river (HYD), *MRS* 309 (DD); Near Hussainsagar (HYD), *KMS* 8039 (CAL); Chityal (NLG), *BVRS* 414 (NU).

ISACHNE R. Br.

Isachne globosa (Thunb.) Kuntze, Revis. Gen. Pl. 2 : 478. 1891; Bor, Grass. India 580. 1960; Moulik, Grass. Bamb. India 2: 545. 1997. *Milium globosum* Thunb., Fl. Jap. 49. 1784. *Isachne australis* R. Br., Prodr. 196. 1810; Hook. f., Fl. Brit. India 7: 24. 1896. *I. dispar auct.* non Trin. 1828; Fischer, Fl. Madras 3: 1797. 1934; Bor, Grass. India 580. *p.p.* 1960; Hook. f., Fl. Brit. India 7: 26. 1896. *p.p.*

An annual, tufted, marshy grass; culms up to 25 cm high, slender, often ascending from a root stock, leafy, branches below; nodes sparsely villous. Leaf sheaths 2-5 cm, margins densely ciliate; ligule a tuft of long white hairs, leaf blade 2-5 × 0.2-0.3 cm, lanceolate, base rounded, scabrid on both surfaces. Panicle 8 cm long; rhachis angular, smooth; spikelets 2 mm, globose, sessile or pedicelled; glumes similar, ovate, boat-shaped, membranous, scabrid, 7-nerved; lemmas unequal, lower ovate, obscurely nerved, glabrous; lodicules 2; stamens 3; upper crustaceous. 7-nerved, ciliate; palea glabrous. Caryopsis ellipsoid.

In marshy areas of Ranga Reddi district.

Fl. & Fr.: May-December.

ISCHAEMUM L.

1. Margins of the lower glume of the sessile spikelet expanded below the middle .... **I. indicum**
1. Margins of the lower glume of the sessile spikelet narrowly inflexed from base to apex:
    2. Base of the lower blade rounded, or sub cordate, without a petiole:
        3. Perennial, rhizomatous grass; lower glume of the spikelets not rugose with transverse ridges; dorsal surface densely pilose .... **I. pilosum**
        3. Annual tufted grass; lower glume of the spikelets rugose with transverse ridges; dorsal surface sparsely ciliate .... **I. rugosum**
    2. Base of the lower blade sagittate or deeply cordate, with distant petiole .... **I. semisagittatum**

Ichaemum indicum (Houtt.) Merr., J. Arnold Arbor. 19: 320. 1938; Bor, Grass. India 180. 1960; Moulik, Grass. Bamb. India 1: 292. 1997. *Phleum indicum* Houtt. Nat. Hist. 2, 13: 198, t. 90, f. 2. 1782. *Ischaemum aristatum* auct. non. L., Fischer, Fl. Madras 3: 1721. 1934. *I. ciliare* Retz., Observ. Bot. 6. 36. 1791; Hook. f., Fl. Brit. India 7: 133. 1896.

A perennial, tufted grass; culms up to 60 cm high, decumbent at base, branched above; nodes glabrous. Leaf sheaths 3-9 cm long, loose, glabrous auricled; ligule membranous; leaf blade 3-12 x 0.3-0.5 cm, linear-lanceolate, base rounded, hairy, apex acuminate, midrib prominent. Racemes 2, 5 cm long, stout or slender; spikelets 6 mm, 2-nate. *Sessile spikelet:* 4 mm, oblong; lower glume broadly oblong, 2-winged at apex, 10-12-nerved, cuspidate, scabrid; upper glume lanceolate hyaline, 5-7-nerved, bidentate, slightly winged; lower lemma ovate-lanceolate, hyaline, margin scaberulous; palea lanceolate, involute, hairy at apex, upper lemma hyaline, bidentate, awned; lodicules 2; stamens 3, stigmas plumose. Caryopsis oblong. *Pedicelled spikelet:* Similar to sessile spikelet; upper lemma awnless or with short mucro.

Common in Karimnagar, Warangal and Nalgonda districts near the water sources.

Fl. & Fr.: August-January.

Aklaspur (KRN), *GVS* 22489 (CAL); Pakhal RF (WGL), *KMS* 11661 (CAL); Cherukupally (NLG), *BVRS* 660 (NU).

Ischaemum pilosum (Klein ex Willd.) Wight, Madras Lit. Sci. J. 138. 1835; Hook. f., Fl. Brit. India 7: 130 1896; Fischer, Fl. Madras 3: 1723. 1934; Bor, Grass. India 183. 1960; Moulik, Grass. Bamb. India 1: 295. 1997. *Andropogon pilosus* Klein ex Willd. Sp. Pl. 4: 920. 1806.

Vern.: *Kundaranattu.*

A perennial, stoloniferous grass; culms up to 90 cm high, erect; nodes glabrous. Leaf sheaths 5-20 cm long, glabrous; ligule membranous; leaf blade 7-15 × 0.4-0.7 cm, linear, base rounded, apex acuminate, veins prominent, glabrous. Racemes 6, up to 15 cm long, pilose; rhachis hairy; joints and pedicels slender, compressed, sparsely ciliate; callus bearded with long hairs; spikelet 9 mm, 2-nate. *Sessile spikelet:* 6 mm long, narrowly lanceolate; glumes lanceolate, chartaceous, laterally compressed, 7-9-nerved, dorsally hairy in lower; lower lemma linear-oblong, ciliate above; palea nerveless, male; upper lemma cleft to the middle into acute, ciliate lobes; palea lanceolate, nerveless; lodicules 2; stamens 3, stigma plumose. Caryopsis lanceolate, *Pedicelled spikelet:* Similar to sessile spikelet; upper lemma contains short awn.

Common in black cotton soils of Adilabad, Ranga Reddi and Warangal districts.

Fl. & Fr.: October –March.

Keslapur (ADB), *GO & PVP* 4558; Mekavaripalli (RR), *TP* & *MSM* 11051.

Ischaemum rugosum Salisb., Ic. Stirp. Rar. l, t. 1. 1791; Hook. f., Fl. Brit. India 7:127 1896; Fischer, Fl. Madras 3: 1722. 1934; Bor, Grass. India 184. 1960.

An annual, tufted grass; culms up to 60 cm high, decumbent at base; nodes bearded. Leaf sheaths 4-6 cm long, compressed, loose, glabrous, auricled, membranous; ligule membranous; leaf blade 6-12 x 0.3-0.5 cm, linear-lanceolate, glabrous or hairy, base rounded, apex acuminate. Racemes usually 2.5 cm, digitate, spiciform; rhachis trigonous; spikelets 2-nate. *Sessile spikelet,* 5 mm, ovate-oblong; callus villous; lower glume oblong-elliptic, cartilaginous from the base, pale yellow, concave, crossed by 3-6 deep smooth ridges, the upper portion flat, membranous, incurved, winged: upper glume oblong-lanceolate, keeled, keel ciliate; lower lemma ovate-lanceolate, palea containing male floret; upper lemma deeply cleft into 2 acute, lanceolate lobes, cuneate; palea lanceolate; lodicules 2; stamens 3, stigmas plumose. Caryopsis oblong. *Pedicelled spikelet:* similar to sessile spikelet.

Common weed in rice fields and in wet marshy grounds of all districts.

Fl. & Fr.: August-December.

Sone (ADB), *TP & PVP* 4059; Sarvaipet (ADB), *GO* 4502; Bhasar (ADB), *GO & PVP* 4643; Mandora (NZB), *TP & BR* 6225; Gangapur RF (MDK), *TP & MSG* 12071; Rechapalli (KRN), *GVS* 22277 (CAL); Kodimial (KRN), *GVS* 21857 (CAL); Pakhal RF (WGL), *KMS* 11663 (CAL); Rusthempet (MDK), *KMS* 3008 (CAL); Hyderabad, *MRS* 245 (DD).

Ischaemum semisagittatum Roxb., Fl . Ind. 1: 322. 1820; Hook. f., Fl. Brit. India 7: 130. 1896; Bor, Grass. India 185. 1960; Moulik, Grass. Bamb. India 1: 297. 1997.

An annual, tufted grass; culms up to 60 cm high, decumbent at base, slender; nodes glabrous. Leaf sheaths loose, 2-10 cm long, glabrous, upper most spathiform, higher pilose; blade 5-8 x 1-2 cm, oblong- lanceolate, auricled, base cordate-sagittate, margin scabrid, apex acuminate, lower with a filiform petiole, upper sessile, glabrous-sparsely hairy. Racemes 2, 7 cm long, villous with white or grey hairs, fragile. *Sessile spikelet:* 6 mm; callus bearded; lower glume ovate, convex, 9-11-nerved, the lower half coriaceous, the upper half chartaceous, ciliate margins; upper glume ovate- oblong, coriaceous, 1-keeled, 3-nerved; lower lemma ovate; palea linear-oblong, hyaline, male; upper lemma linear, 2-lobed, awned; palea linear- oblong, female; lodicules 2; stamens 3. *Pedicelled spikelet:* short and narrower than the sessile, closely appressed to the sessile spikelet; pedicel broad; lower glume coriaceous, oblong, obscurely nerved, villous, 2-keeled; upper glume coriaceous, boat-shaped, 1-keeled, 3-nerved; lower lemma ovate, hyaline, 2-keeled; upper lemma narrowly oblong, 2- lobed, awned, minute; lodicules 2; stamens 3.

Rare in moist localities in Medak district.

Fl. & Fr: AugustDecember.

Rusthempet (MDK), *KMS* 6797 (CAL & MH).

ISEILEMA Anderss.

1. Glandular tubercles present on the keels of the spatheoles and often on the nerves of the lower glume of the involucral spikelets
   2. Cluster of spikelets in dense bascides, robust species, 60-120 cm tall, hermaphrodite spikelets longer than involucral spikelets .... **I. holei**
   2. Cluster of spikelets not in dense bascides, slender grass, hermaphrodite spikelets as long as involucral spikelets .... **I. prostratum**
1. Glandular tubercles not present on the keel of the spatheoles and on the nerves of the lower glume of the involucral spikelets:
   3. Pedicels of the involucral spikelets slender, longer than broad .... **I. laxum**
   3. Pedicels of the involucral spikelets almost as broad as long .... **I. anthephoroides**

Iseilema anthephoroides Hack, in DC., Monogr. Phan. 6: 683. 1889; Hook. f., Fl. Brit. India 7: 219, 1896; Fischer, Fl. Madras 3: 1748. 1934; Bor, Grass. India 187. 1960; Moulik, Grass. Bamb. India 1: 236. 1997.

An annual, tufted grass; culms up to 50 cm high, decumbent at base; nodes glabrous. Leaf sheaths 1-4 cm long, glabrous; ligule membranous with ciliate apex; leaf blade 3-10 × 0.1-0.3 cm, linear-lanceolate, flat, base rounded, apex acute, glabrous. Panicle up to 8 cm long; spatheoles 8 mm long, cymbiform, glabrous. *Involucral spikelet:* 4 mm, oblong, enclosing both sessile and pedicelled spikelets, male; glumes lanceolate-oblong, 3-nerved, upper scabrid and pilose on outside; lemma obovate-oblong, nerveless; lodicules 2; stamens 3. *Sessile spikelet:* 3 mm, linear, bisexual; lower glume ovate, acuminate, scabrid outside; upper glume oblong, scabrid outside; lower lemma lanceolate; upper lemma linear, awned; lodicules 2; stamens 3; stigma plumose. Caryopsis ellipsoid. *Pedicelled spikelet:* Similar to involucral spikelet.

Common along the hedges of fields and abandoned plains and in lower hills of Adilabad, Ranga Reddi, Hyderabad, Karimnagar and Nalgonda districts.

Fl. & Fr.: June-October.

Satnella (ADB), *GO* & *PVP* 4220; Tiryani RF (ADB), *PVP* 9499; Malkalcheruvu (RR), *MSM* & *KH* 10569; Rangammagudem (RR), *MSM* 10582; Krishna river bank (NLG), *KMS* 9858 (CAL).

Iseilema holei Haines, Bot. Bihar & Orissa 1055, 1924; Bor, Grass. India 188. 1960.

Perennials; culms robust, tufted, 100 – 150 cm high; nodes bearded; internodes purplish. Leaf blades linear-lanceolate, 20 – 50 × 0.3 – 0.5 cm, flat, acute at apex, glabrous, sparsely ciliate at the base; ligules ciliate, 2 mm long; leaf sheaths 5 – 10 cm long, keeled, glabrous, long ciliate at mouth. Inflorescence dense fascicle, terminal and axillary, subtended by spathe and spatheole, spathe 2 – 5 cm long, spatheole 0.9 – 1.2 cm long, tuberculate. Racemes solitary, bearing three heteromorphous spikelets;

lower 4 involucral spikelets male, 1 fertile sessile spikelet and 2 male pedicelled spikelet; peduncle 3 – 4 mm long, hairy, tuberculate; rachis carrying sessile spikelet 1.2 mm long, sparsely hairy. *Involucral spikelets* lanceolate, 3.7 – 4.5 × 0.7 – 1 mm, scabrid throughout, sparsely hairy; pedicel 0.7 – 1 × 0.2 – 0.3 mm, hairy on upper side. Lower glumes lanceolate 3.7 – 4.5 × 0.7 – 1 mm, scabrid, 5-nerved, 2-keeled, sparsely tuberculate hairy on keels, apex truncate; upper glumes lanceolate, 3.6 – 4.2 × 0.6 – 0.8 mm, glabrous, 3-nerved, 2-keeled, keels winged, wings hairy; upper lemma linear-lanceolate, 2.5 – 3 × 0.2 – 0.3 mm, hyaline; stamens 3, anthers 1.2 – 1.4 mm long. *Sessile spikelets* linear-lanceolate, 3.5 – 4 × 0.8 mm, broader at base and narrowed at apex, awned. Lower glumes linear-lanceolate, 3.5 – 4 × 0.7 – 0.8 mm, 4-nerved, 2-keeled, apex truncate, scabrid on upper half and glabrous on other half; upper glumes linear lanceolate, 3.4 - 3.7 × 0.6 – 0.8 mm, acuminate to arsitate at apex, arista up to 0.5 mm long, 3-nerved, 2-keeled; upper lemma linear, 2.4 – 3 × 0.1 – 0.2 mm, awned; awn geniculate, 13 – 14 mm long; upper paleas, oblong-lanceolate, 1.8 – 2 × 0.7 – 0.8 mm, hyaline nerveless, apex truncate. Caryopsis oblong lanceolate, 1.8 – 2 × 0.8 – 1 mm. *Pedicelled spikelets* linear-lanceolate, 4.3 – 5.5 × 0.5 – 0.8 mm, scabrid and tuberculate throughout; pedicel 2 mm long, ciliate; lower glumes linear-lanceolate, 4 – 5.3 × 0.6 – 1 mm, acute at apex, 9-nerved, scabrid, tuberculate on nerves; upper glumes linear-lanceolate, 4 – 5 × 0.5 – 0.7 mm, acuminate at apex, 3-nerved, glabrous, scabrid and tuberculate on nerves. Upper lemma linear-lanceolate, 3.8 – 4 × 0.2 – 0.3 mm, hyaline.

On moist grassy patches in Adilabad district.

Fl. & Fr.: September – October.

Kawal Tiger Reserve, Indanpalli Range (ADB), *Alok Chorghe, Chandra Mohan & P. S. Annamma* 3937 (BSID).

Iseilema laxum Hack., in DC., Monogr. Phan. 6 : 682. 1889; Hook. f., Fl. Brit. India 7: 219. 1896; Fischer, Fl. Madras 3: 1748. 1934; Bor, Grass. India 188. 1960; Moulik, Grass. Bamb. India 1: 237. 1997.

A perennial, tufted grass; culms up to 30 cm high, decumbent or ascending at base; nodes glabrous. Leaf sheaths 2-4 cm, loose, glabrous; ligule a fimbriate rim; leaf blade cauline, 2-8 × 0.2-0.4 cm, lanceolate, flat, base rounded, margin scaberulous, apex acuminate. Panicle up to 15 cm long, flexuous, interrupted; spatheoles 18 mm, plicate, glabrous. *Involucral spikelet:* 3 mm, linear-oblong, whorled, joints pilose; lower glume chartaceous, linear-oblong, margins inflexed, 3-5-nerved, 2-keeled, keels ciliate, pilose near apex at outside; upper glume oblong, margins inflexed, 3-nerved, glabrous, apex acute; lemma linear, hyaline; lodicules 2; stamens 3. *Sessile spikelet:* 3 mm, elliptic-lanceolate, bisexual; lower glume chartaceous, lanceolate, bidentate, beak elongate, tip scabrid, sparsely hairy; upper glume lanceolate, acuminate, scabrid on outside; lower lemma lanceolate; upper lemma bidentate, awned; stamens 3, stigmas plumose. Caryopsis ellipsoid. *Pedicelled skikelet:* Similar to involucral spikelet.

Common in waste places and in forests of Adilabad, Nizamabad, Medak, Ranga Reddi, Warangal, Karimnagar, Hyderabad and Nalgonda districts.

Fl. & Fr.: August-March.

Mancheriyal (ADB), *GO* & *PVP* 4938; Nasrullabad (NZB), *TP* & *BR* 6147; Pakhal (WGL), *ANH* 15977 (CAL); Aklaspur (KRN), *GVS* 25652 (CAL); Pegarikutta, Narsapur (MDK), *KMS* 6702; Mekavaripalli (RR), *TP* & *MSM* 11053; Yerraguntapalli (NLG), *BVRS* 638.

Iseilema prostratum (L.) Andersson, Nov. Act. Soc. Sci. Upsal. 3, 2; 251. 1856; Fischer, Fl. Madras 3: 1747. 1934; Bor, Grass. India 188. 1960; Moulik, Grass, Bamb. India 1: 237. 1997. *Andropogon prostratum* L., Mant. Alt . 304. 1771. *Iseilema wightii* Andersson *loc.cit.* 1856; Hook. f., Fl. Brit. India 7: 218. 1896.

A perennial, tufted grass; culms up to 70 cm high, prostrate and ascending at base; nodes more or less bearded. Leaf-sheaths 3-5 cm, loose, glabrous; ligule membranous; leaf blade 4-12 × 0.2-0.4 cm, linear, flat, base rounded, apex acute, glabrous. Panicle up to 30 cm, lax, spreading; rhachis bearded at nodes; spathes linear-lanceolate; margin scarious. *Involucral spikelet:* 4 mm, oblong, male, whorled; lower glume oblong, 5-nerved, sulcate on the back, pilose; upper glume lanceolate, 1-nerved, glabrous; lemma oblong, nerveless, glabrous; lodicules 2; stamens 3. *Sessile spikelet:* 5 mm, lanceolate, bisexual; lower glume ovate, acuminate, scabrid on outside near the tip, 4-nerved; upper glume elliptic, acute, 3-nerved; lower lemma lanceolate; upper lemma, bidentate, awned; lodicules 2; stamens 3; stigmas plumose. Caryopsis oblong. *Pedicelled spikelet:* similar to involucral spikelet.

Common in plains and occasional in forests of Adilabad, Nalgonda, Medak, Ranga Reddi, Hyderabad, Warangal, and Mahabubnagar districts.

Fl. & Fr.: November-June.

Bellampalli (ADB), *PVP* 9466; Posanpalle (MDK), *TP* & *MSG* 11962; Mohammadabad (RR), *MSM* 10595; Krishna river bank (NLG), *KMS* 9821 (CAL); Hyderabad (HYD), *MRS* 253 (DD); Karkalpadu, Amangal RF (MBNR), *SRS* 104549 (BSID).

LEERSIA Solander ex Sw. *nom. cons*

Leersia hexandra Sw., Prodr. 21. 1788; Hook. f., Fl. Brit. India 7: 94. 1896; Fischer, Fl. Madras 3: 1845. 1934; Bor, Grass. India 599. 1960; Moulik, Grass. Bamb. India 1: 45. 1997.

An annual, aquatic, floating grass; culms up to 40 cm high; nodes bearded. Leaf sheaths 4-10 cm long, glabrous; ligule membranous; leaf blade 5-15 × 0.3-0.7 cm, lanceolate, base narrowed, apex acute, scabrid on both surfaces. Panicle up to 7 cm long; pedicel apically cupular, 2-lipped; spikelets 4 mm, solitary, oblong, laterally compressed; glumes obscure; lemma boat-shaped, coriaceous, 3-nerved, ciliate, 1-keeled; palea linear, midrib scabrid, 3-keeled; lodicules 2; stamens 6; stigma plumose, greenish to creamy white. Caryopsis oblong.

In ponds and marshy areas of Medak and Hyderabad districts.

Fl. & Fr.: September- February.

LEPTOCHLOA P. Beauv.

1. Spikelet 1-flowered; ligule membranous, eciliate .... **L. uniflora**
1. Spikelet 4-10-flowered; ligule a fringed membrane or ciliate:
   2. Lemmas 3-lobed; racemes loosely spiculate .... **L. fusca**
   2. Lemmas not lobed; racemes densely spiculate .... **L. chinensis**

Leptochloa chinensis (L.) Nees in Syll. Ratisb. 1:4.1824; Hook. f., Fl. Brit. India 7:299. 1896; Fischer, Fl. Madras 3:1820. 1934; Bor, Grass. India 516. 1960; Moulik, Grass. Bamb. India 2:613.1997. *Poa chinensis* L., Sp. Pl. ed. 1:69.1753.

Annuals, culms erect or ascending, tufted, to 90 cm; nodes glabrous. Leaf sheaths glabrous; ligule membranous; leaf blade 10-20 × 0.3-0.6 cm, linear or linear-lanceolate, base rounded, acuminate, glabrous. Panicles open, 25-40 cm long, rhachis stiff; spikelets ovate-oblong, 3-7-flowered; lower glume lanceolate, chartaceous, 1-keeled, keels scaberulous, 1-nerved; upper glume ovate-lanceolate, chartaceous, 1-keeled, keel scaberulous, 1-nerved; lemmas ovate, chartaceous, 3-nerved; paleas ovate or oblong, 2-keeled, 2-nerved, keels ciliate; lodicules 2; stamens 3. Caryopsis oblong.

Common near water sources in Adilabad, Ranga Reddi, Karimnagar, Hyderabad and Nalgonda districts.

Fl. & Fr.: July-December.

Bijjur RF (ADB), *GO* & *PVP* 4877; Kodimial (KRN), *GVS* 25667 (CAL & MH); Chityal (NLG), *BVRS* 419 (NU).

Leptochloa fusca (L.) Kunth, Revis. Gramin. 1: 91. 1829. *Diplachne fusca* (L.) P. Beauv., Ess. Agrost. 80 & 163. 1812; Hook. f., Fl. Brit. India 7: 329. 1896; Fischer, Fl. Madras 3: 1829. 1934; Bor, Grass. India 492. 1960; Gayathri & Pullaiah in Pullaiah, Fl. Andhra Pradesh 3: 1183. 1997. *Festuca fusca* L., Sp. Pl. ed.2. 109.1762.

A perennial, tufted, coarse grass; culms up to 70 cm high, decumbent or geniculately ascending below; nodes glabrous. Leaf sheaths 5-10 cm long, glabrous; ligule membranous; blade 5-15 x 0.2-0.4 cm, linear, convolute, base rounded, apex acuminate. Panicle up to 15 cm, erect; rhachis scabrid; spikelets 7 mm, oblong, 4-10-flowered; pedicel short, scabrid; glumes oblong, 1-nerved, membranous, scabrid on the keels, mucronate; lemma oblong, 1-nerved, 4-toothed, aristate, scabrid, pilose below; palea oblong, truncate, pilose on the keels; stamens 3; stigmas plumose. Caryopsis oblong.

Common in moist and shady localities of the forests, and in streams and ponds in Hyderabad and Nalgonda districts.

Fl. & Fr.: October-April.

Nagarjuna sagar (NLG), *KMS* 9738 & 9869 (CAL & MH).

Leptochloa uniflora Hochst. ex A. Rich, Tent. Fl. Abyss. 2: 409. 1850.

Annual or perennial, short lived, caespitose grass, erect or decumbent, 50-150 cm high. Ligule an eciliate membrane. Leaf-blade lanceolate or oblong, 4-12 × 0.6-1.8

cm, flaccid, apex acute. Racemes numerous, borne along the central axis, 2.5-9 cm long; spikelets solitary, fertile spikelets sessile, spikelets comprising 1 fertile floret, spikelet oblong, laterally compressed; glumes persistent, lower glume lanceolate, upper glume elliptic; fertile lemma oblong, palea keels ciliolate. Caryopsis ellipsoid, trigonous, dark brown or red.

Rare in Medak district.

Fl. & Fr.: September – December.

Medak (MDK), *RG* 104162 (BSID).

LOPHOPOGON Hackel

Lophopogon tridentatus (Roxb.) Hack. in Engl. & Prantl, Pflanzenfam. 2 (2): 26. 1887; Hook. f., Fl. Brit. India 7: 149 1896; Fischer, Fl. Madras 3: 1724. 1934; Bor, Grass. India 190. 1960; Moulik, Grass. Bamb. India 1: 337. 1997. *Andropogon tridentatus* Roxb., Fl. India 1: 261. 1820.

A perennial, tufted grass; culms up to 30 cm high, decumbent; nodes glabrous. Leaf sheaths 1-4 mm, glabrous; ligule hyaline, ciliate; leaf blade 2-5 × 0.1-0.3 cm, linear, flat or convolute, base narrowed, apex acuminate, glabrous. Binate spiciform racemes up to 3 cm long, solitary, closely appressed, enclosed within ventricose sheaths or exserted; spikelets 4 mm, 2-nate. *Sessile spikelet:* oblong, 4 mm, dorsally compressed, brownish, callus bearded with ferruginous hairs; lower glume oblong, 5-nerved, brownish tuft hairs above the middle, margin ciliate, apex truncate with lateral teeth; upper glume oblong, tufted hairs on outer side, apex bidentate, narrowed into a fine arista; lower lemma linear, 1-nerved, keeled; palea linear, male or empty; upper lemma lanceolate, apex bidentate, hyaline, awned; palea bisexual; lodicules absent; stamens 3; stigmas plumose. *Pedicelled spikelets:* similar to sessile spikelet.

Occurs on hill slopes and abandoned plains of Karimnagar, Hyderabad, Nalgonda and Mahabubnagar districts.

Fl. & Fr.: July – October.

Lingampalle (RR), *MSM* 12754; Krishna river bank (NLG), *KMS* 9828 (MH); way to Mysamma temple (MBNR), *SRS* 108628 (BSID); Umamaheswaram (MBNR), *SRS* 108812 (BSID).

MELANOCENCHRIS Nees

1. A tiny, annual grass; rhachilla glabrous; lemmas bidentate ..... **M. jacquemontii**
1. A perennial grass; rhachilla puberulous; lemmas acute, apiculate or aristate .... **M. monoica**

Melanocenchris jacquemontii Jaub. & Spach, Illus. Pl. Orient. 4: 36. t. 325. 1851; Bor, Grass. India 473. 1960; Moulik, Grass. Bamb. India 2: 574. 1997. *M. royleana* Nees ex Steud., Syn. Pl. Glum. 1: 218 1854; Fischer, Fl. Madras 3: 1831. 1934. *Gracilea royleana* Hook. f., Fl. Brit. India 7: 284. 1896.

A tiny annual, tufted grass; culms up to 20 cm high, decumbent; nodes glabrous. Leaf sheaths 1-2 cm long, sparsely hairy; ligule hairy; leaf blade 2-4 × 0.1-0.5 cm, linear, convolute, base rounded, apex acuminate, glabrous. Spikes 6 cm long, filiform, recurved, flexuous; spikelets 8 mm, clustered on a simple axis, 2-flowered; lower glume linear, coriaceous, 1-nerved, densely ciliate on margin, awned; upper glume ovate, densely ciliate on margin, awned; lower lemma and palea ovate, membranous, bidentate, sparsely ciliate, awned; lodicules 2; stamens 3; stigmas plumose. Caryopsis elliptic, obtuse. Upper lemma ovate-oblong, bidentate, sparsely ciliate, awned; upper palea broadly ovate, bidentate, scabrid, male; stamens 3.

Common in dry sandy soils and waste lands in all districts.

Fl. & Fr.: July –December.

Bellampalli RF (ADB), *PVP* 9453; Wankidi RF (ADB), *TP & GO* 5486; Manchippa RF (NZB), *TP & BR* 6008; East Narsapur (MDK), *BR* 11206 11931; Anantagiri (RR), *TP & MSM* 11072; Kodimial (KRN), *GVS* 25672 (CAL); Way to Pagarikutta (MDK), *KMS* 6618 (CAL); Adikmet (HYD), *MRS* 312 (DD); Rachakonda (NLG), *BVRS* 545 (NU); Guttakindipalle (MBNR), *SRS* 107482 (BSID); Edupayalu RF (MDK), *RG* 113453 (BSID); Jhakaram (WGL), *RKP* 105313 (BSID).

Melanocenchris monoica (Rottl.) Fischer. Fl. Madras 1831. 1934; Bor, Grass. India 473. 1960; Moulik, Grass. Bamb. India 2: 574. 1997. *Pommereulla monoica* Rottl., Ges. Natur. Frunde Berlin, Neue Schr. 4: 218. 1803. *Gracilea nutans* Koen. ex Rottl. *loc.cit.;* Hook. f., Fl. Brit. India 7: 283. 1896.

A perennial, tufted grass; culms up to 60 cm high; nodes glabrous. Leaf sheaths 2-5 cm long, sparsely tubercled based hairs; ligule fimbriate membrane; blade 2-10 x 0.2-0.5 cm, linear-lanceolate, flat or convolute, base cordate, apex acuminate, sparsely tubercled based hairs. Spike up to 15 cm long; rhachilla puberulous. Spikelets 6 mm, clustered, subsessile, secund, 2-nate, lower floret bisexual, upper male or neutral; lower glume linear, coriaceous, 1-3-nerved, pilose, awned; upper glume oblong, pilose, margin hyaline; lower lemma broadly ovate, membranous, sparsely ciliate, aristate; palea oblong-ovate, obtuse, sparsely ciliate; lodicules 2; stamens 3; stigmas plumose. Caryopsis oblong. Upper lemma oblong, sparsely aristate, ciliate; palea oblong-ovate, scabrid margins.

Rare grass in Hyderabad and Nalgonda districts.

Fl. & Fr.: August-December.

Nagarjuna konda valley (NLG), *KT* 9658 (CAL); Hyderabad, *MRS* 205 (DD).

MICROCHLOA R. Br.

Microchloa indica (L.f.) P. Beauv., Ess. Agrost. 20. 1812; Fischer, Fl. Madras 3: 1830. 1934; Bor, Grass. India 473. 1960; Moulik, Grass. Bamb. India 1: 569. 1997. *Nardus indica* L.f., Suppl. Pl. 105. 1781. *Microchloa setacea* R. Br., Prodr. 208.1810; Hook. f., Fl. Brit. India 7: 283. 1896.

An annual, tufted grass; culms up to 20 cm high, slender. Leaf sheaths sparsely hairy; ligule a minutely ciliate rim; leaf blade 1.5-4 cm long, linear, apex acute, glabrous. Spikes solitary, terminal 2-8 cm long, curved; rhachilla minutely ciliate; spikelets 2 mm, 1-flowered, dorsally compressed; glumes subequal, lanceolate, 1-nerved, glabrous; lemmas lanceolate-oblong, ciliate along the nerves; lodicules 2; stamens 3, stigmas plumose. Caryopsis oblong, subfusiform.

Common in open grasslands, at hill slopes in Karimnagar, Hyderabad, Mahabubnagar and Warangal districts.

Fl. & Fr.: August-November.

OPHIUROS Gaertn.

Ophiuros exaltatus (L.) Kuntze, Revis. Gen. Pl. 2: 780. 1891; Fischer, Fl. Madras 3: 1758. 1934; Bor, Grass. India 199. 1960; Moulik, Grass. Bamb. India 1: 198. 1997. *Aegilops exaltata* L., Mant. 2: 575. 1771. *Ophiuros corymbosa* Gaertn. f., Fruct. 3. 4. t. 181. 1805; Hook. f., Fl. Brit. India 7: 160. 1896.

Vern.: *Peddapanuku.*

A perennial, stoloniferous grass; culms up to 70 cm high, erect; nodes glabrous. Leaf sheaths 7-15 cm long, glabrous; ligule membranous; leaf blade 10-20 × 0.5-1 cm, lanceolate, base cordate, margins minutely tubercled at base, apex acute, glabrous, midrib prominent. Spikes 3-6, corymbose, terminal and axillary, up to 15 cm long; rhachis fragile; joints completely fused with pedicels, hollow; spikelets 3 mm, awnless; lower glume oblong, coriaceous, 5-7-nerved; upper glume oblong, boat-shaped, membranous, margins inflexed, 3-nerved; lower lemma elliptic, hyaline; palea lanceolate, hyaline; upper lemma lanceolate; palea linear; lodicules 2; stamens 3, stigmas plumose. Caryopsis oblong slightly compressed.

Occasional grass particularly in black cotton soils of Adilabad, Karimnagar and Warangal districts.

Fl. & Fr.: July-November.

Boath (ADB), *GO & PVP* 4637; Pakhal (WGL), *KMS* 11574 (MH).

OPLISMENUS P. Beauv.

1. Perennial, diffuse grass; leaf blade elliptic – lanceolate; awns reddish, stout, smooth .... **O. compositus**
1. Annual, slender branching grass; leaf blade linear- lanceolate; awns pale, capillary, minutely scaberulous .... **O. burmannii**

Oplismenus burmannii (Retz.) P. Beauv., Ess. Agrost. 54, 168, 169. 1812; Hook. f., Fl. Brit. India 7: 68. 1896; Fischer, Fl. Madras 3: 1778. 1934; Bor, Gras. India 317. 1960; Moulik, Grass. Bamb. India 1: 106. 1997. *Panicum burmannii* Retz., Observ. Bot. 3: 10. 1783.

An annual, slender grass; culms up to 50 cm high, creeping, diffusely branched at base; nodes glabrous-sparsely hairy. Leaf sheaths 15 cm long, below the nodes, margins densely ciliate; ligule a ridge of hairs; leaf blade 2-5 x 0.5-1 cm, ovate-lanceolate, base amplexicaul, apex acuminate, sparsely pilose on both sides. Spiciform secund racemes up to 6 cm long, terminal; rhachis slender, trigonous; branches distant; spikelets elliptic-ovate, 3 mm, solitary, 2-nate, awned; lower glume membranous, ovate, hairy, 3-5-nerved, awned. upper glume membranous, ovate, dorsally pubescent,

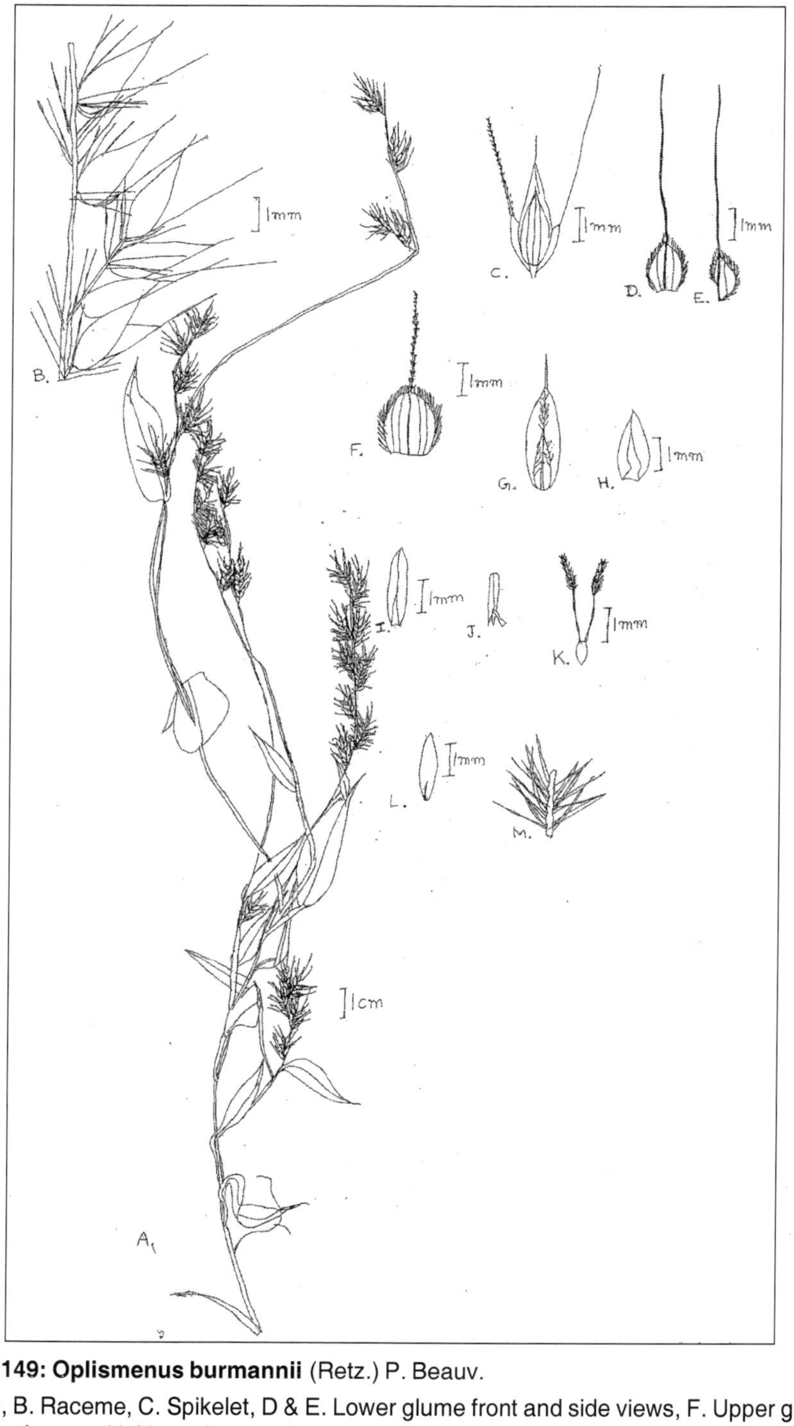

Figure 149: Oplismenus burmannii (Retz.) P. Beauv.

A. Twig, B. Raceme, C. Spikelet, D & E. Lower glume front and side views, F. Upper glume, G. Lower lemma, H. Upper lemma, I. Palea, J. Stamen, K. Pistil, L. Caryopsis, M. Pedicel.

5-nerved, margins ciliate, apex with capillary, 5-7 mm long, scabrid awn; lower lemma ovate, margins curved, 7-9-nerved, hairy on margins, aristate; upper lemma ovate, margins curved, 3-5-nerved; palea linear; lodicules 2; stamens 3; stigmas plumose. Caryopsis linear, brown.

Common grass in cultivated fields and shades of tall trees in the forests of Adilabad, Nizamabad, Medak, Hyderabad, Warangal and Mahabubnagar districts.

Fl. & Fr.: August-December.

Sattenapalli RF (ADB), *TP & PVP* 4083; Wankidi (ADB), *GO* 4474; Bersaipet RF (ADB), *GO & PVP* 4600; Manchippa RF (NZB), *BR* 6043; Narsapur RF (MDK), *TP & MSG* 11995; Gangapur RF (MDK), *TP & MSG* 12069; Pakhal RF (WGL), *KMS* 11640 (CAL); Amrabad (MBNR), *BR* & *BSS* 32205 (BSID).

Oplismenus compositus (L.) P. Beauv., Ess. Agrost. 54, 168, 169. 1812; Hook. f., Fl. Brit. India 7: 66 1896; Fischer, Fl. Madras 3: 1778. 1934; Bor, Grass. India 317. 1960; Moulik, Grass. Bamb. India 1: 109. 1997. *Panicum compositum* L., Sp.Pl. 57.1753.

Vern.: *Kodijuttugaddi.*

A perennial, stoloniferous grass; culms up to 60 cm high, creeping, rooting at nodes; nodes glabrous. Leaf sheaths 2-9 cm long, margin densely ciliate; ligule fimbriate membrane; leaf blade 3-10 x 1-2 cm, elliptic-lanceolate, base amplexicaul, apex acuminate, sparsely pubescent. Spiciform racemes 4-10, up to 20 cm long, terminal; branches distant; spikelets 4 mm, elliptic, 2-flowered, awned; lower glume ovate, membranous, scaberulous above, 5-nerved, awned; upper glume oblong, membranous, 7-nerved, margins ciliate at apex, 7-9-nerved, aristate; lower lemma ovate-oblong, membranous, ciliate on margin, 5-nerved, shortly aristate; palea oblong; upper lemma ovate; palea ovate; lodicules 2; stamens 3, stigmas plumose. Caryopsis linear.

Common in damp, shady places in the forests of Medak and Karimnagar districts (Naqvi, 2001).

Fl. & Fr.: November-January.

Gangapur (MDK), *TP & MSG* 12094; Umamaheswaram (MBNR), *SRS* 111365 (BSID); Mallelatheertham (MBNR), *SRS* 108853, 111401 (BSID).

OROPETIUM Trin.

Oropetium roxburghianum (Steud.) S.M.Phillips, Kew Bull. 36:469. 1975. *Lepturus roxburghianus* Steud., Syn. Pl. Glum. 1:357. 1854; Hook. f., Fl. Brit. India 7:365. 1896. *Tripogon roxburghianus* (Steud.) Bhide, J. Proc. Asiat. Soc. Beng. 1911 (n.s.) 7:515. 1912; Fischer, Fl. Madras. 3: 1834. 1934; Bor, Grass. India 524. 1960.

Annual, culms erect, tufted, slender to 15 cm high; nodes glabrous. Leaf sheaths glabrous; ligules membranous; leaf blades 5-8 cm long, filiform, margin ciliate. Spike solitary; spikelets 1-2-flowered; lower glume oblique, hyaline, 2-lobed, 1-nerved; upper glume coriaceous, 3-nerved;lower lemma membranous, bidentate, awned in the cleft, ciliate in lower part, callus bearded; upper lemma bisexual or imperfect or 0; palea 2-keeled, scabrid; stamens 3. Caryopsiss terete, oblong-ellipsoid, pointed at one side.

Rare in dry rocky areas in Warangal district.

Fl. & Fr.: July – December.

Madikonda (WGL), *CSR* 1116 (KU).

Oropetium thomaeum (L.f.) Trin., Fund. Agrost. 98. t.3. 1820; Hook.f., Fl. Brit. India 7:336.1896; Fischer, Fl. Madras 3: 1830. 1934; Bor, Grass. India 474. 1960; Moulik, Grass. Bamb. India 2: 617. 1997. *Nardus thomaea* L.f., Suppl. 105.1781.

Perennial, culms erect, tufted, to 6-12 cm high; nodes glabrous. Leaf sheaths 0.5-2.5 cm long, compressed; ligules membranous; leaf blade 1.5-2.5 cm long, linear, filiform, base rounded, acute, ciliate on margins. Spikes solitary, terminal, slender; spikelets sessile, lanceolate, solitary, sunk in the alternatingly distichous hollows in the rhachis, 1-4-flowered; lower glume minute, hyaline, nerveless; upper glume linear-lanceolate, puberulous on dorsal side, 3-nerved; lemmas ovate-oblong, hyaline, 1-3-nerved; paleas lanceolate, 2-keeled; lodicules 2; stamens 3. Caryopsis oblong.

Common in forest plains in dry places in Ranga Reddi, Hyderabad and Karimnagar districts.

Fl. & Fr.: July-November.

Vikarbad (RR), *MRS* 87 (DD); Kodimial (KRN), *GVS* 20048 (MH).

ORYZA L.

1. Ligule of the lower leaf blades less than 0.6 cm long .... **O. meyeriana** subsp. **granulata**
1. Ligule of the lower leaf blades 1-5 cm long:
    2. Spikelets persistent, unawned, cultivated .... **O. sativa** var. **sativa**
    2. Spikelets caducous, awned, wild .... **O. rufipogon**

Oryza meyeriana (Zollin. & Mor.) Baill., Hist. Pl 12: 166. 1894; subsp. **granulata** Tateoka, Bot. Mag. Tokyo 75: 460. 1962; Fischer, Fl. Madras 3: 1845. 1934. *O. granulata* Nees & Arn. ex Watt., Dict. Econ. Prod. Ind. 5: 500. 1891; Hook. f., Fl. Brit. India 7: 96. 1896; Bor, Grass. India 604. 1960.

A perennial, tufted grass; culms up to 90 cm high. Leaf sheaths 5- 7 cm long, glabrous; ligule short, membranous; blades 7-15 x 1- 1.5 cm, lanceolate, glabrous. Racemes 8 cm long, very slender; spikelets 6 m long, obtuse, unawned; lower glume absent; upper glume linear; lemma densely irregularly granulate, glabrous; palea elliptic-oblong, granulate, glabrous; stamens 6. Caryopsis ellipsoid or narrowly oblong.

Occasional grass in Medak district.

Fl. & Fr.: July-December.

Ramancheru, Perugunda (MDK), *RG* 96296 (BSID).

Oryza rufipogon Griff., Notul. 3: 5. 1851; Bor, Grass. India 605. 1960; Moulik, Grass. Bamb. India 1: 4. 1997.

Tufted, annual, aquatic grass; culms erect or decumbent, up to 1 m high, rooting at the nodes, spongy below; nodes glabrous. Leaf sheaths overlapping, 6-15 cm long, glabrous; ligule membranous; leaf blade 20-30 x 0.2-0.6 cm, linear, flat, base rounded, apex acuminate, scaberulous. Pancile up to 15 cm long, drooping; spikelets linear-oblong, 7 mm, awned; glumes triangular, acuminate, 1-nerved, scaberulous; lemma boat-shaped, 3-nerved, scabrid on the outer surface, awned, coriaceous; palea linear, scabrid on the outer surface, coriaceous; lodicules 2; stamens 6; stigmas plumose. Caryopsis narrowly oblong, free with in the lemma and palea.

Common aquatic grass along margins of ponds, water courses and in wetlands in most districts.

Fl. & Fr.: August-December.

Sirpur (ADB), *GO* 4418; Mosra RF (NZB), *TP & BR* 6108; Lingampet (NZB), *BR* 9296; Pocharam tank (MDK), *TP & MSG* 12050; Sardhana (MDK), *CP* 11769; Anantasagar RF (RR), *MSM* & *KH* 11006; Jakharam (WGL), *RKP* 105289 (BSID).

Oryza sativa L., Sp. Pl. 333. 1753. var. **sativa:** Hook. f., Fl. Brit. India 7: 92. 1896; Fischer, Fl. Madras 3: 1844. 1934; Bor, Grass. India 605. 1960; Moulik, Grass. Bamb. India 1: 48. 1997.

Vern.: *Vadlu.*

An annual, tufted grass; culms up to 1 m high, creeping or floating at base; nodes glabrous. Leaf sheaths 15-25 cm long, glabrous; ligule membranous. 2-partite; leaf blade 20-50 x 4-7 cm, linear, flat, base rounded, margin scabrid, apex acuminate, glabrous, midrib prominent. Panicle up to 30 cm long, lax, compound; spikelets 5 mm, linear-oblong, numerous, unawned; glumes oblong, scaly, 1-nerved; lemma boat-shaped, coriaceous, punctate, dorsally ciliate, 3-nerved; palea linear, scabrid on the outer surface; stamens 6; stigma plumose. Caryopsis narrowly oblong, compressed.

Widely cultivated in all districts for its grains and straw.

Fl. & Fr.: August-February.

Narsapur town (MDK), *KMS* 6672 (MH); Pocharam Wild Life Sanctuary (MDK), *RG* 116407 (BSID).

PANICUM L.

1. Upper lemma transversely rugose .... **P. maximum**
1. Upper lemma smooth:
    2. Lower glume more than half as long as the spikelet:
        3. An annual grass; leaf blade base rounded or truncate; caryopsis ovoid:
            4. Spikelet < 3 mm long .... **P. trypheron**
            4. Spikelet >4 mm long .... **P. miliaceum**

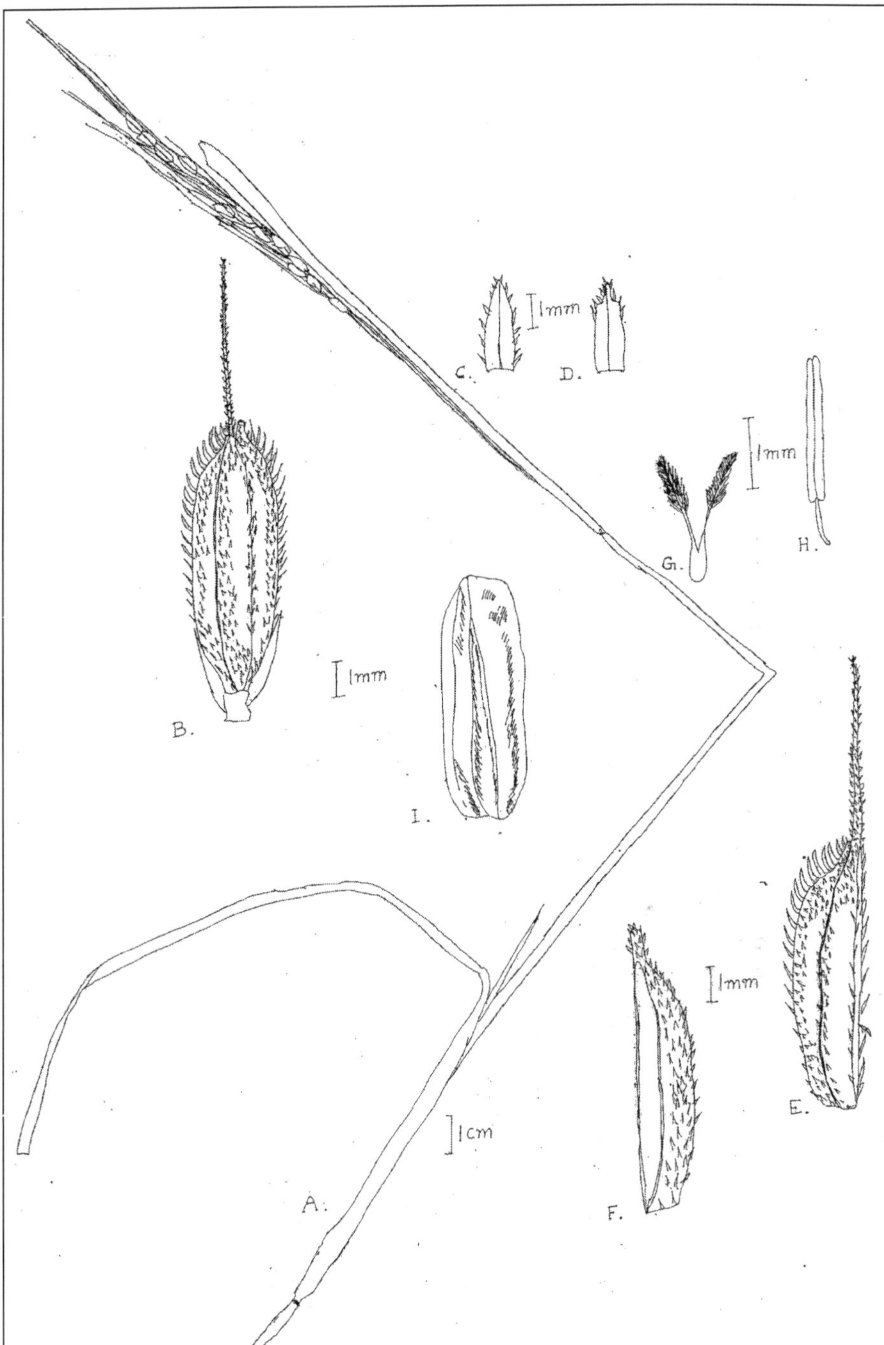

Figure 150: Oryza rufipogon Griff.

A. Twig, B. Spikelet, C. Lower glume, D. Upper glume, E. Lemma, F. Palea, G. Pistil, H. Stamen, I. Caryopsis.

3. A perennial grass; leaf blade base amplexicaul; caryopsis ellipsoid .... **P. notatum**

2. Lower glume less than half as long as the spikelet:

5. Culms spongy, lower glume nerveless .... **P. paludosum**

5. Culms not spongy, lower glume prominently nerved:

6. Leaf-sheaths margins densely ciliate; lower glume suborbicular, apex truncate .... **P. repens**

6. Leaf-sheath margins glabrous; lower glume ovate, apex acute or acuminate:

7. Panicle effuse; caryopsis ellipsoid; wild:

8. An annual; leaf blade linear-lanceolate, upper glume 3 or 10-nerved .... **P. walense**

8. A perennial, tall grass; leaf blade elliptic-lanceolate; upper glume 9-nerved .... **P. psilopodium**

7. Panicle contracted; caryopsis ovoid; cultivated .... **P. sumatrense**

Panicum maximum Jacq., Collectanea t. 1: 76. 1786; Hook. f., Fl. Brit. India 7: 49 1896; Fischer, Fl. Madras 3: 1783. 1934; Bor, Grass. India 327. 1960; Moulik, Grass. Bamb. India 1: 119. 1997.

A perennial, tufted grass; culms up to 1 m high, stout, branching from the base; nodes bearded. Leaf sheaths 10-25 cm long, ciliate near the mouth; ligule membranous, usually dense hairs from behind it; leaf blade 10-60 x 4-10 cm, linear, flat, base cordate, margin scaberulous, apex acuminate, glabrous, midrib prominent. Panicle up to 25 cm long, lax; spikelets 4 mm, linear-oblong, dense; lower glume orbicular, half the length of spikelet, faintly 3-nerved, glabrous to ciliate; upper glume broadly oblong, margins incurved, 5-nerved, glabrous to ciliate; lower lemma oblong, 5-nerved; palea oblong, 2-nerved, male; upper lemma ovate-oblong, 3-nerved, transversely rugose, pallid; palea similar to upper lemma, nerveless, bisexual; lodicules 2, stamens 3; stigma plumose, purplish. Caryopsis narrowly ellipsoid.

A tropical African grass cultivated for its valuable fodder and occasionally found as an escape.

Fl. & Fr.: July-October.

Kamareddy (NZB), *BR & KH* 9627.

Panicum miliaceum L., Sp. Pl. 58. 1753; Hook. f., Fl. Brit. India 7: 45. 1896; Fischer, Fl. Madras 3: 1782. 1934; Bor, Grass. India 327. 1960.

Vern.: *Warigalu, Barigalu.*

An annual, tufted grass; culms up to 1 m high, sparingly branched; nodes pilose. Leaf sheaths 7 cm long, loose, hirsute with tuberculate based hairs; ligule a ciliate

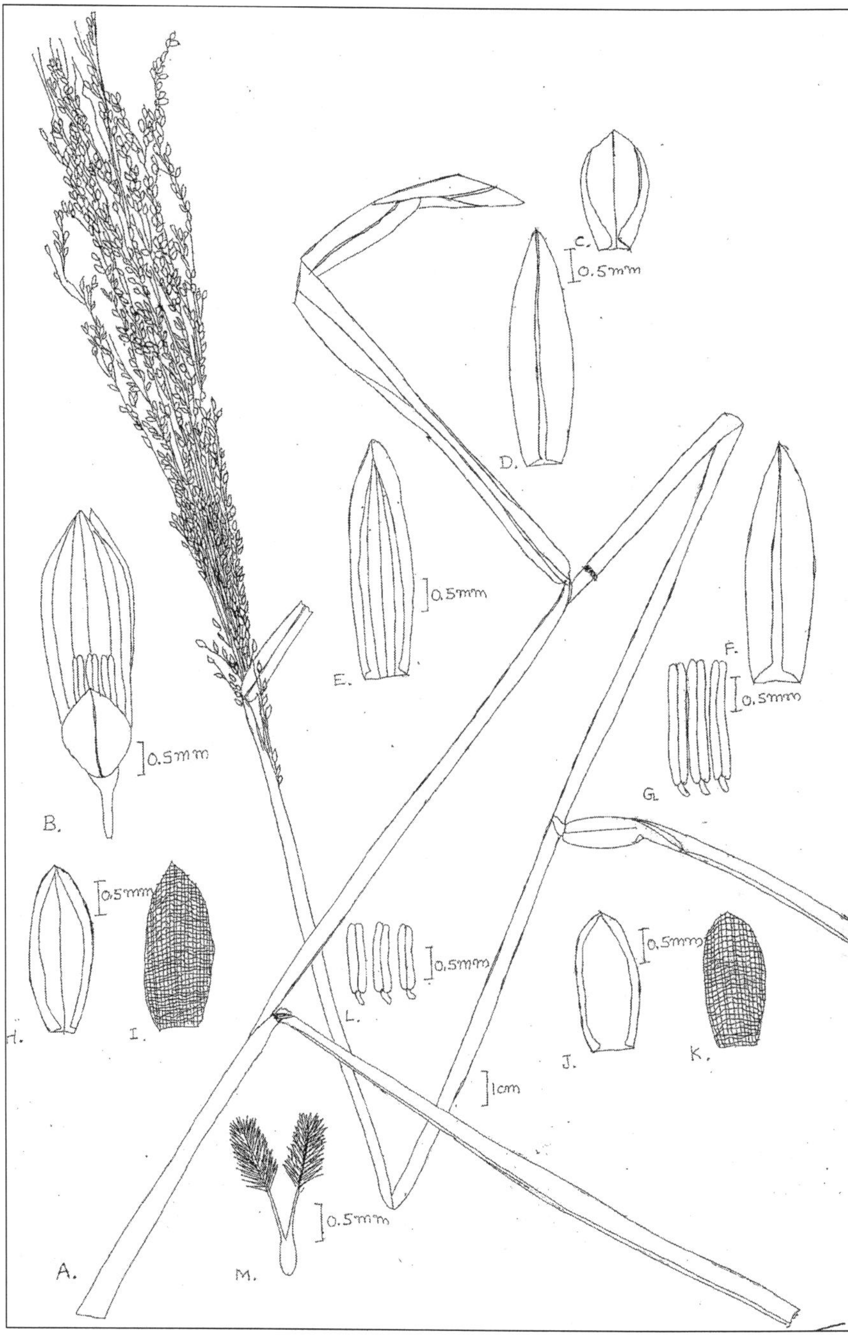

Figure 151: Panicum maximum Jacq.

A. Twig, B. Spikelet, C. Lower glume, D. Upper glume, E. Lower lemma, F. Palea, G. Stamens, H. Upper lemma, front and back views, J & K. Palea front and back views, L. Stamens, M. Pistil.

rim; blades 5-20 x 0.4-1 cm, lanceolate, base rounded, margins involute, apex acuminate, sparsely hairy on both the surfaces. Panicle 20 cm long, narrowly oblong; branches alternate, filiform, scabrid; spikelets 5 mm, ovate-oblong, apiculate; glumes persistent, unequal; lower broadly ovate, 5-nerved; upper ovate-oblong, 11- nerved; lower lemma oblong; palea ovate-oblong; upper lemma elliptic-oblong, crustaceous; palea similar to its lemma, bisexual; stamens 3; stigmas plumose. Caryopsis ovoid.

Occasionally cultivated for its grains and often as an escape in black cotton soils of Karimnagar and Mahabubnagar districts.

Fl. & Fr.: July-November.

Kollapur (MBNR), *BSS* & *SKB* 32915 (BSID).

Panicum notatum Retz., Observ. Bot. 4: 18. 1786; Bor, Grass. India 701. 1960. *P. montanum* Roxb., Fl. Ind. 1:315. 1820; Hook. f., Fl. Brit. India 7: 53. 1896; Fischer, Fl. Madras 3: 1783. 1934; Bor, Grass. India 329. 1960.

A perennial, slender grass; culms up to 1 m high, erect from woody stock, sparsely branched: nodes glabrous-sparsely pilose. Leaf sheaths 3-5 cm, ciliate on margin; ligule of soft long hairs; leaf blade 5-12 x 1-2 cm, linear-oblong, base amplexicaul, margin with tubercle based hairs, apex acuminate, sparsely pubescent on both surfaces. Panicle up to 25 cm long, lax, open; branches spreading, filiform; spikelets 2.5 mm, oblong, few, not gaping at maturity; pedicel very long, 2 cm, filiform; lower glume ovate, 3-5-nerved, pilose 3/4 length of the spikelets; upper glume broadly ovate, 5-nerved; lower lemma ovate, 5-nerved, sparsely hairy on outside, epaleated; upper lemma ovate, crustaceous, indurated; palea coriaceous with involute margins; lodicules 2; stamens 3, stigmas plumose, yellowish purple. Caryopsis ellipsoid, dorsally compressed.

Occasional grass along the forest paths in Adilabad district.

Fl. & Fr.: September – February.

Panicum paludosum Roxb., Fl. Ind. 1 :310. 1820; Fischer, Fl. Madras 3: 1783. 1934; Bor, Grass. India 329. 1960; Moulik, Grass. Bamb. India 1: 122. 1997. *P proliferum* sensu Hook. f., Fl. Brit. India 7: 50 1896. non Lam. 1797.

A perennial, tufted, aquatic grass; culms up to 60 cm high, branched, spongy; nodes glabrous. Leaf sheaths 4-12 cm long, glabrous; ligule a fimbriate membrane; leaf blade 10-20 x 0.3-0.6 cm, oblong- lanceolate, scabrid, apex finely acuminate. Panicle 10-20 cm long, spreading; branches alternate, rhachis slender, angular; spikelets 4 mm, narrowly ellipsoid-lanceolate, shortly acute-acuminate; lower glume orbicular, less than half the length of the spikelet, prominently 3-5-nerved; upper glume lanceolate, 7-9-nerved; lower lemma 9-11-nerved, acuminate; palea male or empty; upper lemma pallid, smooth, 3-nerved; palea similar to its lemma; stamens 3, stigmas plumose, purplish. Caryopsis ellipsoid.

Occasional near marshy areas of Medak district.

Fl. & Fr.: September-March

Rusthampet (MDK), *KMS* 6800 (MH).

Note: It is often confused with *Panicum repens,* but can be distinguished by the spongy culms, broad, flat leaf blade, lax panicle and narrowly lanceolate spikelets which are appressed to the branches. *P. paludosum* does not posses rhizomes unlike the other ones.

Panicum psilopodium Trin., Gram. Panic. 217. 1826; Hook. f., Fl. Brit. India 7: 46. 1896; Fischer, Fl. Madras 3: 1782. 1934; Bor, Grass. India 329. 1960; Moulik, Grass. Bamb. India 1: 122. 1997.

A perenniah tufted grass; culms up to 50 cm high; nodes glabrous. Leaf sheaths 4-8 cm long, glabrous; ligule a fimbriate membrane; leaf blade 4-10 × 0.3-0.9 cm, elliptic-lanceolate, base cordate, apex acuminate, glabrous. Panicle up to 12 cm long, ovate-oblong, effuse; branches numerous, alternate; spikelets 3 mm, elliptic-oblong, deciduous; lower glume less than half the length of the spikelet, ovate, obscurely 3-nerved; upper glume broadly ovate, acute, 9-nerved, margins incurved; lower lemma broadly ovate, 10-nerved; palea oblong, obtuse, scabrid on margin, male; upper lemma oblong, acute; palea oblong, obtuse, bisexual; lodicules 2; stamens 3; stigmas plumose, purplish. Caryopsis ellipsoid.

Common in shady and wet or marshy areas of Adilabad, Medak, Karimnagar, Warangal, Hyderabad, Khammam and Nalgonda districts.

Fl. & Fr.: July-September.

Bheemaram RF (ADB), *TP & GO* 5460; Bellampalli (ADB), *PVP* 9458; Poosampalle (MDK), *TP & MSG* 11959; Ramayampet (MDK), *BR* & *CP* 11521; Upparapalli (HYD), *MRS* 366 (DD); Boddupalle (NLG), *BVRS* 650 (NU); Hamsanpalli to Medak (MDK), *RG* 104087 (BSID); Ramanagutta (KMM), *RR* 108009 (BSID); way to Gangavaram (WGL), *PVS* 84033 (BSID).

Panicum repens L., Sp. Pl. ed. 2. 87. 1762; Hook. f., Fl. Brit. India 7: 49. 1896; Fischer, Fl. Madras 3: 1783. 1934; Bor, Grass. India 330. 1960; Moulik, Grass. Bamb. India 1: 123. 1997.

A perennial, stoloniferous grass; culms up to 45 cm long, creeping at base, rooting from the nodes; nodes glabrous. Leaf sheaths 2-6 cm, densely ciliate on margin; ligule membranous; leaf blade 5-15 x 0.4-0.8 cm, linear-lanceolate, flat, base rounded, apex acute, sparsely pubescent. Panicle 7-20 cm long, lax, branches subfiliform, pedicels capillary, scabrid; spikelets 3 mm, ovate-lanceolate; lower glume sub orbicular, apex truncate, 3-nerved; upper glume lanceolate, 9-nerved; lower lemma lanceolate, 10-nerved, paleated, male; upper lemma ovate, pallid, indurated; palea similar to upper lemma; stamens 3; stigma plumose, greenish to purplish. Caryopsis ellipsoid.

Common in fields and waste lands of Adilabad, Medak, Hyderabad, Karimnagar and Ranga Reddi districts.

Fl. & Fr.: March –December.

Sone (ADB), *TP & PVP* 4032; Choutkoor (MDK), *TP & MSG* 11907; Central University campus (RR), *MSM* & *TP* 15142; Hyderabad, *MRS* 3111 (DD); Near Hussainsagar dam (HYD), *KMS* 8040 (CAL).

Panicum sumatrense Roth ex Roem. & Schult., Syst. Veg. 2;434. 1817; Bor, Grass. India 701. 1960; Moulik, Grass. Bamb. India 1: 123. 1997. *P. miliare* Lam., Tabl. Encycl. 1: 173. 1791; Hook. f., Fl. Brit. India 7: 46. 1896; Fischer, Fl. Madras 3: 1782. 1934; Bor, Grass. India 329. 1960.

Vern.: *Gangasamulu.*

An annual, tufted grass; culms up to 1 m high; nodes glabrous. Leaf sheaths 4-10 cm, glabrous; ligule membranous; leaf blade 7-15 x 0.5-1 cm, flat, base rounded. Panicle 10-25 cm long, thyrsiform or contracted; lower branches alternate; spikelets paired, 5 mm, ellipsoid, glabrous; lower glume broadly ovate, 3-5-nerved, obtuse; upper glume ovate-lanceolate, membranous, 9-13-nerved, obtuse; lower lemma 9-nerved; palea equal to the length of its lemma, neuter; upper lemma elliptic-oblong, 3-5-nerved, glabrous; palea similar to its lemma; lodicules 2; stamens 3; stigma plumose. Caryopsis ovoid.

Occasionally cultivated grass in some districts.

Fl. &Fr.: October-December.

Mohammadabad (RR), *MSM* 10552 & 10555.

Panicum trypheron Schult., Syst. Veg. 2. Mant. 244. 1824; Hook. f., Fl. Brit. India 7; 47. 1896; Fischer, Fl. Madras 3: 1783. 1934; Bor, Grass. India 331. 1960; Moulik, Grass. Bamb. India 1: 124. 1997.

An annual, tufted grass; culms up to 60 cm high; nodes glabrous. Leaf sheaths 2-8 cm long, sparsely ciliate-glabrous; ligule membranous, with long pilose tufts; leaf blade 2-12 x 0.2-0.6 cm, linear-narrowly lanceolate, base truncate, apex acuminate, pilose. Panicle lax, 15 cm long, open; branches alternate; spikelets 3 mm, ellipsoid, gaping at maturity; glumes ovate, acuminate-cuspidate; lower glume membranous, 3-5-nerved, scabrid, apex cuspidate-acuminate; upper glume membranous, 7-9-nerved, apex acuminate; lower lemma ovate, acute; palea male or empty; upper lemma pallid, smooth; palea similar to its lemma; lodicules 2; stamens 3; stigmas plumose. Caryopsis ovoid.

Common grass in moist plains and hill slopes of most districts.

Fl. & Fr.: October – March.

Vempalli (ADB), *GO* & *PVP* 4760; Pocharam (NZB), *BR* 9300; Pocharam RF (MDK), *TP & MSG* 12035, 12036. 12041; Hyderabad (HYD), *MRS* 238 (DD).

Panicum walense Mez., Bot. Jahrb. 34: 146. 1904. (*"watense"*); T.A. Cope in Nasir & Ali, Fl. Pak. 143; 166. 1982; Moulik, Grass. Bamb. India 1: 124. 1997. *P. humile* Nees ex Steud., Syn. Pl. Glum. 1: 84. 1854 non Thunb. ex Trin. 1826; Hook. f., Fl. Brit. India 7: 48. 1896; Fischer, Fl. Madras 3: 1782.1934. *P. austroasiaticum* Ohwi, Act. Phytotax. Geobot. 2: 45. 1942; Bor, Grass. India 324. 1960.

An annual, tufted grass; culms up to 30 cm high, dwarf, slender, geniculately ascending and branched below. Leaf sheaths 0.5-2 cm long, glabrous, sparsely ciliate at mouth; leaf blade linear, 5-10 x 0.3-0.8 cm, flat, base cordate, apex acuminate, glabrous. Panicle 7 cm long, lax, diffusely branched; branchlets capillary; spikelets 2 mm, ovate, acute, glabrous; glumes unequal; linear, small, ovate, acuminate, 3-nerved; upper larger, ovate-lanceolate, acuminate, 3-nerved; lower lemma ovate, acuminate, 3-nerved; upper lemma shorter, broadly oblong, obtuse; palea similar to its lemma, shining. Caryopsis ellipsoid.

Occasional in moist areas and hill slopes of Hyderabad district.

Fl. & Fr.: October- February.

Hyderabad (HYD), *MRS* 252 (DD).

PASPALIDIUM Stapf

1. Leaves blunt; spikes shorter than the internodes .... **P. flavidum**
1. Leaves acute or acuminate; spikes longer than the internodes:
   2. Upper glume broadly oblong; upper lemma rugulose .... **P. punctatum**
   2. Upper glume broadly ovate; upper lemma granulose .... **P. geminatum**

Paspalidium flavidum (Retz.) A. Camus in Lecomte., Fl. Gen. de 1' Indo-chine 7:419. 1922; Fischer, Fl. Madras 3: 1744. 1934; Bor, Grass. India 333. 1960; Moulik, Grass. Bamb. India 1: 125. 1997. *Panicum flavidum* Retz., Observ. Bot. 4. 15. 1786; Hook. f., Fl. Brit. India 7: 28. 1896.

Vern.: *Udagaddi.*

An annual, tufted grass; culms up to 70 cm high, geniculately ascending; nodes glabrous. Leaf sheaths 6-15 cm, glabrous; ligule fringed, membranous; leaf blade 5-20 x 0.5-0.8 cm, linear-lanceolate, base rounded, margins scaberulous, apex acuminate, slightly pubescent. Paniculate racemes up to 20 cm long; branches distant, secund; spikelets 3 mm, broadly ovate, secund in two rows, 2 florets, lower usually neutral or sometime male, upper bisexual; lower glume orbicular, 3-nerved, keeled, glabrous; upper glume broadly ovate, 7-nerved, keeled; lower lemma broadly ovate, 5-nerved, paleated, empty; upper lemma broadly ovate or orbicular, convex, coriaceous, mucronate; palea similar to its lemma; lodicules 2; stamens 3; stigmas plumose, brownish. Caryopsis ovate-oblong, tightly enclosed in the hardened lemma and palea.

Common in cultivated fields and waste lands of most districts.

Fl. & Fr.: July-October.

Sattenapalli beat (ADB), *TP & PVP* 4097; Vempalli (ADB), *GO & PVP* 4781; Bijjur RF (ADB), *GO & PVP* 4868; Yeddapally (NZB), *BR & KH* 6475; Marrikunta (MDK), *CP* 11391; Rudraram RF (RR), *TP* & *MSM* 12719; Hyderabad (HYD), *MRS* 89 (DD); Bhupatipur (KRN), *GVS* 25710 (MH); Pakhal (WGL), *KMS* 13158 (MH); Koriguttalu (KMM), *RR* 106036 (BSID).

Paspalidium geminatum (Forssk.) Stapf in Prain, Fl. Trop. Afr. 9. 583 1920; Fischer, Fl. Madras 3: 1774. 1934; Bor, Grass. India 333. 1960; Moulik, Grass. Bamb. India 1: 125. 1997. *Panicum geminatum* Forssk., Fl. Aegypt-Arab. 18. 1715. *P. paspaloides* Pers., Syn. 1, 81. 1805; Hook. f., Fl. Brit. India 7: 30. 1896.

A perennial, marshy grass; culms up to 50 cm high, spongy, creeping below; branches from lower nodes; nodes glabrous. Leaf sheaths 4-10 cm, loose, glabrous; ligule a ciliate rim; leaf blade 5-15 x 0.3-0.6 cm, linear, base rounded, apex acuminate, glabrous. Spike like panicle up to 15 cm long, racemes 2 cm, 10- 20, alternate on the rhachis, compressed; spikelets 2 mm, ovoid, 2 florets, lower male or neutral, upper bisexual; lower glume orbicular, nerveless; upper glume ovate, 5-7-nerved, concave; lower lemma ovate-oblong, membranous, 5-7-nerved, paleated, male; upper lemma coriaceous, broadly ovate, mucronate; palea similar to its lemma, margins inflexed; lodicules 2; stamens 3, stigma plumose, brownish. Caryopsis ellipsoid.

Occasional grass near water courses, usually near marshy areas of Adilabad, Nizamabad, Medak, Ranga Reddi, Hyderabad, Karimnagar and Mahabubnagar districts.

Fl. & Fr.: December-July.

Kadam dam (ADB), *TP & PVP* 4139; Kammarpally (NZB), *TP & BR* 6235; Narsapur tank (MDK), *TP & MSG* 12006; Anantasagar RF (RR), *MSM* & *KH* 11002; Moosi river (HYD), *MRS* 273 (DD).

Paspalidium punctatum (Burm. f.) A. Camus in Lecomte, Fl. Gen. de. 1' Indo-chine 7: 41. 1922; Fischer, Fl. Madras 3: 1774.1934; Bor, Grass. India 333. 1960; Moulik, Grass. Bamb. India 1: 126. 1997. *Panicum punctatum* Burm. f., Fl. Ind. 26. 1768; Hook. f., Fl. Brit. India 7: 29. 1896.

A perennial, tufted grass; culms up to 1 m high, prostrate, rooting at lower nodes, spongy. Leaf sheaths 10 cm long, more or less ribbed; ligule a ridge of hairs; leaf blade 10-15 × 0.5-1 cm, linear, flat or convolute, apex acuminate, glabrous. Spike like panicle 30 cm long; racemes 20 cm long, distant, sessile, compressed, appressed to the rhachis; spikelets 3 mm long, 18-40 per raceme, ovoid; lower glume truncate, more or less cuspidate, faintly 0-3-nerved; upper glume broadly oblong, prominently 5-nerved; lower lemma acuminate, prominently 5-nerved; palea empty; upper lemma punctate, apiculate; palea coriaceous, margins incurved; lodicules 2; stamens 3; stigmas plumose, purplish. Caryopsis ovoid.

Common in wet places in Adilabad, Mahabubnagar and Warangal districts.

Fl. & Fr.: December- July.

Kadthala (ADB), *GO & DAM* 5027; Swarna project (ADB), *MHR* 14574.

PASPALUM L.

1. Stoloniferous grass; spikelets elliptic; upper glume pubescent .... **P. paspalodes**
1. Tufted grass, non-stoloniferous; spikelets orbicular; upper glume glabrous .... **P. scrobiculatum**

Paspalum paspalodes (Michx.) Scribn., Mem. Torr. Bot. Cl. 5: 29. 1894. (*'paspaloides'*); Bor., Grass. India App. 11. 1973; Moulik, Grass. Bamb. India 1: 133. 1997. *Digitaria paspalodes* Michx., Fl. Bor. Amer. 1 : 46. 1803. *Paspalum distichum* auct non L. 1759, *nom. confus.*; Hook. f., Fl. Brit. India 7: 12. 1896; Bor, Grass. India 338. 1960.

A perennial, stoloniferous grass; culms up to 50 cm, ascending from creeping base, rooting at nodes, branches from lower nodes; nodes glabrous. Leaf sheaths 4-10 cm long, glabrous; ligule membranous, divaricate; leaf blade 5-14 × 0.4-0.8 cm, linear-lanceolate, base rounded, acute, glabrous. Spikes up to 9 cm long, axis flattered; spikelets 3 mm, elliptic-oblong, subsessile, dorsally compressed, 2-flowered; lower glume suppressed; upper glume elliptic-oblong, membranous, pubescent, margins inflexed, 5-nerved; lower lemma similar to upper glume; upper lemma elliptic-concave, apex acuminate, glabrous; palea elliptic; lodicules 2; stamens 3; stigmas plumose. Caryopsis ovoid-ellipsoid.

Occasional grass in marshy areas of Hyderabad, Mahabubnagar, Nizamabad and Warangal districts.

Fl. & Fr.: August-November.

Dichpally (NZB), *BR & KH* 6442; Kachiguda (HYD), *MRS* 107 (DD).

Paspalum scrobiculatum L., Mant. Pl. 29. 1767; Hook. f., Fl. Brit. India 7: 10. 1890; Fischer, Fl. Madras 3: 1772. 1934; Moulik, Grass. Bamb. India 1: 134. 1997. *P commersonii* Lam., Tabl. Encycl. 1: 175. t. 43. f. 1. 1791; Bor, Grass. India 335. 1960. *P. orbiculare* Forst. f., Insul. Austr. Prodr. 7. 1786; Fischer, Fl. Madras 3: 1772. 1934.

Vern.: *Arikelu.*

An annual, tufted grass; culms up to 70 cm high, leafy; nodes glabrous. Leaf sheaths 5-15 cm long, glabrous; ligule a ciliate rim; leaf blade 10-40 x 0.3-0.7 cm, linear, base rounded, apex acute, midrib prominent, glabrous. Racemes 2-5, up to 10 cm long, axis flattened; spikelets 2 mm, globose, in two rows; lower glume obscure; upper glume ovate, boat-shaped, chartaceous, 5-7-nerved. Florets 2, lower barren, epaleate, upper bisexual; lower lemma ovate, membranous, 5-7-nerved; upper lemma ovate, crustaceous, 3-nerved; palea orbicular; lodicules 2; stamens 3; stigmas plumose. Caryopsis plano-convex, free with in the lemma and palea.

Occasionally cultivated in black cotton soils and rarely as an escape in most districts.

Fl. & Fr.: Throughout the year.

Bellampalli (ADB), *PVP* 9443; Laxmapur (NZB), *TP & BR* 6284; Pocharam tank (MDK), *TP & MSG* 12037; Kondapur RF (RR), *MSM* & *KH* 10590; Rudraram RF (RR), *MSM* 12729; Rajawaram (NLG), *BVRS* 449 (NU); Mallelatheertham (MBNR), *BSS* & *BKS* 32211 (BSID).

PENNISETUM Rich. ex Pers.

1. Anther cells not bearded at the apex:
    2. A low, copiously branched perennial, forming mats with numerous stolons with short internodes, culms very short, 10 cm or less high .... **P. clandestinum**
    2. Annual or perennial, erect, culms >70 cm high:

3. Involucral bristles scabrid; lemmas homomorphous, not 3-lobed:
   4. An annual; upper glume broadly ovate, upper lemma transversely rugose; leaf blades flat .... **P. glaucum**
   4. A perennial; upper glume elliptic-oblong, upper lemma not transversely rugose; leaf blades convolute .... **P. hohenackeri**
3. Involucral bristles densely villous or ciliale below the middle; lemmas heteromorphous, 3-lobed:
   5. Spikelets solitary, pedicellcd or in groups of 2-5, with one sessile and other pedicelled, lower glume of sessile spikelet 1.7-3 mm long .... **P. pedicellatum**
   5. Spikelets solitary, sessile, lower glume of sessile spikelet reduced, 0.2-0.8 mm long:
      6. Involucral bristles densely plumose, annual grass .... **P. polystachyon**
      6. Involucral bristles outer not ciliate, the inner long with long silky hairs below the middle, perennial grass .... **P. setosum**

1. Anther cells bearded at the apex:
   7. Wild, perennial; spikelets subsessile, not gaping .... **P. purpureum**
   7. Cultivated, annual; spikelets pedicelled, gaping .... **P. americanum**

Pennisetum americanum (L.) Leeke, Z. Naturwiss. 79: 52. 1907; Bor in Rech. f., Fl. Iran. 70: 503. 1970. *Panicum americanum* L., Sp. Pl. 56. 1753. *Alopecurus typhoides* Burm. f., Fl. Ind. 27. 1768. *Pennisetum typhoideum* Rich., in Pers., Syn. Pl. 1: 72. 1805; Hook. f., Fl. Brit. India 7: 82. 1896. *P. typhoides* (Burm. f.) Stapf & C. E. Hubb., Kew Bull. 1933: 271. 1933; Fischer, Fl. Madras 3: 1792. 1934; Bor, Grass. India 350. 1960.

Vern.: *Sajjalu.*

An annual, tufted grass; culms up to 2 m high; nodes bearded. Leaf sheaths overlapping, sparsely tubercled based hairy; ligule a ciliate rim; leaf blade 30-90 x 0.5-5 cm, linear-lanceolate, base rounded, apex acuminate, pilose below, midrib prominent. Panicle up to 25 cm long, spiciform, cylindric, dense; rhachis stout, villous below the inflorescence; involucral bristles 4 mm, often purplish-coloured; spikelet 4 mm, oblong; lower glume absent; upper glume minute, truncate, 3-nerved; lower lemma oblong, obtuse, 5-nerved, epaleated; upper lemma ovate-oblong, membranous, 2-nerved; palea very broad, truncate, ciliate at the tip; stamens 3, apex penicillate, stigmas plumose. Caryopsis oblong-obovoid or fusiform; glabrous, top exposed.

Widely cultivated for its grains in all districts.

Fl. & Fr.: August-December.

Hyderabad (HYD), *MRS* 229 (DD).

Pennisetum clandestinum Hochst. ex Chiov., Ann. Reale 1st. Bet. Roma 8: 41. t. 5 f. 2. 1903; Fischer, Fl. Madras 3: 1793. 1934; Bor, Grass. India 344. 1960; Moulik, Grass. Bamb. India 1: 137. 1997.

A perennial, stoloniferous grass; culms very short, up to 10 cm high, spread on the floor like mats; nodes glabrous. Leaf sheaths overlapping, 1.5-1.6 cm, terete, mouth bearded, sparsely hairy dorsally and above the middle on one margin; ligule a fringe of hairs; leaf blade 5-15 x 0.2-0.5 cm, linear, convolute, base rounded, apex obtuse. Inflorescence axillary, of 2 cleistogamous spikelets; spikelets in pairs, 1 sessile and other pedicelled, bristles few; lower glume absent; upper glume obtuse to truncate; lemmas similar, ovate to oblong-lanceolate; stamens 3; stigma one, plumose.

Excellent tropical African grass, cultivated to improve the grazing.

Fl. & Fr.: May – October.

Pennisetum glaucum (L.) R. Br., Prodr. 1: 195. 1810: T.A. Cope in Nasir & Ali, Fl. Pak. 143: 234, 1982. *Panicum glaucum* L. Sp. Pl. 56. 1753. *Setaria glauca* (L.) Beauv. Ess. Agrost. 51. 109. 178. 1812; Hook. f., Fl. Brit. India 7: 78. 1896; Bor, Grass. India 360. 1960.

Annual, tufted grass; culms up to 60 cm high, ascending; nodes glabrous. Leaf sheaths glabrous; ligule a ridge of hairs; blades 10-30 x 0.5-1 cm, linear, flat, base rounded, margins scabrid, apex acuminate, glabrous, sparsely hairy. Spike like raceme 13 cm long, cylindric, densely-flowered; involucral bristles 6-12, 4-8 mm long, scabrid; spikelets 3 mm, ellipsoid, numerous, glabrous; lower glume sub-equal. broadly ovate, 3-nerved, hyaline, shortly apiculate; upper glume broadly ovate, hyaline, 5-nerved; lower lemma thinly membranous, 5-nerved; palea hyaline, male or empty; upper lemma coriaceous, broadly elliptic, convex, transversely rugose.

In open places and along road sides in Hyderabad district.

Fl. & Fr.: June-October.

Pennisetum hohenackeri Hochst. ex Steud., Syn. Pl. Glumac. 1: 103. 1854; Fischer, Fl. Madras 3: 1792. 1934; Bor, Grass. India 344. 1960; Moulik, Grass. Bamb. India 1: 138. 1997. *P. alopecuros* Steud., Syn. Pl. Glumac. 1: 102. 1854; Hook. f., Fl. Brit. India 7: 84. 1896.

A perennial, densely tufted grass; culms up to 70 cm high; nodes glabrous. Leaf sheaths overlapping, 5-15 cm long, glabrous; ligule a small hairy ring; blade 15-40 × 0.2-0.5 cm, linear, coriaceous, convolute, base truncate, margins hairy near base, apex acuminate, glabrous. Spikelike raceme up to 12 cm long, solitary; rhachis below the inflorescence pilose; involucral bristels 1.5 cm; spikelet 8 mm, elliptic-lanceolate; lower glume reduced, scale like; upper glume elliptic-oblong, acuminate, 5-nerved; lower lemma elliptic-oblong, margins incurved, 9-nerved, palea absent; upper lemma lanceolate, acuminate, 5-nerved, palea subcoriaceous, 2-keeled, 2-nerved; lodicules absent; stamens 3; stigma plumose, purplish to white. Caryopsis narrowly oblong.

Occasional in sandy soils near streams and bunds of rice fields of Adilabad, Nizamabad, Medak, Ranga Reddi, Hyderabad, Karimnagar and Mahabubnagar districts.

Fl. & Fr.: August- October.

Birsaipet (ADB), *GO & PVP* 4622; Mosra - Malkapur road (NZB), *TP & BR* 6093; Patharajampet railway gate (NZB), *BR & CPR* 7161; Rayanapalli (MDK), *TP & MSG* 12111; Ramayyagudem RF (RR), *MSM* & *KH* 10579; Katasintha (HYD), *KMS* 6010; Mannanur (MBNR), *MRS* 54 (DD).

Pennisetum pedicellatum Trin., Mem. Acad. Sci. Petersb. Ser. 6: 3: 184. 1934; Hook. f., Fl. Brit. India 1: 86. 1896; Fischer, Fl. Madras 3: 1792. 1934; Bor, Grass. India 346. 1960; Moulik, Grass. Bamb. India 1: 139. 1997.

An annual, erect, tufted grass; culms up to 90 cm high, branches from the base, leafy. Leaf sheaths 5-20 cm long, glabrous; ligule a shortly ciliate membrane; leaf blade 10-20 x 3-9 cm, linear, flat, glabrous-sparsely hairy, margin scabrid, apex acuminate;. Racemes 13 cm long, cylindric, densely-flowered; involucral bristles outer few, slender, short, 3 mm, inner long, 1 cm, numerous, densely villous below the middle, unequal, free from the base; spikelets 4 mm, solitary in the involucral, sub sessile; lower glume very small, woolly, nerveless; upper glume oblong-lanceolate, membranous, hyaline, apiculate; lower lemma oblong, truncate, 3-lobed, hyaline; upper lemma ovate-oblong, obtuse, apex fimbriate ciliate, coriaceous, shining; palea lanceolate, toothed; lodicules absent; stamens 3. Caryopsis elliptic.

In plains and hill slopes of Medak, Ranga Reddi, Hyderabad and Warangal districts.

Fl. & Fr.: October-November.

Narsapur RF (MDK), *TP & MSG* 12014; Patanchervu (MDK), *BR* 11316; Lingampalle (RR), *MSM* 12734; Pakhal RF (WGL), *KMS* 11596 (CAL).

Note: It can be distinguished from all the other species of this genus by its fluppy - white or purplish panicles, densely woolly, plumose involucral bristles and pedicelled spikelets. This is similar to *P. polystachyon*, but differs by the same character mentioned above.

Pennisetum polystachyon (L.) Schult., Mant. 2 : 146. 1824; Fischer, Fl. Madras 3: 1792. 1934; Bor, Grass. India 346. 1960 . *Panicum polystachyon* L., Syst. Nat. ed. 10. 870. 1759.

An annual, tufted grass; culms up to 40 cm high, much branched. Leaf blades 7-35 x 0.5-1.5 cm. Panicle 20 cm long, linear, slender; rachis slender, puberulous; bristles densely plumose, especially the inner. Spikelets solitary, rarely 2 in an involucre, pedicelled; lower glume minute or suppressed; upper glume puberulous; lower lemma similar to upper glume, 3-toothed; upper chartaceous, fimbriate; palea similar to its lemma.

Frequent along margins of ponds and ditches of Hyderabad district (Ramana, 2010).

Fl. & Fr.: July-January.

Pennisatum purpureum Schumach., Beskr. Guin. Pl. 44. 1827; Bor, Grass. India 348. 1960; Moulik, Grass. Bamb. India 1: 140. 1997.

A perennial, stoloniferous grass; culms up to 2 m high, erect, branching from lower nodes; nodes bearded. Leaf sheaths 10-15 cm long, glabrous; ligule a dense fringe of hairs; leaf blade 30-60 x 0.7-1.5 cm, linear - lanceolate, base rounded, margins scaberulous, apcx acuminate, scabrid, midrib prominent. Spikelike raceme up to 18 cm long, solitary, cylindric, yellowish or purplish; involucral bristles numerous of unequal length, one usually very much longer, 8 mm, scabrid, ciliate, inner most scabrid, 6 mm, sparingly plumose towards the base; spikelet solitary, sessile, if in fascicles 2-4, lateral pedicelled, all lanceolate, 5 mm; lower glume membranous, nerveless; upper glume membranous, narrowly ovate-elliptic, glabrous, 1-nerved; lower lemma oblong - lanceolate, scabrid on outside surface, 5-nerved, male or barren; upper lemma lanceolate, rolled, scabrid at apex, 2-nerved; stamens 3, penicillate at apex; style united throughout; stigmas plumose. Caryopsis ovoid.

Occasional in hill slopes and forest areas of Adilabad, Karimnagar and Medak districts.

Fl. & Fr.: July – January.

Sathenapalle (ADB), *TP & PVP* 4157.

Note: The 'Napeur grass' is a native of Tropical Africa. But now introduced into many tropical countries including India as a valuble fodder grass. Like *Themeda cymbaria,* it is also known as 'elephant grass.'

Pennisetum setosum (Sw.) Rich. in Pers., Syn. Pl. 1: 72. 1805; Hook. f., Fl. Brit. India 7: 87. 1896; Bor, Grass. India 340. 1960. *Cenchrus setosum* Sw., Prodr. Veg. Ind. Occ. 26. 1788.

A perennial, tufted grass; culms up to 80 cm high, often fastigiately branched at the nodes. Leaf sheaths glabrous; ligule with fringed long, soft hairs; leaf blade 10-20 x 5-10 cm, linear, apex acuminate, slightly hairy. Racemes 10 cm long, purplish brown; involucral bristles unequal, the outer not ciliate, short, 3 mm, the inner longer, long silky hairy below the middle, 1.3 cm; spikelet 4 mm, solitary within the involucral; lower glume minute or completely absent; upper glume ovate-oblong, cuspidate, hyaline; lower lemma oblong, obtuse, bidentate, palea narrowly oblong, hyaline, male; upper lemma ovate-oblong, truncate, coriaceous, shining; palea oblong, truncate, toothed or ciliate at the apex; stamens 3.

In plains and hill slopes of Warangal district.

Fl. & Fr.: July-January.

Pakhal (WGL), *KMS* 11596 (MH).

PEROTIS Aiton

Perotis indica (L.) Kuntze, Revis. Gen. Pl. 2:787. 1891; Fischer, Fl. Madras 3: 1814. 1934; Bor, Grass. India 611. 1960; Moulik, Grass. Bamb. India 2: 504. 1997. *Anthoxanthum indicum* L., Sp. Pl. 28. 1753. *Perotis latifolia* Aiton, Hort. Kew 1: 96. 1789; Hook. f., Fl. Brit. India 7: 98. 1896.

A perennial, tufted grass; culms up to 50 cm high, geniculately ascending from base; nodes glabrous. Leaf sheaths 1-7 cm long, glabrous; ligule absent; leaf blade 1-4 x 0.3-0.6 cm, ovate-lanceolate, base amplexicaul, margin cartilaginous, spinulose, apex acute, glabrous. Spikelike raceme up to 15 cm long, terminal; spikelets 2 mm, linear, subsessile; glumes subequal, linear, 1-nerved, margins scabrid in lower, awned; lemma linear, hyaline, 1-nerved; palea narrow, 2-nerved; stamens 3; stigmas plumose. Caryopsis linear.

Very common in sandy tracts and plains of most districts.

Fl. & Fr.: July-January.

Vempalli (ADB), *GO & PVP* 4782; Jaipur RF (ADB), *GO & PVP* 4960; Ankusapuram (ADB), *GO* 4306; Kotpalli RF (ADB), *GO* 4519; Rudraram RF (RR), *TP* & *MSM* 12722; Nagarjunakonda valley (NLG), *KT* 9655 (CAL); Cherukupally (NLG), *BVRS* 173 (NU); Hyderabad (HYD), *MRS* 269 (DD); Pasra (WGL), *RKP* 108122 (BSID).

PHRAGMITES Adans.

Phragmites vallatoria (L.) Veldkamp, Blumea 37: 233. 1992. *Arundo vallatoria* Pluk. ex L., Herb. Amb. 15. 1754. *Phragmates karka* (Retz.) Trin. ex Steud. Nom. Bot. ed. 2. 2: 324. 1841; Hook. f., Fl. Brit. India 7: 304. 1896; Fischer, Fl. Madras 3: 1807. 1934; Bor, Grass. India 416. 1960; Moulik, Grass. Bamb. India 2: 433. 1997; Gayathri & Pullaiah in Pullaiah, Fl. Andhra Pradesh 3: 1240. 1997. *Arundo karka* Retz., Observ. Bot. 4: 21. 1786.

A perennial shrub; culms up to 5 m high, stout woody below, stoloniferous, creeping; nodes glabrous. Leaf sheaths 1.5-20 cm long, glabrous; ligule a ciliate rim; leaf blade 15-60 x 0.7-4 cm, linear-lanceolate, stiff, base rounded, apex acuminate, glabrous. Panicle up to 60 cm long, lax, open, decompounds, branches filiform, scabrid; spikelets 1.3 cm long, lanceolate, 4-flowered, laterally compressed, pedicelled; callus pilose, it covers the spikelet; lower glume oblong, 1-keeled, keel minutely scabrous, 3-nerved; upper glume linear-lanceolate, 3-nerved, apex acuminate; lower lemma linear, 1-keeled, 1-3-nerved, apex acuminate; palea oblong, obtuse, 2-keeled, 2-nerved; lodicules 2; stamens 3; stigmas plumose. Caryopsis oblong, terete.

Occasional robust grass near the water courses of Adilabad, Nizamabad and Medak districts.

Fl. & Fr.: November-March.

Kunthala water falls (ADB), *GO & DAM* 5131; Laxmapur (NZB), *TP & BR* 6296; Pochammaralu (MDK), *BR* & *CP* 11470.

Note: It is very similar to *Arundo donax,* but can be distinguished by the narrower leaves, glabrous lemma and densely villous spikelets. This grass said to be poisonous to cattle.

POGONATHERUM P. Beauv.

Pogonatherum paniceum (Lam.) Hack, in Allg. Bot. Z. Syst. 12: 178. 1906; Fischer, Fl. Madras 3: 1714. 1934; Bor, Grass. India 202. I960. *Saccharum paniceum* Lam., Encycl. 1: 595. 1785. *Pogonatherum saccharoideum* P. Beauv., Ess. Agrost. 176. 177. 1812; Hook. f., Fl. Brit. India 7: 141. 1896.

A perennial, densely tufted grass; culms up to 60 cm high, leafy, woody; nodes on the culms sparsely hairy. Leaf sheaths bearded; blades 2.5-7 x 0.25-1 cm, linear, broad, apex acuminate. Spikes 5 cm long, spatheolate, terminal; rachis compressed, bearded; spikelets 2 mm long, 2-nate. *Sessile spikelet:* lower glume narrow-oblong, apex ciliate; upper glume conduplicate, apex densely ciliate, awned; lower lemma sometimes absent; palea of the upper lemma broadly ovate-oblong; ovary minute. *Pedicelled spikelets:* similar to the sessile spikelet.

Cultivated in Mahabubnagar district.

Fl. & Fr.: September-April.

Mallelatheertham (MBNR), *SRS* 110703 (BSID).

PSEUDORAPHIS Griff.

Pseudoraphis spinescens (R.Br.) Vickery, Proc. Roy. Soc. Queensl. 62, n. 7. 69. 1952; Bor, Grass. India 353. 1960; Moulik, Grass. Bamb. India 1: 147. 1997. *Panicum spinescens* R. Br., Prodr. Fl. Nov. Holl. 193. 1810. *Chamaeraphis spinescens* (R.Br.) Poir. in Lam., Encycl. 2: 189. 1811; Hook. f., Fl. Brit. India 7: 62. 1896. *Pseudoraphis aspera* (Koen.) Pilger, Notizbl. Bot. Gart. Berlin 10: 210. 1928; Fischer, Fl. Madras 3: 1790. 1934.

A perennial, marshy, floating grass; culms up to 60 cm high, decumbent, much branched; nodes woolly. Leaf sheaths 4-8 cm long, loose, glabrous; ligule membranous; leaf blade 5-10 x 0.2-0.5 cm, linear, flat or convolute, base truncate, margins spinulose, apex acuminate, tuberculate pubescent above, glabrous below. Panicle up to 14 cm long, effuse or contracted; branches filiform, scabrid; spikelets narrowly linear, 6 mm, 2-flowered; lower glume ovate, truncate, membranous, nerveless; upper glume ovate, acuminate, aristate, 11-nerved, scabrid along the margins; lower lemma ovate-oblong, 7-nerved; palea hyaline, male; upper lemma oblong, 2-nerved; palea linear, hyaline, bisexual; lodicules 2; stamens 3, stigma plumose, brown. Caryopsis linear-oblong.

Occasional grass in ponds and marshy areas of Medak, Warangal and Hyderabad districts.

Fl. & Fr.: October-March.

Narsapur tank (MDK), *TP & MSG* 12012; Ooracheru lake (MDK), *KMS* 6755 (MH).

PSEUDOSORGHUM Camus

Pseudosorghum fasciculare (Roxb.) Camus, Bull. Mus. Hist. Nat. (Paris) 26: 662. 1920; Fischer, Fl. Madras 3: 1734. 1934; Bor, Grass. India 205. 1960. *Andropogon fascicularis* Roxb., Fl. Ind.1: 269. 1820;Hook. f., Fl. Brit. India 7: 177. 1896.

An annual, tufted grass; culms up to 75 cm high, slender, decumbent below; nodes villous. Leaf blades 3-20 x 0.2-0.8 cm, linear, sparsely tuberculate-based hairs. Panicle 25 cm long, dense, compound; branches erect, verticillate; rachis fragile; joint and pedicels hairy;spikelets 2-nate. *Sessile spikelet:* 6 cm long, bisexual; callus bearded;glumes subequal, chartaceous; lower 2-keeled; upper boat-shaped; lemmas

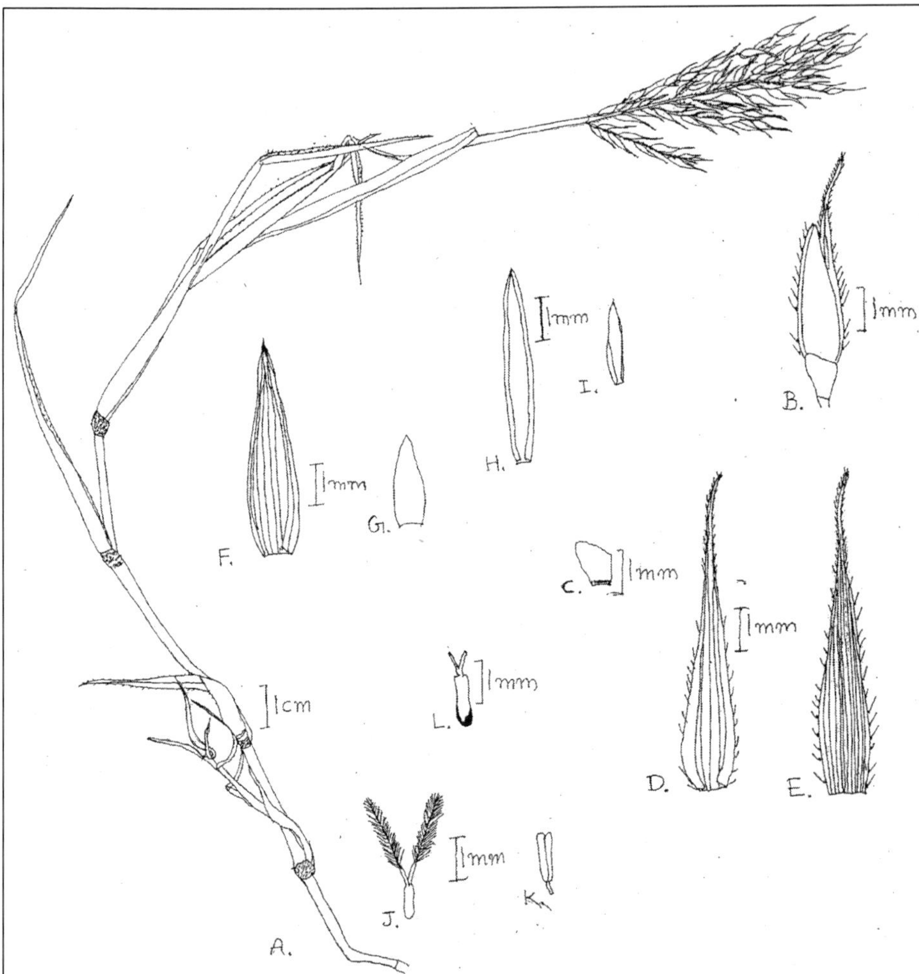

Figure 152: Pseudoraphis spinescens (R.Br.) Vickery

A. Twig, B. Spikelet, C. Lower glume, D & E. Upper glume front and back view, F. Lower lemma, G. Palea, H. Upper lemma, I. Palea, J. Pistil, K. Stamen, L. Caryopsis.

hyaline, lower similar to upper, narrow, short, bi-dentate, awned, paleated; stamens 3. *Pedicelled spikelet:* narrower than sessile spikelet, male, similar to sessile spikelet.

Rare in forests in Warangal district (C.S. Reddy, 2001).

Fl. & Fr.:October-January.

Pakhal (WGL), *KMS* (MH).

RHYNCHELYTRUM Nees

Rhynchelytrum repens (Willd.) C.E. Hubb., Kew Bull. 1934. 110: 1934; Bor, Grass. India 355. 1960; Moulik, Grass. Bamb. India 1: 148. 1997. *Saccharum repens* Willd., Sp. Pl. 1; 372 1798. *Tricholaena wightii* Nees & Arn. ex Steud., Syn. Pl. Glum. 1: 98. 1854; Hook. f., Fl. Brit. India 7: 65. 1896. *Rhynchelytrum villosum* (Parl.) Chiov., Annuario Reale Ist. Bot. Roma 8: 310. 1908; Fischer, Fl. Madras 3: 1791. 1934; Bor *loc. cit. Monachyron villosum* Parl. in Hook., Niger Fl. 191. 1849.

An annual, tufted grass; culms up to 60 cm high; nodes pubescent. Leaf sheaths 2-4 cm long, tuberculately pilose; ligule ciliate; leaf blade 3-15 x 0.2-0.5 cm, lanceolate, base truncate, apex acuminate, pubescent - glabrous on both surfaces. Panicle up to 10 cm long, contracted; branches filiform; pedicels capillary, pilose; spikelets 5 mm, ovate-obovate, laterally compressed, 2-flowered, villous; callus bearded; lower glume linear, pilose, 1-nerved; upper glume ovate, acuminate, 5-nerved, densely pilose, awned; lower lemma oblong, densely pilose, aristate; palea linear, male; upper lemma retuse, arista absent; palea hyaline, bisexual; lodicules 2; stamens 3; stigmas plumose, purplish to brownish. Caryopsis obovoid.

Occasional in plains and open forests of Hyderabed and Karimnagar districts.

Fl. & Fr.: August-December.

Upperpalli (HYD), *MRS* 318 (DD).

ROTTBOELLIA L. f. *nom. cons.*

Rottboellia cochinchinensis (Lour.) Clayton, Kew Bull. 35; 817. 1981; Moulik, Grass. Bamb. India 1: 199. 1997. *Stegosia cochinchinensis* Lour., Fl . Cochinch. 1: 51. 1790. *Rottboelia exaltata* L.f., Suppl. Pl. 114. 1781. non (L.) L. f. 1779; Hook. f., Fl. Brit. India 7: 156. 1896; Fischer, Fl. Madras 3: 1759. 1934; Bor, Grass. India 206. 1960.

A perennial, coarse grass; culms up to 3 m high, supported by stilt roots, branching from the base; nodes glabrous. Leaf sheaths up to 12 cm long, loose, glabrous; ligule a short ciliate rim; leaf blade 10-35 × 0.2-0.7 cm, linear, flat, base rounded, margins scabrid, apex acuminate, glabrous, midrib prominent. Spiciform racemes up to 12 cm long, solitary, cylindric; rhachis fragile; spathe clasping the racemes at base; spikelets 4 mm, 2-nate, one sessile and other pedicelled. *Sessile spikelet*: 2 mm, ovoid; glumes subequal, crustaceous, keeled; lower ovate, close to the cavity; upper boat-shaped; lemmas hyaline, oblong- ovate, paleated; lower male, upper bisexual; paleas hyaline; stamens 3; stigmas plumose. Caryopsis ellipsoid. *Pedicelled spikelet:* 4 mm, ovate; glume ovate, crustaceous; lemmas hyaline, ovate-oblong, upper palea red; palea oblong, empty or male.

Occasional grass in Medak, Karimnagar and Hyderabad districts.

Fl. & Fr.: August-December

Kodimial (KRN), *GVS* 25665 (CAL); Nagsanpally RF (MDK), *RG* 104178 (BSID).

SACCHARUM L.

1. Cultivated; peduncle glabrous below the panicle; glumes glabrous .... **S. officinarum**
1. Wild; peduncle hairy below the panicle; glumes ciliate .... **S. spontaneum**

Saccharum officinarum L., Sp. Pl. 54. 1753; Hook. f., Fl. Brit. India 7:118. 1896; Fischer, Fl. Madras 3: 1709. 1934; Bor, Grass. India 212. 1960; Moulik, Grass. Bamb. India 1: 321. 1997.

Vern.: *Cheruku.* English: Sugarcane.

Annual, tufted grass; culms up to 7 m high, solid; nodes many, glabrous, pubescent below the panicle, coated with white wax below the nodes. Leaf sheaths up to 25 cm, terete, glabrous; ligule membranous, ciliate; leafblade 100-150 × 4-7 cm, linear-lanceolate, midrib stout, margins scabrid. Panicle up to 1 m long, densely branched, silvery; primary branches fragile, glabrous; racemes 12 cm; joints and pedicels filiform, ciliate. *Sessile spikelet :* 5 mm, lanceolate; callus with a tuft of long silky hairs; glumes subequal, lanceolate, chartaceous below and hyaline upwards; lower 4-nerved; upper 1-3-nerved; lower lemma oblong, hyaline, empty; upper lemma ciliate; lodicules 2; stamens 3. Caryopsis oblong. *Pedicelled spikelet:* similar to sessile spikelet.

Widely cultivated in all districts for sugar extraction.

Fl. & Fr.: October-March.

Saccharum spontaneum L., Mant. Alt. 2: 183. 1771; Hook. f., Fl. Brit. India 7: 118. 1896; Fischer, Fl. Madras 3: 1709. 1934; Bor, Grass. India 214. 1960; Moulik, Grass. Bamb. India 1: 322. 1997.

Vern.: *Rellugaddi.*

A perennial, tufted grass; culms up to 6 m high, stout, solid, terete; nodes waxy. Leaf sheaths 5-20 cm long, glabrous; ligule membranous with fringe of hairs; leaf blade 15-90 x 0.3-0.8 cm, linear, flat or convolute; base narrowed, margins scaberulous, apex acuminate, midrib prominent, coriaceous, glabrous. Panicle up to 60 cm long, contracted; rhachis below the inflorescence villous; spikelets linear, 5 mm, 2-nate; callus with silky hairs. *Sessile spikelet:* 4 mm, linear-lanceolate; glumes equal, lanceolate, chartaceous below, membranous above, acuminate, hairy in the middle, 1-nerved; lower lemma linear, hyaline, ciliate, epaleate; upper lemma linear, ciliate; palea ovate; lodicules 2; stamens 3, plumose. Caryopsis oblong. *Pedicelled spikelet:* 4 mm, linear-lanceolate, falling at maturity, similar to sessile spikelet.

Very common along streams and banks of rivers in most districts.

Fl. & Fr.: October-February.

Sone (ADB), *TP & PVP* 4041; Nizamsagar project (NZB), *BR & CPR* 9244; Gangapur-Pupplevagu (MDK), *TP & MSG* 12085; Chityal (NLG), *BVRS* 415 (NU).

SACCIOLEPIS Nash

1. Culms creeping at base; spikelets in interrupted crowded panicle, glabrous; lower glume less than half the length of the spikelets:
   2. Spikelets linear – lanceolate; lower glume broadly ovate, 5-9-nerved .... **S. interrupta**
   2. Spikelets ovoid; lower glume ovate, 3-nerved .... **S. myosuroides**
1. Culms slender; spikelets in continuous crowded panicle, tuberculate-pilose; lower glume half as long as the spikelets .... **S. indica**

Sacciolepis indica (L.) Chase, Proc. Biol. Soc. Wash. 21: 8. 1908; Fischer, Fl. Madras 3: 1787. 1934; Bor, Grass. India 357. 1960; Moulik, Grass. Bamb. India 1: 149. 1997. *Aira indica* L., Sp. Pl. 1753 in "Errata" *Panicum indicum* L., Mant. 2: 184. 1771 non Mill. 1768; Hook. f., Fl. Brit. India 7: 41. 1896.

An annual, marshy, tufted grass; culms up to 70 cm high, slender; nodes glabrous. Leaf sheaths 2-5 cm long, glabrous; ligule absent; leaf blade 5-15 x 0.1-0.4 cm, linear, filiform, base truncate, apex acute, glabrous. Spikelike panicle spiciform, up to 7 cm long, contracted, cylindric, dense; spikelets linear, 3.5 mm; lower glume elliptic, 5-nerved; upper glume obovate, keeled, 9-nerved, tuberculate based hairs at apex of outer surface; lower lemma oblong, 7-nerved, keeled, hairs at apex of outer surface; upper lemma ovate, hyaline; palea linear; lodicules 2; stamens 3; stigma plumose. Caryopsis linear.

Occasional marshy grass in Medak, Ranga Reddi, Hyderabad, Karimnagar and Mahabubnagar districts.

Fl. & Fr.: August-October.

Pocharam tank (MDK), *TP & MSG* 12038; Mohammadabad (RR), *MSM* 10553; Aklaspur (KRN), *GVS* 22486 (CAL & MH); Yenkur (KMM), *RR* 108570 (BSID).

Sacciolepis interrupta (Willd.) Stapf in Prain Fl. Trop. Afr. 9: 757. 1920; Fischer, Fl. Madras 3: 1787. 1934; Bor, Grass. India 358. 1960; Moulik, Grass. Bamb. India 1: 150. 1997. *Panicum interruptum* Willd., Sp. Pl. 1: 341. 1798; Hook. f., Fl. Brit. India 7: 40. 1896.

A perennial, aquatic, tufted grass; culms up to 1 m high, spongy, creeping at base; nodes glabrous, with thick coloured patch. Leaf sheaths 5-15 cm long, loose, glabrous; ligule membranous; leaf blade 10-30 x 0.5-1.5 cm, linear-lanceolate, flat, base truncate, margins scaberulous, apex acuminate, midrib prominent, sparsely hairy. Spikelike panicle up to 20 cm long, cylindric, contracted, interrupted; spikelet up to 6 mm, elliptic-lanceolate; lower glume broadly ovate, 5-9-nerved; upper glume elliptic-lanceolate, 9-11-nerved, scabrid; lower lemma 7-9-nerved, scabrid on outer surface; upper lemma indurated, 5-nerved; lodicules 2; stamens 3, stigmas plumose. Caryopsis ovoid.

Occasional grass in marshy and swamp areas of Warangal, Nizamabad, Medak, Ranga Reddi and Mahabubnagar, districts.

Fl. & Fr.: June-December.

Mosra RF (NZB), *TP & BR* 6109; Narayanpur RF (MDK), *BR* & *CP* 11272; Anantasagar (RR), *MSM* & *KH* 11002; Banks of Pakhal lake (WGL), *KMS* 1168 & *ANH* 15916 (CAL & MH); Narsapur (MDK), *RG* 116460 (BSID).

Sacciolepis myosuroides (R. Br.) A. Camus in Lecomte Fl. Gen. de 1' Indo-chine 7: 460. 1922; Fischer, Fl. Madras 3: 1787. 1934; Bor, Grass. India 358. 1960; Moulik, Grass. Bamb. India 1: 150. 1997. *Panicum myosuroides* R. Br., Prodr. 189. 1810.

An annual, tufted grass; culms up to 90 cm high, shortly creeping, leafy. Leaf sheaths 10-13 cm long, margins naked; ligule absent; leaf blade 10-25 x 0.3-0.6 cm, oblong-lanceolate, flat, base narrow, apex acuminate. Panicle 20 cm long, spiciform, often interrupted; spikelets 2-5 in a cluster, 2 mm, ovoid, subsessile; lower glume ovate, 3-nerved; upper glume ovate, 9-11-nerved; lower lemma broadly ovate, 7-nerved; palea hyaline, empty; upper lemma ovate-oblong, coriaceous; palea similar to its lemma; lodicules 2; stamens 3; stigma plumose. Caryopsis ellipsoid,

Marshy areas of Warangal and Medak districts.

Fl. & Fr.: October- January.

Pocharam tank (MDK), *TP* & *MSG* 12040.

Note: It can be easily distinguished from all other species of the genus by its long narrow leaves, smaller ovate or subglobose, blunt spikelets and long, narrow whip-like spiciform panicles.

SCHIZACHYRIUM Nees

Schizachyrium exile (Hochst.) Pilger in Engler, Bot. Jahrb. 54: 284. 1917; Fischer, Fl. Madras 3: 1752. 1934; Bor, Grass. India 216. 1960; Moulik, Grass. Bamb. India 1: 218. 1997. *Andropogon exilis* Hochst. in Flora 27: 241. 1844; Hook. f., Fl. Brit. India 7: 166. 1896.

An annual, tufted, robust grass; culms up to 15 cm high, decumbent; nodes glabrous. Leaf sheaths 1-3 cm long, glabrous; ligule membranous; leaf blade 2- 8 x 0.05-0.1 cm, linear, filiform, base truncate, apex acuminate, glabrous or covered with densely bulbous based hairs near mouth. Racemes spatheolate, up to 6 cm long, slender, spathe 2 cm long; spikelets 4 mm, lanceolate, 2-nate. *Sessile spikelet:* 4 mm, dorsally compressed; callus hairy; lower glume lanceolate, 2-keeled, obscurely 2-nerved, densely villous; upper glume subcoriaceous, boat-shaped, linear, 3-nerved; lemmas hyaline, upper bidentate, awned; palea hyaline, narrow; lodicules 2; stamens 3, stigmas plumose. *Pedicelled spikelet*: 3 mm long, empty.

Rare grass in Medak and Mahabubnagar districts.

Fl. & Fr.: August-December.

Pocharam tank (MDK), *TP & MSG* 12053.

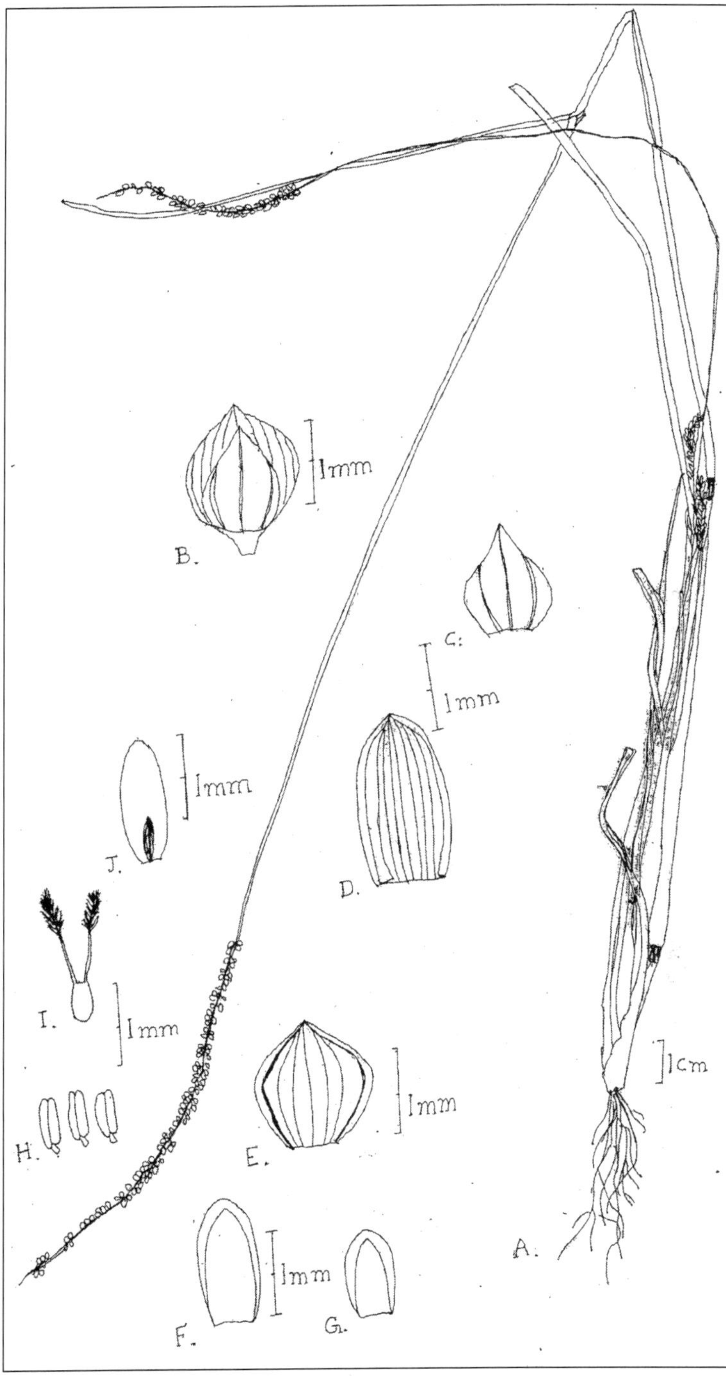

Figure 153: Sacciolepis myosuroides (R.Br.) A. Camus

A. Twig, B. Spikelet, C. Lower glume, D. Upper glume, E. Lower lemma, F. Upper lemma, G. Palea, H. Stamens, J. Pistil.

SEHIMA Forssk.

1. Ciliate on one margin only of pedicels; lower glume of sessile spikelets rounded on the back with a deep longitudinal slit-like groove .... **S. sulcatum**
1. Ciliate on both edges of pedicels; lower glume of sessile spikelets flat on the back, without a slit, but grooved deeply or shallowly..... **S. nervosum**

Sehima nervosum (Rottl.) Stapf in Prain, Fl . Trop. Afr. 9: 36. 1917; Fischer, Fl. Madras 3: 1724. 1934; Bor, Grass. India 218; Moulik, Grass. Bamb. India 1: 301. 1997. *Andropogon nervosus* Rottl. apud Willd., Verh. Ges. Naturf. Freund. Berlin. Neue Schr. 4: 218. 1806. *Ischaemum laxum* R. Br., Prodr. 205. 1810; Hook. f., Fl. Brit. India 7: 136. 1896.

A perennial, tufted grass; culms up to 1 m high; nodes glabrous. Leaf sheaths 5-9 cm long, glabrous; ligule membranous; leaf blade 5-30 x 0.3-0.5 cm, linear, flat, coriaceous, base rounded, apex acuminate, glabrous. Racemes up to 10 cm long, solitary, densely ciliate along the angles; rhachis jointed, fragile; spikelets 12 mm, laterally compressed, 2-nate. *Sessile spikelet:* 3mm, linear; lower glume chartaceous, linear, keeled, apex truncate, winged on the keels with a transculent longitudinal depression in the centre, 5-6-nerved; upper glume boat-shaped, 3-nerved, apex scabrid, awned; lower lemma hyaline, lanceolate; palea lanceolate, margin ciliate; upper lemma hyaline, bidentate, awned; palea linear, apex scabrid; lodicules 2; stamens 3, stigmas plumose. Caryopsis oblong. *Pedicelled spikelet:* 6 mm, lanceolate; lower glume lanceolate, chartaceous, winged on the keels, ciliate; upper glume hyaline, lanceolate; lower lemma linear, margins ciliate; palea linear, male; upper lemma oblong, obtuse; palea oblong, margins ciliate; stamens 3.

Common in open forests and occasional in plains of Karimnagar, Hyderabad and Nalgonda districts.

Fl. & Fr.: August-December.

Kodimial (KRN), *GVS* 25673 (CAL); Nagarjuna konda valley (NLG), *KT* 9669 (CAL); Hyderabad (HYD), *MRS* 213 (DD); Moosi river (HYD), *MRS* 214 (DD).

Sehima sulcatum (Hack.) A. Camus, Bull. Mus. Hist. Nat. Paris 27: 373. 1921; Fischer, Fl. Madras 3: 1724. 1934; Bor, Grass. India 219. 1960; Moulik, Grass. Bamb. India 1: 301. 1997. *Ischaemum sulcatum* Hack., in DC., Monogr. Phan. 6: 248. 1889; Hook. f., Fl. Brit. India 7: 137. 1896.

A perennial, tufted grass; culms up to 50 cm high, geniculately ascending, much branched below; nodes glabrous. Leaf sheaths lax, glabrous, constricted at the mouth; ligule ciliate membrane; leaf blade 10-20 x 0.3-0.4 cm, linear-lanceolate, apex acuminate, glabrous. Racemes solitary, 8 cm long. *Sessile spikelet:* linear, 1 cm long; callus shortly hairy at the base; lower glume lanceolate, subchartaceous, 7-9-nerved, 2-keeled, bidentate, margins incurved; upper glume shorter than the lower, chartaceous, 3-5-nerved, shortly bidentate, apex ciliate, awned; lower lemma linear, hyaline, margins incurved, palea oblong; upper lemma bidentate, awned; palea scaberulous towards apex, lodicules 2; stamens 3. Caryopsis oblong. *Pedicelled spikelet:* smaller than the sessile, awnless; lower glume lanceolate, bidentate, margin incurved; upper glume unawned; lemmas similar to sessile but unawned.

Rare in open forest and plains in Medak district.

Fl. & Fr.: August- December

SETARIA P. Beauv. *nom. cons.*

1. Involucral bristels retrorsely barbellate .... **S. verticillata**
1. Involucral bristles antrorsely barbellate:
   2. Cultivated; upper lemma faintly transversely rugose, upper glume equal to the length of upper lemma .... **S. italica**
   2. Wild; upper lemma distinctly transversely rugose, upper glume smaller than the upper lemma;
      3. Panicle not continuous, interrupted; involucral bristles 3-4; spikelets up to 2 mm long .... **S. intermedia**
      3. Panicle continuous, dense; involucral bristles 6-11; spikelets more than 2 mm long .... **S. pumila**

Setaria intermedia Roem. & Schult., Syst. Veg. 2. 489. 1817; Hook. f., Fl. Brit. India 7: 79. 1896; Fischer, Fl. Madras 3: 1789. 1934; Moulik, Grass. Bamb. India 1: 155. 1997. *Panicum tomentosum* Roxb., Fl. Ind. 1: 303. 1820. *Setaria tomentosa* (Roxb.) Kunth, Revis. Gram. 1: 47. 1829; Bor, Grass. India 365. 1960.

An annual, tufted grass; culms up to 80 cm high, decumbent; nodes glabrous. Leaf sheaths 4-10 cm long, loose, ciliate at mouth; ligule ciliate; leaf blade 7-15 x 0.4-1 cm, linear-lanceolate, flat, flaccid, base rounded, apex acuminate. Panicle up to 10 cm long, lax, interrupted; involucral bristles antrorsely barbed; lower glume membranous, broadly ovate, glabrous, 3-nerved; upper glume broadly ovate, hyaline, glabrous, 5-7-nerved; lower lemma lanceolate-ovate, 5-7-nerved, thinly membranous, paleated, empty; upper lemma broadly ovate, crustaceous, 5-nerved, transversely rugose; palea elliptic; crustaceous; lodicules 2; stamens 3, stigma plumose, purplish. Caryopsis ellipsoid.

Common grass in moist localities of all districts.

Fl. & Fr.: July-November.

Manchippa RF (NZB), *TP* & *BR* 6048; Nadipally (NZB), *BR & KH* 6455; Ramayampet (MDK), *BR* & *CP* 11515; *ICRISAT* (MDK), *LJGV* 26631 (CAL); Central University campus (RR), *MSM* 15171; Along moosi river bank (HYD), *KMS* 5931 (CAL); Madikonakutta (MDK), *KMS* 6725 (CAL); Nagarjuna konda (NLG), *KT* 9830 (CAL); Mysamma temple (MBNR), *SRS* 108633 (BSID).

Setaria italica (L.) P. Beauv., Ess. Agrost. 51. 170. 178. 1812; Hook. f., Fl. Brit. India 7: 78. 1896; Fischer, Fl. Madras 3: 1789. 1934; Bor, Grass. India 362. 1960; Moulik, Grass. Bamb. India 1: 155. 1997. *Panicum italicum* L., Sp. Pl. 56. 1753.

Vern.: *Korralu.*

An annual, tufted grass; culms up to 1 m high; nodes glabrous. Leaf sheaths 3-10 cm long, ciliate on the margins; ligule membranous; leaf blade 10-35 × 0.5-2 cm, linear, flat, base cordate, apex acuminate, minutely scaberulous. Panicle up to 10 cm long, contracted; involucral bristles antrorsely barbed; spikelets 2 mm, ovoid, compressed; lower glume ovate, 3-nerved, hyaline; upper glume broadly ovate, prominently 5-nerved, hyaline; lower lemma ovate, hyaline and its palea scaly, empty; upper lemma ovate, coriaceous, transversely rugose; palea elliptic, coriaceous; lodicules 2; stamens 3, stigmas plumose. Caryopsis ellipsoid.

Rarely cultivated for its grains.

Fl. & Fr. : September-November.

Setaria pumila (Poir.) Roem. & Schult., Syst. Veg. 2: 891. 1817. *Panicum pumilum* Poir., Encycl. Suppl. 4: 273. 1816; Moulik, Grass. Bamb. India 1: 156. 1997. *P. pallidefuscum* Schumach., Beskr. Guin. Pl. 58. 1827. *Setaria pallidefusca* (Schumach.,) Stapf & C. E. Hubb., Kew Bull. 1930: 25. 1930; Fischer, Fl. Madras 3: 1789. 1934; Bor, Grass. India 363. 1960.

Vern.: *Nakkathoka gaddi.*

An annual, tufted grass; culms up to 60 cm high, ascending; nodes glabrous. Leaf sheaths 2-5 cm long, glabrous; ligule a ridge of hairs; leaf blade 5-15 × 0.2-0.6 cm, linear-lanceolate, flat, base cordate, apex acuminate, flaccid. Panicle spiciform, up to 8 cm long, cylindric, densely-flowered; involucral bristles antrorsely barbed; spikelets 2 mm, ovoid; lower glume ovate, glabrous, 3-nerved; upper glume ovate, glabrous, 5-nerved, hyaline; lower lemma ovate, 5-nerved, hyaline, paleated and its palea hyaline, empty; upper lemma ovate, crustaceous, obscurely 5-nerved, transversely rugose; palea similar to its lemma; lodicules 2; stamens 3, stigmas plumose, violet to purplish. Caryopsis ovoid-ellipsoid.

Common in open grass lands, waste places and along roadsides of all districts.

Fl. & Fr.: July-November.

Jannaram (ADB), *TP & PVP* 4198; Ankusapuram (ADB), *GO* 4317; Gupalapatnam (ADB), *GO* 4410; Mosra forest Nursery (NZB), *TP & BR* 6068; Kammarpally (NZB), *TP & BR* 6199; Velutla (NZB), *BR* 9520; Choutkoor fields (MDK), *TP & MSG* 11929; Madikona kutta (MDK), *KMS* 6728 (CAL); Mohammadabad (RR), *MSM* & *KH* 10534, 10592; Pakhal lake (WGL), *KMS* 13119 (CAL); Adikmet (HYD), *MRS* 276 (DD); Koriguttalu (KMM), *RR* 106047 (BSID); Kambalpally (NLG), *BVRS* 187 (NU); Mysamma temple (MBNR), *SRS* 108630 (BSID).

Note: It is a polymorphic species, where the length of leaves and panicles show a gradation in variation. This highly variable grass occurs in all sorts of localities.

Setaria verticillata (L.) P. Beauv., Ess. Agrost. 51: 178. 1812; Hook. f., Fl. Brit. India 7: 80. 1896; Fischer, Fl. Madras 3; 1789. 1934; Bor, Grass. India 365. 1960; Moulik, Grass. Bamb. India 1: 159. 1997. *Panicum verticillatum* L., Sp. Pl. ed. 2. 82. 1762.

An annual, tufted grass; culms up to 60 cm high; nodes glabrous. Leaf sheaths 3-7 cm long, glabrous; ligule a fringe of hairs; leaf blade lanceolate, 3-10 × 0.3-1 cm, flat, base rounded, sparsely pubescent; apex acuminate. Panicle up to 10 cm long, spiciform; involucral bristles retrorsely barbed; spikelets oblong-ovate, 2 mm; lower glume broadly ovate, 3-nerved; upper glume membranous, broadly ovate, glabrous, 5-7-nerved; lower lemma oblong-ovate, shortly apiculate, 5-nerved, margins incurved; palea linear; upper lemma oblong, crustaceous, transversely rugose, 3-nerved; palea similar to its lemma; lodicules 2; stamens 3, stigmas plumose. Caryopsis ellipsoid.

Occasional in moist and shady places of Nizamabad, Ranga Reddi, Karimnagar, Nalgonda, Warangal, Hyderabad, Mahabubnagar districts.

Fl. & Fr.: July-January.

Yedapally (NZB), *BR* & *KH* 6467; Central University campus (RR), *MSM* 15183; Aklaspur (KRN), *GVS* 25652 (CAL); Kishna river bank (NLG), *KMS* 9805 (CAL); Pakhal RF (WGL), *KMS* 13169 (CAL); Amberpet (HYD), *MRS* 270 (DD); Appanapally (MBNR), *SRS* 110735 (BSID).

SORGHUM Moench *nom. cons.*

1. Panicle branches simple; awn long, well developed; nodes bearded .... **S. nitidum**
1. Panicle branches divided; awn short, poorly developed; nodes glabrous or shortly pilose;
    2. Annual; cultivated; sessile spikelet broadly obovate, without rhizomatous culms .... **S. bicolor**
    2. Perennial; wild; sessile spikelet linear-lanceolate; with well developed creeping rhizomes .... **S. halepense**

Sorghum bicolor (L.) Moench, Meth. Pl. Hort. Hot. Merburg. 207. 1794. var. **bicolor;** Bor, Grass. India 228. 1960. *Holcus sorghum* L., Sp. Pl. 1047. 1753.

Vern.: *Jonnalu.*

An annual, robust, tufted grass; culms up to 2 m high, glabrous, polished. Leaf blade 10-30 x 2.5-5 cm, broadly lanceolate, flat, apex acuminate, glabrous. Panicle 25 cm long, densely thyrsiform, decompound; branches verticillate; spikelets 5 mm, in groups at the ends of the branches. *Sessile spikelets:* broadly ovate-lanceolate, hairy; glumes coriaceous, nerves obscure; upper lemma awned. Caryopsis broadly obovate.

Cultivated for its grains at some places.

Fl. & Fr.: October-March.

Sorghum halepense (L.) Pers., Syn. 1: 101. 1805; Fischer, Fl. Madras 3: 1735. 1934; Bor, Grass. India 222. 1960; Moulik, Grass. Bamb. India 1: 249. 1997. *Holcus halepensis* L., Sp. Pl. 1047. 1753. *Andropogon halepensis* (L.) Brot., Fl. Lusit, 1: 89. 1804; Hook. f., Fl. Brit. India 7: 182. 1896. *Sorghum miliaceum* (Roxb.) Sw., J. Linn. Soc. Bot. 55: 207. 955 incl. var. *parvispiculum* Sw. *loc.cit.* 209; Bor, Grass. India 223. 1960.

Vern.: *Gaddi jonna.*

A perennial, tufted grass; culms up to 2 m high, stout, sparsely branched; nodes minutely silky. Leaf sheaths 15-20 cm long, glabrous; ligule short, membranous, ciliate; blade 15-60 x 1-4 cm, linear-lanceolate, base cordate, margins scaberulous, apex acuminate, midrib prominent, glabrous. Panicle effuse, up to 20 cm long, decompound, branches scaberulous. Spikelets 2-nate. *Sessile spikelet:* 6 mm, linear-lanceolate, hairy; callus short, bearded; glumes subequal, coriaceous; lower ovate, 5-7-nerved, silky hairs outside; upper lanceolate, 5-nerved, apex acuminate; lower lemma triangular-lanceolate, hyaline, 3-nerved; upper hyaline, margins ciliate, 2-lobed, awned; lodicules 2; stamens 3, stigmas plumose. Caryopsis ovoid. *Pedicelled spikelet:* similar to sessile spikelet, but narrowly lanceolate, awnless, male.

Occasional near water courses of Adilabad and Karimnagar districts.

Fl. & Fr.: Throughout the year.

Penganga (ADB), *GO & PVP* 4280; Rechapalli (KRN), *GVS* 22244 (MH).

Note: The introduced 'Johnson grass' is now running wild and rapidly spreading all over India. Snowden (J. Linn. Soc. Bot. 55 : 207, 1856) believed that *S. halepense* is not indigenous to India and treated the Indian materials under *S. miliaceum*. But, later on Cope (1982) has merged the later with *S. halepense.*

Sorghum nitidum (Vahl) Pers., Syn. 1: 101. 1805; Fischer, Fl. Madras 3: 1735. 1934; Bor, Grass. India 245. 1960; Moulik, Grass. Bamb. India 1: 254. 1997. *Holcus nitidum* Vahl, Symb. Bot. 2: 102. 1791. *Andropogon serratus* Thunb., Fl. Jap. 41. 1784; Hook. f., Fl. Brit. India 7: 185. 1896. *Sorghum serratum* (Thunb.) Kuntze, Revis. Gen. Pl. 2: 792. 1891, non Roem. & Schult. 1817.

A perennial, tufted grass; culms up to 4 m high; nodes densely villous. Leaf sheaths terete below, keeled, silky-villous near the mouth; ligule truncate; leaf blade 10-50 x 0.8-2 cm, base rounded, apex acuminate, midrib prominent. Panicle 30 cm long, lax; branches capillary; spikelet margins shortly villous. *Sessile spikelet:* broadly ellipsoid; lower glume broadly oblong-elliptic, coriaceous, 7-9-nerved, keels hispid; upper glume broadly cymbiform, 5-7-nerved, ciliate upwards; lower lemma hyaline, margins incurved, 2-nerved, 2-keeled, ciliate; upper lemma linear-oblong, bidentate, awned or unawned. Caryopsis elliptic. *Pedicelled spikelet:* linear-oblong; lower glume oblong, dorsally depressed, keeled; upper glume narrower, margins incurved, keeled; lower lemma hyaline, linear.

Occasional in Hyderabad district.

Fl. & Fr. : October- March.

SPODIOPOGON Trin.

Spodiopogon rhizophorus (Steud.) Pilger in Engl. & Prantl, Nat. Pflanzenfam. ed. 2. 14e: 119. 1940; Bor, Grass. India 246. 1960; S. Moulik, Grass. Bamb. India 1: 326. t. 66. 1997.

Tufted annuals; culms erect, 90 – 150 cm high; nodes glabrous, internodes purplish. Leaf blade broadly lanceolate, 10 – 25 × 2 – 4 cm, acute at apex, apex aristate, sparsely tuberculate hairy on both the sides, margin scabrid, pseudopetiolate at base, pseudopetiole 1 – 12 cm long; ligules membranous, ca. 1 mm; leaf sheath 6 – 15 cm long, apex auricled. Panicle densely villous, 6 – 13 × 2 – 4 cm, with spathe like sheath below panicle; peduncle 10 – 20 cm long; racemes 1 – 3 per nodes, 1 – 4 cm long; spikelets paired, one sessile other pedicelled, homogamous, pedicelled spikelets smaller in size. *Sessile spikelets* oblong-lanceolate, 4 – 5 mm × 1 – 1 – 1.5 mm, acute at apex, hairy throughout, hairs pinkish white. Lower glumes oblong to lanceolate, 4 – 4.8 × 1.5 – 1.7 mm, acuminate at apex, 7 – 9 nerved, hairy, hairs up to 2 mm long pinkish-white in colour; upper glumes oblong-lanceolate, 3.9 – 4.5 × 2 – 2.2 mm, acuminate at apex, sometimes aristate, arista up to 0.3 mm long, 9 – 11-nerved, sparsely hairy; florets 2, lower male and upper bisexual; lower lemma oblong-lanceolate, 3 - 3.5 × 1.6 – 2 mm, hyaline, truncate at apex with minute hairs on it, nerveless; lower paleas linear-lanceolate, 4 - 4.4 × 0.5 -0.6 mm, hyaline, acute at apex with minute hairs, nerveless; upper lemma lanceolate, 2.8 - 3 × 0.7 - 0.8 mm, chartaceous, 2-lobed, 1-nerved, lobes acuminate at apex, awn arising from sinus, geniculate, 6 – 8 mm long. Upper paleas oblong-lanceolate, 2 - 2.3 ×0.7 - 0.8 mm, hyaline, chartaceous, obtuse at apex, sparsely hairy, 5 – 6-nerved; stamens 2, anthers 2.3 – 2.7 mm long; ovary elliptic-oblong, 0.5 × 0.1 mm; stigma plumose, 2.6 – 2.8 mm long. Caryopsis elliptic-lanceolate, 1 – 1.2 × 0.2 – 0.3 mm.

Under forest cover in Adilabad district.

Fl. & Fr.: September - October

Kawal Tiger Reserve, Jannaram Range (ADB), *Alok Chorghe, Chandra Mohan & P. S. Annamma* 3987 (BSID).

SPOROBOLUS R.Br.

1. Lower glume shorter than the spikelet:
    2. Upper glume conspicuously shorter than the lemma:
        3. Panicle effuse:
            4. Lemmas more than 1mm long:
                5. Stamens 2; spikelets crowded; lower glume truncate ..... **S. indicus**
                5. Stamens 3; spikelets scattered; lower glume very short, obtuse .... **S. wallichii**
            4. Lemmas not more than 1mm long .... **S. tenuissimus**
        3. Panicle otherwise .... **S. spicatus**
    2. Upper glume as long as or longer than the lemma:
        7. Panicle contracted; branches not whorled:
            8. Leaf-margins smooth; spikelets 1.5-1.7mm long; lower glume lanceolate .... **S. humilis**
            8. Leaf-margins spinulose; characters not as above .... **S. maderaspatanus**
        7. Panicle effuse; branches truly whorled .... **S. coromandelianus**
1. Lower glume as long as the spikelet .... **S. virginicus**

Sporobolus coromandelianus (Retz.) Kunth, Revis. Gram. 1:68.1829; Hook. f., Fl. Brit. India 7:252. 1896; Fischer, Fl. Madras, 3:1817.1934; Bor, Grass. India 627. 1960; Moulik, Grass. Bamb. India 2:477.1997. *Agrostis coromandeliana* Retz., Observ. Bot. 4:19.1786. *Sporobolus commutatus* (Trin.) Kunth, Enum. Pl. 1:214.1833. *Vilta commutata* Trin., Diss. Bot. 156.1824.

Perennials, culms erect or decumbent, tufted, to 20 cm high; nodes glabrous. Leaf sheaths 1-5 cm long, glabrous; ligule a rim of short hairs; leaf blade linear- lanceolate with sparsely tubercle based hairy, base rounded or cordate, margins spinulosely toothed, acute. Panicles up to 5 cm, pyramidal, decompound, whorled; spikelets ovate; lower glume oblong, small, nerveless; upper glume ovate, chartaceous, 1-nerved; lemma ovate, chartaceous, 1-nerved; palea elliptic, 2-nerved; lodicules 2; stamens 3; stigma plumose, purplish. Caryopsis ellipsoid.

Occasional in forest plains and in dry habitats in Hyderabad, Khammam and Mahabubnagar districts.

Fl. & Fr.: August-December.

Kinnerasani (KMM), *BR* & *AMR* 26816; Golkonda (HYD), *MRS* 110 (DD).

Sporobolus humilis J. Presl., Reliq. Haenk 1: 241. 1830. subsp. **minor** Veldkamp, Blumea 35: 420. 1991. *S. tremulus* auct non (Willd.) Kunth, 1829; Hook. f., Fl. Brit. India 7:250.1896; Fischer, Fl. Madras 3:1817. 1934; Bor, Grass. India 633. 1960; Moulik, Grass. Bamb. India 2:481. 1997.

Annuals, culms tufted, to 30 cm, stoloniferous; nodes glabrous. Leaf sheaths ca 0.6 cm, glabrous; ligules hairy; leaf blade linear-lanceolate 1.3-6 × 0.2-0.3 cm, apex acuminate to acute. Panicles subspiciform, interrupted; spikelets ellipsoid; lower glume lanceolate, 1-nerved, keeled, keel serrulate above middle; upper glume ovate-lanceolate, membranous, acute; lemma oblong, chartaceous, 1-nerved; palea ovate, chartaceous; lodicules 2; stamens 3. Caryopsis ellipsoid.

Occasional in forest peripheries in Hyderabad and Mahabubnagar districts.

Fl. & Fr.: July-December.

Sporobolus indicus (L.) R.Br. Prodr. 170. 1810. var. **flaccidus** (Roem. & Schult.) Veldkamp, Blumea 35: 433. 1991. *Agrostis indica* L., Sp. Pl. 63. 1753. *Agrostis elongata* (R.Br.) Roth ex Roem. & Schult. var. *flaccida* Roth ex Roem. & Schult. Syst. Veg. 2: 368. 1817. *Sporobolous diander* (Retz.) P. Beauv., Ess. Agrost. 26.147.178. 1812; Hook. f., Fl. Brit. India 7:247.1896; Fischer, Fl. Madras 3:1817.1934; Bor, Grass. India 629.1960; Moulik, Grass. Bamb. India 2:478.1997. *Agrostis diandra* Retz., Observ. Bot. 5:19.1789.

Annuals, culms erect, tufted, to 90 cm high; nodes glabrous. Leaf sheaths 5-20 cm long, glabrous; ligule a rim of short hairs; leaf blade 5-12 × 0.1-0.2 cm, linear-lanceolate, base rounded, apex acuminate, glabrous. Panicles up to 25 cm long, effuse, filiform; spikelets obovoid, lanceolate; lower glume lanceolate, truncate, chartaceous, nerveless; upper glume ovate, chartaceous, 1-nerved; lemma lanceolate, chartaceous, 1-nerved; palea oblong, elliptic, 2-nerved; lodicules 2; stamens 2; stigma plumose. Caryopsis obovoid.

Occasional in moist and shady localities of the forests in Adilabad, Nizamabad, Medak, Ranga Reddi and Mahabubnagar districts.

Fl. & Fr.: July – December.

Vempalli RF (ADB), *GO* & *PVP* 4777; Mingaram (NZB), *TP* & *BR* 6300; Yedapally (NZB), *BR* & *KH* 6473; Choutkoor fields (MDK), *TP* & *MSG* 11919; Mohammadabad (RR), *MSM* & *KH* 10535; Appapur (MBNR), *AMR* 25686.

Key to Varieties

1. Racemes slender, open; caryopsis cuneate at base, 0.6-0.9 mm long .... var. **flaccidus**
1. Racemes shorter, contracted; caryopsis 0.9-1.1 mm long .... var. **major**

Sporobolus indicus (L.) R.Br. var. **major** (Buese) Baaijens, Blumea 35: 437. 1991. *Sporobolus diandrus* (Retz.) P. Beauv. var. *major* Buese in Miq., Pl. Jungh. 3: 343. 1854. *S. indicus* var. *fertilis* (Steud.) Jovet & Guedes, Bull. Centre & Rech. Sci. Biarrtiz. 7: 50. 1968 & Taxon 22: 163. 1973; Gayathri & Pullaiah in Pullaiah, Fl. Andhra Pradesh 3: 1259. 1997. *Sporobolus fertilis* (Steud.) Clayton, Kew Bull. 19:291. 1965; Moulik, Grass. Bamb. India 2:478.1997. *Agrostis fertilis* Steud., Syn. Pl. Glumac. 1: 170. 1854. *Sporobolus indicus* sensu Hook. f., Fl. Brit. India 7:247. 1896; Fischer, Fl. Madras 3:1817. 1934; Bor, Grass. India 630.1960.

Perennials, culms erect, tufted, to 1.1m high; nodes glabrous. Leaf sheaths 5-20 cm long, glabrous; ligule hairy or membranous; leaf blade 15-30 × 0.2-0.5 cm, linear, base rounded or cordate, apex acuminate, glabrous. Panicles to 45 cm long, contracted; spikelets obovoid; lower glume ovate, oblong, chartaceous, nerveless, obtuse; upper glume ovate, chartaceous, 1-nerved; lemma oblong-ovate, chartaceous, 1-nerved; palea oblong-elliptic, hyaline, 2-nerved; lodicules 2; stamens 3; stigma plumose. Caryopsis ovate-oblong.

Occasional along road sides and forests plains in Mahabubnagar district.

Fl. & Fr.: September – December.

Kollapur (MBNR), *BSS* & *SKB* 32996.

Sporobolous maderaspatanus Bor, Kew Bull. 1957: 234. 1957 & Grass. India 632.1960; Moulik, Grass. Bamb. India 2:480.1997. *S. orientalis* Kunth, *nom. illegit*; Hook. f., Fl. Brit. India 7:251.1896; Fischer, Fl. Madras 3: 1817. 1934.

Perennials, culms erect, to 50 cm high, stoloniferous; nodes glabrous. Leaf sheaths 3-6 cm long, glabrous; ligule membranous; leaf blade 5-15 × 0.2-0.4 cm, linear, base rounded, acuminate, convolute. Panicles 9 cm long, pyramidal; spikelets oblong; lower glume ovate, hyaline, nerveless; upper glume ovate-lanceolate, 1-nerved, apex acute; lemma ovate, lanceolate, chartaceous, prominently 1-nerved, apex acute; palea ovate-lanceolate, hyaline; stamens 3; stigma plumose. Caryopsis ovoid.

Rare in forest shade and plains in Mahabubnagar district.

Fl. & Fr.: July-November.

Kollapur (MBNR), *BR* & *BSS* 32917.

Sporobolus spicatus (Vahl) Kunth, Revis. Gram. 1: 67. 1829; Hook. f., Fl. Brit. India 7:250.1896; Fischer, Fl. Madras 3:1817.1934; Bor, Grass. India 632.1960; Moulik, Grass. Bamb. India 2:481.1997. *Agrostis spicata* Vahl, Symb. Bot. 1:9.1970.

Perennials, stoloniferous, aquatic, culms erect, tufted, to 30 cm high; nodes glabrous. Leaf sheaths 1-4 cm long, glabrous; ligule membranous; leaf blade 2-6 × 0.05-0.1 cm, oblong, base cordate or truncate, apex acute, glabrous. Panicles spiciform, up to 9 cm long, solitary; spikelets elliptic; lower glume ovate-oblong, truncate, hyaline, 1-nerved; upper glume ovate-oblong, chartaceous, 1-nerved; lemma linear, lanceolate, chartaceous, hyaline; palea oblong, truncate, 2-nerved; lodicules 2; stamens 3; stigma plumose, greenish. Caryopsis obovoid.

Occasional along stream banks and moist places in Mahabubnagar district.

Fl. & Fr.: April – December.

Sporobolus tenuissimus (Schrank) Kuntze, Revis. Gen. Pl. 3: 369. 1893; Bor, Grass. India 633.1960; Moulik, Grass. Bamb. India 2:481.1997. *Panicum tenuissimum* Schrank in Denkschr., Bot. Ges. Regensb. 2:26.1822. *Sporobolus minutiflorus* (Trin.) Link., Hort. Berol. 1:88.1827; Hook. f., Fl. Brit. India 7:248.1896; Fischer, Fl. Madras 3: 1816. 1934.

Annuals, culms erect, tufted, to 1.2 m high; nodes glabrous. Leaf sheaths 5-15 cm long, glabrous; ligules hairy, membranous; leaf blade 3-6 × 0.05-0.15 cm, linear-lanceolate, base rounded, acuminate. Panicle up to 25 cm long effuse; spikelets ovate; lower glume oblong, nerveless, hyaline; upper glume ovate, hyaline, 1-nerved; lemma ovate, chartaceous, 1-nerved; palea oblong, hyaline, 2-nerved; lodicules 2; stamens 2. Caryopsis obovoid.

Occasional near water courses in Khammam, Medak and Mahabubnagar districts.

Fl. & Fr.: August – December.

Narsapur (MDK), *TP* & *MSG* 12019; Kinnerasani (KMM), *BR* & *AMR* 26820.

Sporobolus virginicus (L.) Kunth, Revis. Gram. 1:67. 1829; Hook. f., Fl. Brit. India 7:249. 1896; Fischer, Fl. Madras 3:1818. 1934; Bor, Grass. India 634. 1960; Moulik, Grass. Bamb. India 2:482. 1997. *Agrostis virginica* L., Sp. Pl. ed. 1:63. 1753. *Sporobolus tremulus* (Willd.) Kunth, Revis. Gram. 1: 67. 1829; Hook. f., Fl. Brit. India 7:250.1896; Fischer, Fl. Madras 3:1817. 1934; Bor, Grass. India 633. 1960; Moulik, Grass. Bamb. India 2:481. 1997. *Agrostis tremula* Willd., Sp. Pl. 1:372.1798.

Perennials, culms erect, tufted, to 10 cm high, stoloniferous, nodes glabrous. Leaf sheath 1-8 cm long, terete; ligule row of hairs, leaf blade 2.5-7.5 cm long, linear-lanceolate, glabrous, base rounded, convolute, apex acute. Panicle 10 cm long, sub-spiciform; spikelets lanceolate, crowded; lower glume lanceolate, chartaceous, 1-nerved; upper glume ovate-lanceolate, chartaceous, 1-nerved; lemma ovate-lanceolate, chartaceous, 1-nerved; palea oblong, elliptic, hyaline; lodicules 2; stamens 3; stigma plumose, green. Caryopsis obovoid.

Occasional in river banks, hill slopes and near streams in Hyderabad (Ramana, 2010) and Mahabubnagar district.

Fl. & Fr.: September - December.

Sporobolus wallichii Munro ex Trin., J. Bot. 27: 171. 1889; Hook. f., Fl. Brit. India 7: 248. 1896; Fischer, Fl. Madras 3: 1817. 1934; Bor, Grass. India 634. 1960; Moulik, Grass. Bamb. India 2:482.1997.

Perennials, culms erect, tufted, to 1.2 m high, slender; nodes glabrous. Leaf sheath 2-8 cm long, glabrous; ligule membranous; leaf blade 10-25 × 0.2-0.4 cm, linear, convolute, base cordate or truncate, apex acuminate, glabrous. Panicles up to 10 cm long, effuse, flexuous; spikelets oblong-lanceolate; lower glume ovate, chartaceous, truncate, nerveless; upper glume ovate-lanceolate, acute, 1-nerved; lemma ovate-elliptic, chartaceous, 3-nerved; palea oblong, chartaceous; lodicules 2; stamens 3; stigma plumose, white. Caryopsis oblong.

Common in moist localities of the forests in Mahabubnagar district.

Fl. & Fr.: July-December.

Mannanur (MBNR), *BSS* & *SKB* 30600, 30601 (SKU, BSID).

TETRAPOGON Desf.

Tetrapogon tenellus (Koen. ex Roxb.) Chiov., Ann. Ist. Bot. Roma 8: 352. 1908; Bor, Grass. India 475. 1960; Moulik, Grass. Bamb. India 2: 570. 1997. *Chloris tenella* Koen. ex Roxb., Fl . Ind. 1: 330. 1820; Hook. f., Fl. Brit. India 7: 290. 1896; Fischer, Fl. Madras 3 : 1838. 1934.

Vern.: *Eruvangaddi.*

An annual, tufted grass; culms up to 60 cm high; nodes glabrous. Leaf sheaths 3-7 cm long, glabrous; ligule membranous; leaf blade 2-8 x 0.5-0.15 cm, linear-lanceolate, base rounded, apex acuminate, glabrous. Racemes up to 7 cm long, solitary, terminal; spikelets 5 mm, cuneate, bifarious, 3-4-awned; perfect florets 4; glumes linear, persistent, strongly 1-nerved; upper broader than lower, mid nerve produced a small arista; all perfect lemmas broadly ovate, strongly 3-nerved, truncate; callus densely hairy, awned (shape is same, but size is different in all 4 perfect lemmas); paleas elliptic, 2-nerved, margins ciliated; stamens 3, stigmas plumose. Caryopsis ellipsoid, compressed. Empty lemma broadly ovate, 3-nerved, awnless.

Occasional grass of sandy tracts and waste lands in Nalgonda district.

Fl. & Fr.: August- September.

Cherukupally (NLG), *BVRS* 174 (NU).

THELEPOGON Roth ex Roem. & Schult.

Thelepogon elegans Roth ex Roem. & Schult., Syst. Veg. 2: 788. 1817; Hook. f., Fl. Brit. India 7: 148. 1896; Fischer, Fl. Madras 3: 1892. 1934; Bor, Grass. India 247. 1960; Moulik, Grass. Bamb. India 1: 302. 1997.

An annual, tufted grass; culms up to 60 cm high, geniculately ascending, rooting from base; nodes glabrous. Leaf sheaths 3-7 cm, loose, tubercle-based hairs; margins densely ciliate; ligule ciliate; leaf blade 5-15 x 0.7-2 cm, linear-lanceolate, base amplexicaul, margins spinulose, apex acute, tubercle-base pubescent, nerves prominent. Racemes 4-12, up to 15 cm long, digitate, corymbosely fasicled; rhachis fragile, compressed; spikelets ovate-oblong, 5 mm, 2-nate, pedicelled spikelet reduced. *Sessile spikelet;* ovate-oblong, 5 mm; callus bearded; lower glume ovate, crustaceous, apex 2-lobed, transversely rugose, interrupted, strongly ridged, prominent at apex; upper glume ovate, acuminate, coriaceous, keeled, scabrid; lower lemma oblong, hyaline; palea oblong, nerveless, male; upper lemma obovate, 2-lobed, awned; palea narrower, bisexual; stamens 3; stigmas plumose. Caryopsis oblong.

Occasional gregarious grass in Nizamabad district.

Fl. & Fr.: August-October.

Gandhari (NZB), *BR & CPR* 7231.

THEMEDA Forssk.

1. Branching effuse; clusters few, in few, small superposed fascicles on filiform branches .... **T. laxa**
1. Branching short; clusters of inflorescence numerous, in numerous fascicles:
    2. Perennial; culms robust; clusters globose or fan-shaped .... **T. quadrivalvis**
    2. Annual; culms geniculately ascending; clusters usually narrow, erect .... **T. triandra**

Themeda laxa (Anderss.) Camus, Bull. Mus. Hist. Nat. Paris 26: 423. 1920; Fischer, Fl. Madras 3: 1746. 1934; Bor, Grass. India 251. 1960. *Anthistiria laxa* Anderss., Nov. Act. Sci. Upsal. 3, 2: 243. 1856. Hook. f., Fl. Brit. India 7: 213. 1896.

A perennial, tufted grass; culms up to 40 cm high. Leaf-blades on the main culms up to 30 x 0.7 cm, apex acuminate, upper blades 7.5-12 cm long, filiform. Panicle up to 8 cm long, leafy, few racemes, shortly peduncled outer spathes 2.5 cm; proper spathes glabrous, longer than the spikes. *Involucral spikelet:* 4, all on the same level, 4 mm, male; lower glume glabrous except few bristles near the apex. *Pedicelled spikelets:* similar to involucral spikelet. *Sessile spikelet:* solitary, 4 mm, bisexual, awned; lower glume glabrous, shining below, hispidulous at apex.

Occasional in Medak district.

Fl. & Fr.: August-December.

Medak (MDK), *RG* 104106 (BSID).

Themeda quadrivalvis (L.) Kuntze, Revis. Gen. Pl. 2: 794. 1891; Fischer, Fl. Madras 3: 1746. 1934; Bor, Grass. India 252. 1960; Moulik, Grass. Bamb. India 1: 233. 1997. *Andropogon quadrivalvis* L. in Murr. Syst. Veg. ed. 13. 758. 1774. *Anthistiria ciliata* L. f., Suppl. 113. 1781; Hook. f., Fl. Brit. India 7: 213. 1896.

An annual, slender grass; culms up to 1 m high, geniculately ascending and rooting from the lower nodes; nodes glabrous. Leaf sheaths 3-7 cm long, glabrous, ligule a narrow, ciliate membrane; leaf blade 7-15 x 0.3-0.5 cm, lanceolate, flat, base cordate, margins scaberulous, apex acuminate, glabrous, midrib prominent. Spatheolate panicle up to 20 cm long, pedunculate racemes up to 5 cm; spatheolate 4 mm, lanceolate. *Involucral spikelets:* 4 mm, oblong, tuberculate pilose, male, whorled; lower glume oblong, tuberculate pilose on outer surface and margins; upper glume linear; lemma linear, hyaline. *Pedicelled spikelet:* 8 mm, linear-lanceolate, similar to above. *Sessile spikelet:* 7 mm, linear, bisexual; callus pungent; lower glume oblong-obtuse; upper glume lanceolate, hairy; lower lemma oblong, glabrous; upper lemma linear, awned, hispid; stamens 3, stigmas plumose. Caryopsis linear, grooved on the front.

Common on hill slopes of Adilabad, Medak and Hyderabad districts.

Fl. & Fr: August- December.

Sone (ADB), *TP & PVP* 4052; Narsapur (MDK), *TP & MSG* 11996; Hyderabad, *MRS* 210 (DD); Pakhal RF (WGL), *KMS* 11672 (CAL).

Themeda triandra Forssk., Fl. Aegypt-Arab. 123 & 178. 1775; Fischer, Fl. Madras 3: 1746. 1934; Bor, Grass. India 254. 1960; Moulik, Grass. Bamb. India 1: 234. 1997. *Anthistiria imberbis* Retz., Observ. Bot. 3: 11. 1783; Hook. f., Fl. Brit. India 7: 211. 1896.

A perennial, tufted grass; culms up to 2 m high, robust, solid; nodes glabrous. Leaf sheaths 4-12 cm long, compressed, glabrous; ligule a ciliate membrane; leaf blade 8-25 x 0.3-0.5 cm, linear, flat, base rounded, margin scaberulous, apex acuminate, scabrid on both surfaces, midrib prominent. Spatheolate panicle up to 30 cm long, lax, bearing clusters of racemes up to 3 cm, distant; spatheolates 1.5 cm, lanceolate, pilose at base. *Involucral spikelet:* 4, 4 mm, oblong, tubercle-based hairs at apex, whorled, all arise at same level, male; lower glume oblong, ridge, keel winged, ciliate, coriaceous; upper glume linear, margins incurved; lemma linear. *Pedicelled spikelet:* 8 mm, linear-lanceolate, similar to above. *Sessile spikelet:* 6.5 mm, linear; callus densely pilose with yellow hairs; lower glume oblong, crustaceous, apex scabrid outside; upper glume oblong, obtuse, keeled; lower lemma oblong, acute, hyaline, ciliate at apex; upper lemma stipiform, awned, hispid, ciliate at apex; lodicules 2; stamens 3; stigmas plumose. Caryopsis linear, grooved on the front.

Occasional grass in hilly areas, slopes and in plains near the streams of Karimnagar, Medak and Mahabubnagar districts.

Fl. & Fr: August-December.

Patancheru (MDK), *BR & CP* 11315; Mallelatheertham (MBNR), *SRS* 108857, 111412 (BSID).

THYSANOLAENA Nees

Thysanolaena latifolia (Roxb. ex Hornem.) Honda, J. Fac. Sci. Univ. Tokyo, Sect. 3, Bot. 3: 312. 1930. *T. maxima* (Roxb.) Kuntze, Revis. Gen. Pl. 2 : 794. 1891; Fischer, Fl. Madras 3 : 1797. 1934; Bor, Grass. India 650. 1960. *T. agrostis* Nees, Edinburgh New Philos. J. 18 : 180. 1835; Hook.f. Fl. Brit. India 7 : 61. 1896.

Perennial tufted grass; culms 1.5-3.5 m high, erect, woody. Leaf-blades flat, 30-60 x 3-7 cm. broadly lanceolate, acuminate, sub-amplexicaul, coriaceous, glaucous below. Panicles 30-60 cm long, branches mostly straight, erect or sub-erect, capillary; rachis flexuose. Spikelets lanceolate, 12-1.8 mm long, lower florets sterile; median one bisexual; upper one male or sterile; glumes subequal, 0.5-0.8 mm long, ovate, subacute to sub-rotund; sterile lemma glabrous; fertile lemma ciliate on the margins with hairs up to 0.5 mm long. Caryopsis elliopsid, 0.5 mm long.

In moist places in Karimnagar district (Naqvi, 20001), also planted in gardens (Ramana, 2010).

Fl. & Fr.: September-December.

TRACHYS Pers.

Trachys muricata (L.) Pers. ex Trin., Mem. Acad. St. Petersb. Set. 6. 3: 189. 1834; Fischer, Fl. Madras 3:1813. 1934; Bor, Grass. India 369. 1960; Moulik, Grass. Bamb. India 1: 165. 1997. *Cenchrus muricatus* L., Mant. 2: 302. 1771. *Trachys mucronata* Pers., Syn. Pl. 1: 85. 1805; Hook. f., Fl. Brit. India 7: 96. 1896.

An annual, diffuse grass; culms up to 60 cm high, weak, ascending or prostrate, rooting from lower nodes; nodes villous. Leaf sheaths 3-5 cm, loose, villous; ligule a thin membrane; leaf blade 4-10 x 0.5-1.5 cm, lanceolate, flat, base rounded, apex acute, villous on both surfaces. Spiciform raceme paired, up to 7 cm long; rhachis angular, clusters of spikelets sunk in cavity of rhachis; spikelets intermixed with small scale like empty glumes; spikelets 6 mm, globose; glumes unequal, lanceolate; lower coriaceous; upper hyaline; lower lemma broadly ovate, coriaceous, many-nerved, cuspidately acute; palea lanceolate, hyaline, nerveless, empty; upper lemma ovate-lanceolate, chartaceous, 3-nerved; palea lanceolate, nerveless, bisexual; lodicules 2; stamens 3; stigmas plumose. Caryopsis broadly ovate.

Common weed in waste lands and in cultivated fields of Nalgonda district.

Fl. & Fr.: August-November.

Nagarjuna konda valley (NLG), *KT* 7116 (CAL); Nagarjunasagar (NLG), *KMS* 9732 (CAL); Kambalpally (NLG), *BVRS* 224 (NU)

Note: Persoon (Syn. Pl. 1:85. 1805) made the combination *Trachys mucronata* by citing *Cenchrus mucronatus.* L. as basionym. There is no Linnean species C. *mucronatus* but, there is a C. *muricatus* L. Trinius (1834) corrected this mistake.

TRAGUS Heller *nom. cons.*

Tragus roxburghii Panigr., Kew Bull. 29:495, 1974; Moulik, Grass. Bamb. India 2: 506. 1997. *T. racemosus* Hook. f., Fl. Brit. India 7: 97. 1896 non Scop. 1777. *T. biflorus auct.* non Schult. 1824; Fischer, Fl. Madras 3: 1813. 1934; Bor, Grass. India 682. 1960.

A perennial, tufted grass; culms up to 25 cm high, decumbent or geniculately ascending from the base; nodes glabrous. Leaf sheaths 2-4 cm long, glabrous; ligule membranous ciliate rim; leaf blade 1-3 x 0.2-0.4 cm, lanceolate, flat, base rounded, margin pectinately ciliate, apex acute, glabrous. Spiciform racemes up to 4 cm long, solitary; spikelets 3 mm, lanceolate, 2, in a deciduous subsessile clusters facing each other, 1-flowered; glumes unequal; lower 0 or scarious, hyaline; upper glume lanceolate, prickled, prickles terete at base, hooked at the apex; lemma oblong-lanceolate, chartaceous; palea linear, hyaline; lodicules 2; stamens 3; stigmas plumose. Caryopsis ellipsoid, free within the lemma and palea.

Common in waste plains and lower slopes of Adilabad, Medak, Ranga Reddi, Karimnagar, Hyderabad and Nalgonda districts.

Fl. & Fr.: July-September

Vempalli RF (ADB), *GO & PVP* 4805; Choutkoor (MDK), *TP & MSG* 11945; Rudraram RF (RR), *TP* & *MSM* 12711; Nagarjuna konda valley (NLG), *KT* 9682 (CAL); Peddavagu (NLG), *BVRS* 518 (NU); Vempalli (KRN), *GVS* 20178 (MH).

TRIPOGON Roem. & Schult.

1. Lemmas 4-lobed .... **T. bromoides**
1. Lemmas 2-lobed:
    2. Leaf blade filiform, convolute .... **T. jacquemontii**
    2. Leaf blade linear to linear-lanceolate .... **T. purpurascens**

Tripogon bromoides Roem. & Schult., Syst. Veg. 2:600.1817; Hook. f., Fl. Brit. India 7:287. 1896; Fischer, Fl. Madras 3:1834.1934; Bor, Grass. India 521. 1960; Moulik, Grass. Bamb. India 2:622.1997.

Perennials, culms erect, tufted, to 30 cm high, nodes glabrous. Leaf sheaths 2-6 cm long, glabrous; leaf blade 5-25 × 0.2-0.4 cm, linear, base cordate, acuminate, hairy. Spike up to 20 cm long, secund; spikelets 6-13-flowered; lower glume oblong-lanceolate, chartaceous, 1-nerved; upper glume lanceolate, ovate-oblong, 2-lobed, 1-nerved; lemmas ovate, chartaceous, 3-nerved, 3-awned; paleas elliptic-oblong, keeled, 2-nerved, hyaline; lodicules 2; stamens 3; stigma plumose. Caryopsis oblong.

Common on hill slopes and forest plains in Mahabubnagar districts.

Fl. & Fr.: August-February.

Irlapenta (MBNR), *BR* & *BSS* 32289.

Tripogon jacquemontii Stapf, Kew Bull. 85.1892; Hook. f., Fl. Brit. India 7:286. 1896; Fischer, Fl. Madras 3:1833. 1934; Bor, Grass. India 522.1960; Moulik, Grass. Bamb. India 2:622.1997.

Perennials, culms erect, tufted, to 90 cm high; nodes glabrous. Leaf sheaths 2-8 cm long, glabrous; ligules fimbriate; leaf blade 5-20 cm long, linear, base narrow, apex acuminate, glabrous. Spikelets oblong, 8-20-flowered; lower glume ovate-lanceolate, 1-nerved, lobed on one side, sub coriaceous; upper glume ovate-lanceolate, 1-nerved, chartaceous; lemma elliptic-obovate, awned; palea lanceolate, hyaline, 2-keeled, winged; lodicules 2; stamens 3. Caryopsis oblong.

Rare in barren lands and forest plains in Mahabubnagar district. Endemic to Peninsular India.

Fl. & Fr.: September–December.

Mallelateertham (MBNR), *BR* & *BSS* 32217.

Tripogon purpurascens Duthie, Ann., Roy. Bot. Gard. Cal. 9:75. 1901; Bor, Grass. India 522. 1960; Moulik, Grass. Bamb. India 2:623. 1997; B.R.P. Rao & *al.*, J. Econ. Taxon. Bot. 33: 435.2009. *T. abyssinicus* Hook. f., Fl. Brit. India. 7:287. 1896 *p.p.* non Nees (1854).

Perennials, culms erect, tufted, slender, to 15 cm high; nodes glabrous. Leaf sheaths glabrous; ligules ciliate, membranous; leaf blade filiform, convolute, tip obtuse with purple colour, densely hairy near the base. Inflorescence terminal, contracted; spikelets 2-3-flowered; lower glume small, triangular, 1-nerved, slightly awned; upper glume lanceolate, 1-nerved, membranous, central thickened, coriaceous, margin hyaline, ciliate upper portion; lemmas 3, 3-nerved, cleft, median nerve slightly awned in the cleft, margin hyaline; paleas 3, 2-keeled, tip 2-lobed except third palea; stamens 2-3. Caryopsis linear.

Occasional in dry and moist rocky habitats of forests in Mahabubnagar district.

Fl. & Fr.: July-November.

Amrabad (MBNR), *BSS* & *SKB* 32201; Appapur (MBNR), *BSS* & *SKB* 32220.

TRITICUM L.

1. Rhachis fragile at maturity; caryopsis entirely enclosed between the lemma and palea and not removed by threshing .... **T. dioccum**
1. Rhachis tough at maturity; caryopsis separating from the gaping lemma and palea .... **T. aestivum**

Triticum aestivum L., Sp. Pl. 85. 1753 *nom. cons*; Bor, Grass. India 670. 1960. *T. vulgare* Vill., Hist. Pl. Bauph. 2:153. 1789; Hook. f., Fl. Brit. India 7: 367. 1896.

Vern.: *Godhumalu*. English: Wheat.

An annual, tufted grass; culms up to 1 m high. Leaf-blade linear-lanceolate, flat, apex acuminate. Spike up to 20 cm long, erect, curved at maturity, compact; spikelets solitary, laterally compressed, 3-5-flowered; glumes obtuse, shortly awned; lemma muticous, or shortly 1-3-awned, bisexual. Caryopsis oblong, ventraly grooved, free within the lemma and palea.

Occasionally cultivated in some districts.

Fl. & Fr.: November-March.

Triticum dicoccum (Schrank) Schubler, Char. & Descr. Cerealium in Hort. Tuling. 28. 1818; Fischer, Fl. Madras 3:1856. 1934; Bor, Grass. India 679. 1960.

Vern.: *Godhumalu*. English: Wheat.

An annual, cultivated, tufted grass; culms up to 50 cm high, leafy; nodes glabrous. Leaf sheaths 6-15 cm long, glabrous; ligule membranous; blade 10-30 × 0.5-1 cm, base rounded, apex acuminate, glabrous. Spikes up to 15 cm long, erect, compact; spikelets 1.2 cm, laterally compressed, ellipsoid; rhachis fragile at maturity; spikelets in two rows, few-flowered; glumes broadly ovate, persistent, coriaceous, often unequal, shortly awned; lower lemma coriaceous, keeled, shortly awned; upper lemma boat-shaped, geniculately awned; palea elliptic, 2-keeled, keels ciliate, the lower enclosing a bisexual floret; the upper male or empty; stamens 3. Caryopsis oblong, often hairy, free within the lemma and palea.

Occasionally cultivated crop in all districts.

Fl. & Fr: January-March.

Kadthala fields (ADB), *GO & DAM* 5015.

UROCHLOA P. Beauv.

Urochloa panicoides P. Beauv., Ess. Agrost. 53, 6. 11, t. 1. 1812; Fischer, Fl. Madras 3:1775. 1934; Bor, Grass. India 372. 1960; Moulik, Grass. Bamb. India 1: 167. 1997. *Panicum javanicum* Poir. in Lam., Encycl. 4: 274. 1816; Hook. f., Fl. Brit. India 7:35. 1896.

An annual, stoloniferous grass; culms up to 70 cm high, geniculately ascending and rooting at lower nodes; nodes pubescent. Leaf-sheaths 2-5 cm long, loose, glabrous, keeled; ligule ciliate rim; leaf blade 4-12 x 0.6-1.2 cm, lanceolate, base cordate, margin tuberculate based hairs, apex acute, glabrous. Spiciform racemes of 4-9, up to 5 cm long; rhachis serrate, flat, scabrid; spikelets 4 mm, ovate-elliptic, solitary, dorsally compressed; glumes unequal; lower very small, ovate-orbicular, hairy, 1-nerved, hyaline; upper glume ovate, acute, membranous, 7-nerved, densely ciliate; lower lemma ovate-oblong, acute, 3- nerved, densely ciliate, palea ovate, hyaline, keeled, male or empty; upper lemma ovate-obovate, coriaceous, nerveless, arista at apex, paleated, its palea similar to lemma, bisexual; lodicules 2; stamens 3; stigmas plumose. Caryopsis ovate, compressed.

Common on the bunds of cultivate fields and roadsides of all districts.

Fl. & Fr: August-October.

Kowtal (ADB), *GO & PVP* 4832; Vempalli (ADB), *GO & PVP* 4745; Hyderabad, *MRS* 313 (DD); ICRISAT (MDK), *LJGV* 3865 (CAL); Mallepally (NLG), *BVRS* 534 (NU).

Note: This grass superficially looks like some species of *Brachiaria,* but it can be easily distinguished by the mucronate upper glume. It has glabrous or pubescent and setosely fringed spikelets.

Key to Varieties

1. Leaf sheaths glabrous .... var. **panicoides**
1. Leaf sheaths pubescent .... var. **pubescens**

Urochloa panicoides P. Beauv. var. **pubescens** (Kunth) Bor, Grass. India 372. 1960.

Annual grass, culms 30-60 cm high, geniculately ascending from the rooting base. Leaf sheaths loose, hirsute with tubercle based hairs; ligule a ciliate rim; leaf blade lanceolate, 4-8 ×0.8-1.5 cm, base semiamplexicaul, finely hirsute, margins wavy.

Inflorescence of 4-7 spiciform, second racemes, 5-7 cm long; spikelets ovate-elliptic, 3.5-4 mm long, pubescent, unawned. Caryopsis elliptic, up to 2 mm.

Rare in moist places in Adilabad district.

Fl. & Fr.: August – October.

Kowtal (ADB), *GO* & *PVP* 4832.

Note: This variety is easily distinguishable by its pubescent spikelets.

Urochloa panicoides Beauv. var. **velutina** (Henr.) Bor, Grass. India 373. 1960.

There are four varities of *Urochloa panicoides* in India (Bor, 1960): *panicoides, pubescens, marathensis* and *velutina*. The latter differs from other taxa in having dense pubescence on spikelets and conspicuous bristled lower lemma.

This variety was reported for the first time from Andhra Pradesh and Southern Peninsular India by Sadasivaiah & *al.* (2013).

VETIVERIA Bory de St. Vincent

1. Ligule of a fimbriate membrane with long hairs; lower glume with a row of marginal tubercled spinulose joints and pedicels with a basal tuft of hairs; callus of sessile spikelets bearded .... **V. lawsonii**
1. Ligule a scarious ring; lower glume muriculate, joints and pedicels glabrous; callus of sessile spikelets glabrous .... **V. zizanioides**

Vetiveria lawsonii (Hook, f.) Blatt. & McCann, J. Bombay Nat. Hist. Soc. 32: 409, 1928; Fischer, Fl. Madras 3:1733. 1934; Bor, Grass. India 258. 1960; Moulik, Grass. Bamb. India 1: 256. 1997. *Andropogon lawsonii* Hook. f., Fl. Brit. India 7:187. 1896.

A perennial, tufted, robust grass; culms up to 2 m high; nodes glabrous. Leaf sheaths 5-20 cm long, glabrous; ligule a small ring of fimbriate membrane, apex with long hairs; leaf blade 10-20 x 0.4-0.6 cm, linear, base truncate, apex acuminate, glabrous. Panicle up to 20 cm long; branches capillary, whorled; racemes up to 6 cm long; spikelets lanceolate, 2-nate. *Sessile spikelets:* 5 mm, linear, laterally compressed, bisexual; glumes equal, coriaceous; lower linear, spinulose; upper boat-shaped, keels ciliate, spinulose at apex, awned; lemma hyaline; lower lemma linear, margins ciliate; upper lemma linear, 2-lobed, awned; palea linear, ciliate; lodicules 2; stamens 3, stigmas plumose. Caryopsis linear-oblong. *Pedicelled spikelet:* 6 mm, lanceolate, male; lower glume lanceolate, chartaceous; upper glume lanceolate, awned, chartaceous; lemma lanceolate hyaline; stamens 3.

Occasional grass in marshy areas and in cultivated fields of Warangal, Nizamabad and Medak districts.

Fl. & Fr: August- March.

Ghanapur (NZB), *BR & KH* 7112; Jalalpur RF (NZB), *BR & KH* 9564; Gangapur fields (MDK), *TP & MSG* 12093.

Vetiveria zizaniodes (L.) Nash in Small, Fl. South South-east U.S. 67. 1903; Fischer, Fl. Madras 3: 1733. 1934; Bor, Grass. India 258. 1960; Moulik, Grass. Bamb. India 1: 255. 1997. *Phalaris zizanioides* L., Mant. Pl. 2:183. 1771. *Andropogon squarrosus* auct. non. L. f.; Hook. f., Fl. Brit. India 7: 186. 1896; Bor, *loc.cit.*

Vern.: *Vattiveru.*

A perennial, tufted, robust, aquatic grass; culms up to 2 m high: nodes glabrous. Leaf sheaths 4-10 cm long, glabrous; ligule a scarious ring; leaf blade 5-25 x 0,5-1.2 cm long, linear, flat, base truncate, apex acuminate, glabrous. Panicle up to 25 cm long; branches whorled, capillary; racemes up to 10 cm long; spikelets lanceolate, 2-nate. *Sessile spikelet:* 4 mm, lanceolate, bisexual; glumes equal, coriaceous; lower glume lanceolate, spinulose; upper boat-shaped, keeled, keel ciliate, scabrid on back; lower lemma linear, hyaline, keel winged, wings ciliate on margin, epaleated; upper lemma oblong, hyaline, 2-lobed, lobes mucronate, awned, paleated; palea linear, hyaline, nerveless; lodicules 2; stamens 3; stigmas plumose. Caryopsis oblong, slightly oblique at apex. *Pedicelled spikelet:* 4 mm, lanceolate, male; glumes equal, coriaceous; lower glume lanceolate, spinulose; upper glume boat-shaped, keeled, keels ciliate, scabird at apex; lower lemma lanceolate, hyaline, margins ciliate; upper lemma oblong, 2-lobed, lobes mucronate; lodicules 2; stamens 3.

Common in marshy plains and near ponds of Adilabad and Warangal districts.

Fl. & Fr.: August-October.

Sarvaipet (ADB), *GO* 4512; Pakhal RF (WGL), *KMS* 13148 & 11683 (CAL); Vatavarlapally (MBNR), *KI* 18729; Thupakulagudem (WGL), *PVS* & *NRR* 76993 (BSID).

ZEA L.

Zea mays L., Sp. Pl. 971. 1753; Hook. f., Fl. Brit. India 7:102. 1896; Fischer, Fl. Madras 3:1704. 1934; Bor, Grass. India 482. 1960; Moulik, Grass. Bamb. India 1: 205. 1997.

Vern.: *Mokkajonna.* English: Maize, Corn.

An annual, cultivated, monoecious grass; culms up to 2 m high, robust; nodes bearded. Leaf sheaths 5-20 cm long, terete, densely pubescent along the margins; ligule a pubescent ring; leaf blade 20-90 x 3-10 cm, lanceolate; flat, broad, base cordate, margin ciliate, apex acute, sparsely hairy on both surfaces. Panicle monoecious; male panicle up to 40 cm long, terminal, racemosely arranged; female panicle up to 2.5 cm long, solitary, axillary; spikelets 2-nate, 2-flowered.

Cultivated in all districts for its grains.

Fl. & Fr.: July-October.

Indanapally fields (ADB), *GO & DAM* 5411; Bellampalli (ADB), *PVP* 9442.

Gymnosperms

The following species are often seen in gardens: *Araucaria columnaris* (Forst.) Hook.f. (Araucariaceae), *Cycas revoluta* Thunb., *Cycas rumphii* Miq. (Cycadaceae), *Callitris cupressiformis* Vent., *Callitris glauca* R.Br. ex Baker & Smith, *Cupressus sempervirens* L., *Cryptomeria japonica* D.Don, *Juniperus communis* L., *Thuja occidentalis* L., *Thuja orientalis* L. (Cupressaceae), *Pinus caribaea* Morelet, *Pinus excelsa* Wall. ex D. Don, *Pinus roxburghii* Sarg. and *Pinus rudis* Lindl. (Pinaceae).

Bibliography

Ahmedulla, M. and M.P. Nayar. 1987. Endemic Plants of the Indian region. Vol. I. Peninusular India.Bot. Surv. India, Calcutta.

Aiton, W. 1789. Hortus Kewensis (ed. 1). (2 vols.) London.

Aiton, W.T. 1810-1813. Hortus Kewensis. (ed.2). London.

Ali, S.I. 1967. The genus *Goniogyna* DC. Taxon **16**: 463- 464.

Ali, S.I. 1968. Revision of the genus *Indigofera* L. from West Pakistan and north west Himalayas. Bot Notiser **3**: 543-577.

Anandakumar. 1984. A new combination in *Drimia* Jacq. ex Willd. (Liliaceae). J. Econ. Taxon. Bot. **5:** 962.

Anonymous 1948-76. The Wealth of India - Raw materials. (11 vols.) CSIR New Delhi.

Ansari, A.A. 1991. Critical notes on the genus *Crotalaria* L. IV. – Taxonomic status of a few taxa. J. Econ. Taxon. Bot. **15**: 455-456.

Ansari, A.A. 2008.*Crotalaria* L. in India. Bishen Singh Mahendra Pal Singh, Dehra Dun.

Ansari, R. and N.P. Balakrishnan. 1994. The Family Eriocaulaceae in India. Bishen Singh Mahendra Pal Singh, Dehradun. (revised ed. 2009).

Babu, C.R. 1974. The genus *Zanthoxylum* L. (Rutaceae) in India. Bull. Bot. Surv. India **16** : 48-71.

Babu, P.S.P., D. Ali Moulali and T. Pullaiah. 1994. *Jatropha tanjorensis* Ellis et Saroja – a new record for Andhra Pradesh. J. Bombay Nat. Hist. Soc. **91(1):** 163.

Bahadur, B. 1975. *Neanotis montholoni* (Hook.f.) W.H.Lewis: A new record for Andhra Pradesh. Curr. Sci. **44:** 208.

Bahadur, B., M. Bhanja, S. Karuppusamy and T. Pullaiah. 2007. *Holmskioldia sanguine* Retz. f. *citrina* Mold. – A new record for Andhra Pradesh. Proc. A.P.Akademi Sci.**11:** 132-133.

Bahadur, B., M. Sujatha and N. Carels. 2013. *Jatropha*, Challenges for a new energy crop. Springer, New York.

Bahadur, B., T. Pullaiah and G.V.S. Murthy. 2013. Systematics of Indian Jatrophas. In: Bahadur, & *al.* (eds.) *Jatropha*, Challenges for a new energy crop. Springer, New York. pp. 153-186.

Baken, C.V. 1960. Die Cactaceae. Band IV. Jena

Bailey, L.H. 1949. Manual of cultivated Plants. New York.

Baker, C.A. and R.C. Bakhuizen Van der Brink. 1963, 1965 & 1968. *Flora of Java.* (3 vol.). The Netherlands.

Balakrishnan, N.P. 1966. Nomenclatural notes on some flowering plants. J. Bombay Nat. Hist. Soc. **63** : 327-331.

Balakrishnan, N.P. 1980. Nomenclature of some flowering plants. Bull. Bot. Surv. India **22**: 173-179.

Balakrishnan, N.P. and T. Chakrabarty. 2007.The Family Euphorbiaceae in India. A synopsis of its profile, taxonomy and Bibliography. Bishen Singh Mahendra Pal Singh, Dehra Dun.

Banerjee, S.P. and P.K. Mukherjee. 1970. Studies in the Rhamnaceae III. A taxonomic revision of Indian Ventilaginae. Indian Forester **96**: 35-36.

Basak, R.K. 1997. Simaroubaceae. In: P.K. Hajra & *al.* (eds.). Flora of India. Vol. 4, pp. 409-421, Bot. Surv. India, Calcutta

Beddome, R.H. 1868-74. Icones Plantarum India Orientalis. Madras.

Beddome, R.H. 1869-74. The Flora Sylvatica for Southern India. (3 vols.). Madras.

Bennet, S.S.R. 1971. The taxonomic status of the genus *Pongamia* Vent. (Papilionaceae). J. Bombay Nat. Hist. Soc.**68**: 302-303.

Bennet, S.S.R. 1987. Name changes in Flowering plants of India and adjacent regions. Dehradun.

Bentham, G. and J.D. Hooker 1862-83. Genera Plantarum. (3 vols.). London.

Bhandari, M.M. 1964. *Artabotrys hexapetalus:* Correct name for *A. odoratissimus* R. Br. Baileya **12**: 147-150.

Bhandari, M.M. 1978. Flora of the Indian Desert. Jodhpur.

Bhirava Murthy, P.V. 1984. Proceedings of the seminar on Resources. Development and Environment of the Eastern Ghats. Andhra University, Waltair.

Biju, S.D. 1997. Taxonomic and morphologic studies on the family Convolvulaceae of southern peninsular India. Ph. D. thesis, Univ. Calicut.

Biksham, G., C.T. Hash, M.M. Sharma, F. Thampal and S.M.M. Javed. (eds.) 2010. Biodiversity @ ICRISAT, Patancheru, India (version 2.0). WWF-ICRISAT, Hyderabad.

Binojkumar, M.S. and N.P. Balakrishnan. 2010.The genus *Euphorbia* L. (Euphorbiaceae) in India: A taxonomic revision.Bishen Singh Mahendra Pal Singh, Dehra Dun.

Biswas, K. 1943. Systematic and taxonomic studies on the flora of India and Burma. Proc. 30th Indian Sci. Congr. Assoc. Calcutta. 101-152.

Biswas, K. and C.C. Calder, 1937. Handbook of Common Water and marshy plants of India and Burma.Delhi. (ed. 2. 1955). Calcutta.

Biswas, S.N. 1993. Hypericaceae. In: B.D. Sharma and M. Sanjappa (eds.). Flora of India. Vol. 3. pp. 43-85. Bot. Surv. India, Calcutta.

Blatter, E. 1918. A revision of the Indian species of *Rotala* L. and *Ammannia* L. J. Bombay Nat. Hist. Soc. **25** : 701-722.

Blatter, E. and C. McCann. 1934. Revision of the flora of Bombay Presidency. J. Bombay Nat. Hist. Soc. **37**: 15-35, 265-277,764- 779.

Blatter, E. and W.S. Millard. 1954. Some Beautiful Indian Trees (ed.2.) revised by W.T. Stearn. Bombay.

Bor, N. L. 1960. The Grasses of Burma, Ceylon, India and Pakistan. Pergamon Press. Oxford.

Bradley, H. 1849. Statistics of the circar of Dowlutabad, Hyderabad, Deccan. Madras J. Lit. & Sci. 15: 481-551. (Plant products pp 489-499).

Brown, R. 1810. Prodomus florae novae Hollandiae, Vol.1. R Taylor, London, pp. i – viii. 145 – 590.

Brandis, D. 1906. Indian Trees. New Delhi.

Burkhill, I.T. 1965. Chapters on the history of Botany in India. Calcutta.

Burman, N.L. 1768. Flora Indica. Leiden, Amsterdam.

Borssum Walkes, J.V. 1966. Malesian Malvaceae revised. Blumea **14**: 1-213.

Buchanan-Hamilton, F. 1807. A. journey from Madras through the countries of Mysore, Canara, Malbar. (3 vols.) London (repr. ed. in 2 vols. 1870).

Calder, C.C., V. Narayanaswamy and M.S. Ramaswamy 1926. List of species and genera of Indian Phanerogams not included in J.D. Hooker's flora of British India. Rec. Bot. Surv. India. **11**: 1- 157.

Candolle, A.L.P.P. de and A.C.P. de Candolle 1878-1896. Monographiae phanerogamarum. (9 vols.). Paris.

Candolle, A.P. de 1824-1874. Prodromus systematis Naturalis Regni vegetabilis. 17 vols. & Index, Paris. (vols. 8-17 by A.L. P.P. de Candolle).

Carter, S. 1984. New taxa and notes on herbaceous species of *Euphorbia* from East and North east Africa. Kew Bull. **39:** 650.

Chakravarthy, A. and S.K. Jain. 1973. On the genus *Zornia* Gmel. (Fabaceae) in India and Ceylon. Curr. Sci. **42**: 744-745.

Chakravarthy, H.L. 1959. Monograph of the Indian Cucurbitaceae. Rec. Bot. Surv. India. **17**: 1-234.

Chakravarthy, H.L. 1982. Cucurbitaceae. Fascicles of Flora of India. No. 11, Howrah.

Chakravarthy, H.L. and D.K. Hore 1979. Role of seed structure in cucurbit classification. Indian J. Forest. **2**: 323-329.

Chakrabarty, T. andN.P. Balakrishnan. 2006. The genera *Megistostigma* Hook.f. and*Tragia* L. (Euphorbiaceae) in India. Rheedea **16:** 15-28.

Champion, H.G. and S.K. Seth. 1968. A revised survey of the forest types of India. Govt. of India, New Delhi.

Chandra, S.K. 1966. A note of *Trianthema triquetra* Rottl. ex Willd. J. Bengal Nat. Hist. Soc. **35**: 40-42.

Chandrabose, M. 1976. Notes on some rare and interesting plants from South India. Bull. Bot. Surv. India.**18**:236-238.

Chandrabose, M. 1970. *Abutilon indicum* (L.) Sweet. ssp. *albescens* (Miq.) Borss. var. *australiense* Hochr. - A new record for India. Bull. Bot. Surv. India. **12**: 276-277.

Chandrabose, M. and N.C. Nair. 1981. The genus *Polygala* L. (Polygalaceae) in Andhra Pradesh, Kerala and Tamilnadu (South India). Proc. Indian Acad. Sci. (Plant Sci.) **90**: 107-127.

Chandrabose, M. and S.R.Srinivasan. 1975. *Leucas lavandulifolia* Rees and its varieties (Lamiaceae) in South India. Bull. Bot. Surv. India **17**: 164-167.

Chandra Mohan Reddy, N., R. Padmavathi, R. Shailaja, M.K. Prasad, K. Kalyani, K.B. Gopalam, P. Sridhar Rao and S. Jaltare. 2002. Our trees. Municipal Corporation of Hyderabad and Centre for Environment Education, Hyderabad.

Chatterjee, D. 1949. Indian Species of *Canavalia* DC.

Chaudhary, L.B., J.V. Sudhakar, A. Kumar, O. Bajpai, R. Tiwari and G.V.S. Murthy. 2012. Synopsis of the genus *Ficus* L. (Moraceae) in India.Taiwania **57(2):** 193-216.

Chennakesavulu, M. and R.R. Venkata Raju 1999. Note on two rare and little known aquatic angiosperms from Andhra Pradesh, India. Indian J. Forest **22**: 87 – 90.

Chopra, R.N. 1956. Glossary of Indian Medicinal Plants. New Delhi.

Chorge, A., K. Chandramohan, P.S. Annamma, M. Bhanja and E. Venkatesh. 2014. Additions to the grasses (Poaceae) of Andhra Pradesh from Kawal Tiger Reserve, Adilabad district. J. Threatened Taxa (In press).

Clayton, W.D and S.A. Renvoize. 1986. Genera Graminum – Grasses of the world. Kew Bulletin Additional series XIII, RBG, Kew; HMSO, London.

Clayton, W.D., M. Vorontsova, K.T. Harman and H. Williamson. 2013. Grass Base – The online World Grass Flora. Royal Botanic Gardens, Kew. www.kew.org/data/grass-db/www/imp05781,htm accessed on 17-6-2013.

Cook, C.D.K. 1996. Aquatic and wetland plants of India: A reference book and identification manual for the vascular plants found in permanent or seasonal fresh water in the subcontintnet of India south of the Himalayas. Oxford University Press, New Delhi.

Cooke, T. 1901-08. The Flora of the Presidency of Bombay. London. (3 vol.) repr. ed. 1958. Calcutta.

Cronquist, A.J. 1968. The Evolution and classification of Flowering Plants. Boston.

Cronquist, A.J. 1981. An Integrated system of classification Flowering Plants. London.

Dahlgren, R.M.T. 1980. A revised system of classification of the Angiosperms. Bot. J. Linn. Soc. **80** : 91-124.

Dalzell, N.A. and A. Gibson. 1861. The Bombay Flora. (2 vols.) Bombay.

Daniel, P. 1980. The fast vanishing umbrella thorn tree (*Acacai planifrons* Wt. & Arn.). Indain J. Forest. **3**: 349-352.

Dayananda, P. 1981. Conserving the flora of Peninsular hills. Bull. Bot. Surv. India. **23**: 250-253.

Deb, D.B. 1965. Plants new to India. Indian Forester **91**: 193-194.

Delile, A.R. 1813. Description de 1'Egypte. Historoire naturelle. (2 vols.). Paris.

Desfontaines, R.L. 1798-1799. Flora Atlantica etc. (2 vols.) Paris.

Deshpande, U.R. 1984. Poaceae: Andropogoneae. Flora of India, Fascicles. Fascicle 15. BSI, Calcutta.

Dey, S. and P.V. Prasanna. 2009. Exclusion of *Schoenus nigricans* L., *Rhynchospora triflora, Scleria foliosa* Hochst. & *Scleria reticulata* (Holtt.) Kern. (Cyperaceae) from Flora of India. Indian J. Forestr. **32:** 685-686.

Dinesh, V. and P.P. Sharma. 2012 Additions to the flora of Nizamabad district; Andhra Pradesh, India. Indian Forester **138(11):** 103-1032.

Ding Hou, L. 1962. Celastraceae-I. inFlora Malesiana Ser. 1. **6**: 227-291.

Don, D. 1825. Prodromus Florae Nepalensis. London.

Don, G. 1831-1838. A general history of Dichlamydeous Plants. London.

Duthie, J.F. 1903-1929. Flora of the upper gangetic plains and of the adjacent Siwaliks and Sub-Himalayan tracts. (3 vols.) Calcutta.

Endlicher, S.L. 1836-1840. Genera planatarum. Vienna.

Engler, H.G.A. 1900-1937. Das Pflanzenreich Regni vegetabilis conspectus. Leipzig.

Engler, H.G.A. and K.A.E. Prantl. 1897-1915. Die Naturlichen Pflanzen familien. (23 vols.) ed. 1. Leipzig.

Engler, H.G.A. and E. Gilg. 1924. Syllabus der Pflanzenfamilien. Berlin.

Forster, J.R. and J.G.A. Forster. 1775. Characters Genera Plantarum etc. London.

Fosskal, P. 1775. Flora Aegyptiaco-Arabica. Copenhagen.

Gaertner, J. 1788-1807. De Fructibus et seminibus plantarum. (3 vols.) Stuttgart.

Gamble, J.S. and C.E.C. Fischer. 1915-1935. Flora of the Presidency of Madras. London. Rep. ed. 1957. Calcutta.

Ganesh Babu, N.M., R. Vijaya Kumar, K. Ravikumar and D.K. Ved. 2006. Notes on some interesting and noteworthy plants from southern India. J. Econ. Taxon. Bot.**30:** 390-393.

Gayathri, M.S. and T. Pullaiah. 1996. On the occurrence of *Dimeria kanjirapalliana* K.C.Jacob (Poaceae) in Andhra Pradesh. J. Bombay Nat. Hist. Soc.**93(1):** 122-123.

Gayathri, M.S., T. Pullaiah and P.V. Prasanna. 1994. *Neanotis lancifolia* (Hook.f.) Lewis – An addition to the Flora of Andhra Pradesh. J. Bombay Nat. Hist. Soc. **91(3):** 478480.

Gmelin, J.F. 1791-1792. Linnaeus systema Nature. ed. 13. Leipzig.

Goetghebeur, P. and D.A. Simpson. 1991. Critical notes on *Actinoscirpus, Bolboschoenus, Isolepis, Phylloscirpus* and *Amphiscirpus* (Cyperaceae). Kew Bull. **46(1):** 169-178.

Gopalan, R., V. Chitra and C. Murugan. 2006. Two distributional records for Andhra Pradesh. J. Econ. Taxon. Bot.**30:** 980-983.

Govaerts, R. 1995. World checklist of seed plants. MIM, Deurne.

Govaerts, R. and D.A. Simpson 2007. World checklist of Cyperaceae. Sedges. The Board of Trustees of the Royal Botanic Gardens, Kew.

Gray, A. 1848. A Manual of Botany of the northern United States.Boston & Cambridge (ed. 2 New York).

Hackel, E .1889. MonographiaAndropogonearum. Paris.

Haines, H.H. 1924. The botany of Bihar and Orissa. Adlard & Son & West Newman Ltd., London.

Haines, R.W. and K.A. Lye. 1971. Studies on African Cyperaceae. IV. *Lipocarpha*,R.Br., *Hemicarpha* Nees and *Isolepis* R.Br. Bot. Notiser**124:** 473-482.

Hajra, P.K, V.J. Nair and P. Daniel 1997. Flora of India. Vol. 4. Malpighiaceae – Dichapetalaceae. Bot. Surv. India. Calcutta.

Hajra, P.K., R.R. Rao, D.K. Singh and B.P. Uniyal 1995. Flora of India. Vol. 12 and 13 Asteraceae. Bot. Surv. India. Calcutta.

Haworth, A.H. 1812. Synopsis plantarum succulentarum. London.

Heipko, P. 1984. Opiliaceae. In Flora Malesiana Ser. 1. **10**: 31-52.

Hemambara Reddy, M. and R.R. Venkata Raju 1998. Taxonomic study of the family Amaranthaceae in South India. J. Econ. Taxon. Bot.**21**: 577 – 586.

Henry, A.N., and Bela Roy, 1968. Nomenclatural notes on Indian flowering plants. Bull. Bot. Surv. India. **10**: 274.

Henry, A.N., K. Vivekanathan and N.C. Nair, 1978. Rare and threatened flowering plants of South India. J. Bombay Nat. Hist. Soc.**75**: 684-697.

Hoenselaar, K., B. Verdcourt and H. Beentje. 2010. Cyperaceae. Flora of Tropical East Africa.

Hooker, J.D. (ed.) 1872-1897. Flora of British India. (7 vols.) London.

Hooker, J.D. and T. Thomson 1855. Flora Indica. London.

Hooker, W.J. and G.A.W. Arnott. 1830-1841. The Botany of Captain Beechey's voyage. London.

Hooper, S.S. 1976. A concept of cyperoid spikelet and its terminology, illustrated by new species of *Lipocarpha* (Cyperaceae). Kew Bull. **41**: 423-428.

Huber, H. 1973. In: Abeywickrama. B.A. (ed.) A revised handbook of Flora of Ceylon. Vol. 1. University of Ceylon, Peradenia.

Humboldt, F.H.A., A.J.A. Bonpland and C.S. Kunth. 1816-1825. Nova Genera et species plantarum etc. Paris.

Hutchinson, J. 1959. Evolution and Phylogency of Flowering plants. London.

Hutchinson, J. 1969. The Families of flowering plants. Vol. I. Dicotyledons. Oxford.

Iltis, H.H. 1960. Studies in the Capparidaceae VII. Old world *Cleomes* adventive in the new world. Brittonia **12**: 279-294.

Indira, K. and R.R. Venkata Raju. 1998. Note on taxonomy and distribution of two rare and little known aquatic angiosperms from Andhra Pradesh. J. Indian Bot. Soc. **77:** 135-137.

Indira, K. and R.R.Venkata Raju. 2001. Two new additions to the sedges, family Cyperaceae of Andhra Pradesh. J. Bombay Nat. Hist. Soc. **98**: 497-499.

Indira, K. and R.R.Venkata Raju. 2001. Note on *Fimbristylis microcarya* F.V.Muell. (Cyperaceae) a new distributional record to South India. J. Bombay Nat. Hist. Soc.**98(1):** 157-158.

IUCN. 1994. IUCN Red List Categories. IUCN, Gland, Switzerland.

Jacobs, M. 1960. Capparidaceae. In Flora Malesiana Ser. 1. **6** : 61-226.

Jacobs, M. 1781-1795. Icones plantarum rariorum.Wien.

Jadhav, S.N., D.K. Ved, U. Ghate, K.N. Reddy and C.S. Reddy. 2001. Proceedings of the workshop on Conservation Assessment and Management Planning of Medicinal plants of Andhra Pradesh.FRLHT, Bangalore.

Jain, S.K. and R.R. Rao. 1976. A Hand Book of field and Herbarium methods. New Delhi.

Jain, S.K. and A.R.K. Sastry, 1980. Threatened plants of India, A State of the Art Report. New Delhi.

Jacquin, N.J. 1787-1797. Collectanae ad Botanicam, Chemiam etc. Historium naturalem spectamtia. 5 vols. Wien.

Javed, S.M.M. and T. Farida. 2008. Study of the Biological diversity of the Hussainsagar lake and its environs (Report). WWF-India, A.P., Hyderabad.

Jeffrey, C. 1980. Further notes on Cucurbitaceae of the Indian subcontinent. Kew Bull.**34**: 789-810.

Jussieu, A.L. de 1789. Genera plantarum secundum Ordines Naturales Disposita. (ed. 2, 1791). Paris.

Kapoor, S.L. and L.D. Kapoor, 1973. Further contribution to the flora of the Karimnagar district of A.P. Bull. Bot. Surv. India **15**: 76-84.

Karthikeyan, S., S.K. Jain, M.P. Nayar and M. Sanjappa. 1989. Florae Indicae Enumeratio: Monocotyledonae. Fl. Ind. Ser. 4. Bot. Surv. India, Calcutta.

Karthikeyan, S.and B.D. Sharma 1983. A catalogue of species added to Gamble's Flora of the Presidency of Madras. J. Bombay Nat. Hist. Soc.**80**: 63-70.

Karthikeyan, S., M.P. Nayar and R.S. Raghavan 1981. An annotated Bibliography of Taxonomic Botany of Peninsular India. 1959-1978. Botanical Survey of India. Howrah.

Karuppusamy, S., A. Ugraiah and T.Pullaiah. 2013. *Caralluma* (Sensu lato). Regency Publications, New Delhi.

Kashyap, G. 1961. *Shorea talura* Roxb. a synonym of *S. roxburghii* G. Don. J. Bombay Nat. Hist. Soc. **58**: 543-544.

Kelkar, S.S. and C.S. Latto, 1979. Further notes on the identification of the Genus *Tephrosia* Pers. J. Bombay Nat. Hist. Soc.**76**: 560-563.

Khan, A.A. 1954. The Andhra Forest. Indian Forester **80**: 753-758.

Khan, M.S. 1953. Forest Flora of Hyderabad State, being revised and enlarged edition of Patridge's "Flora of Hyderabad". Government Press, Hyderabad.

Konda Reddy, C.V. 1978. *Prosopis juliflora* (SW.) DC. the precocious child of the plant world. Indian Forester **104**: 14-18.

Krishna Murthy, A.V.R.G. 1960. Forest types of Andhra Pradesh. Mss.

Kumari G.R. and G.V. Subba Rao. 1982. The genus *Picrasma* Bl. (Simaroubaceae) in South India. J. Econ. Taxon. Bot.**3**: 249-250.

Kunth, C.S. 1833. 1850. Enumeratio plantarum ombium hucusgue cognitarium. Stuttgart.

Kuntze, C.E.O. 1891-98. Revisio Generum Plantarum. 3 vols. Leipzig.

Kurz. S. 1877. Forest Flora of British Burma. (2 vols.) Calcutta.

Lamarck, J.B. A.P.M. de 1783-1817. Encyclopaedic Methodique Botanicque. 8 vols. & 5 Suppl., Paris. (Vol. 5-13 byJ.L.M. Poiret).

Lamarck, J.B. A.P.M. de 1791-1823. Tableau Encyclopedique et Methodique; Botanique. 3 vols. & supp. 1. Paris (Vol. 2 part 5, 2 & Vol. 3 by J.L.M. Poiret.)

Lawrence, G.H.M. 1951. Taxonomy of Vascular Plants. New York.

Lazarrides, M. 1980. Tropical Grasses of Southeast Asia. J. Cramer, Hirschberg.

Leong-Skornickova, J., O. Sida and K.Marhold. 2010. Back to types ! Towards stability of names in Indian *Curcuma* L. (Zingiberaceae). Taxon **59(1):** 269-282.

Linnaeus, C. 1753. Species Plantarum (2 vols.) Stockholm.

Linnaeus, C. 1755-56. Centuria Plantarum. (2 vols.) Stockholm.

Linnaeus, C. 1758-59. Systema Naturae. ed. 10. Vol. 2. Botany. Stockholm.

Linnaeus, C. 1767. Mantissa Plantarum.pages 1-142. Stockholm.

Linnaeus, C. 1771. Mantissa Plantarum Altera.pages 143- 587. Stockholam.

Linnaeus, fil., 1781. Supplementum Plantarum Systamatis vegetabilium etc. Brunsvigae.

Lushington, A.W. 1915. Vernacular list of trees, shrubs and woody climbers in the Madras Presidency.Madras.

Maheshwari, J.K. 1960. Taxonomic studies on Indian Guttiferae - the genus *Calophyllum* L. Bull. Bot. Surv. India 2: 139- 148.

Maheshwari, J.K. 1962. Studies on the naturalised flora of India. Proc. Summer School, Botany, Darjeeling, 156-170.

Maheshwari, J.K. 1976. Plant resources in India. Sci. & Cult.**42** : 18-23.

Majumdar, N.C. 1979 a. Dilleniaceae. In Fascicles of Flora India 2: 1-16.

Majumdar, N.C. 1993. Caryophyllaceae. In: B.D. Sharma and N. P. Balakrishnan (eds.) Flora of India. Vol. 2. pp. 502-595, Bot. Surv. India, Calcutta.

Malick, K.C. and B. Safui. 1979. Field characters of two species of *Biophytum sensitivium* and *B. peterisianum*.Bull. Bot. Surv. India **21**: 209-210.

Malick, K.C. and B. Safui. 1980. A review of the androceium in Sterculiaceae with a key to the Genera. Bull. Bot. Surv. India **22**: 213-216.

Mamgain, S.K. and R.R.Rao. 1995. Asteraceae tribe 5. Cichorieae. In: Flora of India 12: 226-345.

Mathew, K.M. 1981-83. The Flora of the Tamilnadu Carnatic. (3 vols.) Tiruchirapalli.

Mathew, K.M. 1999. The Flora of the Palni hills, S. India. The Rapinat Herbarium, St. Joseph's College, Tiruchirapalli, India.

McCann. C. 1935. Some observations on *Nymphaea pubescens* Willd. J. Bombay Nat. Hist. Soc.**37**: 895-901.

McCann. C. 1952. A note on the Genus *Ludwigia* L. J. Bombay Nat. Hist. Soc.**50**: 892-894.

Mehra, K.L. 1965-66. Portuguese introduction of fruit plants in India I-III, Indian Hort.**10**: 8-12, 23-25.

Merril, E.D. 1917. An interpretation of Rumphius Herbarium Amb oinensc. Bur. Sci. Publ. Manila **9**: 1-595.

Meve, U. and Liede-Schumann, S. 2012. Taxonomic dissolution of *Sarcostemma* (Apocynaceae: Asclepiadoideae). Kew Bull. **67:** 751-758.

Meve, U. and Liede-Schumann, S. 2013. Erratum to :Taxonomic dissolution of *Sarcostemma* (Apocynaceae: Asclepiadoideae). Kew Bull. **68:** 187-188.

Michael, P.W. 1978. Notes on *Echinochloa* in the Philippines. Philipp. J. Weed Sci. **5:** 16-18.

Michaux, A. 1803. Flora Borealiamericana. (2 vols.) Paris.

Millard, W.S. 1911. The varieties of *Hibiscus* cultivated in Gardens. J. Bombay Nat. Hist. Soc. **20**: 892-894.

Miller, P. 1768. The Gardener's Dictionary. (ed. 8). London.

Misra, S., Panda, S.P. and Sahoo, D. 2008. Orchid flora of Andhra Pradesh, India. Bull. Bot. Surv. India **50 (1-4):** 129-146.

Mooney, H. 1938. Supplement to the Botany of Bihar and Orissa. Ranchi.

Moulali, D.A. and T. Pullaiah. 1991. A new combination in *Leucas*.J. Econ. Taxon. Bot.**15:** 454.

Moulik, S. 1997. Grasses and Bamboos of India. 2 vols. Scientific Publishers, Jodhpur.

Mukherjee, P.K. and Rita Gupta. 1970. Field identification of *Tephrosia* Pres. J. Bombay Nat. Hist. Soc. **67**: 135.

Mukherjee, S.K. 1968. The genus *Alangium* Lam. in India. Bull. Bot. Surv. India **10** : 330-331.

Mukherjee, S.K. 1972 a. Indian *Allophyllus* (Sapindaceae). Indian Forester **98**: 489-496.

Mukherjee, S.K. and Dali Chandra. 1983. An outline of the revision of Indian Anacardiaceae. Bull. Bot. Surv. India**25**: 52-61.

Murthy, E.N., C.S. Reddy and V.S. Raju. 2010. Tree wealth of Sivaram wildlife sanctuary, Andhra Pradesh, India. J. Econ. Taxon. Bot. **34(3):** 670-681.

Murthy, E.N., C.S.Reddy and V.S.Raju. 2006. Additions to the Flora of Warangal district, Andhra Pradesh, India. J. Econ. Taxon. Bot. **30(4):** 745-750.

Murthy, G.V.S., K.A.A. Kabeer, J.H.F. Benjamin, R.K. Premnath, B. Bahadur and V.J. Nair. 2008. Additions to the flora of Andhra Pradesh since Gamble (1935). Proc. A.P. Akademi Sci. **12 (1&2)**: 76-100.

Murugan, C. and R.Gopalan. 2006. Four additions to Indian *Memecylon* L. (Melastomataceae) from South India. Indian J. Forest.**29(1):** 105-108.

Nair, N.C., M. Chandrabose and S.R. Srinivasan. 1980. A further contribution to the weed flora of South India. Indian J. Forest.**3**: 56-59, 111-115, 231-237.

Nair, N.C. and A.N. Henry. 1983. Flora of Tamil Nadu, India Ser. 1.Analysis Vol. I. Coimbatore.

Nair, K.N. and M.P. Nayar 1997. Rutaceae. In: P.K. Hajra *et al.*(eds.). Flora of India Vol. 4. pp. 259-408. Bot. Surv. India, Calcutta.

Naithani, H.B. 1993. Contributions to the Taxonomic studies of Indian Bamboos. Vol. II. Ph.D.thesis. HNB Garhwal University, Srinagar, Garhwal.

Naithani, H.B., P.V.C. Rao, B.V.P. Reddy and C.K. Rao. 2011. Note on the rediscovery of *Cephyalostachyum pergracile* in Andhra Pradesh, India. Indian Forester **137(4):** 528-531.

Nampy, S. 2005. Morphologic and Taxonomic studies on Commelinaceae of Peninsular India. Final Technical Report of DST Project.

Naqvi, A.H. 2001. Flora of Karimnagar district, Andhra Pradesh. Ph.D.thesis, Kakatiya University, Warangal.

Naqvi, A.H. and V.S. Raju 1995. Further additions to the Flora of Karimnagar district, Andhra Pradesh, India. J. Econ. Taxon. Bot. **19:** 667-676.

Naqvi, A.H. and V.S. Raju 1998. Some plants new to the flora of Karimnagar district, Andhra Pradesh, India. J. Econ. Taxon. Bot. **22**: 451-458.

Narasimhan, D., N.Rama Rao and T. Ravisankar. 1989. Two rare interesting taxa of Euphorbiaceae from Andhra Pradesh, India. J. Econ. Taxon. Bot. **13:** 56-59.

Narasimha Rao, G. 1986. Studies on Floristics of Angiosperms and Dermotypes of Acanthaceae of Narsapur Taluk (Medak District, A.P.). Ph.D. Thesis, Osmania University, Hyderabad.

Narasimha Rao, G. and N. Ramayya. 1987. Biological spectrum of Narsapur, Medak district (A.P.), India. J. Swamy Bot. Cl. **4**: 183-187.

Nayar, M.P. 1980 a. Endemism and patterns of distribution of endemic genera (Angiosperms) in India. J. Econ. Taxon. Bot.**1**: 99-110.

Nayar, M.P. 1980 b. Endemic Flora of Peninsular India and its significance. Bull. Bot. Surv. India **22**: 12-23.

Nayar, M.P. 1996. 'Hot spots' of endemic plants of India, Nepal and Bhutan. Tropical Botanical Garden and Research Institute, Palode, Trivandrum.

Nayar, M.P.,M. Ahmed and D.C.S. Raju. 1984. Endemic and rare plants of Eastern Ghats. Indian J. Forest.**7**: 35-42.

Nayar, M.P. and G.S. Giri. 1982. A synopsis of the genus *Tribulus* L. (Zygophyllaceae) in India. Bull. Bot. Surv. India. **24**: 160-164.

Nayar, M.P., D.C.S. Raju and M. Ahmedulla. 1986. The dwindling flora of Andhra Pradesh. A call for conservation. Indian J. Forest. **9**: 283- 286.

Nayar, M.P. and A.R.K. Sastry 1987-1990. Red data book of Indian Plants. Bot. Surv. India. Calcutta. 3 vols.

Nayar, M.P., K. Thothathri and M. Sanjappa. (ed.) 1989. Fascicles of Flora of India. No. 19. Howrah.

Noltie, H.J. 2000. Flora of Bhutan. Vol. 3 part 2. The Grasses of Bhutan. Royal Botanic Garden, Edinburgh and Royal Government of Bhutan.

Oza, G.M. 1966. Correct name for *Ventilago calyculata* Tulasne. J. Bombay Nat. Hist. Soc. **63**: 455.

Panigrahi, G. 1974. A note on *Euphorbia heyneana* (Euphorbiaceae). Kew Bull. **29:** 695-697.

Panigrahi, G. 1977. Proposal to reject the name *Limonia acidissima nom. ambia rejc.* in favour of *L. elephantum* (Corr.) *comb. nov.* Taxon **26** : 576-577.

Panigrahi, S. 1976. Nomenclatural notes on 2 species of *Rotala* L. (Lythraceae) from the Indian region. Indian Forester **102**: 766-767.

Patridge, E.A. 1911. Forest Flora of H.E.H. The Nizam's Dominions, Hyderabad, Deccan. Hyderabad.

Prabhakar Raju and R.R.V. Raju. 1988. Additions to the genus *Blumea* in Andhra Pradesh. Indian J. Forest. **21:** 270-272.

Prabhakar Raju, C. and R.R. V. Raju 1995. *Bidens cynapifolia* H.B.K. (Asteraceae). A new record for peninsular India. Rheedea **5**: 43 – 45.

Prabhakar Raju, C. and R.R. V. Raju 1996. Some rare and interesting Asteraceous taxa from the forests of Andhra Pradesh. J. Econ. Taxon. Bot. **20**: 261 – 265.

Prabhakar Raju, C. and R.R. V. Raju 1998. Some interesting plant records from Andhra Pradesh. In: Higher Plants of Indian Subcontinent (Additional Series of Indian J. Forest. **10**: 223 – 228.

Prabhakar Raju, C. and R.R. V. Raju 1999. *Cyathocline manilaliana* (Asteraceae) – a new species from Andhra Pradesh, India .Rheedea **9**: 151 – 154.

Prain, D. 1890-1905. Bengal Plants. 2 vols. (repr. ed. 1963). Calcutta.

Prain, L.C. 1909. Botany in 'Imperial Gazetter of India' provincial series, Hyderabad *state.* Calcutta.

Priyadarsini, P. and B.R.P.Rao. 2011. Cultivated plants of Andhra Pradesh. In: T.Pullaiah (ed.) Biodiversity in India. Vol. 5. Daya Publishing House pp. 292-314.

Pullaiah, T. 1997. Flora of Andhra Pradesh. Vol.3. Scientific Publishers, Jodhpur.

Pullaiah, T and D.Ali Moulali 1997. Flora of Andhra Pradesh. Vol.2. Scientific Publishers, Jodhpur.

Pullaiah, T and E. Chennaiah. 1997. Flora of Andhra Pradesh. Vol.1. Scientific Publishers, Jodhpur.

Pullaiah, T and S. Karuppusamy 2008. Flora of Andhra Pradesh. Vol. 5. Scientific Publishers, Jodhpur.

Pullaiah, T., S. Karuppusamy and S. Sandhya Rani. 2008. The district Floras of Andhra Pradesh. Proc. Andhra Pradesh Akad. Sci. **12**: 34-47.

Pullaiah, T. and D. Muralidhara Rao. 2002. Flora of Eastern Ghats. Vol. 1. Regency Publications, New Delhi.

Pullaiah, T., C. Prabhakar and B. Ravi Prasad Rao. 1998. Flora of Medak District. Daya Publishing House, Delhi.

Pullaiah, T., P.V. Prasanna and G. Obulesu. 1992. Flora of Adilabad district, Andhra Pradesh. CBS Publishers and Distributors, New Delhi.

Pullaiah, T. and B. Ravi Prasad Rao. 1994. Flora of Nizamabad. Bishen Singh Mahendra Pal Singh, Dehra Dun.

Pullaiah, T. and S. Sandhya Rani 1999. Trees of Andhra Pradesh. Regency Publications, New Delhi.

Pullaiah, T., S. Sandhya Rani and S. Karuppusamy. 2010. Flora of Eastern Ghats.Vol. 4. Regency Publications, New Delhi.

Pullaiah, T. and M. Silar Mohammed. 2000. Flora of Ranga Reddi District. Regency Publications, New Delhi.

Pullaiah, T., and K. Sri Ramamurthy. 2001. Flora of Eastern Ghats. Vol.2. Regency Publications, New Delhi.

Pullaiah, T., K. Sri Rama Murthy and S. Karuppusamy. 2007. Flora of Eastern Ghats. Vol. 3. Regency Publications, New Delhi.

Pullaiah, T., K.Sri Rama Murthy and S.Sandhya Rani. 2004. Floristic analysis of the Flora of Andhra Pradesh, India. In: N.Ramaswamy, N.Pratap Reddy, T. Pullaiah, G.V.S. Murthy and G.P. Rao (eds.) Gleanings in Plant Sciences. Regency Publications. New Delhi pp. 34-42.

Pullaiah, T. and P. Surya Prakash Babu. 1997. Flora of Andhra Pradesh Volume 4. Illustrations. Scientific Publishers, Jodhpur.

Radford, A.E., W.C. Dickison, J.R. Massey and C.R. Bell. 1974. Vascular plant systematics. New York.

Ragan, A., S. Geetha and V.S. Raju 1998. The genus *Lipocarpha* R. Br. (Cyperaceae) in Andhra Pradesh, India. Rheedea **8**: 97 – 102.

Ragan, A. and V.S. Raju. 1990. *Eleiocharis setifolia* (A. Rich) Raynal and *Scleria multilacunosa* T. Koyama : Two unreproted sedges from Peninsular India. Indian J. Bot. **13**: 199- 203.

Ragan, A., A. Samatha and V.S. Raju 2004. *Flemingia involucrata* (Fabaceae): A new record for Andhra Pradesh. Indian J. Forest. **28(2):** 164-165.

Raghava Rao, S.R. 1989. Flora of Mahabubnagar district. Ph.D. Thesis, Osmania University, Hyderabad.

Raghava Rao, S.R., T. Rajagopal and S.T. Ramachandra Chary. 1985. *Fuirena wallichiana* Kunth (Cyperaceae) - A new record for Andhra Pradesh State. Indian bot. Reptr. **4**: 189-190.

Raghava Rao, S.R., T. Rajagopal and S.T. Ramachandrachari. 1989. *Alysicarpus mahabubnagarensis* sp. nov. (Fabaceae) from India. Asian J. Plant Sci. **1**: 97-99.

Raghava Rao, S.R., T. Rajagopal and S.T. Ramachandrachari. 1990. New records of flowering plants from Andhra Pradesh, India. Indian J. Bot. **13**: 197-198.

Raghava Swamy, B.V. and P.N. Rao 1987. Further additions to the flora of Andhra Pradesh. J. Econ. Taxon. Bot. **9**: 489-492.

Raghava Swamy, B.V. and P.N. Rao 1985. *Corchorus urticifolius* Wt. & Arn. (Tiliaceae) a new record for Andhra Pradesh. J. Econ. Taxon. Bot.**7**: 556-558.

Rajagopal, T. 1973. Flora of Hyderabad including study of the foliar epidermal characters of species as an aid to taxonomy. Ph.D. Thesis, Osmania University, Hyderabad.

Rajagopal, T. and N. Ramayya. 1968. A little known flowering plant from India *Rotala illecebroides* Kohne (Lythraceae). Curr. Sci. **37:** 336-338.

Rajagopal, T. and N. Ramayya. 1970. New record from Hyderabad *Bacopa floribunda* (R.Br.) West. Curr. Sci. **39:** 201-203.

Rajendran, A. and P. Daniel. 2002. The Indian Verbenaceae. Bishen Singh Mahendra Pal Singh. Dehra Dun

Raju, R.R.V. and M. Chennakesavulu. 2003. Two new additions to the sedge (Cyperaceae) flora of Andhra Pradesh, India.J. Econ. Taxon. Bot. **23**: 1181-1185.

Raju, R.R.V. and C. Prabhakar Raju 1997. Two new records of Asteraceae to Andhra Pradesh. J. Bombay Nat. Hist. Soc. **94**: 180 – 181.

Raju, R.R.V and C. Prabhakar Raju 2000. On the genus *Sonchus* L. (Asteraceae) in Andhra Pradesh, India. J. Econ. Taxon. Bot. **24**: 93 – 98.

Raju, V.S. 1983. Some additions to the Flora of Andhra Pradesh. J. Econ. Taxon. Bot. **4**: 295-297.

Raju, V.S. 1984. New Angiospermous taxa described from the state of Andhra Pradesh after its formation. J. Econ. Taxon. Bot. **5**: 469- 472.

Raju, V.S. 1985. On the name *Euphorbia laeta* Heyne ex Roth. J. Econ. Taxon. Bot. **6:** 413.

Raju, V.S. 1986. Three interesting plant records for Andhra Pradesh. Ind. Bot. Reptr. **5**: 113-114.

Raju, V.S. 1986. A Taxonomic note on *Salomonia oblongifolia* DC. (Polygalaceae). Indian J. Bot.**9**: 74-76.

Raju, V.S. 2004. New combinations for the Indian *Chamaesyce* (Euphorbiaceae). J. Econ. Taxon. Bot. **28(1):** 91-92.

Raju, V.S. and A. Ragan. 1988. Additions to the Cyperaceae of Andhra Pradesh. J.Econ. Taxon. Bot. **14:** 467-468.

Raju, V.S. and A. Ragan. 1989. The genus *Eleocharis* in Andhra Pradesh India. In : M.L. Trivedi, B.S. Gill & S.S. Saini (ed.) Plant Science Research in India. Part I. 213-216. Today &Tomorrow's, New Delhi.

Raju, V.S. and A. Ragan. 1990. Additions to the genus *Cyperus* L. in Andhra Pradesh. J. Econ. Taxon. Bot. **14:** 467-468.

Raju, V.S., A. Ragan, K.Omkar and S.Geetha. 2008. Sedges of Andhra Pradesh: Biodiversity at species and Ecosystem level, along with economic and ethnobotanical value. Proc. A.P.Akademi Sci. **12:** 214-226.

Raju, V.S. and N. Rama Rao. 1986. On the identify and distribution of *Cassia obtusifolia* in Southern India. J. Econ. Taxon. Bot. **8**: 485-486.

Raju, V.S. and P.N. Rao. 1976. *Cuscuta approximata* Bab. – a new record for South India. Geobios **3:** 180.

Raju, V.S., C.S. Reddy and A. Ragan. 2006. *Curcuma* L. (Zingiberaceae) in Andhra Pradesh. A preliminary study. J. Econ. Taxon. Bot. **30(4):** 773-775.

Raju, V.S., C.S. Reddy and K.G. Rajarao. 2007. The myth of "*minima* and *maxima*" the species of *Physalis* in the Indian Subcontinent. Acta Phytotaxonomica Sinica **45(2**): 239-245.

Raju, V.S., C.S. Reddy, K.N. Reddy, K.S. Rao and B. Bahadur. 2008. Orchid wealth of Andhra Pradesh, India. Proc. A.P. Akademi Sci. **12(1&2):** 180-192.

Rakshit, S.C. and B.C. Kundu. 1970. Revision of the Indian species of *Hibiscus* L. Bull. Bot. Surv. India. **12**: 151-175.

Ramachandrachari, S.T. 1980. Floristic studies of Achampeta taluk (Mahaboobnagar dt.) Ph.D. Thesis, Osmania University, Hyderabad.

Ramachandrachari, S.T. and J.J. Wood. 1981. A new species of *Habenaria* (Orchidaceae) from India. Kew Bull. **36**: 235-237.

Ramachandran, A. 1975. Inaugural address. 125th Anniversary Souvenir of Madras Herbarium, Botanical Survey of India, Southern Circle, Coimbatore, pp. 9-12.

Ramana, M.V. 2010. Flora of Hyderabad district (Andhra Pradesh). Ph.D.thesis, Osmania University, Hyderabad.

Ramana, M.V., V. Hanumantha Rao and V. Vasudeva Rao. 2011. Note on the distribution of *Ceropegia bulbosa* var. *lushii* (Apocynaceae) in Andhra Pradesh. Indian J. Forest.**34**: 471-472.

Ramana, M.V., P.V. Prasanna and P. Venu. 2011. *Hybanthus stellaroides* (Violaceae), a new record for India. Rheedea 21: 10-12.

Ramana, M.V., P.V. Prasanna and P. Venu. 2012. *Ledebouria hyderabadensis* (Hyacinthaceae), a new species from India. Kew Bull. **67(3):** 561-564.

Ramana, M.V., P.V. Prasanna, P. Venu and A. Bhattacharjee. 2011. Some critical notes on *Leonotis nepetifolia* (L.) R.Br. var. *africana* (P.Beauv.) J.K.Morton (Lamiaceae). Taiwania **56(3):** 254-256.

Ramana, M.V., C.S. Reddy and C. Srinivasulu. 2012. Vascular plants of Greater Hyderabad, Andhra Pradesh, India. In: C.Srinivasulu and B.Srinivasulu (eds.) Glimpses of Biodiversity of Greater Hyderabad. GHMC, Hyderabad, Zoo Outreach Organisation, Coimbatore.

Rama Rao N. and A.N. Henry 1996. The ethnobotany of Eastern Ghats in Andhra Pradesh, India. Bot. Surv. India, Kolkata.

Rama Rao, N., D. Narasimhan, T. Ravishankar and A.N. Henry. 1987. The genus *Adenia* Forsskal (Passifloraceae) in Southern India. J. Econ. Taxon. Bot.**11**: 241-245.

Ramayya, N. and T. Rajagopal. 1971. Systematics, distribution and anatomy of two Indian species of the genus *Elatine* L. Bull. Bot. Surv. India. **13**: 328-337.

Ramayya, N. and T. Rajagopal. 1973. Taxonomy, forest ecology and anatomy of the flowering plants of the erstwhile state of Hyderabad A review. Indian Nat. Sci. Acad.**45:** 302-313.

Ramayya, N. and T. Rajagopal. 1975. *Pseudowolfia hyalina* (Dell) Harp. Plas (Lemnaceae). A new record for Asia from Hyderabad, India. Aquatic Botany **1:** 71-73.

Rani, N. and K.M. Mathew 1983. Labiatae (Lamiaceae). In: The Flora of Tamil Nadu Carnatic. (Ed.) K.M. Mathew. 2: 1244-1284.

Rao, B.R.P., A.Lakshmaiah, P.M.Rajeswaramma, M.Sridhar Reddy, B.Sadasivaiah and S.Khader Basha. 2008. Agrestals of Southern Andhra Pradesh region. J. Econ. Taxon. Bot. **32:** 694-709.

Rao, B.R.P. and T.Pullaiah. 1992. *Vetiveria lawsonii* (Hook.f.) Blatter & McCann and *Potamogeton crispus* L. – Additions to the flora of Andhra Pradesh. J.Bombay Nat. Hist. Soc.**89(1):** 150-151.

Rao, B.R.P., T.Pullaiah and K.Hanumanthappa. 1991. Two new plant records from Andhra Pradesh. J. Bombay Nat. Hist. Soc. **88(2):** 307-308.

Rao, B.R.P. and A. M. Reddy. 2010. Two little known grasses of India. Indian Forester **136**: 12301233.

Rao, B.R.P., A. M. Reddy, B. Sadasivaiah, K.V.Subbaiah and S. Sunitha. 2009. Two new distributional records of Poaceae for Peninsular India. J. Econ. Taxon. Bot. **33**(2): 434-436.

Rao, B.R.P., A. M. Reddy, S.K. Basha and B. Sadasivaiah. 2010. Two endemic grasses of India: Additions to Flora of Andhra Pradesh. J. Econ Taxon. Bot. **34(2):** 394-397.

Rao, B.R.P., M. S. Reddy and T. Pullaiah. 2008. Flora and Vegetation of Andhra Pradesh. Proc. A.P. Akademi of Sciences **12** (1&2): 1-13.

Rao, B.R.P., B. Sadasivaiah, P. Priyadarsini, A. M. Reddy, S. Sunitha, A. Lakshmaiah, K.V. Subbaiah, J.N. Prasad, S.K. Basha, M.V. Suresh Babu and P.V. Prasanna. 2011. A conspectus of the Tribe Eragrostideae (Poaceae) in Andhra Pradesh, India. In: T. Pullaiah (ed.) Biodiversity in India. Vol. 5. Daya Publishing House, New Delhi. P.p. 259-291.

Rao, E.S. and V.S. Raju 1994. The genus *Jatropha* L. in Andhra Pradesh. J. Econ. Taxon. Bot.**18**: 585-589.

Rao, M.A. 1981. A review of Indian Ranunculaceae. Bull. Bot. Surv. India **23**: 213-216.

Rao, P.N., B.V. Raghava Swamy and T. Pullaiah. 2001.Flora of Nalgonda District. Shipra Publications, Delhi

Rao, P.N., V.S. Raju and R.S. Rao, 1983. Floristic Research work in Andhra Pradesh. Bull. Bot. Surv. India **9**: 493-494.

Rao, Rolla S. and R.S. Raghavan. 1965. Critical notes on three species of *Capparis* Linn. from Peninsular India. J. Bombay Nat. Hist. Soc.**62**: 412-424.

Ravi, N., N.Mohanan, T. Shaju, M.S. Kiran Raj and R. Rajesh. 2001. Three more new species of *Ischamum* L. (Poaceae) from Kerala, India. Bot. Bull. Acad. Sin. **42**: 223-230.

Reddy, A.M. 2002. Grasses of Eastern Ghats of Andhra Pradesh. Ph.D., Thesis, Sri Krishnadevaraya University, Anantapur.

Reddy, A.M., B. Sadasivaiah, K.V. Subbaiah, M.V. Suresh Babu and B. Ravi Prasad Rao 2009. Three new distributional records of Poaceae for Andhra Pradesh, India. J. Econ. Taxon. Bot. **33**(2): 411-413.

Reddy, C.S. 2001a. *Hybanthus vatsavayii* (Violaceae) – A new species from Andhra Pradesh, India. J. Econ. Taxon. Bot.**25**: 219-220.

Reddy, C.S. 2001 b. Floristic studies in Warangal district, Andhra Pradesh, India. Ph.D.thesis, Kakatiya University, Warangal.

Reddy, C.S. 2007. Forest types in Andhra Pradesh. Paryavaran. EPTRI-ENVIS (SoE-AP)) Newsletter. **1**: 1-8. EPTRI, Hyderabad.

Reddy, C.S. 2008. Flora of Andhra Pradesh. Paryavaran, EPTRI-ENVIS SoE-AP)) Newsletter. **3**: 1-4. EPTRI, Hyderabad.

Reddy, C.S. 2008. Floral diversity of Andhra Pradesh. Biodiversity News 1(3&4). 8. Andhra Pradesh Biodiversity Board, Hyderabad.

Reddy, C.S. 2008. Gap analysis for protected areas of Andhra Pradesh, India for conserving biodiversity. J. Amer. Sci. 6(11): 472-484.

Reddy, C.S., G.Bhagyanarayana, K.N.Reddy and V.S.Raju. 2008. Invasive alien flora of India. US National Biological Information Infrastructure, USHS, USA (http://www.isisinetwork.org/).

Reddy, C.S., M.R. Bhanja and V.S. Raju 1999a. *Nicotiana plumbaginifolia* Viv. and *Youngia japonica* (L.) DC. – New records for the flora of Andhra Pradesh, India. J. Econ. Taxon. Bot.**23**: 685-686.

Reddy, C.S., M.R. Bhanja and V.S. Raju. 1999b. Angiospermic flora and Biological spectrum of Jakaram Reserve Forest, Warangal district, Andhra Pradesh. Indian Forester **125:** 1152-1166.

Reddy, C.S., M.R. Bhanja and V.S. Raju. 2000. *Cassia uniflora* Miller: A new record for Andhra Pradesh, India. Indian J. Forest **23(3):** 324-325.

Reddy, C.S., M.R. Bhanja, P.S. Rao and V.S. Raju. 2000. A contribution to the Tree Flora of Hyderabad, Andhra Pradesh, India. Add. Ser. Indian J. Forest, **11**: 33-48.

Reddy, C.S., M.S.R. Murthy and C.B.S. Dutt. 2002. Vegetational diversity and endemism in Eastern Ghats, India. Proc. National Seminar on Conservation of Eastern Ghats, EPTRI, Hyderabad, pp. 109-134.

Reddy, C.S., A.H. Naqvi and V.S. Raju 2000. *Rorippa dubia* (Pers.) Hara (Brassicaceae) – a new record for southern India. J. Econ. Taxon. Bot. **24**: 287 – 288.

Reddy, C.S. and C. Pattnaik. 2007. *Argemone ochroleuca* Sweet – a new invasive species in Andhra Pradesh State. Zoos print Journal **11(12):** 2949.

Reddy, C.S., C. Pattnaik, M.S.R. Murthy and K.N. Reddy. 2006. Floristic census of Orchids of Eastern Ghats, India. Botanica **56:** 79-96.

Reddy, C.S., C. Pattnaik, M.S.R. Murthy and K.N. Reddy. 2006. Orchids of Eastern Ghats, India.EPTRI – ENVIS News letter11(4): 6-12. EPTRI, Hyderabad.

Reddy, C.S., G.S. Pujar, S. Sudhakar. B. Shilpa, K. Sudha, B. Gharai and M.S.R. Murthy. 2008. Mapping the vegetation types of Andhra Pradesh, India using Remote sensing. Proc. A.P.Akademi Sci. **12(1&2):** 14-23.

Reddy, C.S. and V.S. Raju 1999. *Gnaphalium coarctatum* Willd. (Asteraceae): A new record for southern India. J. Econ. Taxon. Bot. **23:** 687-688.

Reddy, C.S. and V.S. Raju. 2001a. A new variety of *Cleome chelidonii* (Cleomaceae). J. Econ. Taxon. Bot. 25: 217-218.

Reddy, C.S. and V.S.Raju. 2001b. *Physalis pruinosa* L. (solanaceae): A new record for India. Rheedea.

Reddy, C.S. and V.S. Raju 2001d. Endemic flowering plants of Andhra Pradesh, Peninsular India. Indian J. Forest.**24**:

Reddy, C.S. and V.S. Raju 2002. Additions to the weed flora of Andhra Pradesh. J. Econ. Taxon. Bot.**26**: 195-198.

Reddy, C.S. and V.S.Raju. 2008. Endemic Spermatophytes of Andhra Pradesh, India. Proc. A.P. Akademi Sci. **12(1&2):** 48-75.

Reddy, C.S. and V.S.Raju. 2009. *Aeschynomene americana* L. and *Mikania micrantha* Kunth – New invasive weeds in Flora of Andhra Pradesh, India. J. Econ. Taxon. Bot.**33(3):** 540-541.

Reddy, C.S., V.S.Raju and Y.N.R.Verma. 2008. A new variety of *Polycarpaea corymbosa* (L.) Lam. (Caryophyllaceae) from Andhra Pradesh, India. J. Econ. Taxon. Bot. **32(3):** 519-521.

Reddy, C.S. and K.N. Reddy. 2004. *Cassia rotundifolia* (Caesalpiniaceae): a new record for India. J. Econ. Taxon. Bot. **28(1):** 73-74.

Reddy, C.S., K.N. Reddy, M.R. Bhanja and V.S. Raju 1999. On the identity of *Physalis minima* L. (Solanaceae) in Southern India. J. Econ. Taxon. Bot.**23**: 709-710.

Reddy, C.S., K.N. Reddy, C.Pattnaik and V.S.Raju. 2006. Ethnobotanical observations of some endemic plants of Eastern Ghats, India. Ethnobotanical leaflets,http://www.siu.edu/-ebl/leaflets/html.

Reddy, C.S., K.N. Reddy and V.S. Raju 2000. *Cardamine trichocarpa* Hochst. ex A. Rich and *Cleome rutidosperma* DC. as alien weeds in Andhra Pradesh, India. J. Econ. Taxon. Bot. **24**: 291-292.

Reddy, C.S., K.N. Reddy and V. S. Raju. 2008. Supplement to Flora of Andhra Pradesh, India. Deep Publications, New Delhi.

Reddy, C.S., K.N. Reddy and S.N. Jadhav 2001. Threatened medicinal plants of Andhra Pradesh. Medicinal Plants Conservation Centre (MPCC). Hyderabad.

Reddy, C.S., K.Thulsi Rao, I.Siva Rama Krishna and S.M.M. Javed. 2008. Vegetation and Floristic studies in Nallamalais, Andhra Pradesh, India. J.Plant Sci. **3(1):** 85-91.

Reddy, K.N. 2002. Ethnobotany in Khammam district of Andhra Pradesh, India. Ph.D.thesis, Kakatiya University, Warangal.

Reddy, K.N., C. S. Reddy and V.S Raju 2000. On the occurrence of *Taraxacum javanicum* Soest (Asteraceae) in Andhra Pradesh. J. Econ. Taxon. Bot. **24**: 289 –290.

Reddy, K.N. C.S. Reddy and S.N. Jadhav 2001. *Heterostemma deccanense* (Talb.) Swarup & Mangaly (Asclepiadaceae): An endangered and endemic taxon in Andhra Pradesh. Indian Forester **127**: 1403-1404.

Reddy, K.N., C.S. Reddy and S.N. Jadhav 2002. *Dendrobium macrostachyum* Lindl. (Orchidaceae): A new record for Andhra Pradesh, India. J. Non-Timber Forest Products **9:** 91-92.

Reddy, K.N., C.S. Reddy and V.S. Raju 2004. *Memecylon jadhavii* (Melastomataceae): A new species from Andhra Pradesh, India. J. Econ. Taxon. Bot. **28**: 165-166.

Reddy, K.N., G.V. Subbaraju, C.S.Reddy and V.S.Raju. 2006. *Reinwardtia indica* (Linaceae): a new record for Andhra Pradesh. Indian J. Forest. **29:** 379-380.

Reddy, M.H. and R.R.Venkata Raju. 1996. Aquatic angiosperm diversity of Andhra Pradesh. J. Swamy Bot. Cl. **13:** 35-37.

Reddy, M.H. and R.R.Venkata Raju. 1998. Taxonomic study of the family Amaranthaceae in South India. J. Econ. Taxon. Bot.**21:** 35-37.

Reddy, R.D., M.K. Prasad and K. Venkaiah. 1991. Forest Flora Andhra Pradesh (Vernacular names). Hyderabad.

Renner, S.S. and A.K.Pandey. 2012. The Cucurbitaceae of India: Accepted names, synonyms, geographic distribution and information on imags and DNA sequences. Phytokeys 20: 53-118. doi: 10.3897/phytokeys.@.3948

Retzius, A.J. 1779-91. Observations Botanicae. Leipzig.

Ridley H.N. 1930. The dispersal of Plants throughout the world. Kent.

Roemer, J.J. and J.A. Schultes. 1817-30. Carolia Linne Equitis systemas vegetabilicum Secundum classes etc. (7 vols.) Stuttgart.

Roth, A.W. 1821. Novae Plantarum Species Prasertium Indiae Orientalis Collectione Doct. B. Heynes cum descrip-tionibus et observationibus. Halberstedt.

Roth, A.W. 1827. Enumeratio Plantarum Phanerogamarum etc. Leipzig.

Roxburgh, W. 1795-1810. Plants of the coast of Coromandal. (3 vols.) London.

Roxburgh, W. 1814. Hortus Benghalensis. Serampore.

Roxburgh, W. 1832. Flora Indica.(3 vols.) Serampore.

Royle, J.F. 1839. Illustrations of the Botany of the Himalayan Mountains. Vol. 2. London repr. 1971. New Delhi.

Sadasivaiah, B., K. Prasad, S.K. Basha, M.V.S. Babu, V.S. Rao and B.R.P. Rao. 2010. *Eulophia graminea* Lindl., *E. ochreata* Lindl. and *Habenaria barbata* Wight ex Hook.f. – Relocated in Andhra Pradesh eight decades. Indian J. For. **33(2):** 211-214.

Sadasivaiah, B., P. Priyadarsini and B.R.P. Rao. 2013. Three rare grasses (*Eragrostis ciliaris* var. *clarkei*, *E. tenella* var. *insularis* and *Urochloa panicoides* var. *velutina*) from Andhra Pradesh, India. J. Sci. Trans. Environ. Technov. **6(3):** 166-167.

Sadasivaiah, B., K. Venkata Subbaiah, S. Khadar Basha, M.V. Suresh Babu and B. R. P. Rao. 2008. *Arundinella tuberculata* Munro ex Lisboa- A new distributional record for Eastern Ghats, India. Proc. of A.P. Akademi of Sciences **12 (3):**289-291.

Salam, M.A. 1940. On the occurrence of *Cichorium intybus* Linn. (Chicory) in Hyderabad. J. Bombay Nat. Hist. Soc. **41**: 680.

Saldhana, C.J. 1984-96. Flora of Karnataka. Vol. 1 & 2. Oxford & IBH, New Delhi.

Saldhana, C.J. and D.H. Nicolson. 1976. Flora of Hassan district, Karnataka. India. New Delhi.

Sandhya Rani, K. Sri Rama Murthy, D. Muralidhara Rao and T. Pullaiah. 2004. On the occurrence of *Dimeria connivens* Hack. in Andhra Pradesh. J. Bombay Nat. Hist. Soc. **101 (3)**: 478-479.

Sanjappa, M. 1984. Additions to the genus *Indigofera* L. (Fabaceae) of India and Bhutan. Bull. Bot. Surv. India **26**: 38- 42.

Sanjappa, M. 1984. *Indigofera trita* L.f. (Fabaceae) complex in India. Bull. Bot. Surv. India **26**: 114-119.

Santapau, H. 1953. The Flora of Khandala on the Western Ghats of India. Rec. Bot. Surv. India **16:** 1-396. (3rd rev. ed. 1967).

Santapau, H. 1954. Annual excursion of the Indian Botanical Society, Hyderabad, Deccan. J. Indian Bot. Soc. **33**: 78-92.

Santapau, H. 1955. Botanical collector's Manual. Calcutta.

Santapau, H. 1956. The Botanical exploration of India. J. Indian. Bot. Soc. **35**: 1-6.

Santapau, H. and S.K. Wagh. 1964. Nomenclatural changes in Indian plants. Bull. Bot. Surv. India **5**: 105-109.

Santapau, H. and A.N. Henry. 1973. A dictionary of flowering plants in India. New Delhi.

Sardesai, M.M. and S.Y.Chavan. 2010. *Andrachne telephoides* L. (Phyllanthaceae) – An addition to the Flora of Peninsular India. J. Bombay Nat. Hist. Soc. **107(1):** 73-74.

Sarkar, P.K. 1995. Rare, endangered and endemic orchids in India. J. Econ. Taxon. Bot. (Addl. Ser.) 11: 33-47.

Satyanarayana Reddy, P. 1985. Floristic studies of Warangal and epidermology of Cyperaceae as an aid to taxonomy.Ph.D. Thesis, Osmania University.

Satyanarayana Reddy, P. and T.Rajagopal. 1993. New records of flowering plants for the Telangana region of Andhra Pradesh. Indian Bot. Reptr.**12:** 17-22.

Saxena, H.O. and M.Brahmam. 1996. The Flora of Orissa. Orissa Forest Development Corporation, Bhubaneswar.

Sayeeduddin, M. 1935. Some of the common flowering plants of the Hyderabad state; their distribution, economic and medicinal importance. J. Asiat. Soc. Bengal Sci. **1:** 9-22.

Sayeeduddin, M. 1936. Some of the common flowering plants of the Hyderabad state; their distribution, economic and medicinal importance. Osmania Univ. Journal **2:** 73-94.

Sayeeduddin, M. 1938. A further contribution to some of the common flowering plants of Hyderabad state; their distribution and economic importance. J. Bombay Nat. Hist. Soc. **40** : 190-212.

Sayeeduddin, M. 1941 a. Preliminary notes of a recent tour to Amrabad forest reserve. J. Bombay Nat. Hist. Soc.**41**: 907-910.

Sayeeduddin, M. 1941 b. Additions to our knowledge of the flowering plants of H.E.H. The Nizam's Dominions, Hyd. Deccan. J. Bombay Nat. Hist. Soc.**42**: 903-924.

Sayeeduddin, M. 1954. A brief sketch of the flora of Hyderabad, Deccan, Souvenir, 41st Session, Indian Science Congress, Hyderabad.

Sebastine, K.M. and A.N. Henry. 1966. A contribution to the flora of Pakhal and surrounding regions of Narasampet taluk, Warangal district, Andhra Pradesh. Bull. Bot. Surv. India **8:** 304-311.

Sebastine, K.M., K. Thothathri and N.P. Balakrishnan. 1960. Observations on the flora of Narsapur, Medak district. Andhra Pradesh. Bull. Bot. Surv. India **2:** 275-285.

Sebastine, K.M.,K. Thothathri, N.P. Balakrishnan and K. Rama Murthy 1966. Some additions to the Flora of the Presidency of Madras II. Bull. Bot. Surv. India **8:** 80-81.

Sedgwick, L.J. 1919. On the *Alysicarpus rugosus* DC. and its allied species. J. Indian Bot. Soc.**1:** 14-18.

Seshagiri Rao, K. 2012. University of Hyderabad Flora. (online version): http://ks-lab.in/uhflora/. Dated 25 July 2012.

Sharma, B.D. and N.P. Balakrishnan (eds). 1993. Flora of India. Vol.2. Papaveraceae – Caryophyllaceae. Bot. Surv. India. Calcutta.

Sharma, B.D., N.P. Balakrishnan, R.R. Rao and P.K. Hajra (eds.). 1993. Flora of India Vol. 1. Ranunculaceae – Barclayaceae. Bot. Surv. India. Calcutta.

Sharma, B.D. and M. Sanjappa (eds.) 1993. Flora of India Vol. 3. Portulacaceae – Ixonanthaceae. Bot. Surv. India. Calcutta.

Sharma, K.P. and B. Gopal. 1980. A note on identity of *Typha elephantina* Roxb. Aquatic Bot. **9:** 381-387.

Shivakumar, S. and M. Prabhakar. 2002. *Oxalis pubescens* H.B.K. – a new record to Andhra Pradesh. J. Econ. Taxon. Bot.**26:** 119-120.

Siddiq, E.A. 1962. Foliar variations in *Naravelia zeylanica* DC. J. Bombay Nat. Hist. Soc.**59**: 325-327.

Singh, K.K., S.K. Palvi and H.B. Singh. 1981. Survey and Biological activity of some medicinal plants of Mannanur forest, Andhra Pradesh. Indian J. Forest. **4**: 115-118.

Singh, Nagendra. 1978. Critical taxonomic notes on some species of *Cassia* L. found in India. J. Bombay Nat. Hist. Soc. **67**: 614-615.

Singh, P., G.S. Giri and V. Singh. 1983. A new species of *Tribulus* L. (Zygophyllaceae) from South India. Bull. Bot. Surv. India**25**: 197-198.

Singh, V. 1987. The genus *Cassia* L. (Caesalpiniaceae)- Some new taxa and combinations from India. J. Econ. Taxon. Bot. **10**: 321-327.

Sing, V. 1992. Some new combinations in *Chamaecrista* Moench and *Senna* Mill. J. Econ. Taxon. Bot. **16**: 590 – 600.

Singh, V. 2001. Monograph on Indian subtribe Cassinae (Caesalpinaceae). Scientific Publishers, Jodhpur.

Singh, V. 2001. Monograph on Indian *Leucas* R.Br. (Dronapushpi) Lamiaceae. Scientific Publishers (India), Jodhpur.

Sivarajan, V.V. 1981. Taxanomic notes on the genus *Portulaca* L. in India. J. Bombay Nat. Hist. Soc.**78**: 256- 260.

Sprengal, K. 1824-27. Carol Linnaeus Systema vegetabilitam. ed. 16. Gottingae.

Sri Rama Murthy, K., Sandhya Rani, S. and T. Pullaiah, 1998. *Desmodium* Desv. (Faboideae) in Eastern Ghats, India – a systematic survey. J. Bombay Nat. Hist. Soc. **95(3):** 536-541.

Sri Rama Murthy, K., Sandhya Rani, S. and Pullaiah, T. 1999. Genus *Dalbergia* in Eastern Ghats, J. Econ. Taxon. Bot.**23**: 133 –139.

Sri Rama Murthy, K., Sandhya Rani, S. and T. Pulliaiah. 2000. A contribution to the Flora of Eastern Ghats, Indian Genus *Rhynchosia* Lour. (Faboideae – Fabaceae) In: Higher Plants of Indian sub continent **9:** 1-15.

Srinivasan, S.R. and D. Narasimhan. 1987. A new variety of *Polycarpaea corymbosa* (L.) Lam. (Caryophyllaceae) from southern India. J. Econ. Taxon. Bot.**10:** 347-349.

Srivastava, S.K. and V.J.Nair. 2010. Genus *Ischaemum* L. (Poaceae) in India. Nelumbo **52:** 63-92.

Stapf, O . 1917-1934. Gramineae. in D.Prain (ed.), Flora of Tropical Africa, vol.9. L. Reeve and co., Ltd., Kent.

Stearn, D.A. and D.J. Mabberly. 1998. *Rotheca* (Lamiaceae) revived. Novon **8(2):** 204-206.

Stearn, W.T. 1976. Union of *Chionanthus* and *Lonicera* (Oleaceae). Ann. Mo. Bot. Gard. **63:** 355-357.

Steenis, C.G.G.J. Van. 1950-74. Flora Malesiana. Noordhoff & T.D. Netherlands.

Subbaiah, K.V. 2008. Grass diversity of Arid Zone of Peninsular India. Ph.D. thesis submitted to Sri Krishnadevaraya University, Anantapur.

Subba Rao, T.V. 1960. Forest types of Andhra Pradesh. Thesis.

Subramanyam, K. 1962. Aquatic Angiosperms. New Delhi.

Subramanyam, K. and C.P. Sreemadhavan. 1969. A conspectus of the families Podostemaceae and Tristichaceae. Bull. Bot. Surv. India **11**: 161-168.

Sudhakar, J.V. and G.V.S. Murthy 2012. *Ficus variegata* (Moraceae), a new record for Peninsular India. Rheedea **22(1):** 62-65.

Sudhakar, J.V. and G.V.S. Murthy. 2012. Additions to *Ficus* L. (Moraceae) species to the South Indian states. Indian J. Forest. **35 (3):** 345-350.

Sundara Raghavan, R. 1993. Capparaceae. In: B.D. Sharma and N. P. Balakrishnan (eds.). Flora of IndiaVol. 2. pp 248-335. Bot. Surv. India, Calcutta.

Sunitha, S. and B. R. P. Rao 2003. *Fimbristylis litoralis* Gaudich (Cyperaceae) a new record to Andhra Pradesh. J. Econ. Taxon. Bot.**25**: 24 – 25.

Sunitha, S., A. M. Reddy and B. R. P. Rao 2002. *Kaempferia rotunda* L., a new report to Andhra Pradesh. J. Econ. Taxon. Bot.**26**: 156-158.

Sur, P.R. 2001. A revision of the genus *Ischaemum* Linn. (Poaceae) in India. J. Econ. Taxon. Bot.**25**: 407-438.

Suxena, M.R. 1947. A study of the grasses of the H.E.H., the Nizam's Dominions, Hyderabad, Deccan. Osmani Univ. J. Res. **12**: 28-65.

Swarupanandan, K., N.Sasidharan and J.K.Manglay. 1989. A reconsideration of the generic circumscription of *Heterostemma* Wight & Arn. (Asclepiadaceae) and a new species from India. J. Linn. Soc.**101**: 249-259.

Syam Prasad, N., D. Vasantha Kumar, T. Samba Murthy and G. Maheswara Rao 1997. *Nothopegia beddomei* a new species to Eastern Ghats. Indian Forester**123**: 260.

Tadulingam, M.C. and G. Venkata Narayana. 1932. A Handbook of some South Indian weeds. Madras.

Takhtajan, A. 1969. Flowering plants : Origin and Dispersal. U.K.

Takhtajan, A. 1980. Outline of the classification of the flowering plants (Magnoliphyta). Bot. Rev. **46** : 225-359.

Talbot, W.A. 1909-1911. Forest Flora of the Bombay Presidency and Sind.

Thothathri, K. 1964. The flora of Nagarjunakoda valley and the surrounding hills. Indian Forester**90**: 539-546.

Thothathri, K. 1964. Studies in Leguminosae IV. New varieties of *Crotalaria paniculata* Willd. and *Derris brevipes* Baker. Bull. Bot. Surv. India **6:** 67-68.

Thothathri, K. 1964. Studies in Leguminosae 4. New varities of *Crotalaria panicualta* Willd. var. *nagarjunakondensis.* Bull. Bot.Surv. India. **6**: 67-68.

Thothathri, K. and A.A. Ansari 1978. Critical notes on the genus *Crotalaria* II. Bull. Bot. Surv. India **20**: 179-182.

Thothathri, K. and A.A. Ansari 1979. Critical notes on the genus *Crotalaria* L. Indian J. Forest. **2:** 280-283.

Thulsi Rao, K., S.M.Javed and C.S.Reddy. 2006. *Taverniera cuneifolia* (Roth) Arn. (Papilionaceae): New record for Andhra Pradesh. J. Econ. Taxon. Bot. **30(4):** 912.

Thunberg, 1794-1800. Prodromous plantarum capensium. Upsala.

Thwaites, G.H.W. 1858-64. Enumeratio plantarum Zeylaniae. London.

Trimen, H. 1893-1931. A Hand Book of the Flora of Ceylon. London.

Trimen, H. and J.D. Hooker. 1893-1900. A Hand Book to the Flora of Ceylon. Vol. 6 (suppl.) London (in 1931 by A.H.G. Alston).

Torrey, J. 1843. A Flora of the State of New York.(2 vols.) Albany.

Tuntin *et al.* (ed.) 1964-1980. Flora Europaea. (5 vols.) Cambridge.

Vahl, M. 1879-94. Symbole botanicae. Hawanieae.

Veeranjaneyulu, D., A.Lakshmaiah and B.R.P.Rao. 2013. *Fimbristylis salbundia* and *Queenslandiella hyalina* (Cyperaceae), new distributional records for Andhra Pradesh. Indian J. Forest. **36(2):** 277-278.

Venkata Ramana, M. 2010. Flora of Hyderabad District (Andhra Pradesh). Ph.D.Thesis, Osmania University, Hyderabad.

Venu, P. 2006. *Strobilanthus* Blume (Acanthaceae) in Peninsular India. Bot. Surv. India, Kolkata

Venu, P. and P. Daniel 2000. The identity of *Leucas wightiana* Benth. (Lamiaceae). J. Swamy Bot. Cl.**17**: 69-74.

Verdcourt, B. 1985. A synopsis of the Moringaceae. Kew Bull.**40**: 1-23.

Verma, D.M. amd Veena Chandra. 1980. The genus *Rhynchospora* Vahl in India. Bull. Bot. Surv. India **22:** 126-135.

Vijayakumar, B.K. and N. Ramayya. 1986. Phytogeography of the genus *Indigofera* L. in South India. J. Swamy Bot Club **3**: 31-36.

Vijayakumar, B.K., and M. Prabhakar. 1986. Poorly collected and presumably rare taxa of *Indigofera* L. frm South India.J. Swamy Bot. Club **3**: 167-168.

Voigt, J.O. 1845. Hortus suburbanus calcuttensis, Calcutta.

Wadoodkhan, M.A. and P.Lakshminarasimhan. 2008. Two new species of Cyperaceae from Peninsular India. J. Bot. Res. Inst. Texas **2(1):** 379-384.

Wadoodkhan and S.N.Solanke. 2008. Novelties in Cyperaceae VII – A new variety and a new record. Bioinfolet **5(2):** 103-107.

Wadoodkhan, M.A., Solanke, S.N. and D.P.Chavan. 2007. Genus *Scleria* Berg. (Cyperaceae) in Peninsular India, rest of Maharashtra and Andamañ. J. Econ. Taxon. Bot. **31:** 598-612.

Wadoodkhan, M.A., R.D.Taur, P.Lakshminarasimhan, M.M.Sardesai and R.I.Shaikh. 2011. Novelties in Cyperaceae IX – New specific and intraspecific taxa of *Eleocharis* R.Br. Bioinfolet **8(1):** 5-10.

Wagh, S.K.1964. The genus *Zornia* Gmel. in India. J. Bombay Nat. Hist. Soc. **61**: 213-215.

Walker, A.L. 1849. Statistical report on the circar of Warangal (Hyderabad, Deccan). Madras J. Lit & Sci. **15**: 219-301.

Wallich, N. 1829-32. Plantae Asiaticae Rarires. (3 vols.) London.

Wallich, N. 1829-49. A numerical list of dried specimens of plants in the East India company Museum. London.

Wight, R. 1834. Contributions to the Botany India. London.

Wight, R. 1838-1853. Icones plantarum Orientalis. (6 vols.) Madras.

Wight, R. 840-1850. Illustrations to Indian Botany. London.

Wight, R. and G.A. Walker-Arnott. 1834. Prodromus Florae Penninsuale Indiae Orientalis. London.

Willdenow, C.L. 1797-1830. Species Plantarum. (ed. 5.) (5 vols.) Berlin.

Willis, J.C. 1973. A dictionary of flowering plants and ferns. (rev. ed. 8th by H.K. Airy Shaw) Cambridge.

Yadav, S.R. and S.P.Gaikwad. 2003. A revision of the Indian Aponogetonaceae. Bull. Bot. Surv. India **45:** 39-76.

Index to Families, Genera and Species

Italicized names indicate Synonymy

Index to Common Names

Index to English Names